METHODS IN MOLECULAR BIOLOGY™

Series Editor
John M. Walker
School of Life Sciences
University of Hertfordshire
Hatfield, Hertfordshire, AL10 9AB, UK

For further volumes:
http://www.springer.com/series/7651

Nanoparticles in Biology and Medicine

Methods and Protocols

Edited by

Mikhail Soloviev

School of Biological Sciences, Royal Holloway University of London, Egham, Surrey, UK

Humana Press

Editor
Mikhail Soloviev, Ph.D.
School of Biological Sciences
Royal Holloway University of London
Egham, Surrey, UK

ISSN 1064-3745 ISSN 1940-6029 (electronic)
ISBN 978-1-4939-6247-1 ISBN 978-1-61779-953-2 (eBook)
DOI 10.1007/978-1-61779-953-2
Springer New York Heidelberg Dordrecht London

Softcover re-print of the Hardcover 1st edition 2012

Cover illustration: Image from Chapter 18, Figure 3.

Printed on acid-free paper

Humana Press is a brand of Springer
Springer is part of Springer Science+Business Media (www.springer.com)

Preface

Despite being used for centuries in crafts such as pottery and stained glass, nanoparticles never before attracted as much attention as they have since the term "nanoparticles" came into use in the early 1990s. This term refers to a very special intermediate category between atoms or molecules and bulk material. While the term "nanoparticles" technically refers to any particle with a dimension of less than 1 μm, it is more commonly used to describe particles ranging from approximately 1–100 nm.

Physical properties of nanoparticles differ from both bulk and molecular properties of the same material. As the surface-to-volume ratio of a particle increases with the shrinking of particle size, an increasing number of individual atoms become exposed on the surface of a particle compared to the number of atoms or molecules remaining buried inside it. The uniquely small size of nanoparticles, which is of the same order as the wavelength of the electron wave function, results in quantum confinement effects and consequently in the unique chemical, physical, and optical properties of nano-scaled materials.

Since the early 1990s, the number of publications on nanoparticles, their properties, and their applications in chemistry, material sciences, physics, medicine, and life sciences has grown exponentially with no sign of slowing down. The modern fascination with micro- and nano-sized materials can actually be traced back further to the 1960s and 1970s when the first few reported attempts were made to use nanoparticles for controlled drug delivery. Since then the use of nanoparticles in biology and medicine has flourished, with a number of drug-containing nanoparticle products eventually reaching the market in the past decade.

This volume presents a wide range of methods for synthesis, surface modification, characterization, and application of nano-sized materials (nanoparticles) in life science and medical fields (mostly for drug delivery). The methods presented cover all stages of nanoparticle manufacturing, modification, analysis, and applications.

The methods and techniques covered in this volume encompass a number of materials including noble metal nanoparticles (Chapters 1, 2, 3, 4, 5, 6, 7), semiconductor nanoparticles (Chapters 8, 9, 10, 11, 12, 13, 14, 15, 16), magnetic nanoparticles (Chapters 17, 18, 19, 20), and organic, polymeric, biomimetic, and biocompatible nanoparticles (Chapters 21, 22, 23, 24, 25, 26, 27, 28, 29, 30, 31). Nano-toxicity and nano-safety aspects are covered in Chapters 32, 33, 34, 35, 36, while Chapters 37, 38, 39, 40, 41, 42, 43 focus on nanoparticle analysis. This volume will help the beginner to become familiar with this fascinating field of research and will provide scientists at all levels of expertise with easy-to-follow practical advice needed to make, modify, and analyze nanoparticles of their choice and to use them in a wide range of biomedical and pharmaceutical applications, including functional protein studies, drug delivery, immunochemistry, imaging, and many others.

I would like to express my gratitude to all authors who have contributed invaluably to this volume.

Egham, Surrey, UK *Mikhail Soloviev*

Contents

Part IV Organic, Polymeric, Biomimetic, and Biocompatible Nanoparticles and Their Applications

Part V Nano-Toxicity and Nano-Safety

Part VI Nanoparticles Characterization

Contributors

Shrikant Anant • *The University of Kansas Medical Center, Kansas City, KS, USA*
Maria Andreou • *Department of Biological Sciences, University of Cyprus, Nicosia, Cyprus*
Ioanna Antoniades • *Department of Biological Sciences, University of Cyprus, Nicosia, Cyprus*
Yoshinobu Baba • *FIRST Research Center for Innovative Nanobiodevices, Nagoya University, Nagoya, Japan; Department of Applied Chemistry, Graduate School of Engineering, Nagoya University, Nagoya, Japan; Health Technology Research Center, National Institute of Advanced Industrial Science and Technology, Takamatsu, Japan*
Julian Bailes • *School of Biological Sciences, Royal Holloway, University of London, Egham, Surrey, UK*
Anil K. Bajpai • *Department of Chemistry, Bose Memorial Research Laboratory, Government Autonomous Science College, Jabalpur, MP, India*
Jaya Bajpai • *Department of Chemistry, Bose Memorial Research Laboratory, Government Autonomous Science College, Jabalpur, MP, India*
Pedro Viana Baptista • *CIGMH, Departamento de Ciências da Vida, Faculdade de Ciências e Tecnologia, Universidade Nova de Lisboa, Caparica, Portugal*
Paulo C. Bessa • *Department of Biology, CBMA–Centre of Molecular and Environmental Biology, University of Minho, Braga, Portugal; 3B's Research Group, Biomaterials, Biodegradables and Biomimetics, University of Minho, Headquarters of the European Institute of Excellence on Tissue Engineering and Regenerative Medicine , Guimarães , Portugal; IBB—Institute for Biotechnology and Bioengineering, PT Associated Laboratory, Guimarães, Portugal*
Tobias Blom • *Department of Engineering Sciences, Division of Applied Materials Science, Uppsala University, Uppsala, Sweden*
Linda Bowman • *Pathology and Physiology Research Branch, Health Effects Laboratory Division, National Institute for Occupational Safety and Health, Morgantown, WV, USA*
Ana Maria Carmona-Ribeiro • *Departamento de Bioquímica, Biocolloids Lab, Instituto de Química, Universidade de São Paulo, São Paulo, SP, Brazil*
Antonio Carlos Campos de Carvalho • *Instituto de Biofísica Carlos Chagas Filho, Universidade Federal do Rio de Janeiro, Rio de Janeiro, RJ, Brazil; Department of Neuroscience, Albert Einstein College of Medicine, Bronx, NY, USA*
Margarida Casal • *Department of Biology, CBMA–Centre of Molecular and Environmental Biology, University of Minho, Braga, Portugal*
Wayne E. Cascio • *U.S. Environmental Protection Agency, Durham, NC, USA*
Onofre Castell • *Servei de Microscòpia, Universitat Autònoma de Barcelona, Bellaterra, Barcelona, Spain*

Vincent Castranova • *Pathology and Physiology Research Branch, Health Effects Laboratory Division, National Institute for Occupational Safety and Health, Morgantown, WV, USA*
Anna Charalambous • *Department of Biological Sciences, University of Cyprus, Nicosia, Cyprus*
Xianfeng Chen • *Department of Physics and Materials Science, City University of Hong Kong, Kowloon, Hong Kong*
Yun Chen • *Wallenberg Laboratory, Department of Molecular and Clinical Medicine/Clinical Physiology, The Sahlgrenska Academy and University Hospital, University of Gothenburg, Gothenburg, Sweden*
Zhongpin Chen • *Experimental Center of Biotechnology and Biomaterials, BME, Southeast University, Nanjing, China*
Neophytos Christodoulou • *Department of Biological Sciences, University of Cyprus, Nicosia, Cyprus*
Zuny Cifuentes • *IFAPA, Centro Alameda del Obispo, Ãrea de Mejora y Biotecnología, Córdoba, Spain*
Maria C. Collins • *Department of Cardiovascular Sciences, East Carolina Heart Institute, East Carolina University Brody School of Medicine, Greenville, NC, USA*
João Conde • *CIGMH, Departamento de Ciências da Vida, Universidade Nova de Lisboa, Caparica, Portugal; Instituto de Nanociencia de Aragón, Universidad de Zaragoza, Zaragoza, Spain*
Kaitlin Marie Daniel • *Office of Research and Development, National Health and Environmental Effects Research Laboratory, Toxicology Assessment Division, U.S. Environmental Protection Agency, Research Triangle Park, NC, USA*
Diksha • *Department of Chemistry, University of Delhi, Delhi, India*
Konstantinos Dimas • *Laboratory of Pharmacology, University of Thessaly, Biopolis, Larissa, Greece*
Min Ding • *Pathology and Physiology Research Branch, Health Effects Laboratory Division, National Institute for Occupational Safety and Health, Morgantown, WV, USA*
Peter J. Dobson • *University of Oxford, Kidlington, Oxon, UK*
Gonçalo Doria • *CIGMH, Departamento de Ciências da Vida, Faculdade de Ciências e Tecnologia, Universidade Nova de Lisboa, Caparica, Portugal*
Dietmar Eberbeck • *Physikalisch-Technische Bundesanstalt, Berlin, Charlottenburg, Germany*
Pedro Fevereiro • *Plant Cell Biotechnology Laboratory, Instituto de Tecnologia Química e Biológica, Universidade Nova de Lisboa, Oeiras, Portugal*
Jesus M. de la Fuente • *Instituto de Nanociencia de Aragón, Universidad de Zaragoza, Zaragoza, Spain*
Sara Gazi • *School of Biological Sciences, Royal Holloway, Faculdade de Ciências e Tecnologia, University of London, Egham, Surrey, UK*
Valérie A. Gérard • *CRANN and the School of Chemistry, Trinity College Dublin, Dublin, Ireland*
Ning Gu • *Experimental Center of Biotechnology and Biomaterials, BME, Southeast University, Nanjing, China*

YURII K. GUN'KO • *CRANN and the School of Chemistry, Trinity College Dublin, Dublin, Ireland*
PHILIP R. GUNST • *Metabolon Inc., Durham, NC, USA*
SANGILIYANDI GURUNATHAN • *Division of Molecular and Cellular Biology, Department of Biotechnology, Kalasalingam University (Kalasalingam Academy of Research and Education), Krishnankoil, Tamilnadu, India*
CLAUDIA GUTIÉRREZ-WING • *Ciencias Aplicadas-Tecnología de Materiales, Instituto Nacional de Investigaciones Nucleares, La Marquesa Ocoyoacac, Edo. de México, Mexico*
MIRIAN A.F. HAYASHI • *Departamento de Farmacologia, Universidade Federal de São Paulo (UNIFESP), São Paulo, SP, Brazil*
YONG HOU • *Experimental Center of Biotechnology and Biomaterials, BME, Southeast University, Nanjing, China*
COURTNEY W. HOUCHEN • *Department of Medicine, Digestive Diseases and Nutrition, University of Oklahoma Health Sciences Center, Oklahoma City, OK, USA*
SABER M. HUSSAIN • *Air Force Research Laboratory, Applied Biotechnology Branch, Human Effectiveness Directorate, Wright Patterson Air Force Base, OH, USA*
M. RICARDO IBARRA • *Departamento de Física de la Materia Condensada, Universidad de Zaragoza, Zaragoza, Spain; Instituto de Nanociencia de Aragón, Universidad de Zaragoza, Zaragoza, Spain*
ROSITSA IVANOVA • *School of Biological Sciences, Royal Holloway, University of London, Egham, Surrey, UK*
SYED HASSAN MUJTABA JAFRI • *Department of Engineering Sciences, Division of Applied Materials Science, Uppsala University, Uppsala, Sweden*
JASMIN • *Instituto de Biofísica Carlos Chagas Filho, Universidade Federal do Rio de Janeiro, Rio de Janeiro, RJ, Brazil; Department of Neuroscience, Albert Einstein College of Medicine, Bronx, NY, USA*
LINDA JELICKS • *Department of Physiology and Biophysics, Albert Einstein College of Medicine, Bronx, NY, USA*
MIGUEL JOSÉ-YACAMÁN • *Department of Physics and Astronomy, University of Texas at San Antonio, San Antonio, TX, USA*
NORITADA KAJI • *FIRST Research Center for Innovative Nanobiodevices, Nagoya University, Nagoya, Japan; Department of Applied Chemistry, Graduate School of Engineering, Nagoya University, Nagoya, Japan*
KALIMUTHU KALISHWARALAL • *Division of Molecular and Cellular Biology, Department of Biotechnology, Kalasalingam University (Kalasalingam Academy of Research and Education), Krishnankoil, Tamilnadu, India*
RICHARD L. KARPEL • *Department of Chemistry and Biochemistry, University of Maryland Baltimore County, Baltimore, MD, USA*
MARIOS KASINOS • *Particle & Interfacial Technology Group, Ghent University, Ghent, Belgium*
IRINA KERKIS • *Laboratory of Genetics, Butantan Institute, São Paulo, SP, Brazil*
ALAN P. KYPSON • *Department of Cardiovascular Sciences, East Carolina Heart Institute, East Carolina University Brody School of Medicine, Greenville, NC, USA*

KLAUS LEIFER • *Department of Engineering Sciences, Division of Applied Materials Science, Uppsala University, Uppsala, Sweden*
SCOTT C. LENAGHAN • *Department of Mechanical, Aerospace, and Biomedical Engineering, The University of Tennessee, Knoxville, TN, USA*
HAO-YING LI • *Biomanufacturing Research Centre, School of Mechanical and Electronic Engineering, Soochow University, Suzhou, China*
HONG LIANG • *Materials Science and Mechanical Engineering, Texas A&M University, College Station, TX, USA*
MAIK LIEBL • *Physikalisch-Technische Bundesanstalt, Berlin, Charlottenburg, Germany*
JONATHAN B. LIN • *Applied Biotechnology Branch, Human Effectiveness Directorate, Air Force Research Laboratory, Wright Patterson Air Force Base, XXX, OH, USA*
YINGXUN LIU • *Experimental Center of Biotechnology and Biomaterials, BME, Southeast University, Nanjing, China*
RAUL MACHADO • *Department of Biology, CBMA–Centre of Molecular and Environmental Biology, University of Minho, Braga, Portugal*
CLARA MARQUINA • *Instituto de Ciencia de Materiales de Aragón (ICMA), CSIC-Universidad de Zaragoza, Zaragoza, Spain; Departamentto de Fisica de la Materia Condensada, Universidad de Zaragoza, Zaragoza, Spain*
CHRISTOPHER MAYCOCK • *Organic Synthesis Laboratory, Instituto de Tecnologia Química e Biológica, Universidade Nova de Lisboa, Av. da República, Oeiras, Portugal;Faculdade Ciências da Universidade de Lisboa, Campo Grande, Lisboa, Portugal*
SAMANTHA MAYOSS • *HFL Ltd., Fordham, Ely, Cambridgeshire, UK*
PAUL VAN DER MEEREN • *Particle & Interfacial Technology Group, Ghent University, Ghent, Belgium*
SERGIO MEJÍA-ROSALES • *Center for Innovation and Research in Engineering and Technology and CICFIM, Universidad Autónoma de Nuevo León, San Nicolás de los Garza, NL, Mexico*
ROSALIA MENDEZ-OTERO • *Instituto de Biofísica Carlos Chagas Filho, Universidade Federal do Rio de Janeiro, Rio de Janeiro, RJ, Brazil*
ARBEN MERKOÇI • *Nanobioelectronics & Biosensors Group, Institut Català de Nanotecnologia, Campus Universitat Autónoma de Barcelona, Bellaterra, Barcelona, Spain; ICREA, Barcelona, Spain*
ANA SOFIA MIGUEL • *Organic Synthesis Laboratory, Instituto de Tecnologia Química e Biológica, Universidade Nova de Lisboa, Av. da República, Oeiras, Portugal; Biomolecular Diagnostics Laboratory, Instituto de Tecnologia Química e Biológica, Universidade Nova de Lisboa, Oeiras, Portugal*
S. MOEIN MOGHIMI • *Centre for Pharmaceutical Nanotechnology and Nanotoxicology, University of Copenhagen, Copenhagen Ø, Denmark*
HELENA MONTÓN • *Nanobioelectronics & Biosensors Group, Institut Català de Nanotecnologia, Campus Universitat Autónoma de Barcelona, Bellaterra, Barcelona, Spain*
BARBARA J. MULLER-BORER • *East Carolina University, Greenville, NC, USA*

Carme Nogués • *Departament de Biologia Cellular, Fisiologia i Immunologia. Universitat Autònoma de Barcelona, Bellaterra, Barcelona, Spain*
Enda O'Connell • *National Centre for Biomedical Engineering Science, National University of Ireland, Galway, Ireland*
Yukihiro Okamoto • *FIRST Research Center for Innovative Nanobiodevices, Nagoya University, Nagoya, Japan*
Abel Oliva • *Biomolecular Diagnostics Laboratory, Instituto de Tecnologia Química e Biológica, Universidade Nova de Lisboa, Av. da Republica, Oeiras, Portugal*
Eduardo B. Oliveira • *Departamento de Bioquímica e Imunologia, Universidade de São Paulo, Ribeirão Preto, Brazil*
Panayotis Pantazis • *ADNA, Inc., Oklahoma City, OK, USA*
Jayanth Panyam • *Department of Pharmaceutics, College of Pharmacy, University of Minnesota, Minneapolis, MN, USA*
Ladan Parhamifar • *Centre for Pharmaceutical Nanotechnology and Nanotoxicology, University of Copenhagen, Copenhagen Ø, Denmark*
Yeon-Su Park • *FIRST Research Center for Innovative Nanobiodevices, Nagoya University, Nagoya, Japan*
Alejandro Pérez-de-Luque • *IFAPA, Centro Alameda del Obispo, Área de Mejora y Biotecnología, Córdoba, Spain*
Arturo Ponce • *Department of Physics and Astronomy, University of Texas at San Antonio, San Antonio, TX, USA*
Babu R. Prasad • *National Centre for Biomedical Engineering Science, National University of Ireland, Galway, Ireland*
Rama P. Ramanujam • *ADNA, Inc., Oklahoma City, OK, USA; ADNA, Inc, Columbus, OH, USA*
Rui L. Reis • *3B's Research Group, Biomaterials, Biodegradables and Biomimetics, University of Minho, Headquarters of the European Institute of Excellence on Tissue Engineering and Regenerative Medicine, Guimarães, Portugal; IBB—Institute for Biotechnology and Bioengineering, PT Associated Laboratory, Guimarães, Portugal*
Aracely Rocha • *Materials Science and Mechanical Engineering, Texas A&M University, College Station, TX, USA*
Yury Rochev • *National Centre for Biomedical Engineering Science, National University of Ireland, Galway, Ireland*
José Carlos Rodriguez-Cabello • *Dpto. Física de la Materia Condensada, BIOFORGE Research Group, E.T.S.I.I, Universidad de Valladolid, Valladolid, Spain; CIBER BBN, Valladolid, Spain*
Mónica Roldán • *Servei de Microscòpia, Universitat Autònoma de Barcelona, Bellaterra, Barcelona, Spain*
Emma Rossinyol • *Servei de Microscòpia, Universitat Autònoma de Barcelona, Bellaterra, Barcelona, Spain*
Indrajit Roy • *Department of Chemistry, University of Delhi, Delhi, India*

RAJESH SAINI • *Department of Chemistry, Bose Memorial Research Laboratory, Government Autonomous Science College, Jabalpur, MP, India*
CARLOS J. SANCHEZ • *Materials Science and Mechanical Engineering, Texas A&M University, College Station, TX, USA*
ANA RAQUEL SANTOS • *Diagnostic, Instituto de Tecnologia Química e Biológica, Universidade Nova de Lisboa, Av. da República, 2780-157 Oeiras, Portugal*
HANS SAVEYN • *Particle & Interfacial Technology Group, Ghent University, Ghent, Belgium*
AMANDA M. SCHRAND • *Applied Biotechnology Branch, Human Effectiveness Directorate, Air Force Research Laboratory, Wright Patterson Air Force Base, OH, USA*
PARIS A. SKOURIDES • *Department of Biological Sciences, University of Cyprus, Nicosia, Cyprus*
TERRY J. SMITH • *National Centre for Biomedical Engineering Science, National University of Ireland, Galway, Ireland*
MIKHAIL SOLOVIEV • *School of Biological Sciences, Royal Holloway, University of London, Egham, Surrey, UK*
DAVID C. SPRAY • *Department of Neuroscience, Albert Einstein College of Medicine, Bronx, NY, USA*
MUTHU IRULAPPAN SRIRAM • *Division of Molecular and Cellular Biology, Department of Biotechnology, Kalasalingam University (Kalasalingam Academy of Research and Education), Krishnankoil, Tamilnadu, India*
PHIL TEALE • *HFL Ltd., Fordham, Ely, Cambridgeshire, UK*
ANA LUIZA MACHADO TORRES • *Instituto de Biofísica Carlos Chagas Filho, Universidade Federal do Rio de Janeiro, Rio de Janeiro, RJ, Brazil*
MANABU TOKESHI • *FIRST Research Center for Innovative Nanobiodevices, Nagoya University, Nagoya, Japan; Department of Applied Chemistry, Graduate School of Engineering, Nagoya University, Nagoya, Japan*
J. JESÚS VELÁZQUEZ-SALAZAR • *Department of Physics and Astronomy, University of Texas at San Antonio, San Antonio, TX, USA*
NEERAJ VIJ • *Department of Pediatrics, Pulmonary Medicine & Institute of NanoBiotechnology, Johns Hopkins University School of Medicine, Baltimore, MD, USA*
JINKE WANG • *Experimental Center of Biotechnology and Biomaterials, BME, Southeast University, Nanjing, China*
KEN WELCH • *Department of Engineering Sciences, Division of Nanotechnology and Functional Materials, Uppsala University, Uppsala, Sweden*
MATTHEW WRIGHT • *NanoSight Ltd, Amesbury, UK*
JAMES H. WYCHE • *Howard University, Washington, DC, USA*
LIJIN XIA • *Department of Mechanical, Aerospace, and Biomedical Engineering, The University of Tennessee, Knoxville, TN, USA*
SADAKI YOKOTA • *Division of Functional Morphology, Nagasaki International University, Sasebo, Nagasaki, Japan*
FAN ZHANG • *Medical Research Council, Laboratory of Molecular Biology, Cambridge, UK*

MINGJUN ZHANG • *Department of Mechanical, Aerospace, and Biomedical Engineering, The University of Tennessee, Knoxville, TN, USA*
YAN ZHOU • *Materials Science and Mechanical Engineering, Texas A&M University, College Station, TX, USA*
QIN ZHU • *Department of Mechanical, Aerospace, and Biomedical Engineering, The University of Tennessee, Knoxville, TN, USA*
ROBERT MARTIN ZUCKER • *Toxicology Assessment Division, Office of Research and Development, National Health and Environmental Effects Research Laboratory, U.S. Environmental Protection Agency, Research Triangle Park, NC, USA*

Part I

Noble Metal Nanoparticles: Synthesis and Applications

Chapter 1

Procedures for the Synthesis and Capping of Metal Nanoparticles

Claudia Gutiérrez-Wing, J. Jesús Velázquez-Salazar, and Miguel José-Yacamán

Abstract

The increasing impact of metallic nanoparticles in life sciences has stimulated the development of new techniques and multiple improvements of the existing methods of manufacturing nanoparticles with tailored properties. Nanoparticles can be synthesized through a variety of physical and chemical methods. The choice of preparation procedure will depend on the physical and chemical characteristics required on the final product, such as size, dispersion, chemical miscibility, optical properties, among others. Here we review basic practical procedures used for the preparation of protected and unprotected metallic nanoparticles and describe a number of experimental procedures based on colloidal chemistry methods. These include gold nanoparticle synthesis by reduction with trisodium citrate, ascorbic acid, or sugars in aqueous phase; nanoparticle passivation with alkanethiols, cetyltrimethylammonium bromide, or bovin serum albumin. We also describe microwave-assisted synthesis, nanoparticle synthesis in ethylene glycol, template-assisted synthesis with dendrimers and briefly describe how to control nanoparticle shape (star-shaped and branched nanoparticles).

Key words: Nanoparticles, Nanostructures synthesis, Passivated particles, Nanomaterials, Nanotechnology, Colloidal particles

1. Introduction

The synthesis of nanoparticles has two main avenues, physical methods and chemical methods. In this chapter we will describe the basic procedures used in nanotechnology for the preparation of metallic nanoparticles. It should be mentioned that in many cases these methods have been known for a long time, and the modern investigator is using or perfecting what was developed before (1).

Mikhail Soloviev (ed.), *Nanoparticles in Biology and Medicine: Methods and Protocols*, Methods in Molecular Biology, vol. 906, DOI 10.1007/978-1-61779-953-2_1, © Springer Science+Business Media, LLC 2012

1.1. Physical Approaches to the Synthesis of Nanoparticles

In this case a target of the metal or semiconductor material is placed under vacuum between 10^{-4} and 10^{-8} Pa, and then a saturated vapor of the elements on the target is produced using direct heating, impact by an electron beam, or impact by an ion beam, among others (2).

If the vacuum chamber is filled with an inert gas (such as argon), see Fig. 1, the atoms released from the collisions of the metal (semiconductor) atoms with the inert gas induce rapid growth of the particles. The growth mechanism is well established and can be explained with aid of Fig. 2. The atoms in the super saturated gas will tend to form nuclei of a few atoms (Fig. 2a). The nuclei will have the probability of being dissolved or to have more atoms attached. This is known as the nucleation stage (3, 4).

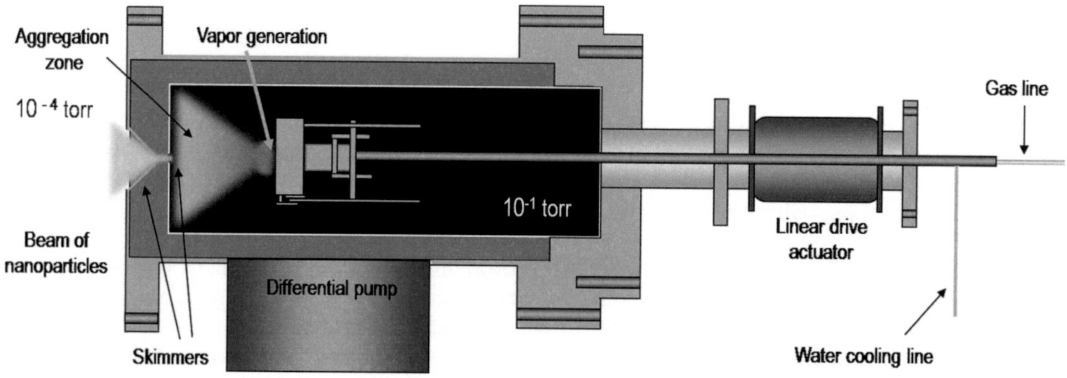

Fig. 1. Schematic of the preparation of metal nanoparticles by evaporation.

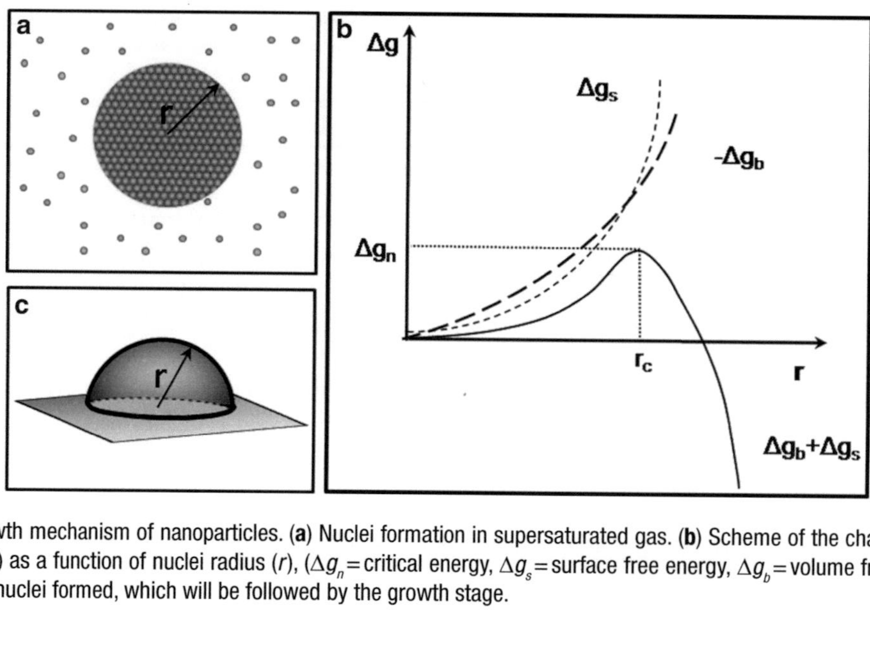

Fig. 2. Growth mechanism of nanoparticles. (**a**) Nuclei formation in supersaturated gas. (**b**) Scheme of the change in free energy (Δg) as a function of nuclei radius (r), (Δg_n = critical energy, Δg_s = surface free energy, Δg_b = volume free energy). (**c**) Critical nuclei formed, which will be followed by the growth stage.

Fig. 3. Critical nuclei of 13 atoms corresponding to an icosahedron.

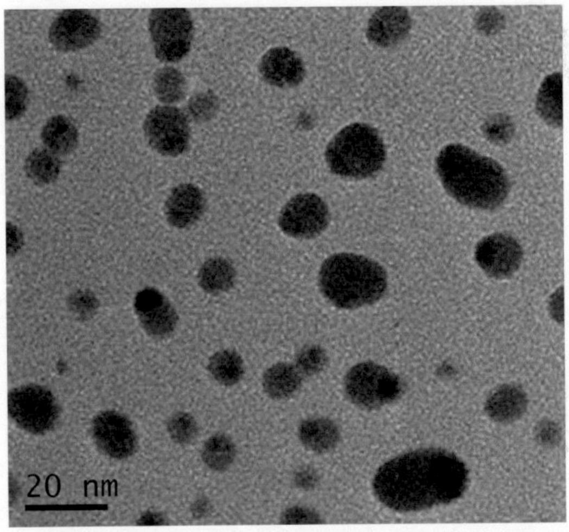

Fig. 4. Gold nanoparticles produced by evaporation.

However at this given point, a cluster size will be reached at which continuing addition of atoms becomes energetically favorable (Fig. 3). This nuclei size is the critical nuclei (r_c) and the new stage is the growth stage (Fig. 2b, c). All new atoms arriving at the gas will be attached to existing nuclei. A typical sample of gold nanoparticles produced by evaporation is shown in Fig. 4.

Nanoparticle growth is usually very fast. If the particles are deposited onto a substrate, they normally cool quickly. This means that in most cases the structures produced by physical methods are metastable structures, i.e., do not correspond to an equilibrium that has the conservancy that gold nanoparticles produced by vacuum evaporation tend to have icosahedral (Ic) or decahedral (Dh)

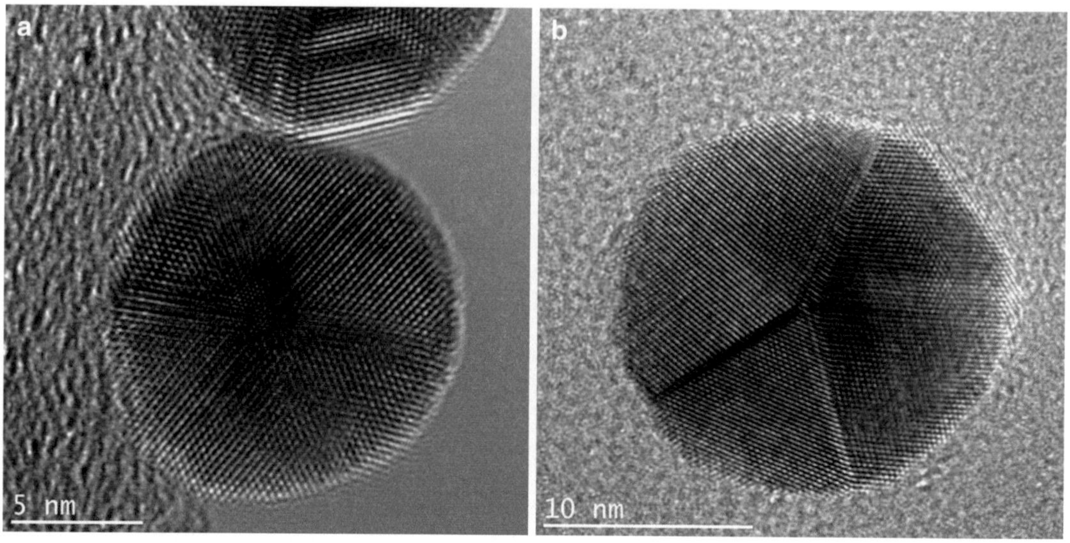

Fig. 5. High resolution TEM. (**a**) Au icosahedral (Ic) and (**b**) Au decahedral (Dh) spheres produced by evaporation.

spheres (5), as shown in Fig. 5. This is a remarkable property of gold nanoparticles since the Ic and Dh are fivefold crystal symmetries that cannot exist in bulk crystals. However they are very common at the nanoscale. Many important properties of gold can be linked to structure.

Not all the spheres obtained are Ic and Dh, also many other structures are observed. In general, physical methods provide the following advantages:

- Good control of particle size.
- Good control of crystal structure of the nanoparticles.
- Cleaner fabrication conditions.

The drawbacks are that particles cannot be easily transferred into a solution (they have to be deposited on a substrate) and that it is difficult to protect the particles with a passivating layer.

1.2. Chemical Methods

This approach is based on the chemical reduction of a salt or acid of the metal with a reducing agent such as citrate, hydrides, hydrazine, ethylene glycol, as it will be described later. For example, in gold chloride the gold ions are reduced in such a way that we have

$$Au^{+3} + 3e^{-} = Au^{0}$$

Then, once the metal atoms are formed in the solution, nucleation and growth will occur, resulting in formation of nanoparticles. A major difference to evaporation, however, is that the growth ratio tends to be much slower than in the evaporation case. This results in nanoparticles having structure resembling the bulk (equilibrium) structure, which, in the case of gold it corresponds to the FCC structure.

Reducing agent plays an important role in determining the size and crystal structure of the nanoparticles. However, many other factors influence the synthesis such as the temperature of the solution, the pH, type and relative concentrations of reactants, and other factors that have made nanoparticle synthesis sometimes difficult to control in general, when nanoparticles are not protected there is a strong tendency towards aggregation. A way to avoid this is to protect their surface with a stabilizing compound (a ligand, e.g., organic molecules such as alkanethiols, alkyl amines, citrate ions, polymers) that will reduce their tendency to aggregate by either electrostatic or steric stabilization (4). Such surface capping will serve as a diffusion barrier in the subsequent growth of nanoparticles. This implies that the reaction occurs in two steps: one in which the salt is reduced and another in which the metal is covered by the stabilizing compound. This reaction proceeds as follows:

$$M^{+} + e^{-} = M^{0}$$

$$(M^{0})_{n} + mL \rightarrow (M^{0})_{n} L_{m}$$

where (L) represents the ligand used as a stabilizing compound, (n) is the number of atoms in the particle, and (m) is the number of ligand molecules surrounding the nanoparticle.

In the following section, detailed procedures for the synthesis of metallic nanoparticles based on colloidal chemistry methods are presented.

2. Materials

2.1. Synthesis by Reduction with Trisodium Citrate in Aqueous Phase (Two Separate Techniques Described)

Deionized water should be used for making all solutions, unless specified otherwise.

1. 1 mM $HAuCl_4$ aqueous solution (use $HAuCl_4 \cdot 3H_2O$, 99.9%).
2. 1 % Solution of trisodium citrate dihydrate (use $Na_3C_6H_5O_7 \cdot 2H_2O$, 99+%).
3. Hot plate magnetic stirrer.
4. Second method: a mixture containing 0.25 mM $HAuCl_4$ and 0.25 mM trisodium citrate.
5. 0.1 M $NaBH_4$ solution (use $NaBH_4$, 99%), keep on ice.

2.2. Synthesis by Reduction with Ascorbic Acid in Aqueous Phase

1. 0.1 N aqueous solution of $HAuCl_4$ (use $HAuCl_4 \cdot 3H_2O$, 99.9%).
2. 0.1 N L-Ascorbic Acid ($C_6H_8O_6$, 99+%).
3. Magnetic stirrer plate.

2.3. Sugars as Reducing Agents

1. 0.25 mM $HAuCl_4$ in water (use $HAuCl_4 \cdot 3H_2O$, 99.9 %).
2. 1 M D-glucose aqueous solution.
3. 1 M NaOH aqueous solution (use NaOH, 98+ %).
4. Hot plate magnetic stirrer.

2.4. Alkanethiol Passivated Nanoparticles

1. Toluene ($C_6H_5CH_3$, 99.8 %).
2. Tetraoctylammonium bromide ($C_{16}H_{36}NBr$, 98 %).
3. Hydrogen tetrachloroaurate trihydrate ($HAuCl_4 \cdot 3H_2O$, 99.9 %).
4. 0.4 M Sodium borohydride, prepare fresh form $NaBH_4$, 99 %.
5. 1-Dodecanethiol ($CH_3(CH_2)_{11}SH$, ≥98 %).
6. Ethanol (≥99.5 %).

2.5. Cetyltrimethylammonium Bromide Stabilized Nanoparticles

1. 0.2 M Cetyltrimethylammonium bromide, solution in water (use CTAB, 99 %).
2. 0.5 mM $HAuCl_4$, solution in water ($HAuCl_4 \cdot 3H_2O$, 99.9 %).
3. 0.01 M $NaBH_4$ solution (use $NaBH_4$, 99 %), keep on ice.
4. Magnetic stirrer plate.

2.6. Bovin Serum Albumin-Capped Nanoparticles

1. 0.12 M $HAuCl_4$ aqueous solution (use $HAuCl_4 \cdot 3H_2O$, 99.9 %).
2. Bovin serum albumin (BSA).
3. 1 M $NaBH_4$, aqueous solution ($NaBH_4$, 99 %).
4. Ethanol (≥99.5 %).

2.7. Synthesis of Nanoparticles Through a Microwave-Assisted Process

1. 1.2 mM $HAuCl_4$, aqueous solution (use $HAuCl_4 \cdot 3H_2O$, 99.9 %).
2. High pressure tetrafluoromethaxil vessel for microwave oven.
3. ETHOS EZ Digestion System MicroWave (Milestone, 2.5 GHz, sensor ATC400).

2.8. Synthesis of Nanoparticles in Ethylene Glycol

1. Ethylene glycol (EG).
2. Hydrogen tetrachloroaurate trihydrate: dissolve 0.08 g of $HAuCl_4 \cdot 3H_2O$, 99.9 % in 5 mL EG.
3. Polyvinylpyrrolidone (PVP Mw 55,000): dissolve 0.208 g of PVP in 5 mL EG.

4. Ethanol (≥99.5 %).
5. Acetone.
6. Three-neck round-bottom flask with a reflux condenser.
7. Hot plate magnetic stirrer.

2.9. Use of Dendrimers as Templates for the Synthesis of Nanoparticles

1. 0.01 M $HAuCl_4$, aqueous solution (use $HAuCl_4 \cdot 3H_2O$, 99.9 %).
2. Fourth generation (G4) amine terminated PAMAM dendrimer 1.7 mM solution in methanol (e.g., from Dendritech, Inc.).
3. 0.15 M $NaBH_4$, aqueous solution (use $NaBH_4$, 99 %).
4. Magnetic stirrer plate.

2.10. Controlling the Shape of Nanoparticles—Star-Shaped and Branched Nanoparticles

1. 0.25 mM $HAuCl_4$ ($HAuCl_4 \cdot 3H_2O$, 99.9 %).
2. 1 M L-Ascorbic acid in water ($C_6H_8O_6$, 99+ %).
3. 0.25 mM $AgNO_3$ aqueous solution (use $AgNO_3$, 99.9999 % metal basis).
4. 0.25 mM Trisodium citrate (use $Na_3C_6H_5O_7 \cdot 2H_2O$, 99+ %).
5. 0.1 M Sodium borohydride (use $NaBH_4$, 99 %).
6. Hydrogen tetrachloroaurate trihydrate ($HAuCl_4 \cdot 3H_2O$, 99.9 %).
7. 1.25 mM Cetyltrimethylammonium bromide (CTAB, 99 %) aqueous solution.
8. 0.125 mM Pluronic F-127 aqueous solution.
9. L-Ascorbic acid ($C_6H_8O_6$, 99+ %).

3. Methods

3.1. Reduction with Trisodium Citrate in Aqueous Phase

The classic method reported by Turkevich et al. (6) and then refined by Frens (7, 8) describes the reduction of gold chloride by sodium citrate at 100 °C, producing colloidal gold of excellent stability and uniform particle size of 20–40 nm diameter (9). In this case, aggregation of the dispersed gold nanoparticles in water is prevented by electrostatic stabilization. Two typical approaches are detailed here.

1. Under constant stirring with a magnetic bar, heat 10 mL of a 1.0 mM $HAuCl_4$ aqueous solution in a round bottom flask to boiling point. Add 1 mL of a 1 % solution of trisodium citrate dihydrate to the boiling solution of $HAuCl_4$. The nanoparticles' formation is achieved when the solution turns deep red color. Figure 6 shows a typical image of isolated particles of an average size of 30 nm.

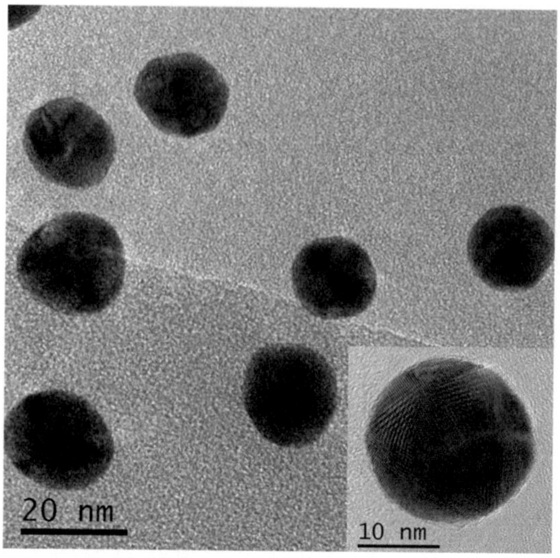

Fig. 6. TEM image of gold nanoparticles synthesized by citrate reduction. Inset shows a higher magnification HRTEM image of one of these gold nanoparticles.

2. Add 20 mL aqueous solution containing 2.5×10^{-4} mol of $HAuCl_4$ and 2.5×10^{-4} mol of trisodium citrate to a round bottom flask. Add 0.6 mL of an ice cold 0.1 M $NaBH_4$ solution under vigorous stirring with a magnetic bar over a stirring plate. The solution immediately turns orange-red, indicating the formation of gold nanoparticles. Typical average particle size is 35 nm.

3.2. Reduction with Ascorbic acid in Aqueous Phase

Gold nanoparticles can be produced using ascorbic acid as reducing agent in an aqueous system (10).

Add 10 mL of a 0.1 N ascorbic acid (vitamin C) aqueous solution to 2 mL of a 0.1 N aqueous solution of $HAuCl_4$ at room temperature in a round bottom flask, under vigorous stirring with a magnetic bar. Stir the solution for 15 min and centrifuge to precipitate the nanoparticles. The precipitate should be washed with deionized water several times. Figure 7 shows a SEM image of nanoparticles obtained by reducing $HAuCl_4$ with ascorbic acid following these experimental conditions.

3.3. Sugars as Reducing Agent

Typical glucose-capped Au nanoparticles obtained through this method are presented in TEM and SEM images in Fig. 8a, b. Nanoparticles produced through this method (11) can have nonuniform size distribution. Two typical approaches are described.

1. Add 2 mL of 1 M D-glucose to 18 mL of 0.25 mM $HAuCl_4$ in a round bottom flask under vigorous stirring with a magnetic bar. Heat the solution to 60 °C. Add 40 μL of a 1 M sodium hydroxide and keep under vigorous stirring for 10 s. An immediate change of the color of the solution to ruby red indicates the formation of gold nanoparticles.

Fig. 7. SEM image of gold nanostructures obtained by reduction with ascorbic acid.

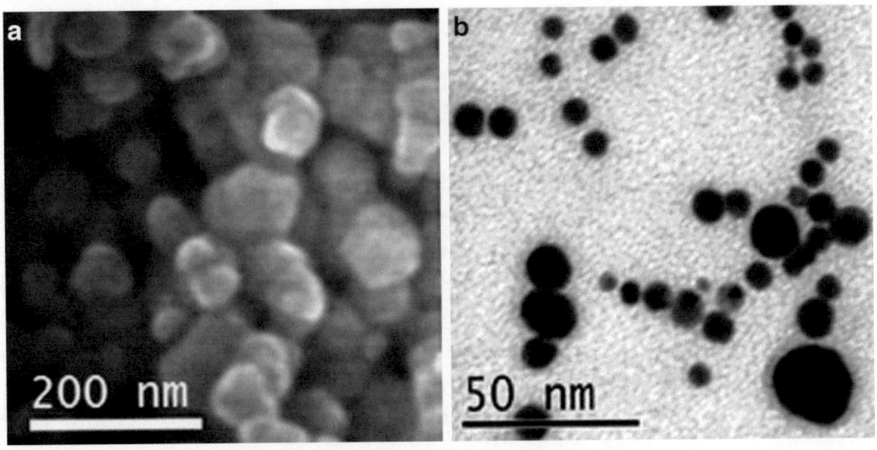

Fig. 8. Glucose-capped gold nanoparticles. (**a**) SEM and (**b**) TEM images.

2. Transfer required volume of 0.25 mM $HAuCl_4$ into a round bottom flask. Add glucose to final concentration of 0.1 M. Boil the solution under vigorous stirring with a magnetic bar for 30 min. The color of the solution turns purple, indicating that $AuCl_4^-$ has been reduced forming colloidal gold particles.

3.4. Alkanethiol Passivated Nanoparticles

Nanoparticles passivated by alkanethiol molecules are prepared by a phase transfer process through the method developed by Brust et al. (12), where metallic clusters are grown with a simultaneous attachment of alkanethiol molecules on the growing nuclei.

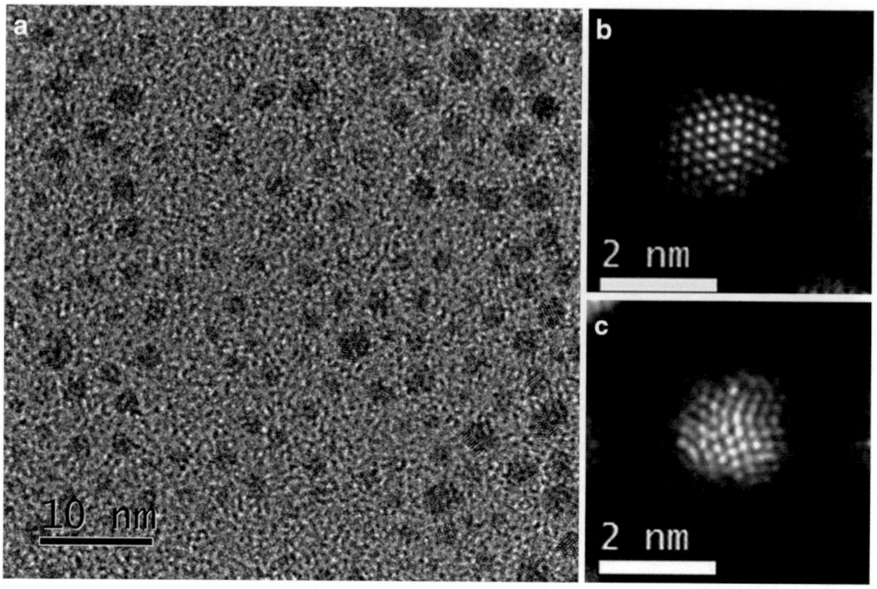

Fig. 9. Gold nanoparticles passivated with 1-dodecanethiol. (**a**) TEM and (**b**, **c**) HAADF-STEM images of these nanoparticles.

Prepare a 50 mM solution of tetraoctylammonium bromide in toluene. Mix 80 mL of this solution with 30 mL of 30 mM solution of $HAuCl_4$ under vigorous stirring until all the $AuCl_4^-$ transfers into the organic layer. Add 170 mg of 1-dodecanethiol to the organic phase. Then slowly add 25 mL of a freshly prepared 0.4 M aqueous solution of sodium borohydride under continuous vigorous stirring. Continue the reaction for 3 h. Separate the organic phase, reduce its volume to 10 mL by evaporating it in a rotatory evaporator and mix with 400 mL ethanol. Passivated nanoparticles will precipitate as a dark brown compound from ethanol after 4 h at −18 °C. Finally, this precipitate is filtered off and washed with ethanol. TEM image in Fig. 9a shows an example of alkanethiol-passivated gold nanoparticles obtained using this method. Figure 9b, c show images in dark-field of particles with sizes of about 2 nm.

3.5. CTAB Stabilized Nanoparticles

Gold nanoparticles can be synthesized using cetyltrimethylammonium bromide as surfactant (13). Typical TEM image of an Au nanocrystal seed obtained through this method is shown in Fig. 10a, exhibiting a spherical shape and a uniform size of 12 nm. HRTEM image in Fig. 10b evidences single crystal cuboctahedron seeds.

Mix 5 mL of a 0.2 M CTAB with 5 mL of a 0.5×10^{-3} 0.5 mM hydrogen tetrachloroaurate trihydrate in a round bottom flask. Add 0.6 mL of 0.01 M sodium borohydride under continuous stirring with a magnetic bar. Continue vigorous stirring of the nanoparticle suspension for 2 min. This will result in the formation of a brownish-yellow solution. Store the solution at 25 °C.

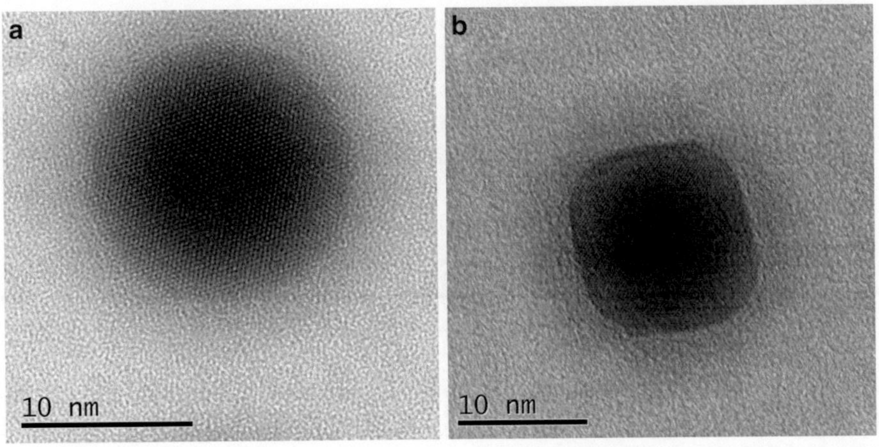

Fig. 10. HRTEM images of gold nanoparticles passivated with CTAB. (**a**) Single crystal nanoparticle, (**b**) Au single crystal cuboctahedrons.

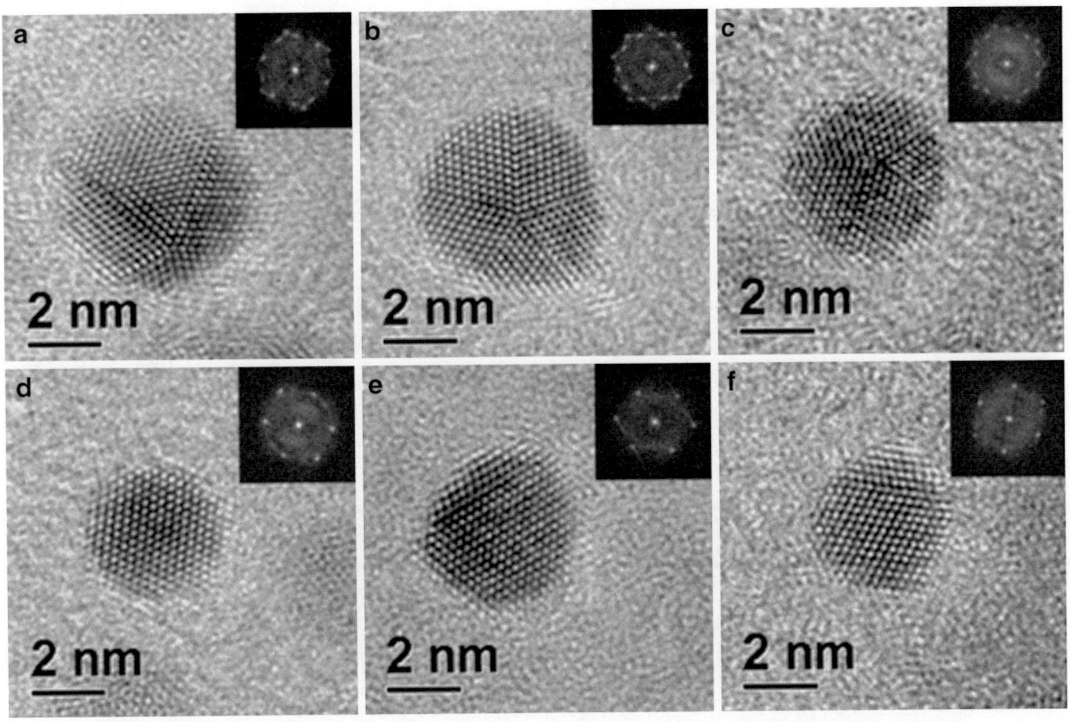

Fig. 11. HRTEM images of gold nanoparticles synthesized with proteins. (**a–c**), Multiple-twinned crystals with fivefold symmetry. (**d–f**) FCC single crystals. The fast Fourier transform (FFT) of each image is included as an inset.

3.6. BSA-Capped Nanoparticles

BSA-conjugated gold nanoparticles can be synthesized by reduction of gold ions in an aqueous phase system at ambient conditions (14). Different structures of gold nanoparticles synthesized with proteins can be obtained. Figure 11 illustrates typical HRTEM images of these particles, identifying multiple-twinned crystals with fivefold symmetry (Fig. 11a, c) and FCC single crystals (Fig. 11d, f).

Dissolve 0.57 g of BSA in 35.5 mL of water in a round bottom flask. Add 2 mL of a 0.12 M $HAuCl_4$ followed by 2.5 mL of 1 M sodium borohydride under vigorous stirring (reaching a total reaction volume of 40 mL). The molar ratio of BH_4^- to $AuCl_4^-$ has to be maintained at 10:1. When $NaBH_4$ solution is added to the metal acid solution, immediate protein foaming occurs and the solution rapidly turns golden brown and eventually takes a deep brown color. After 3 h of reaction, the product is purified by precipitation in ethanol at −5 °C and then re-dispersed in water. The obtained Au nanoparticles should be less than 2 nm in diameter.

3.7. Synthesis of Nanoparticles Through a Microwave-Assisted Process

This method is based on the use of a microwave radiating source, eliminating the need for an additional reducing agent. Kinetically or thermodynamic driven nanoparticles can be obtained, depending on the choice of reagents used and processing conditions (15). Figure 12 shows HRTEM images of Au nanoparticles synthesized using microwave-assisted process.

Transfer 15 mL of a 1.2 mM $HAuCl_4$ to a high pressure tetrafluoromethaxil vessel. Heat the sample using fast temperature program until a temperature of 200 °C is reached and let them cool to room temperature, without the use of any cooling system.

3.8. Synthesis of Nanoparticles in Ethylene Glycol

This method relies on a PVP-assisted reaction, where the solvent, ethylene glycol (EG), has two functions, as a solvent and as a reducing agent. This process is suitable for the fabrication of noble metal nanoparticles, yielding narrow particle size distribution and high colloid stability (16). This method can yield Au nanoparticles with different morphologies as can be seen in the SEM image in Fig. 13.

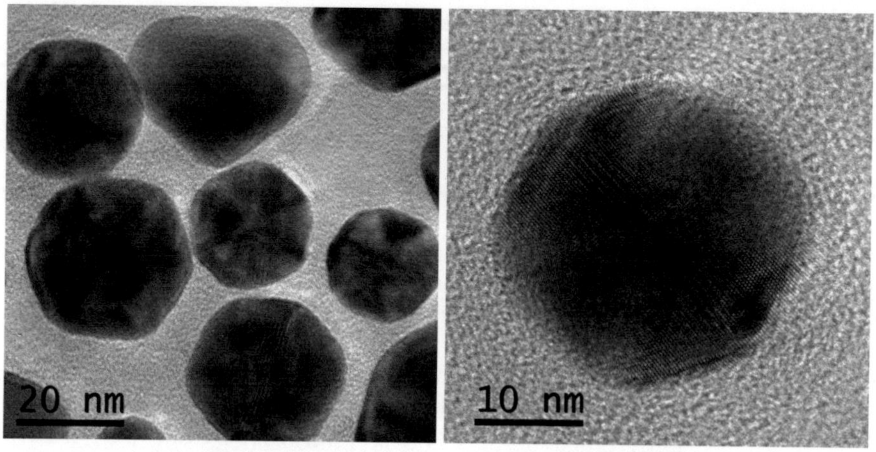

Fig. 12. HRTEM images of Au nanoparticles synthesized using microwave-assisted process. General TEM image (*left panel*) and HRTEM image of one of these nanoparticles (*right panel*).

Fig. 13. SEM image of gold nanoparticles obtained in the presence of PVP as a capping agent.

Transfer 5 mL of EG to a three-neck round-bottom flask with a reflux condenser. Heated to 250 °C under constant stirring with a magnetic bar, incubate for 10 min. Under constant stirring add the solutions of $HAuCl_4$ (5 mL) and of PVP (5 mL). Both solutions should be added simultaneously in a drop-wise fashion. The reaction should continue for about 10 min until the solution turns orange-purple indicating formation of gold nanoparticles. Collect the products by centrifugation at 4,000 rpm for 4 min and wash them with acetone once and with ethanol three times to remove the excess of PVP. The nanoparticles can be dispersed in 10 mL ethanol.

3.9. Use of Dendrimers as Templates for the Synthesis of Nanoparticles

The use of templates has been used to the growth of metal nanoparticles. Dendrimers have been used as templates and stabilizing agents for making nanoparticles. Dendrimers are highly functionalized, hyperbranched polymers, with an almost spherical shape, with a closed packing on the exterior containing functional groups and cavities between the polymeric chains in the interior (Fig. 14). These cavities allow the growth control when the metal precursors are complexed and reduced in situ (17, 18). This method should yield nanoparticles with the size range of 2–3 nm.

In a round bottom flask mix 50 mL of a 0.01 M $HAuCl_4$ and 50 mL of PAMAM dendrimer solution under vigorous stirring with a magnetic bar. The molar ratio of Au to the number of primary amine groups of the dendrimer should be 1:11 (5.9 Au atoms per G4 dendrimer). Slowly add 50 mL of 0.15 M $NaBH_4$ to reduce the Au:PAMAM complexes. The end of the reaction is indicated by formation of a wine-red dispersion, containing the nanoparticles.

Fig. 14. Schematic of the synthesis of metallic nanoparticles within a dendrimer template.

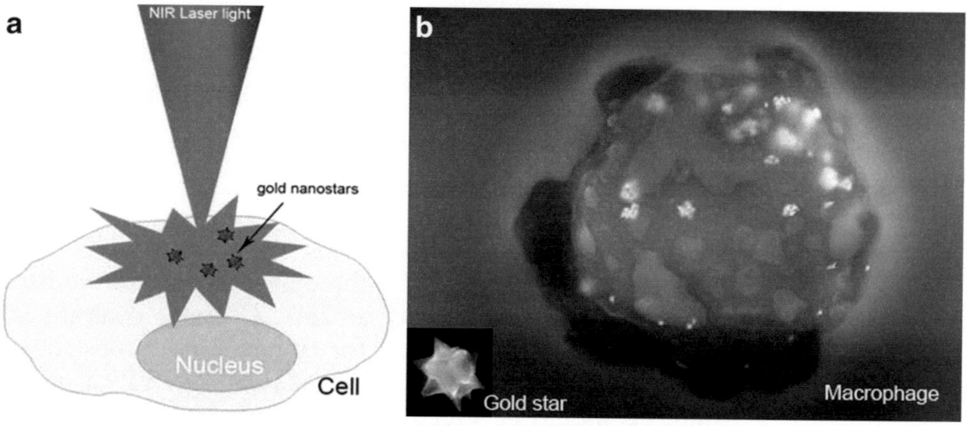

Fig. 15. Star-shaped nanoparticles. (**a**) Schematic showing the experimental setup for NIR laser irradiation, (**b**) SEM image of gold nanostars inside macrophage.

3.10. Controlling the Shape of Nanoparticles—Star-Shaped Nanoparticles

Shape is one of the most important parameters which define the properties of nanoparticles. In nanotechnology one of the most significant applications of nanoparticles is related to photo thermal therapy. An example is illustrated in Fig. 15. If a gold nanoparticle is placed inside a cancer cell and irradiated with a near IR light, then because of plasmon resonance, the particle will then be heated and energy released destroying the malignant cell. However in order to have a significant amount of energy released, the aspect ratio of the particle has to be large. Most often this was achieved using gold nanorods (19) or nanoshells (20). However, a significant improvement can be obtained by using gold nanostars (Fig. 16). In these shapes, every peak acts as a nano-antenna thus enhancing the electromagnetic field around it. Here we illustrate some of the procedures used to grow nanostars (21) and branched nanoparticles (22). Several other shapes can also be obtained such as nanorods, cubes, and spikes, among others (23–25). The shape will depend on the reaction conditions, including type and molar ratios of reagents, time, and temperature.

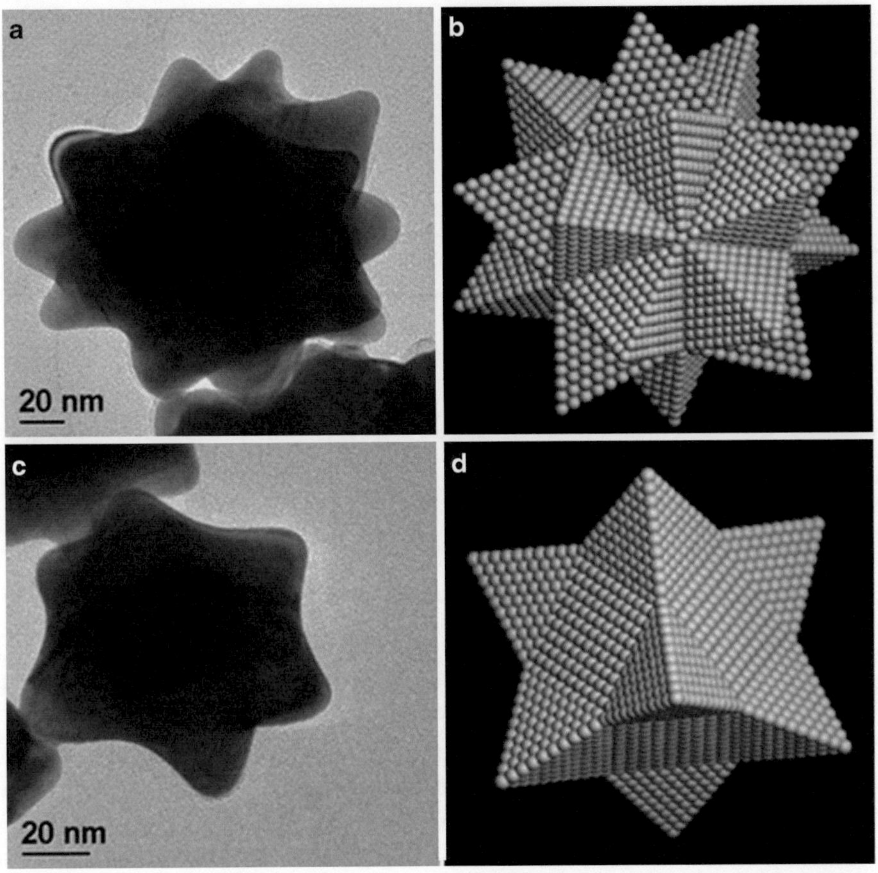

Fig. 16. Two classes of star polyhedral gold nanocrystals. (**a**) TEM image of a multiple-twinned star nanocrystal, and (**b**) proposed model, an icosahedron with each of its 20 {111} surfaces replaced by a tetrahedral pyramid. (**c**) TEM image of a monocrystalline star, and (**d**) proposed model, a cuboctahedron with each of its 8 {111} truncations replaced by a tetrahedral pyramid. (Reprinted with permission from J. M. Cabrera-Trujillo et al. J. Phys. Chem C, 2010, 114(49) 21051-21060. Copyright 2010 American Chemical Society).

1. Star-shaped polyhedral gold nanocrystals are synthesized by colloidal reduction in aqueous solution at ambient conditions. Under vigorous stirring with a magnetic bar, quickly add 10 mL of 1.6 mM ascorbic acid to a round bottom flask containing 20 mL of 5×10^{-5} M $HAuCl_4$ bringing the total volume to 30 mL. Addition of ascorbic acid causes an immediate color change to an opaque orange-red. Leave the reaction for 15 min under vigorous stirring. Different types of star polyhedral gold nanocrystals can be obtained under these conditions, including the two shown in Fig. 16. These correspond to a multiple-twinned star nanocrystal (Fig. 16a, b) and a monocrystalline star (Fig. 16c, d). Observe that the proposed model in Fig. 16b corresponds to an icosahedron with each of its 20 {111}

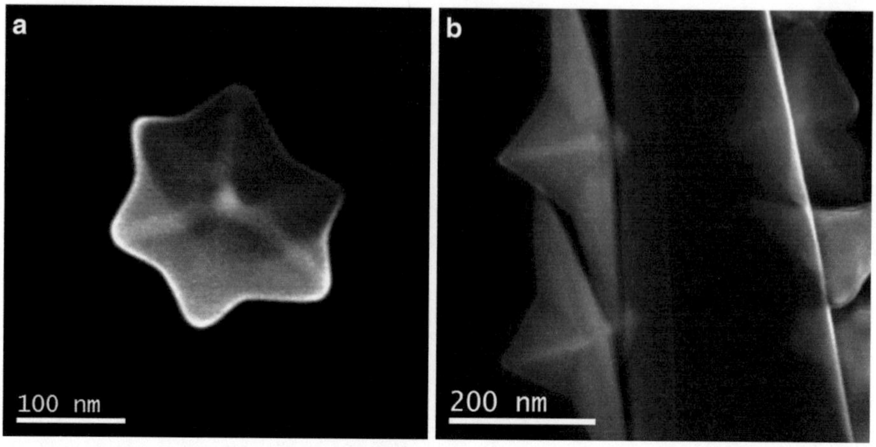

Fig. 17. SEM images of two gold nanoparticles with different branched structures. (**a**) Eight branches; (**b**) four branches.

surfaces replaced by a tetrahedral pyramid, and that in Fig. 16d to a cuboctahedron with each of its 8 {111} truncations replaced by a tetrahedral pyramid.

2. Branched gold nanoparticles can be produced using two-step seed-mediated growth technique. To prepare Silver Seed suspension, mix 25 mL of 0.25 mM $AgNO_3$ with 25 mL of 0.25 mM trisodium citrate in a round bottom flask under vigorous stirring with a magnetic bar. Quickly add 0.6 mL of 0.1 M $NaBH_4$. The color should change immediately to a dark-green/yellow. Keep this silver suspension under static conditions for few hours before using it for nanoparticles growth. To prepare gold nanoparticles, transfer 10 mL of a 1.25 mM CTAB to a round bottom flask, stir with a magnetic bar and incubate at 30 °C until CTAB is completely dissolved. Remove the heat source and immediately add 5 mL of a 0.25 mM $HAuCl_4$ solution; an orange-red mixture should form. Add 10 mL of a 0.125 mM F-127 Pluronic solution, keep stirring vigorously for 5 min. Reduce the gold ions by adding 0.7 mL of 1 M ascorbic acid, followed by the addition of 12.5 μL of the silver seeds suspension. The seeds serve as nucleation sites for the Au nano branched growth. Continue stirring for 30 min and the color of the solution gradually changes to orange-red. Figure 17 shows an example of the type of nanostructures obtained by this method. A gold nanostructure in Fig. 17a displays eight branches, while gold nanocrystal in Fig. 17b has only four branches.

References

1. Siegel RW (1990) Nanophase materials assembled from atomic clusters. MRS Bull 15:60–7
2. Phillips J, Chou CH (1992) Plasma production of metallic nanoparticles. J Mat Res 7:2107–13
3. Porter DA, Easterling KE (1992) Phase Transformations in Metals and Alloys. CRC Press
4. Cao G (2004) Nanostructures and nanomaterials. Synthesis, properties and applications. Imperial College Press, London
5. Elechiguerra JL, Reyes-Gasga J, Jose-Yacaman M (2006) The role of twinning in shape evolution of anisotropic noble metal nanostructures. J Mat Chem 16:3906–19
6. Turkevich J, Stevenson PC, Hillier J (1953) The Formation of Colloidal Gold. J Phys Chem 57:670–673
7. Kimling J, Mainer M, Okenve B, Kotaidis V, Ballot H, Plech A (2006) Turkevich method for gold nanoparticle synthesis revisited. J Phys Chem B 110(32):15700–7
8. Frens G (1973) Controlled nucleation for regulation of particle-size in monodisperse gold suspension. Nat Phys Sci 241:20–2
9. Cademartiri L, Ozrin GA (2009) Concepts of nanochemistry. Wiley-VCH, Germany
10. Nadagouda MN, Varma RS (2007) A greener synthesis of core (Fe, Cu)-shell (Au, Pt, Pd, and Ag) nanocrystals using aqueous vitamin C. Cryst Growth Des 7:2582–7
11. Panigrahi S, Kundu S, Ghosh SK, Nath S, Pal T (2004) General method of synthesis for metal nanoparticles. J Nanopart Res 6:411–4
12. Brust M, Walker M, Bethell D, Schiffrin DJ, Whyman R (1994) Synthesis of thiol-derivatised gold nanoparticles in a two-phase liquid-liquid system. J Chem Soc Chem Commun 7(7):801–2
13. Nikoobakht B, El-Sayed MA (2003) Preparation and growth mechanism of gold nanorods (NRs) using seed-mediated growth method. Chem Mater 15:1962–75
14. Burt JL, Gutierrez-Wing C, Miki-Yoshida M, Jose-Yacaman M (2004) Noble-metal nanoparticles directly conjugated to globular proteins. Langmuir 20:11778–83
15. Vargas Hernandez C, Mariscal MM, Esparza R, Jose-Yacaman M (2010) A synthesis route of gold nanoparticles without using a reducing agent. Appl Phys Lett 96:213115–1
16. Kim F, Connor S, Song H, Kuykendall T, Yang P (2004) Platonic gold nanocrystals. Angew Chem Int Ed Engl 43:3673–7
17. Zhao M, Crooks RM (1999) Dendrimer-encapsulated pt nanoparticles: Synthesis, characterization, and applications to catalysis. Adv Mater 11:217–20
18. Garcia ME, Baker LA, Crooks RM (1999) Preparation and characterization of dendrimer-gold colloid nanocomposites. Anal Chem 71:256–8
19. Bauer LA, Birenbaum NS, Meyer GJ (2004) Biological applications of high aspect ratio nanoparticles. J Mater Chem 14:517–26
20. Wang H, Goodrich GP, Tam F, Oubre C, Nordlander P, Halas NJ (2005) Controlled texturing modifies the surface topography and plasmonic properties of au nanoshells. J Phys Chem B 109:11083–7
21. Burt JL, Elechiguerra JL, Reyes-Gasga J, Montejano-Carrizales JM, Jose-Yacamán M (2005) Beyond archimedean solids: Star polyhedral gold nanocrystals. J Cryst Growth 285:681–91
22. Mayoral A, Vazquez-Duran A, Heinze SG, Jose-Yacamán M (2010) Synthesis and characterization of branched gold nanoparticles. Mater Sci Forum 644:57–60
23. Jana NR, Gearheart L, Murphy CJ (2001) Wet chemical synthesis of high aspect ratio cylindrical gold nanorods. J Phys Chem B 105:4065–7
24. Chen J, Saeki F, Wiley BJ, Cang H, Cobb MJ, Li ZY et al (2005) Gold nanocages: bioconjugation and their potential use as optical imaging contrast agents. Nano Lett 5:473–7
25. Wei Q, Song HM, Leonov AP, Hale JA, Dongmyung Oh, Ong QK et al (2009) Gyromagnetic imaging: dynamic optical contrast using gold nanostars with magnetic cores. J Am Chem Soc 131:9728–34

Chapter 2

Preparation of Concentrated Colloids of Gold Core-Silica Shell Nanoparticles for Biomedical Applications

Yeon-Su Park

Abstract

Encapsulation of gold nanoparticles within a silica shell is highly beneficial to the preparation of highly concentrated gold nanoparticles that can strongly absorb X-ray and hence be used as an X-ray contrast agent. This chapter describes a method for preparing highly concentrated colloidal gold nanoparticles suitable for an X-ray contrast agent application. It describes specific details of procedures for preparing spherical gold nanoparticles, forming thin silica shell on each gold nanoparticle, and enriching the silica-encapsulated gold nanoparticles.

Key words: Contrast agent, Core, Gold, Shell, Silica, Nanoparticle, X-ray absorption

1. Introduction

Gold NPs have attracted great interest because of their biocompatibility, optical and electrical properties, and feasibility for surface modification (1–5) which makes them suitable for various biological applications. Relatively large spherical Au Nanoparticles (NPs) (i.e., $d \geq 10$ nm) have been extensively studied for biological applications owing to their strong surface plasmon resonance (SPR) in the visible region (1–5) as well as their strong X-ray absorption (6–9). Usually such Au NPs are prepared by the reduction of Au^{3+} ions to metallic Au atoms using citrate ions as a reducing and capping agent (10–12). Such methods allow to control the size of Au NPs reasonably well, from a few nm to several tenth nm, by changing concentration of their precursors (i.e., Au^{3+} and citrate) or their ratio.

Au NPs are biocompatible and have large X-ray mass attenuation coefficients (9, 13, 14). Thus, they have high potential as an X-ray contrast agent for computed tomography which is one of the

Mikhail Soloviev (ed.), *Nanoparticles in Biology and Medicine: Methods and Protocols*, Methods in Molecular Biology, vol. 906, DOI 10.1007/978-1-61779-953-2_2, © Springer Science+Business Media, LLC 2012

most indispensable modern diagnostic tools. Currently tri-iodobenzene derivatives are widely used as the contrast agents but they are suffered from some serious drawbacks such as limited imaging time, due to their short vascular circulation time, and renal toxicity (6, 15). Although Au NPs have large X-ray mass attenuation coefficients, their colloidal concentration is typically very low and hence an additional enrichment process is required for their practical application as an X-ray contrast agent. However, it is practically infeasible to prepare stable, highly concentrated colloids of relatively large Au NPs because of their coagulation tendency increasing with their concentration. One practical and bio-friendly solution to the problem is to form a silica shell on each Au NP because silica is chemically inert, biocompatible, and highly flexible for further surface modifications (14, 16, 17), and bears strong negative charge in physiological media (18). The silica shell on the Au NP can act as a physical barrier against other Au NPs and that can enhance the steric stabilization of the NPs (16). In addition, the silica shell can enhance the electrostatic stabilization of the NPs due to its high surface charge density (16). The enhanced steric and electrostatic stabilizations by the silica shell permit to prepare stable, highly concentrated colloids of Au NPs. Here, we report detailed procedures for the synthesis of Au NPs and for making stable, highly concentrated Au-SiO_2 NPs colloids suitable for use as efficient X-ray contrast agent (Fig. 1).

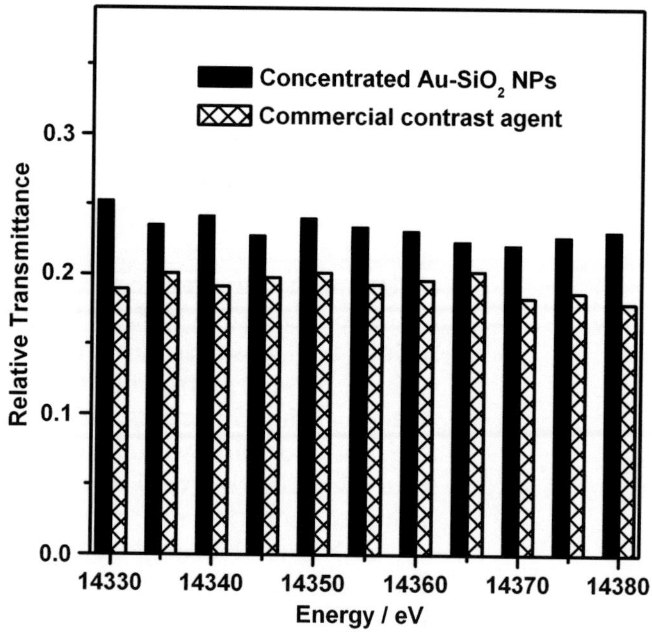

Fig. 1. Transmittances of the concentrated colloid of Au core. Relative X-ray transmittances of the concentrated colloid of Au core (d ~15 nm)-SiO_2 shell (thickness ~4 nm) NPs and a commercial X-ray contrast agent (Iopamiron® 300, Schering) near Au L_1 edge [Au] = 100 mM. Relative transmittance is the ration of sample transmittance to the transmittance of water.

2. Materials

All chemical reagents should be of analytical grade. Use deionized water (i.e., Milli-Q water with $R > 18.2$ MΩ cm) for all solutions and for Au-SiO_2 NPs purification. Strictly follow all waste disposal regulations and guidelines when disposing waste materials. Use personal protective equipment at all times.

2.1. Preparation of Colloidal Au NPs

1. Hydrogen tetrachloroaurate (III) tetrahydrate solution (Au^{3+} stock solution) (see Note 1): 25 mM aqueous solution of $HAuCl_4{\cdot}4H_2O$ in deionised water (see Note 2). Store in a brown glass bottle. Cap the brown bottle tightly and store at 4 °C. Typically 20 mL of the stock solution should be sufficient for a few experiments.
2. Sodium citrate solution: 1 wt % aqueous solution of trisodium citrate (Na_3Cit). Make 15 mL and store in a glass vial.
3. Two magnetic hot plate stirrers.
4. Volumetric flasks (20, 250, 500 mL).
5. Erlenmeyer flask (500 mL).
6. Brown glass vials with caps for Au NP storage.

2.2. Thin Silica Shell Formation

1. 1 mM 3-Aminopropyl trimethoxysilane (APS) solution: add 11.377 mL of water to a 30 mL-capacity glass vial and 2 μL of APS, cap the vial, mix the solution by shaking. Prepare fresh before use.
2. Sodium silicate solution (0.54 wt % Na_2SiO_3): add 0.108 g of commercially available sodium silicate solution (ca. 27 wt % SiO_2 in 14 % NaOH) and 20 mL of water to a 30 mL-capacity glass vial and stir the sodium silicate solution at 500 rpm on a hot plate magnetic stirrer. Check pH using a pH meter or a piece of pH paper. If pH of the solution is outside the pH range 10.5 and 11, add a small amount of DOWEX® 50WX4-400 ion-exchange resin to the sodium silicate solution and then check pH. Keep adding the ion-exchange resin until pH of the sodium is 4. Turn off the stirring and leave the solution at least 30 min at room temperature without any stirring to allow the resin to settle down (see Note 3).
3. DOWEX® 50WX4-400 ion-exchange resin (Sigma-Aldrich, USA).
4. Two magnetic hot plate stirrers.
5. A pH meter.
6. Glass vials (30 mL)

2.3. Purification and Enrichment

1. Plastic centrifuge tubes with 12 mL-capacity ("12 mL-tubes").
2. Plastic centrifuge tubes with 50 mL-capacity ("50 mL-tubes").
3. A centrifuge rotor ("large rotor") with six fixed holes for placing the 50 mL-tubes.
4. A centrifuge rotor ("small rotor") with twelve fixed holes for placing the 12 mL-tubes.
5. A centrifuge (maximum relative centrifugal force, RCF = 40,000 × g or higher) which can handle 12 and 50 mL-tubes.
6. A vortex shaker.

3. Methods

3.1. Preparation of Colloidal Au NPs

All procedures described here should be performed in a fume hood.

1. Place the capped vial containing 15 mL of 1 wt % Na_3Cit solution on a magnetic hot plate stirrer, set the hot plate temperature to 80 °C, and incubate for 20 min or longer.
2. Place a 500 mL-capacity Erlenmeyer flask on another magnetic hot plate stirrer and put a magnetic stirring bar into the flask.
3. Fill the Erlenmeyer flask with 279 mL of water (see Note 4). After that, add 6 mL of the 25 mM Au^{3+} stock solution to the Erlenmeyer flask.
4. Set magnetic stirring speed at 1,500 rpm and hot plate temperature at 160 °C. After that, start stirring and heating. Keep stirring and heating until observing intense boiling of the Au precursor solution in the Erlenmeyer flask (see Note 5).
5. Transfer all of the Na_3Cit solution (15 mL, see step 1) into the Erlenmeyer flask containing boiling Au precursor solution (see Note 6).
6. After 1 min of continuous stirring at 1,500 rpm, decrease the stirring speed to 150 rpm. When the color of the solution changes to dark purple, increase the stirring speed to 1,500 rpm and continue stirring for another 20 min.
7. Remove the Erlenmeyer flask with Au NPs from the hot plate stirrer and place it onto another stirrer kept at room temperature (see Note 7). Continue stirring at 500 rpm until the solution cools down to room temperature.

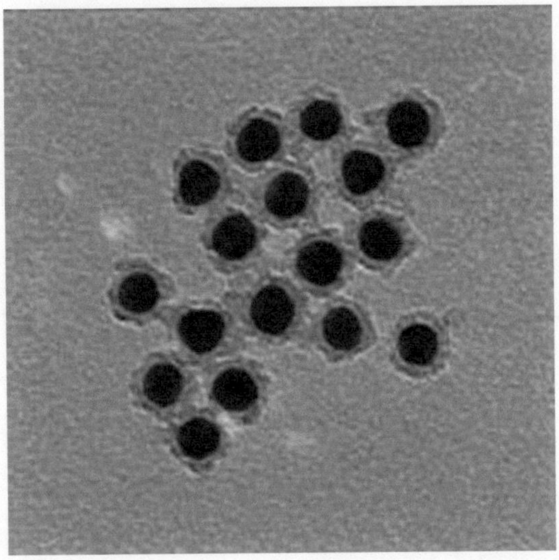

Fig. 2. TEM imaging. TEM image of typical Au core (d~15 nm)-SiO_2 shell (thickness ~4 nm) nanoparticles. Reproduced from (16) with permission from American Chemical Society.

8. The total volume of Au NP solution should be adjusted to compensate for the water which evaporated during the previous stages. Add water to the solution of Au NP to make the total volume of 300 mL (see Note 8). The final concentrations of Au^{3+} and citrate are 0.5 and 0.17 mM, respectively. The prepared Au NPs have diameter of ~15 nm (Fig. 2) and show strong SPR peak at 519 nm (Fig. 3).
9. Store Au NPs in a tightly closed dark glass bottle at 4 °C.

3.2. Thin Silica Shell Formation

This section exemplifies the procedures using 100 mL of the freshly prepared Au NPs (see previous section). This method should yield ~4 nm-thick silica shells (Fig. 2).

1. Transfer 100 mL of the Au NP solution (see Subheading 3.1, step 8) into a 200 mL glass bottle and place it on a magnetic hot plate stirrer, stir the solution at 1,500 rpm.
2. Add 0.25 mL of 1 mM APS solution under constant stirring.
3. Immediately after that, add 4 mL of 0.54 % sodium silicate solution (pH 10.5–11) into the bottle and cap the bottle tightly, continue stirring at 1,500 rpm for 10 min (see Note 9).
4. Transfer the bottle from the magnetic stirrer to a dark place and keep it there for 3 days. The prepared Au-SiO_2 NPs have silica shells approximately ~4 nm think (Fig. 2) and show their SPR at 523 nm (Fig. 3) (see Note 10).

3.3. Purification and Enrichment

Purification and enrichment of Au-SiO_2 NPs should be performed at 4 °C in ambient atmosphere. This section exemplifies the procedures using 100 mL of the freshly prepared Au-SiO_2 NPs (see previous

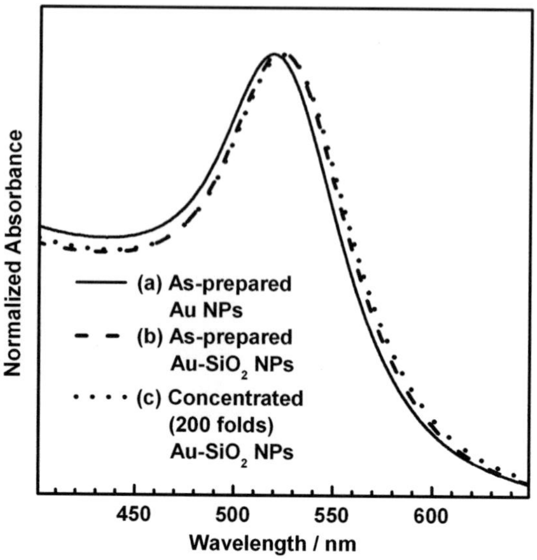

Fig. 3. UV-visible absorption spectra of nanoparticles. UV-visible absorption spectra of Au ($d \sim 15$ nm) and Au core ($d \sim 15$ nm)-SiO_2 shell (thickness ~4 nm) NPs: (**a**) citrate-capped Au NPs ([Au] = 0.5 mM), (**b**) Au core-SiO_2 shell NPs ([Au] = 0.5 mM), and (**c**) Au core-SiO_2 shell NPs ([Au] ~0.5 mM) diluted from corresponding concentrated NPs ([Au] = 100 mM). All spectra were normalized by taking maximum absorbance for each spectrum to be unity.

section). This method should yield 200× fold concentrated NPs solution (equivalent to 100 mM Au).

1. Place four 50 mL centrifuge tubes in a rack on the experimental bench and uncap them.
2. Transfer 25 mL of the sample (Au-SiO_2 NPs) solution into 50 mL centrifuge tubes using a pipette, add 20 mL of water, and cap the tube. Prepare four such sample tubes in total.
3. Weigh each of the capped tubes and balance their weights using water (see Note 11).
4. Centrifuge the large sample tubes at RCF of 25,000 × *g* for 15 min at 4 °C.
5. Immediately after the centrifugation, take all the tubes out of the rotor and put them in a rack on a bench, check that the volume of the bottom layer (dark-colored layer at the bottom of the tubes formed by NP precipitates) is 20 ± 4% of the total volume (see Note 12).
6. Uncap all the tubes and then discard supernatants (see Note 13).
7. Add 40 mL of water into each large sample tube which contains ~5 mL of the Au-SiO_2 NP precipitate (the bottom layer remaining after the centrifugation and subsequent supernatant removal, see steps 5 and 6), tightly cap all the tubes, and re-disperse the nanoparticles by vigorous vortexing.

8. Weigh the capped, large sample tubes and then balance their weights using water (see Note 11).
9. Re-disperse the solutions by using a vortex shaker (see Note 14).
10. Repeat steps 4–9 twice and then steps 4–6 once. This will accomplish Au NPs washing.
11. Tightly cap all the tubes and then re-disperse the Au-SiO_2 NP in the solution remaining after the centrifugation and subsequent supernatant removal (~9 mL/tube) (see step 10) using a vortex shaker.
12. Combine the content of all four tubes and transfer evenly into twelve 12 mL-tubes using a pipette (see Note 15). Top up tubes with water to achieve ~10 mL total volume in each tube and cap the tubes.
13. Weigh the capped tubes and balance their weights using water (see Note 11).
14. Centrifuge the capped tubes at 4 °C for 15 min at RCF of 25,000 × *g* or use the alternative RCF/time settings selected in step 5 above (see Note 12).
15. Immediately after the centrifugation, take all the tubes out of the rotor and put them in a rack on a bench, and discard supernatant (see Note 13).
16. Tightly cap all the tubes and then re-disperse the Au-SiO_2 NP in the solution remaining after the centrifugation and subsequent supernatant removal (~2 mL/tube) (see steps 14 and 15) using a vortex shaker.
17. Combine the content of tubes by transferring content of six of the tubes into the remaining six tubes. This procedure yields six tubes. The empty tubes may be washed as described in the Note 15 to minimize nanoparticles losses. Top up these tubes with water to a final volume of ~10 mL each and cap the tubes.
18. Repeat centrifugation exactly as described in steps 13–16 above.
19. Combine the content of three tubes by transferring all nanoparticles into one tube, do the same with the remaining three tubes. This procedure yields two tubes with nanoparticles. The empty tubes may be washed as described in the Note 15 to minimize nanoparticles losses. Top up these tubes with water to a final volume of ~10 mL each and cap the tubes.
20. 18. Repeat centrifugation exactly as described in steps 13–16 above.
21. Combine the content of the two remaining tubes by transferring all nanoparticles into one tube.

22. Weigh the capped, small sample tube and place it into one of the fixed holes of the centrifuge. Balance the centrifuge using a capped 12 mL-tube containing water.
23. Centrifuge the tubes at the same RCF as in step 14 above at 4 °C until the volume of the dark-colored Au-SiO_2 NP precipitate layer in each tube decreases to 0.50 mL or slightly less (see Note 16).
24. Immediately after the centrifugation, take all the tubes out of the rotor and put them in a rack on a bench, and discard supernatant (see Note 13).
25. Tightly cap all the tubes and then re-disperse the Au-SiO_2 NP in the solution remaining after the centrifugation and subsequent supernatant removal (~0.5 mL/tube) (see steps 21 and 22) using a vortex shaker.
26. Transfer the nanoparticles solution into 1 mL microcentrifuge tube, note the solution volume. Add water to the nanoparticles up to the total volume of 0.5 mL (see Note 17).
27. Wrap the tube using 2–3 layers of aluminium foil and then store it in a refrigerator at 4 °C.

4. Notes

1. Typically manufacturers provide $HAuCl_4{\cdot}4H_2O$ in an airtight glass ampoule housed in a plastic container in a light-protecting box because the chemical is light-sensitive and very hygroscopic. Store it as-received in a refrigerator at 4 °C. To take the chemical from the ampoule, make a circle of a shallow score on the surface of the ampoule top using a glass cutter, clean the ampoule surface using water (or ethanol) and Kimwipes®, and then cut the ampoule with your hands. Be careful not to contaminate the chemical inside the ampoule when cutting the ampoule. After use, tightly seal the ampoule with a few layers of Parafilm®, place back it inside the plastic container, cap the container, and then seal the container using a few layers of Parafilm® to protect the chemical from light and moisture. Store it in a refrigerator at 4 °C.
2. Due to hygroscopic nature of $HAuCl_4{\cdot}4H_2O$, prolonged exposure to ambient air will cause errors in determining accurate mass of the salt. The mass of the $HAuCl_4{\cdot}4H_2O$ should be noted as soon as possible and the salt should be washed off the weighing dish directly into a volumetric flask with small amount of water. Shake gently to dissolve all solids in the solution and fill the flask to the required volume with water.

We typically use 20 mL volumetric flask and 0.206 g of $HAuCl_4 \cdot 4H_2O$ to make 20 mL of 25 mM stock solution.

3. The resin particles start sinking down onto the bottom of the vial as soon as the stirring stops. Do not stir or shake the vial to prevent disturbing the resin particles. Alternatively, centrifuge the sodium silicate solution and transfer the supernatant from the centrifugation tube to a suitable volume glass vial, cap the vial and store it at room temperature.
4. An easy but accurate way is to use a volumetric flask and a pipette. For example, measure 250 mL of water in a 250 mL-capacity volumetric flask, deliver the water completely to the Erlenmeyer flask, and then add 29 mL of water into the Erlenmeyer flask using a pipette.
5. Use an elevated temperature setting (170 °C or higher) if the solution does not boil within 20 min.
6. Upon transferring the Na_3Cit solution into the boiling Au precursor solution, solution color in the flask changes from pale yellow to pale black.
7. The flask is very hot. Wear additional heat-protect gloves and then handle it carefully.
8. The total volume of the as-prepared Au NP solution is less than 300 mL owing to loss of water into atmosphere during the preparation. This procedure makes up for the water loss. This can be achieved for example using 250 and 50 mL volumetric flasks as follows: Fill a 250 mL flask with 250 mL of the Au NP solution. Transfer all of the remaining solution to a 50 mL flask and top it up with water up to 50 mL mark. Combine both of the above Au NP fractions by transferring them into a larger volume dark glass flask. This will yield 300 mL of the Au NP solution with the stated concentration.
9. pH of the solution at this stage should be approximately ~8.5.
10. Longer incubation period may result in thicker SiO_2 shell formation. Thickness can be reached to ~6 nm (16).
11. If weight difference between the heaviest tube and any of the other lighter tubes is larger than 0.001 g, adjust weight of the lighter sample tubes by adding water so that weight difference between the lightest tube and the heaviest tube is within 0.001 g. The weight difference should be 0.001 g or smaller. Much larger weight difference may cause large irregular vibration of the equipment and lead to serious damage to the centrifuge. When setting up the centrifuge, pay attention to the locations of the sample tubes. Each pair of the sample tubes should be located across each other to maintain balance during centrifugation.

12. Before uncapping tubes, check the volume of the bottom layer (dark color due to the NP precipitates) in the tubes. If the layer volume is not within 20 ± 4% (5 ± 1 mL for 25 mL samples, 2 ± 0.4 mL for 10 mL samples) of the solution volume before centrifugation, adjust centrifugation conditions (RCF and time).
13. Carefully decant the supernatant onto a waste bottle or carefully use a pipette (with a 10 mL-capacity tip) to discard the supernatant. In any cases, give a special care to minimize agitation of the solutions which results in spreading the NPs into the supernatant layer.
14. If a vortex shaker is not available, just shake the capped, weight-adjusted sample tubes vigorously until the sample solutions becomes homogeneous.
15. To minimize concentration errors after the sample solution transfer from the 50 mL-tubes to the 12 mL-tubes, the following series of processes are strongly recommended. First, cap the tubes and mix Au NP solutions in each tube by vigorous shaking. Transfer the content of three 50 mL tubes into the forth 50 mL tube, mix vigorously and distribute this mixture to twelve 12 mL tubes using a pipette. Typically there will be a small amount of the sample solution left in each of the 50 mL tubes following the first transfer. To improve yield and minimize concentration errors, use small amount of water to wash sequentially all four tubes and then distribute this solution equally between the twelve 12 mL-tubes. Such washing procedure may be repeated if nanoparticle transfer appears incomplete. Finally top up volume in 12 mL tubes with water to 10 mL.
16. If the volume of the dark-colored Au-SiO_2 NP precipitate layer in each tube is larger than 0.50 mL, run centrifugation further at RCF of 25,000 × *g* at 4 °C until the volume of the layer decreases to 0.50 mL or slightly less. It may take more than 20 min.
17. The desired amount of water added is determined by subtracting the initially delivered sample volume in the tube from 0.5 mL.

Acknowledgments

This chapter is based on author's research work as a postdoctoral researcher and a visiting associate professor at Center for Interdisciplinary Research, Tohoku University, Sendai, Japan. I deeply thank to Professor Atsuo Kasuya and Professor Luis M. Liz-Marzán (Professor at University of Vigo, Spain) for their kind advice and help during my work.

References

1. Jain PK et al (2007) Review of some interesting surface plasmon resonance-enhanced properties of noble metal nanoparticles and their applications to biosystems. Plasmonics 2:107–118
2. Wang Z, Ma L (2009) Gold nanoparticle probes. Coord Chem Rev 253:1607–1618
3. Agasti SS et al (2010) Nanoparticles for detection and diagnosis. Adv Drug Deliv Rev 62:316–328
4. Li Y, Schluesener JS, Xu S (2010) Gold nanoparticle-based biosensors. Gold Bull 43:29–41
5. Khlebtsov N, Dykman L (2011) Biodistribution and toxicity of engineered gold nanoparticles: a review of in vitro and in vivo studies. Chem Soc Rev 40:1647–1671
6. Hainfeld JF et al (2006) Gold nanoparticles: a new X-ray contrast agent. Br J Radiol 79: 248–253
7. Park Y-S et al (2006) X-ray absorption of gold nanoparticles with thin silica shell. J Nanosci Nanotechnol 6:3503–5056
8. Park Y-S et al (2007) Concentrated colloids of silica-encapsulated gold nanoparticles: colloidal stability, cytotoxicity, and X-ray absorption. J Nanosci Nanotechnol 6:3503–3506
9. Hahn MA et al (2011) Nanoparticles as contrast agents for in-vivo bioimaging: current status and future perspectives. Anal Bioanal Chem 399(1):3–27
10. Turkevich J, Stevenson PC, Hillier J (1951) A study of the nucleation and growth processes in the synthesis of colloidal gold. Discuss Faraday Soc 11:55–75
11. Frens G (1973) Controlled nucleation for the regulation of the particle size in monodisperse gold suspensions. Nat Phys Sci 241:20–22
12. Kimling J et al (2006) Turkevich method for gold nanoparticle synthesis revisited. J Phys Chem B 110:15700–15707
13. Connor EE et al (2005) Gold nanoparticles are taken up by human cells but do not cause acute cytotoxicity. Small 1:325–327
14. Xu ZP et al (2006) Inorganic nanoparticles as carriers for efficient cellular delivery. Chem Eng Sci 61:1027–1040
15. Yu S-B, Watson AD (1999) Metal-based X-ray contrast media. Chem Rev 99:2353–2377
16. Liz-Marzán LM, Giersig M, Mulvaney P (1996) Synthesis of nanosized gold-silica core-shell particles. Langmuir 12:4329–4335
17. Barbé C et al (2004) Silica particles: a novel drug-delivery system. Adv Mater 16: 1957–1966
18. Cui X et al (2005) Nonionic triblock copolymer synthesis of SBA-15 above the isoelectric point of silica (pH=2–5). Mater Lett 59: 2257–2261

Chapter 3

Biosynthesis of Silver and Gold Nanoparticles Using *Bacillus licheniformis*

Muthu Irulappan Sriram, Kalimuthu Kalishwaralal, and Sangiliyandi Gurunathan

Abstract

Owing to the wide-ranging applications of noble metal nanoparticles in diverse areas of science and technology, different methods have been proposed for their synthesis. Here, we describe the methods for the intracellular biosynthesis of silver and gold nanoparticles using the bacterium *Bacillus licheniformis* KK2 and this same procedure can be followed for other bacteria as well. The biological synthesis of nanoparticles is highly eco-friendly and possesses distinct advantages such as enhanced stability, better control over the size, shape, and monodispersity of the nanoparticles, when compared with the more traditional physical and chemical methods which often involves the use of hazardous chemicals creating environmental concern.

Key words: Biosynthesis, Bacteria, Silver nanoparticles, Gold nanoparticles, *Bacillus licheniformis*

1. Introduction

The synthesis of noble metal nanoparticles with predefined shape and size is of great importance in diverse applications such as optical devices (1), data storage (2), catalysis (3), biodiagnostics (4), drug delivery system (5), and cancer treatment (6, 7). The highly desirable features for the synthesis of nanoparticles involve the control over size, morphology, composition, and stability of nanoparticles along with utilization of eco-friendly synthesis methods.

One of the most exciting research topics in the field of nanobiotechnology is the synthesis of nanoparticles by biological systems. Although the microorganisms such as bacteria, yeast, and fungi were identified to remediate toxic metals through the reduction of metal ions, very recently they were unveiled to be harnessed

Mikhail Soloviev (ed.), *Nanoparticles in Biology and Medicine: Methods and Protocols*, Methods in Molecular Biology, vol. 906, DOI 10.1007/978-1-61779-953-2_3, © Springer Science+Business Media, LLC 2012

as potential nanofactories. Both the unicellular and multicellular microorganisms were reported to produce inorganic nanomaterials mediated by intracellular and extracellular enzymatic constituents (8–11). The environmentally benevolent synthesis of nanoparticles by biological system is comprised of the process that takes place at neutral pH, ambient temperature and pressure. Considering the bacterial mediated synthesis methods, even the heavy metal salts which are normally toxic to the living organisms can be converted into their corresponding nanoparticles (12–14).

Although a vast array of biological resources are available in nature comprising algae, yeast, fungi, viruses, plant and plant products for the synthesis of nanoparticles (15, 16), the bacterial biomass exhibit distinct advantages such as ease of handling and enhanced suitability for genetic manipulation (17, 18). Hence, in this chapter, we want to emphasize on the bacterial mediated intracellular synthesis of silver and gold nanoparticles. There are reports evidencing the fact that reduction of silver and gold ions to their corresponding metal nanoparticles is mediated by the intracellular and membrane bound reductive enzymes of the bacteria (19, 20). Here we describe biosynthetic approach to the manufacturing of silver and gold nanoparticles using *Bacillus licheniformis* and the methods for their purification and characterization.

2. Materials

Prepare all reagent solutions with the chemicals of analytical grade and use ultra pure Milli-Q water free from ionic, organic, and bacterial contaminants for dissolving the salts and powders (see Notes 1 and 2). Store all the reagents at room temperature unless stated otherwise. Follow the waste disposal regulations while discarding the waste materials. Wear appropriate personal protective equipment (gloves, laboratory coat, and protect hearing when using sonicator devices).

2.1. Microbial Strain and the Bacterial Biomass Production

This rod-shaped gram-positive bacterium *B. licheniformis* KK2 is used here to exemplify the procedures for synthesis of both silver and gold nanoparticles (see Note 3).

1. Bacterial cell stock—*B. licheniformis* KK2 (see Note 4).
2. Nutrient agar (NA) (HiMedia). Use 20 mL of medium per Petri plate (100 × 15 mm) (see Note 5).
3. Nutrient Broth (HiMedia). Do not exceed one fifth of the volume of the flask for liquid cultures (e.g., use no more than 50 mL per 250 mL flask) (see Note 5).
4. Petri plates (100 × 15 mm).

5. 15 mL test tubes (110 × 16 mm).
6. 250 mL conical flasks.
7. Rotary Shaker capable of 160–200 rpm.
8. 5 °C refrigerator.
9. Centrifuge and rotor capable of spinning 50 mL tubes at 1,725 × *g*, e.g., Heraeus Biofuge stratos rotor 3335.

2.2. Synthesis of Noble Metal Nanoparticles

1. Silver nitrate stock solution (10 mM $AgNO_3$): dissolve 84 mg of $AgNO_3$ in 50 mL of Milli-Q water. Store it in a cool dry place away from light (see Note 1).
2. Gold chloride stock solution (10 mM $AuCl_4$): dissolve 98 mg of $AuCl_4$ in 50 mL of Milli-Q water. Store in a closed container at 4 °C (see Note 2).
3. Phosphate-buffered saline (PBS): 8 g NaCl, 1.14 g $NaHPO_4$, 0.2 g KCl, 0.2 g KH_2PO_4 to 1,000 mL with deionized water. The PBS used for washing biomass must be sterilized to maintain the purity of the culture. The pH 7.0 is required for maintaining the isotonic condition to the bacterial cells.
4. Autoclaved conical flasks with 50 mL of sterile Milli-Q water.
5. Orbital shaking incubator.

2.3. Purification of Silver and Gold Nanoparticles

1. Phosphate-buffered saline (PBS)—same as in Subheading 2.2.
2. Sonicator (Sonics Vibra Cell VC-505/220, Newtown, CT, USA).
3. Centrifuge with rotor capable of spinning 50 mL tubes at 2,500 × *g*, e.g., (Heraeus Biofuge stratos, rotor 3335).
4. Sterile micro filter units of 0.22 μm pore size (Millipore Express (33 mm)) that are able to withstand the pressure of up to 150 psi (see Note 6).
5. Sterile or autoclaved centrifuge tubes.

2.4. Characterization of Silver and Gold Nanoparticles

1. Digital camera (see Note 7).
2. UV–vis Spectrophotometer (UV PharmaSpec 1700, Shimadzu) with a resolution of 0.72 nm. The alternative instruments that can be used are Systronics Double Beam Spectrophotometer 2203, NanoDrop 1000 Spectrophotometer, Thermo Scientific, etc.
3. Quartz cuvette.
4. Fourier Transform Infrared (FT-IR) Spectrometer (Shimadzu, Japan) (see Note 8).
5. Transmission Electron Microscope (TEM) (JEOL model 1200 EX) operated at an accelerating voltage of 80 kV (see Note 9).

3. Methods

Perform all procedures at room temperature unless indicated otherwise.

3.1. Production of Bacterial Biomass

This section describes the method for the production of bacterial biomass suitable for either the production of silver or gold nanoparticles.

1. Maintain the working culture in nutrient agar plates at 4 °C and subculture it for every fortnight (see Note 10).
2. Use a sterile inoculation loop to transfer a single colony of cells on the nutrient agar plate into 3 mL of NB medium and incubate overnight in a rotary shaker at 160 rpm at 37 °C. This overnight grown culture is called as seed culture (see Note 11).
3. Add 1 mL of seed culture to the conical flask containing 50 mL of NB production media (2 % inoculation) with the help of sterile micro tip and incubate it for 24 h at 160 rpm at 37 °C (see Note 12).
4. Centrifuge bacterial culture at 1,725 × *g* for 10 min, so that all the biomass are efficiently separated and then discard the cell free culture supernatant.
5. Resuspend the biomass with PBS and then once again centrifuge it at 1,725 × *g* for 10 min to precipitate the biomass. Repeat this step three times.

3.2. Synthesis of Silver and Gold nanoparticles

3.2.1. Biosynthesis of Silver Nanoparticles

1. Resuspend the bacteria (obtained from the 50 mL of original growth medium) in 45 mL of sterile Milli-Q water directly in the centrifuge tube. Transfer to an autoclaved 250 mL conical flask under aseptic condition. The appearance of pale white color indicates the presence of microbial cells.
2. Add 5 mL of 10 mM $AgNO_3$ stock solution to the 45 mL of biomass, so that the final concentration of 1 mM $AgNO_3$ is achieved. Incubate the conical flask in a closed orbital shaking incubator at 200 rpm at 37 °C for 24 h for the biogenesis of silver nanoparticles. Protect the flask from light during the incubation.
3. The characteristic change of color from the initial whitish-yellow shade of silver nitrate to brown indicates the successful biosynthesis of silver nanoparticles (see Note 13).

3.2.2. Biosynthesis of Gold Nanoparticles

1. Resuspend the bacteria (obtained from the 50 mL of original growth medium) in 45 mL of sterile Milli-Q water directly in the centrifuge tube. Transfer to an autoclaved 250 mL conical

flask under aseptic condition. The appearance of pale white color indicates the presence of microbial cells.

2. Add 5 mL of 10 mM $AuCl_4$ stock solution to the biomass dissolved in 45 mL of Milli-Q water, so that the final concentration of 1 mM $AuCl_4$ is obtained. Incubate the conical flask in an orbital shaking incubator at 200 rpm for 24 h at 37 °C for the synthesis of gold nanoparticles (see Note 14).
3. The characteristic change of color from the initial pale-yellow shade of gold chloride to purple indicates the successful biosynthesis of gold nanoparticles (see Note 15).

3.3. Purification of Nanoparticles

1. Centrifuge the nanoparticles carrying biomass suspension at 1,725 × *g* for 10 min and then discard the supernatant. To wash the microbial cultures containing the nanoparticles, resuspend the cells in PBS and precipitate by centrifugation at the same speed mentioned above. Repeat the washing step for the total of three times at room temperature.
2. Resuspend the biomass in ice-cold PBS and sonicate it on ice with an ultrasonic processor for 20 s, pause for 30 s (keep the cells on ice). Repeat sonication cycle three times with an interval of 30 s between each cycle (see Note 16).
3. Filter the sonicated biomass using 0.22-μm filter, so that the purified silver and gold nanoparticles are obtained (see Note 17).
4. Centrifuge the nanoparticles at 2,700 × *g* for 30 min to precipitate the nanoparticles, discard the supernatant. Do not disturb the pellet to avoid losing the nanoparticles. Resuspend the nanoparticles in sterile deionized water.

3.4. Characterization of Nanoparticles

The nanoparticles could be characterized using UV–vis absorption spectrum, FT-IR spectrum, and TEM analysis. The characteristic colors of silver and gold nanoparticles are brown and purple (see Fig. 1).

1. Measure UV–visible absorption spectra (200–800 nm) for the silver and gold nanoparticles. These should show distinct absorption peaks at 420 and 540 nm, respectively (see Figs. 2 and 3). Use Milli-Q water to obtain the background spectrum, subtract from the nanoparticle spectrum.
2. Thoroughly dry the purified nanoparticles and then subject the powder to FT-IR analysis by scanning it in the range 450–4,000 cm^{-1} to obtain the spectra (see Figs. 4 and 5) from which the surface chemistry of the particles can be understood.
3. Use the liquid sample of purified nanoparticles for TEM analysis from which the exact size and shape of the nanoparticles can be confirmed (see Fig. 6a, b).

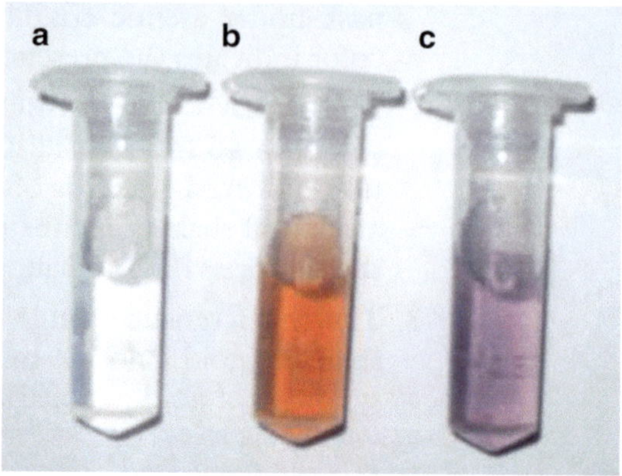

Fig. 1. Nanoparticles resuspended in water. (**a**) Milli-Q water without nanoparticles. The characteristic brown and purple color of finally purified silver (**b**) and gold (**c**) nanoparticles suspended in Milli-Q water.

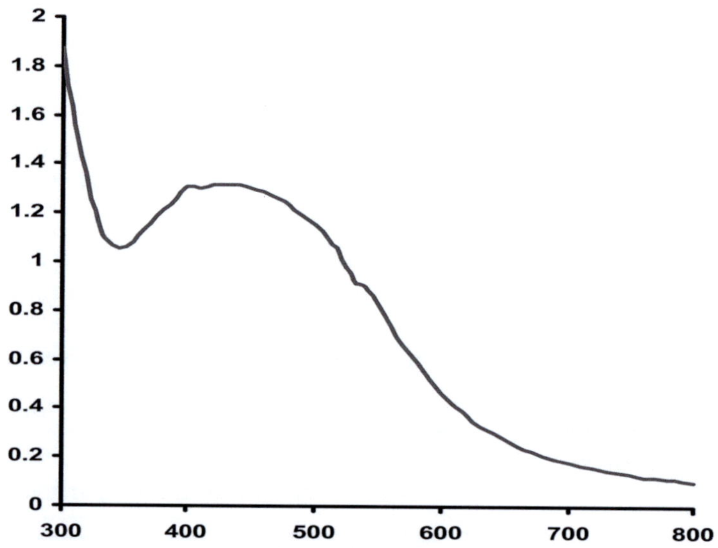

Fig. 2. The UV–vis spectrophotometric analysis of silver nanoparticles. Absorbance maximum is at 430 nm. Vertical axis—absorbance, arbitrary units.

4. Notes

1. Silver nitrate, a transparent, odorless, and colorless crystal salt, is highly corrosive and toxic to humans. Prolonged exposure can lead to a permanent bluish discoloration of skin (Argyria), mucous membranes, and conjunctiva. Lung disease can also

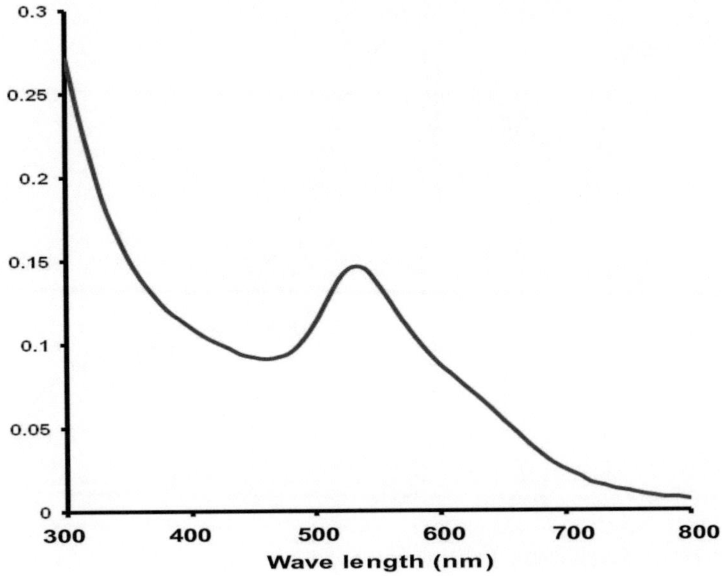

Fig. 3. The UV–vis spectrophotometric analysis of gold nanoparticles Absorbance maximum is at 530 nm. Vertical axis—absorbance, arbitrary units.

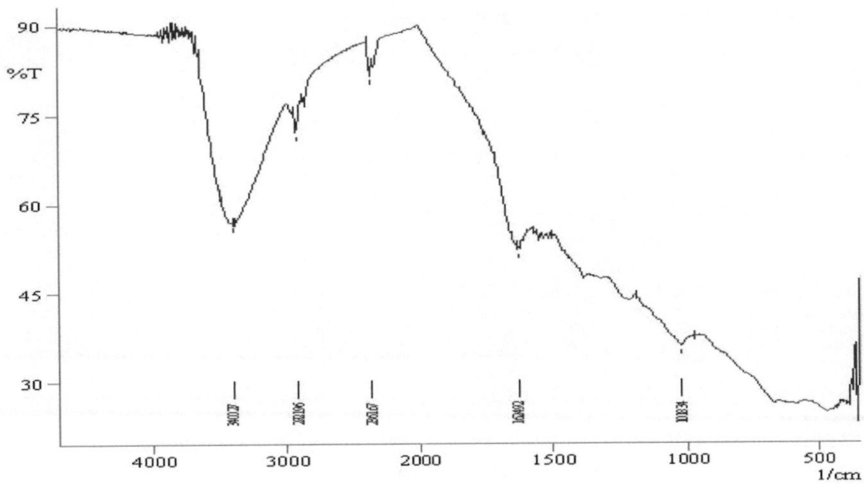

Fig. 4. FT-IR spectrum of silver nanoparticles. The peak values at 3,400, 2,921, 2,361, 1,624, and 1,018/cm correspond to functional groups present on the surface of NPs. These are phenol (O–H), alkane (C–H), nitrile (C = O), amine (N–H), and ester (C–O) groups, respectively.

occur when inhaled repeatedly. Wear appropriate protective lab equipments like goggles and shield, proper gloves, lab coat, and apron before handling silver nitrate. It is extremely annihilative to the mucous membranes and upper respiratory tract when inhaled; always handle this inside the vent hood. The $AgNO_3$ is light sensitive, combustible, and a severe oxidizer.

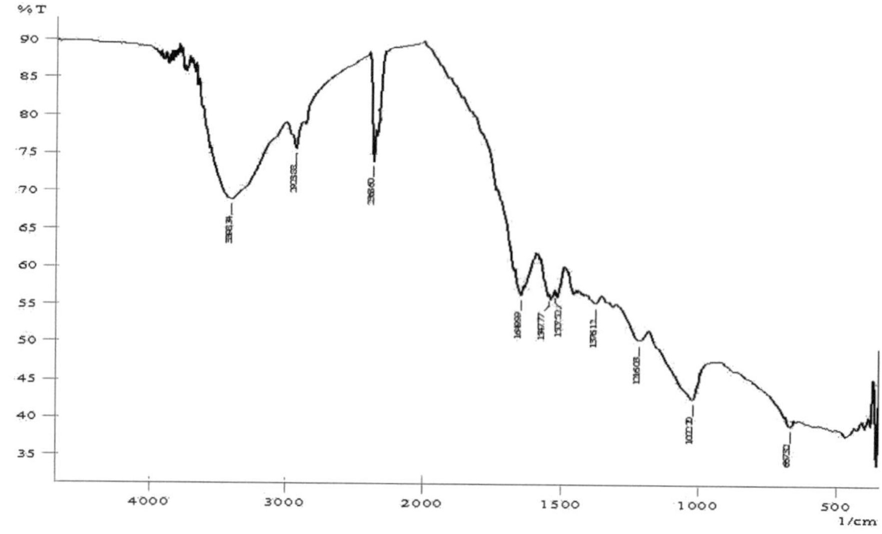

Fig. 5. FT-IR spectrum of gold nanoparticles. The peak values at 3,398, 2,923, 2,368, 1,649, 1,216, 1,020, and 667/cm correspond to functional groups present on the surface of NPs. These are phenol (O–H), alkane (C–H), nitrile (C = O), amide (N–H), acid (C–O), ester (C–O), and alkyne (C–H) groups, respectively. Three peaks values at 1,547, 1,537, and 1,376/cm denote the presence of nitro (N = O) functional group.

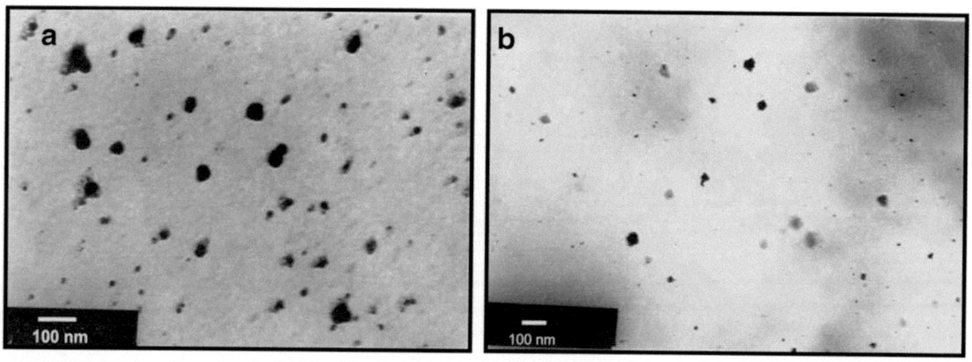

Fig. 6. TEM image of nanoparticles. Silver (**a**) and gold (**b**) nanoparticles. Images indicate spherical shape and size of around 50 nm in each case.

Store it in a dry, cool ventilated area free from light and other readily combustible or oxidizable materials.

2. Gold chloride is not light sensitive. Avoid getting contact with the eyes, skin, or clothing. Never allow the gold chloride solution to come in contact with the other metal objects, since all the metals except platinum will result in the reduction of gold ions into its black, gold powder form.
3. *B. licheniformis* KK2 was isolated from sewage sample and was characterized using 16S rRNA gene sequence analysis and its corresponding gene sequence had been stored in GenBank with the accession number GQ871450.1.

4. When using this or another strain ensure that the bacterial culture is free from contaminant microbes and the purity of the strain can be determined by means of quadrant streak which would result in the isolated single colonies. Further the mother plate culture has to be subcultured for every 14 days in order to sustain the fast and accelerated growth of bacteria.
5. Both nutrient agar and nutrient broth media are highly hygroscopic and hence proper care should be taken to close the caps of reagent bottles tightly for the prolonged use of media.
6. Whereas other low quality filters may have the chance for their filter membrane to get ragged when more pressure is given. Ensure that the syringe is not over pressed as it can often lead to spattering of biomass carrying the nanoparticles.
7. Ordinary digital camera is sufficient; we used Canon Powershot A 640.
8. Depending upon the type of instrument, the sample for FT-IR can be in liquid or powder form. The air-dried nanoparticles can be scrapped off to get the powder sample for FT-IR analysis.
9. Liquid sample of nanoparticles for TEM analysis is prepared on carbon-coated copper TEM grids and then subjected to analysis.
10. Usually overnight incubation is enough for the sufficient growth of bacterial colonies but the culture which is streaked immediately from the glycerol stock takes more time to grow when compared with the normally subcultured bacterial colonies. When the culture plates are taken from refrigerator, it is advisable to bring them to the room temperature before streaking or inoculation.
11. Even 24 h grown culture can be utilized as seed culture but 12 h grown culture is usually preferred as the right choice for the inoculation of production medium since it will have more viable cells when compared with the former one.
12. Direct inoculation of culture from the agar plates to production media is not preferred since it will take long time for the overall growth of organism in the given medium. To avoid such difficulty, it is customary to use seed culture for the inoculation of production media at percentages varying from 2 to 5. In our laboratory, we used to inoculate 2 % of seed culture into the production medium for this particular bacterium *B. licheniformis*.
13. The course of color change during the biosynthesis of silver nanoparticles involves the initial white to pale brown color at the intermediate stage and then finally to dark brown color.

14. As the gold chloride is not light sensitive, the biomass containing gold chloride can be incubated either in open or closed orbital shakers.
15. Considering the color changing trend for the formation of gold nanoparticles, it starts with yellowish color to light purple and eventually to dark purple color.
16. Ensure that the sonicator probe is well immersed in fluid and should be used only for short durations of time (normally around 20 s for each sample). Keep the samples to be sonicated in ice box, to avoid heat damage during the process of sonication.
17. While performing microfiltration wear appropriate lab clothing, gloves, and goggles to avoid contact with the nanoparticles. When Millipore microfilters are used, around 5 mL of sonicated biomass can be filtered. When other brands of microfilters are used only about 2–3 mL of biomass can be filtered. The filters cannot be reused since the biomass will get completely clogged up in the filter membrane.

Acknowledgements

The authors wish to acknowledge the support of Dr. PusphaViswanathan, Professor, Cancer Institute (WIA), Chennai, who helped us with TEM analysis. Prof G. Sangiliyandi was supported by grants from Department of Science and Technology (DST), New Delhi (ProjectNo.SR/NM/NS-31/2010), India. Mr. Kalishwaralal was thankful to Council of Scientific and Industrial Research (CSIR), New Delhi for the award of Senior Research Fellowship (Reference No.9/10/2(0003)/2k10-EMR-I (KK)).

References

1. Han M, Gao X, Su JZ, Nie S (2001) Quantum-dot-tagged microbeads for multiplexed optical coding of biomolecules. Nat Biotechnol 19:631–635
2. Sun S, Murray CB, Weller D, Folks L, Moser A (2000) Monodisperse FePt nanoparticles and ferromagnetic FePt nanocrystal superlattices. Science 287:1989–1992
3. Kim YC, Park NC, Shin JS, Lee SR, Lee YJ, Moon DJ (2003) Partial oxidation of ethylene to ethylene oxide over nanosized Ag/α-Al_2O_3 catalysts. Catal Today 87:153–162
4. Rosi NL, Mirkin CA (2005) Nanostructures in biodiagnostics. Chem Rev 105:1547–1562
5. Mann S, Ozin GA (1966) Synthesis of inorganic materials with complex form. Nature 382:313–318
6. Neal DPO, Hirsch LR, Halas NJ, Payne JD, West JL (2004) Photo-thermal tumor ablation in mice using near infrared-absorbing nanoparticles. Cancer Lett 209:171–176
7. Sriram MI, Selvaraj BMK, Kalimuthu K, Gurunathan S (2010) Antitumor activity of silver nanoparticles in Dalton's lymphoma ascites tumor model. Int J Nanomedicine 5:753–762
8. SimkissK,WilburKM(1989)Biomineralization; cell biology and mineral deposition. Academic, New York, p 337

9. Mann S (1996) Biomimetic materials chemistry. VCH Publishers, Weinheim, p 383
10. Kim YY, Walsh D (2010) Metal sulfide nanoparticles synthesized via enzyme treatment of biopolymer stabilized nanosuspensions. Nanoscale 2:240–247
11. Verma VC, Kharwar RN, Gange AC (2010) Biosynthesis of antimicrobial silver nanoparticles by the endophytic fungus *Aspergillus clavatus*. Nanomedicine 5:33–40
12. Kumar U, Shete A, Harle AS, Kasyutich O, Schwarzacher W, Pundle A, Poddar P (2008) Extracellular bacterial synthesis of protein-functionalized ferromagnetic Co_3O_4 nanocrystals and imaging of self-organization of bacterial cells under stress after exposure to the metal ions. Chem Mater 20:1484–1491
13. Coker VS, Telling ND, Van der Laan G, Pattrick RAD, Pearce CI, Arenholze E, Tuna F, Winpenny REP, Lloyd JR (2009) Harnessing the extracellular bacterial production of nanoscale cobalt ferrite with exploitable magnetic properties. ACS Nano 3:1922–1928
14. Staniland S, Williams W, Telling N, Van der Laan G, Harrison A, Ward B (2008) Controlled cobalt doping of magnetosomes in vivo. Nat Nanotechnol 3:158–162
15. Thakkar NK, Snehit MS, Rasesh PY (2010) Biological synthesis of metallic nanoparticles. Nanomedicine 6:257–262
16. Huang JL, Wang WT, Lin LQ, Li QB, Lin WS, Li M, Mann S (2009) A general strategy for the biosynthesis of gold nanoparticles by traditional Chinese medicines and their potential applications as catalysts. Chem Asian J 4:1050–1054
17. Pugazhenthiran N, Anandan S, Kathiravan G, Kannaian N, Prakash U, Craw Ford S, Ashokkumar M (2009) Microbial synthesis of silver nanoparticles by *Bacillus* sp. J Nanopart Res 11:1811–1815
18. Parikh RY, Singh S, Prasad BLV, Patole MS, Sastry M, Shouche YS (2008) Extracellular synthesis of crystalline silver nanoparticles and molecular evidence of silver resistance from *Morganella* sp.: towards understanding biochemical synthesis mechanism. Chembiochem 9:1415–1422
19. Kalimuthu K, Suresh Babu R, Venkataraman D, Bilal M, Gurunathan S (2008) Biosynthesis of silver nanocrystals by *Bacillus licheniformis*. Colloids Surf B Biointerfaces 65: 150–153
20. Kalishwaralal K, Shubaash G, Ramanathan V, Deepak V, Suresh Babu R, Gurunathan S (2010) Optimization of α-amylase production for the green synthesis of gold nanoparticles. Colloids Surf B Biointerfaces 77:174–180

Chapter 4

Gold Nanoparticle Antibody Conjugates for Use in Competitive Lateral Flow Assays

Julian Bailes, Samantha Mayoss, Phil Teale, and Mikhail Soloviev

Abstract

Gold nanoparticles (GNPs) are widely used in a variety of biomedical diagnostic assays and for imaging. Their popularity stems from key properties such as their low toxicity and high extinction coefficients, as well as straightforward synthesis methods that allow GNPs to be produced quickly and inexpensively. Here we describe the use of GNPs for visual detection in a lateral flow assay using benzodiazepine affinity assay to illustrate the methods

Key words: Lateral flow, Immunochromatography, Test strip, Gold nanoparticles, Au nanoparticles, Colloidal gold, Antibody conjugation

1. Introduction

1.1. Lateral Flow Assays

Immunochromatographic assays are often designed to provide easy-to-use point-of-care test kits. Lateral flow or strip tests are one such example, embodying speed and simplicity over high multiplexity (1, 2). These tests provide rapid results and are relatively inexpensive to produce making them well suited for mass markets and routine assays compared to lab-based assays such as SPR- and QCM-based technologies. Such kits are usually stable for long durations across a wide range of environments and conditions and can be made compatible with all major sample matrices including urine, saliva, serum, plasma, and whole blood. Although immunochromatographic assays were first described in the 1960s, it wasn't until Unipath's Clearblue pregnancy test (www.clearview.com) was released in the late 1980s that they saw their first commercial release. Clearblue provided rapid results for just a single analyte,

Mikhail Soloviev (ed.), *Nanoparticles in Biology and Medicine: Methods and Protocols*, Methods in Molecular Biology, vol. 906, DOI 10.1007/978-1-61779-953-2_4,

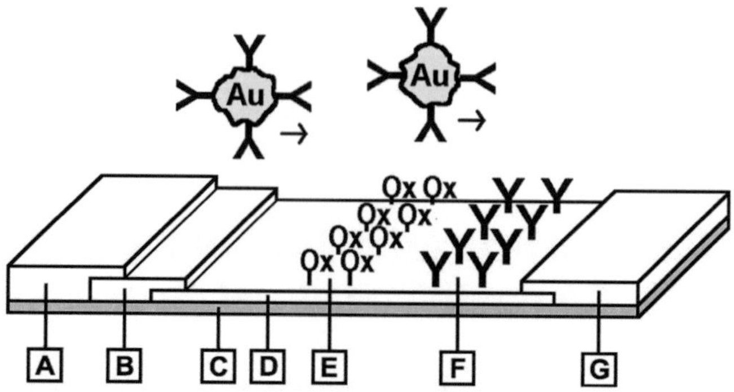

Fig. 1. Schematic diagram of assembled lateral flow components. The sample pad (A) is where the test sample is first deposited and regulates flow rate into the subsequent conjugate pad (B) where it mixes with a labeled reference reagent (represented here as a gold–antibody conjugate) that has been temporarily dried in place. Sample and labeled reference reagent diffuse along a porous membrane (D) coming into contact with capture (E) and control (F) lines (represented here as protein–drug conjugates and antibodies, respectively). An absorbent pad (G) completes the lateral flow strip and serves to maximize the sample volume entering the membrane test strip, thus flushing through unbound detection particles, reducing background and increasing sensitivity. All components are assembled on an adhesive backing layer (C) and usually housed in a plastic test cassette with defined sample application and test result windows.

human chorionic gonadotropin (hCG) (3, 4) and while pregnancy tests remain one of the few mass produced (5), and arguably the best known of lateral flow assays, the technology has been applied in many sectors, from healthcare screening (6, 7), to veterinary and agricultural applications (8), to biodefense and detection of infectious agents (9). Unipath's Clearview product range alone now offers point-of-care diagnostics for pregnancy and ovulation, STDs, infectious disease, respiratory disease, and coagulation/cardiac conditions.

Lateral flow assays are compatible with direct and competitive immunoassay formats, although in practice competitive formats are preferred because they do not require labeling or spotting of the sample prior to use, which is not at all practical in most situations where the tests are to be used and defeats the main attractions of such tests, their simplicity and speed. In their simplest form, lateral flow tests consist of a band of capture antibodies (marked E in Fig. 1) immobilized on a porous membrane strip (marked D in Fig. 1). The assay relies on the diffusion of samples along the length of the strip by capillary forces towards target-specific binding sites. A porous membrane is typically a strip of nitrocellulose although nylon, polyethersulfone, polyethylene, and fused silica can be used. Normally lateral flow test devices include a number of other components such as a sample pad (marked A in Fig. 1) which can serve several critical purposes including regulating flow rate into the subsequent conjugate pad, and filtration when separation of

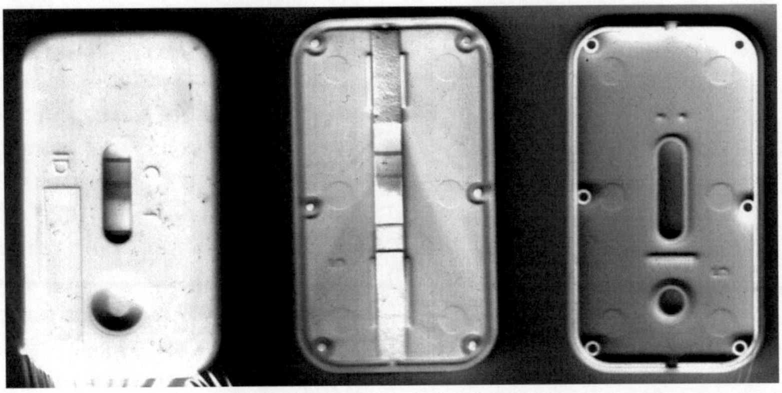

Fig. 2. Photograph of completed sample run shown in fully assembled lateral flow test device complete with designated sample window, and with top casing removed. Capture line (deposited at 13 mm from origin of membrane) and control line (at 16 mm) both clearly visible indicating sample is negative.

plasma from whole blood is required. A conjugate pad (marked B in Fig. 1) is typically included and acts as a zone for labeled reference reagent that is dried in place but becomes mobile as the sample defuses through it. A control line (marked F in Fig. 1) located after the capture line is also a common feature and provides confirmation that the assay has successfully run to completion. An absorbent pad (marked G in Fig. 1), typically cellulose fiber, completes the lateral flow strip and serves to maximize the sample volume entering the membrane test strip, thus flushing through unbound detection particles, reducing background and increasing sensitivity. All components are attached to a backing layer (marked C in Fig. 1) and usually housed in a plastic test cassette with defined sample application and test result windows (see Fig. 2). For direct assay strips the visual presence of a capture band upon completion of the assay indicates a positive sample, whereas in the competitive format reduced color intensity or disappearance of the capture line entirely is indicative of a target analyte's presence in the sample. For a more detailed consideration of lateral flow devices and materials, see ref. (10). The production and reproducibility of lateral flow tests is simplified and increased by using qualified equipment such as the Biodot platforms (www.biodot.com) that can quantitatively dispense reagents onto membranes in bulk and perform precise cutting of batch preparations into individual strips.

1.2. Gold Nanoparticles

While the original Clearblue test used blue latex particles for visual detection, the past 20 years have seen many alternative labeling and visualization techniques employed, including liposomes that contain reporter agents such as colored, fluorescent or bioluminescent dyes (11–13), carbon (14, 15), selenium (16), upconverting phosphors (17–19), quantum dots (20), electrochemiluminescent particles (21), and silica particles (22). More recently, the use of gold nanoparticles (GNPs) for colorimetric detection in lateral flow was reported for testing of troponin 1 (23), bacterial contaminations (24), explosives (25), and disease detection (26). Among the many reasons for their popularity in the field of medical diagnostics and biological imaging is the fact that they are nontoxic, possess extinction coefficients far greater than that of organic dyes (27), and can be produced simply and inexpensively in almost any lab. Synthesis involves chemical reduction of tetrachloroauric acid ($HAuCl_4$) solution with sodium citrate (28). Controlling the amount of reducing solution determines the size of the particles, which in turn determines their color. Although the citrate method is the most commonly used, other methods of synthesis have been reported over the years (29–31). Important considerations for any detection particles to be used in lateral flow are that the population can remain monodisperse and consists of particles with a consistent, spherical shape so as to ensure consistency of migration through the porous membrane. Proteins readily adhere to the nanoparticle surface due to the combination of electrostatic attraction, covalent bonding, and hydrophobic interactions. If insufficient ligand molecules are present on the surface of the nanoparticle then aggregation may occur when in the presence of buffers.

Benzodiazepines are a group of drugs characterized by a benzene ring fused with a diazepine ring and are agonists of the receptor for the neurotransmitter gamma-aminobutyric acid (GABA), which is responsible for regulating neuronal excitability and muscle tone. Here we describe the use of GNPs for visual detection in a benzodiazepine lateral flow assay. We illustrate the lateral flow preparation and assembly methods using anti-benzodiazepine GNP conjugates as an example.

2. Materials

2.1. Antibody Conjugation to GNPs

Use deionized water for all buffers unless stated otherwise.

1. Potassium carbonate buffering reagent (K_2CO_3): make 0.2 M solution in water.
2. Colloidal gold: 40 nm (see Notes 1 and 2).
3. pH paper test strips (pH range 0–14).
4. Hydrogen peroxide (H_2O_2): 30 % solution in water.
5. Mouse anti-benzodiazepine monoclonal antibody: unconjugated, clone BEN1 (Biogenesis, now part of AbD Serotec), protein-A purified in phosphate buffered saline (PBS) buffer (pH 7.2) with 0.05 % sodium azide (see Note 3).

6. Bovine serum albumin (BSA): make 10 % (w:v) solution in water.
7. Benchtop refrigerated centrifuge capable of 10,000 × *g* at 4 °C and 2 mL microcentrifuge tubes.
8. Resuspension solution: 1 % sucrose, 0.1 % sodium azide in water.

2.2. Preparation of Oxazepam Lateral Flow Strips

1. Release agent: 0.5 % Tween-20, 0.5 % human serum albumin, in water.
2. Glass fiber conjugate pad strips: 10 × 300 × 0.41 mm (*G041* from Millipore).
3. Dispensing solution: 20 mM PBS, 5 % methanol, 0.1 % lactose (see Note 4).
4. Sodium chloride (NaCl): make 0.135 M solution in water.
5. Capture reagent: 2.5 mg/mL oxazepam-3-hemisuccinate/BSA conjugate (OX-3HS-BSA) in dispensing solution (see Note 5).
6. Control reagent: 1 mg/mL Goat anti-mouse IgG, whole molecule (Sigma) in 0.135 M sodium chloride. Dilute 1/10 in dispensing solution to a final concentration of 100 μg/mL before use.
7. PBS: 0.01 M phosphate buffer, 0.0027 M potassium chloride, and 0.137 M sodium chloride, pH 7.4, at 25 °C.
8. Nitrocellulose membrane: Hi-Flow Plus 240 SN-HF, mylar-backed nitrocellulose, capillary flow rate 240 s per 4 cm (Millipore UK Ltd.) (see Note 6).
9. Dispensing robot for dispensing control and capture line reagents onto membrane: BioDot XYZ3000 Platform with BioJetQuanti 3000 (BioDot Ltd) (see Note 7).
10. Cutting module for cutting batch product into individual test strips: BioDot CM4000 Guillotine Cutting Module (BioDot Ltd.) (see Note 7).
11. Blocking solution: 1 % dried milk powder, 0.1 % sodium azide in diH_2O.
12. SureWick Cellulose fiber sample pad strips: 17 × 300 mm and 20 × 300 mm, plastic backed (Millipore).
13. Adhesive backing card: 60 × 300 mm (Millipore).
14. Plastic test cassettes for lateral flow assembly, such as shown in Fig. 2 (see Note 8).

2.3. Lateral Flow Assays

1. Diazepam: 1 mg/mL solution of diazepam in methanol.
2. PBS: as in item 7, Subheading 2.2 (see Note 9).
3. Prepare five serial dilution of diazepam in PBS ranging from 1 to 10,000 ng/mL in tenfold increments. Make fresh before use.

3. Methods

3.1. Antibody Conjugation to GNPs

This section describes the conjugation of BEN1 monoclonal antibody to GNPs. The methods are easily adaptable for use with any other immunoglobulins.

1. Adjust the pH of colloidal gold solution and BEN1 monoclonal antibody solution to pH 9.5 using 0.2 M K_2CO_3, check pH using pH paper (see Note 10).
2. Add 30 % H_2O_2 to the colloidal gold solution to a final concentration of 0.1 % (see Note 11).
3. Dilute mouse anti-benzodiazepine monoclonal antibody in water to a final concentration of 50 μg/mL (see Note 12).
4. Mix the antibody and colloidal gold solutions in a ratio of 1:5, respectively, by slowly adding the gold to the protein while mixing briskly (see Note 13).
5. Mix the solution gently at room temperature for 10–15 min.
6. Add 10 % BSA to the antibody–gold conjugate to a final concentration of 1 %. Mix gently at room temperature for 15 min.
7. Purify the conjugate by centrifugation at 4 °C for at least 1 h at 10,000 × *g* (see Note 14).
8. Carefully remove and discard supernatant that contains proteins (see Note 15). Do not disturb the pellet.
9. Resuspend the gold conjugate in resuspension solution (see Note 16).
10. Determine absorbance at 520 nm as a measure of the conjugate concentration (see Note 17).
11. Store nanoparticles at 4 °C until use (add NaN_3 to final concentration of 0.02 % if conjugate is to be stored for over 48 h).

3.2. Preparation of Oxazepam Lateral Flow Strips

1. Take a precut glass fiber strip of 10 × 300 × 0.41 mm. Soak the strip in the release agent for 5 min and dry it at 37 °C for 2 h (see Note 18).
2. Soak the dried, treated glass fiber strip in anti-benzodiazepine monoclonal antibody–gold conjugate (from step 11, Subheading 3.1) (see Note 19). Dry at 37 °C for 2 h.
3. Dispense capture and control lines onto nitrocellulose membrane at 13 and 16 mm from origin (marked E and F, respectively, in Fig. 2). Allow membranes to dry for 10 min at 37 °C (see Note 20).
4. Block membranes by submersion in blocking solution for 30 min at room temperature (see Note 21).

5. Rinse membranes in dH_2O (3× 10 s) and then dry for 30 min at room temperature.
6. Assemble cellulose fiber sample pads, treated membrane strip, and treated glass fiber strip on 300 mm wide adhesive plastic backing card such that there is a 2 mm wide overlap between individual components. Leave to set for 2 h at room temperature (see Fig. 1).
7. Cut the assembled card into 5 mm wide strips and place each into a plastic test cassette (see Fig. 2).
8. Store desiccated at room temperature until use.

3.3. Lateral Flow Assays

1. Apply 3–4 drops (150–200 μL) of diazepam spiked PBS or blank PBS control sample onto sample window of test cassettes (see Fig. 2) and allow the test to run for 15 min (see Note 22).
2. Allow 15 min after adding the sample before reading the assay results. Colored band may form at capture line and will form at control line. Appearance of a band at the control line confirms successful test run. Appearance of two bands (capture and control lines) indicates a negative sample, while the partial or total loss of intensity in color of the capture band indicates a positive sample.

4. Notes

1. We used BBI (British Biocell International) but many other suppliers are currently available, e.g., Aldrich (Sigma-Aldrich) and Fitzgerald (Fitzgerald Industries International).
2. Although readily available from a large number of commercial suppliers, GNPs can be produced relatively easy, see e.g., (28) for the details of nanoparticle manufacturing.
3. Polyclonal antisera can be used but requires purification (e.g., by ammonium sulfate precipitation of IgG followed by immunoaffinity purification) prior to GNP conjugation.
4. Methanol (≤5 %) is added to improve binding of proteins to nitrocellulose (>5 % may damage proteins), and lactose is added to improve stability. Proteins bind best at or near their isoelectric point and so pH 4.5 is used for the BSA–drug conjugate in this instance. For consistency, this solution is also used for control reagents despite not being optimized for the antibodies.
5. To maximize assay sensitivity ensure that the concentration of conjugate used as capture reagent is as low as possible while still yielding a visible signal.

6. When choosing which nitrocellulose membrane to use there are a number of important factors to consider, namely slower membrane flow rates increase the time reagents spend in proximity, thus increasing sensitivity, and that the membrane pore size must be at least ten times the diameter of any particles used as labels. For a more detailed consideration of lateral flow materials, see ref. (10).
7. While manual apparatus (pipettes and scissors) can be used when preparing later flow test strips, its use will compromise consistency. For best results use specialized automated equipment, such as BioDot XYZ3000 Platform and BioDot CM4000 Guillotine Cutting Module to improve reproducibility and output.
8. A variety of similar cassettes are commercially available, e.g., from Millipore (www.millipore.com), Diagnostic Consulting Network (www.dcndx.com), and Advanced Microdevices (www.mdimembrane.com).
9. It is important to develop and optimize the assay for the particular matrix it is intended for, e.g., pooled plasma. Flow characteristics and assay performance will be affected by the rheology of the matrix used. Also consider the color of the sample when choosing a label to work with in order not to compromise visual signal detection.
10. The pH of the colloidal gold solution and the protein to be conjugated should be adjusted to 0.5 pH units above the isoelectric point of the protein. This ensures the overall negative charge on the proteins and their adhesions will not change the surface charge of the NPs, thus avoiding aggregation. Do not use a pH meter when adjusting the pH of colloidal gold to avoid damaging the electrode.
11. H_2O_2 is used to break down residual tannic acid.
12. To determine protein binding capacity of GNPs, prepare a series of BSA dilutions ranging from 1 μg/mL to 10 mg/mL. Add 100 μL of each of BSA dilution to individual 500 μL aliquots of colloidal gold and incubate the samples at 25 °C with constant, gentle agitation. Following the 30 min incubation add 600 μL aliquots of 20 % NaCl to individual GNP–BSA samples. The GNP–BSA sample with the lowest BSA content (100 μL or 1 mg/mL BSA per 500 μL of the GNP preparation) where no color change is observed contains sufficient protein for total coverage of the colloidal gold present. This ratio of total protein to GNPs should be achieved by the conclusion of the blocking stage (step 6, Subheading 3.1) in order to ensure complete coverage of the GNPs. The above assays may be scaled down to reduce reagents usage.

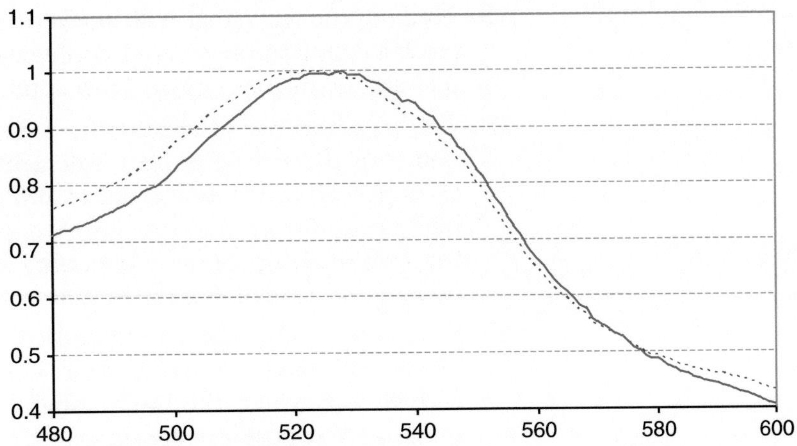

Fig. 3. Absorption spectra of gold nanoparticles. *Dashed line* shows absorption spectrum of nonderivatized NPs. Absorption maximum 522 nm. *Solid line*—same nanoparticles after derivatization with protein (bovine albumin in this example). Absorption maximum 528 nm.

13. Slowly adding the gold to the protein results in a more stable conjugate and minimizes the risk of flocculation (aggregation of gold particles).
14. Unconjugated protein will remain in the supernatant. The conjugate may form a two-phase pellet consisting of looser particles (the conjugate) and very dense, black particles (aggregates of gold particles). Aggregates may be an indication that the amount of protein used in the blocking step was insufficient to achieve complete coverage of the nanoparticles, in which case the conjugation procedure should be restarted.
15. Remove the supernatant carefully as conjugate may go into suspension very easily.
16. Sucrose in the buffer will facilitate the resuspension and will prevent aggregation of conjugate and improve stability during storage.
17. The reported extinction coefficients for gold NPs vary (e.g., $4.2 \times 10^8\ M^{-1}cm^{-1}$ for 16 nm gold NPs (32) or $6.3 \times 10^7\ M^{-1}cm^{-1}$ for 12 nm gold NPs (33)). For simplicity one can assume to be $10^8\ M^{-1}cm^{-1}$, meaning that peak absorbance of 0.1 will correspond to 1 nM NP concentration. Peak wavelength will depend on the nanoparticle size. One can therefore estimate nanoparticle diameter using this formula:

$$\lambda_{max} = 515.04 + 0.3647d$$

where "*d*" is nanoparticle diameter in nanometers (34). Absorbance spectrum of protein–gold conjugate should show a redshift (see Fig. 3).

18. Treating the glass fiber with surfactants, polymers, or proteins prior to application of gold conjugate ensures it accepts and releases the majority of the label applied to it.
19. The glass fiber pad ($10 \times 300 \times 0.41$ mm, *G041* from Millipore) holds 45 μL/cm^2. Ensure a sufficient concentration of gold conjugate is used. The amount of antibody absorbed will eventually be redissolved in the sample applied to the strip. Hence the final working antibody concentration will depend on both the amount loaded and the sample volume.
20. When dispensing capture and control reagents remember that protein will bind to nitrocellulose where it comes into contact with it, not where the buffer might subsequently appear to spread. These positions are suggested as guidelines for a 60 mm test strip (including sample, conjugate, and absorbent pads). Flow rate is a function of the distance from the origin and so altering the location of capture and control lines will in turn affect assay kinetics and sensitivity. Be sure that your chosen positions fall within the viewing window of any test cassette you intend to use.
21. Slowly slide the membrane into the blocking solution at a shallow angle to reduce the risk of air becoming trapped in the three-dimensional structure of the membrane.
22. Orientation and temperature should be standardized, i.e. test cassette placed on flat surface and test run at room temperature.

References

1. Qian S, Bau HH (2003) A mathematical model of lateral flow bioreactions applied to sandwich assays. Anal Biochem 322(1):89–98
2. Qian S, Bau HH (2004) Analysis of lateral flow biodetectors: competitive format. Anal Biochem 326(2):211–224
3. May K (1991) Home tests to monitor fertility. Am J Obstet Gynecol 165(6):2000–2002
4. Chard T (1992) Pregnancy tests: a review. Hum Reprod 7(5):701–710
5. Cole LA (2011) The utility of six over-the-counter (home) pregnancy tests. Clin Chem Lab Med 49(8):1317–1322
6. Wu AH, Smith A, Christenson RH et al (2004) Evaluation of a point-of-care assay for cardiac markers for patients suspected of acute myocardial infarction. Clin Chim Acta 346(2):211–219
7. Rossi AF, Khan D (2004) Point of care testing: improving pediatric outcomes. Clin Biochem 37(6):456–461
8. Liao JY, Li H (2010) Lateral flow immunodipstick for visual detection of aflatoxin B-1 in food using immuno-nanoparticles composed of a silver core and a gold shell. Microchim Acta 171:289–295
9. Peruski AH, Peruski LF Jr (2003) Immunological methods for detection and identification of infectious disease and biological warfare agents. Clin Diagn Lab Immunol 10(4):506–513
10. Posthuma-Trumple GA, Korf J, van Amerongen A (2009) Lateral flow (immuno)assay: its strengths, weaknesses, opportunities and threats. A literature survey. Anal Bioanal Chem 393(2):569–582
11. Gussenhoven GC, van der Hoorn MA, Goris MG et al (1997) LEPTO dipstick, a dipstick assay for detection of Leptospira-specific immunoglobulin M antibodies in human sera. J Clin Microbiol 35(1):92–97

12. Ho JA, Wauchope RD (2002) A strip liposome immunoassay for aflatoxin B1. Anal Chem 74(7):1493–1496
13. Ho JA, Huang MR (2005) Application of a liposomal bioluminescent label in the development of a flow injection immunoanalytical system. Anal Chem 77(11):3431–3436
14. van Amerongen A, Wichers JH, Berendsen LB et al (1993) Colloidal carbon particles as a new label for rapid immunochemical test methods: quantitative computer image analysis of results. J Biotechnol 30(2):185–195
15. Lönnberg M, Carlsson J (2001) Quantitative detection in the attomole range for immunochromatographic tests by means of a flatbed scanner. Anal Biochem 293(2):224–231
16. Lou SC, Patel C, Ching S et al (1993) One-step competitive immunochromatographic assay for semiquantitative determination of lipoprotein(a) in plasma. Clin Chem 39(4): 619–624
17. Corstjens P, Zuiderwijk M, Brink A et al (2001) Use of up-converting phosphor reporters in lateral-flow assays to detect specific nucleic acid sequences: a rapid, sensitive DNA test to identify human papillomavirus type 16 infection. Clin Chem 47(10):1885–1893
18. Niedbala RS, Feindt H, Kardos K et al (2001) Detection of analytes by immunoassay using up-converting phosphor technology. Anal Biochem 293(1):22–30
19. Zuiderwijk M, Tanke HJ, Sam Niedbala R et al (2003) An amplification-free hybridization-based DNA assay to detect *Streptococcus pneumoniae* utilizing the up-converting phosphor technology. Clin Biochem 36(5): 401–403
20. Goldman ER, Clapp AR, Anderson GP et al (2004) Multiplexed toxin analysis using four colors of quantum dot fluororeagents. Anal Chem 76(3):684–688
21. Yoon CH, Cho JH, Oh HI et al (2003) Development of a membrane strip immunosensor utilizing ruthenium as an electro-chemiluminescent signal generator. Biosens Bioelectron 19(4):289–296
22. Xia X, Xu Y, Zhao X et al (2009) Lateral flow immunoassay using europium chelate-loaded silica nanoparticles as labels. Clin Chem 55(1):179–182
23. Choi DH, Lee SK, Oh YK et al (2010) A dual gold nanoparticle conjugate-based lateral flow assay (LFA) method for the analysis of troponin I. Biosens Bioelectron 25(8): 1999–2002
24. Rong-Hwa S, Shiao-Shek T, Der-Jiang C et al (2010) Gold nanoparticle-based lateral flow assay for detection of staphylococcal enterotoxin B. Food Chem 188(2):462–466
25. Girotti S, Eremin S, Montoya A et al (2010) Development of a chemiluminescent ELISA and a colloidal gold-based LFIA for TNT detection. Anal Bioanal Chem 396(2):687–695
26. Yu CY, Ang GY, Chua AL et al (2011) Dry-reagent gold nanoparticle-based lateral flow biosensor for the simultaneous detection of *Vibrio cholerae* serogroups O1 and O139. J Microbiol Methods 86(3):277–282
27. Zhao W, Chiuman W, Brook MA et al (2007) Simple and rapid colorimetric biosensors based on DNA aptamer and noncrosslinking gold nanoparticle aggregation. Chembiochem 8(7):727–731
28. Ferrari E, Darios F, Zhang F et al (2010) Binary polypeptide system for permanent and oriented protein immobilization. J Nanobiotechnology 8:9
29. Brust M, Walker M, Bethell D et al (1994) Synthesis of thiol-derivatised gold nanoparticles in a two-phase liquid–liquid system. J Chem Soc Chem Commun 7(7):801–802
30. Perrault SD, Chan WC (2009) Synthesis and surface modification of highly monodispersed, spherical gold nanoparticles of 50–200 nm. J Am Chem Soc 131(47):17042–17043
31. Martin MN, Basham JI, Chando P et al (2010) Charged gold nanoparticles in non-polar solvents: 10-min synthesis and 2D self-assembly. Langmuir 26(10):7410–7417
32. Demers LM, Mirkin CA, Mucic RC et al (2000) A fluorescence-based method for determining the surface coverage and hybridization efficiency of thiol-capped oligonucleotides bound to gold thin films and nanoparticles. Anal Chem 72(22):5535–5541
33. Browning LM, Lee KJ, Huang T et al (2009) Random walk of single gold nanoparticles in zebrafish embryos leading to stochastic toxic effects on embryonic developments. Nanoscale 1(1):138–152
34. He YQ, Liu SP, Kong L, Liu ZF (2005) A study on the sizes and concentrations of gold nanoparticles by spectra of absorption, resonance Rayleigh scattering and resonance non-linear scattering. Spectrochim Acta A Mol Biomol Spectrosc 61(13–14):2861–2866

Chapter 5

Gold Nanoparticles for Immunogold Localization of Antigens in Plant Tissues

Sadaki Yokota

Abstract

Here we describe postembedding immunoelectron microscopic technique applied to ultrathin sections of plant material. The method relies on the use of gold nanoparticles. The methods include tissue fixation, dehydration, embedding of plant specimens, and staining of semithin and ultrathin sections. The described method is suitable for localization of antigens including proteins (peroxisomal enzymes) and a low molecular substance such as ginsenoide-Rb1.

Key words: Immunofluorescence, Nanogold probes, Protein A-gold, Postembedding immunoelectron microscopy

1. Introduction

Immunoelectron microscopy (IEM) has the potential to provide important information on the subcellular location of macromolecules in cells and tissues in situ. Two main approaches exist for the visualization of antigenic sites by IEM. First group include preembedding methods and the other include postembedding techniques. In the former methods, the immunological probes are allowed to penetrate fixed cells or tissue slices to interact with the antigens prior to embedding. The inevitable main problem in this method is the impermeability of cellular membranes to antibodies. In the plant, the cell walls prevent the antibodies to penetrate the cell. To permeate membranes various procedures have been employed. Fixation conditions such as concentration of glutaraldehyde and fixation time affect markedly the penetration of antibodies (1). It has been shown that the more reliable and repeatable results for animal cells are obtained by the following treatments: the cells or tissue slices are fixed with low concentration of glutaraldehyde (less than 0.01 %) for 5–10 min, treated with 0.1 % Triton X100, and incubated with HRP-conjugated

Mikhail Soloviev (ed.), *Nanoparticles in Biology and Medicine: Methods and Protocols*, Methods in Molecular Biology, vol. 906, DOI 10.1007/978-1-61779-953-2_5, © Springer Science+Business Media, LLC 2012

Fab fragment of specific antibody (2), or by nanogold-labeled antibody in combination with silver enhancement (3). However, this method is not able to be applied to plant cells with cell wall. Alternative labeling postembedding labeling techniques have been developed to overcome these problems. These include postembedding staining of thin sections of acryl resin-embedded material, as in ref. (4). There antibody interact with antigenic sites exposed on the surface of thin sections, which alleviates the problem of cell wall penetration altogether. Therefore, the technique can be applied to plant cells and tissues. The technique visualizes the antibody binding with antigen by colloidal gold particles conjugated to secondary probes such as protein A or G, and antibody, which allows the detection of antigens with high specificity and high resolution. However, in the postembedding technique, sample is prepared through various physicochemical procedures, including fixation with paraformaldehyde or glutaraldehyde, dehydration by ethanol, substitution of ethanol with acryl resin, and polymerization of resin, which may affect or even abolish antigenicity (5). Thus, this technique is not versatile for all antigens but nevertheless remains the method of choice for IEM of plant cells and tissues.

Semithin sections of resin-embedded materials are able to be applied to immunofluorescence staining (IF) of antigens. This is very helpful because it makes possible to observe antigenic sites in wider area of the sections. The combination of IF and IEM in a single method may provide more information on the localization of antigens in cells and tissues. For this purpose, nanogold particles conjugated with fluorescein (FluroNanogold) are a useful tool by which we can visualize the antigens on the semithin sections under a fluorescence microscope (6). In addition, ultrathin sections cut out next to the semithin sections can be observed by electron microscope after staining by the postembedding labeling technique. In this chapter, the postembedding immunoelectron microscopic technique for plant antigens, including peroxisomal enzymes of pumpkin cotyledons and ginsenoide-Rb1 of *Panax ginseng*, is described. The methods mentioned earlier can be applied to animal tissues without major modifications (see Note 1).

2. Materials

2.1. Preparation of Tissue Slices

1. *Panax ginseng* grown for 3 years. We believe that many other tissue materials can be studied using the method with little or no modification.
2. Pumpkins cotyledons grown for 5 days.
3. 25 % Glutaraldehyde (EM grade, Electron Microscopy Sciences). Store at 4 °C.
4. Paraformaldehyde.

5. Phosphate-buffer saline (PBS): To make 500 mL PBS, dissolve 4.0 g of NaCl, 0.1 g of KCl, 0.72 g of Na_2HPO_4, and 0.12 g of KH_2PO_4 in 400 mL of distilled water. Store at 4 °C.
6. Low gelling temperature agarose (36 °C): add 5 g of agarose to 100 mL of PBS and heat in a microwave oven until the agarose melts completely.
7. Instant glue (Cyanoacrylate adhesive).
8. Fixative: 4 % paraformaldehyde, 5 % glutaraldehyde, or 0.2 % glutaraldehyde in 0.05 M HEPES-KOH buffer, pH 7.4 (see Note 2). To make 100 mL of the fixative, dissolve 4 g of paraformaldehyde in 50 mL of distilled water, heat the solution to 80 °C on a hot plate with a magnet stirrer, and add a small amount of 0.1 M KOH until the solution becomes clear (see Note 2). Cool the solution in an ice bath to ~20 °C, add 1.19 g of HEPES to the solution. Adjust the pH to 7.2 with 0.1 M KOH, add 20 mL of 25 % glutaraldehyde (for final concentration of 5 %) or 0.8 mL of that (for final concentration of 0.2 %), and adjust the volume to100 mL, see Note 2.
9. Glass vial (5 mL).
10. Vibratome (Vibratome® Company).
11. Razor.
12. 50, 70, 80, 90, 95, and 100 % ethanol for dehydrating the samples.
13. Vacuum desiccator.

2.2. Embedding of Tissue Slices in LR White Resin

1. 50 mL polystyrene Conical Tube (Becton Dickinson Labware).
2. TAAB capsule 8 mm flat, polyethylene (TAAB Laboratories Equipment Ltd.).
3. LR White embedding resin (hard grade kit) (London Resin Company Ltd.). Store at 4 °C.
4. Embedding resin mixture: Prepare in 50 mL polystyrene Conical Tube by mixing 40 mL of the LR White resin and 5 μL of LR White accelerator (see Note 3). Attach and close the cap, store at −20 °C.
5. Tweezers with flat round tips (Style 2A, Electron Microscopy Sciences).
6. UV rays polymerization device (TUV-200, DOSAKA EM Co.) (see Note 4).
7. TAAB Rotator (TAAB Laboratories Equipment Ltd.).
8. Freezer at −20 °C.

2.3. Preparation of Semithin and Ultrathin Sections

1. Silicon-coated glass slides.
2. Glass stick with round-ground tip (see Note 5).
3. Hot plate.

4. 1 % toluidine blue: Dissolve 1 g of toluidine blue O in 100 mL of 1 % borax.
5. 300 nickel grids (Electron Microscopy Sciences).
6. Cleaning of grids: Wash the grids twice in 1 mL glacial acetic acid and ten times with distilled water. Sonicate the grids in water for 5 min and wash twice with acetone, 5 min each wash. Store the grids in acetone in glass (or acetone-insoluble plastic) vials with tightly closed screw caps (see Note 6).
7. Histodiamond knife (HistoDiatome, Diatome AG).
8. Eyelash probe (see Note 7).
9. Whatman #1 filter paper.
10. Antimagnetic Tweezers (DUMONT Biology, Ted Pella Inc.) or similar.
11. Petri dishes 50 and 10 mm in diameter.

2.4. Immunofluorescence Staining with FluoroNanogold Nanoprobes

1. Cover glass (1.5 × 1.5 mm).
2. Liquid-repellent slide marker pen Pap-pen (Daido Sangyo).
3. Alexa Fluor®*546 FluoroNanogold (Nanoprobes)-conjugated secondary antibody against the primary antibody used.
4. 1 M NaN_3: Dissolve 6.5 g in 100 mL of distilled water.
5. 0.05 % $NaBH_4$ solution: Dissolve 50 mg of $NaBH_4$ in 10 mL of distilled water.
6. 0.1 M Tris–HCl buffer (pH 8.0) containing 0.1 M glycine: To make 100 mL solution, dissolve 1.21 g of Trizma base, 0.75 g of glycine in 50 mL of distilled water. Adjust the pH to 8.0 with 1 N HCl and adjust the volume to 100 mL. Store at 4 °C.
7. 2 % Fish gelatin-15 mM NaN_3 in PBS: Add 4.4 mL of fish gelatin and 1.5 mL of 1 M NaN_3 to 94.1 mL of PBS. Store at 4 °C.
8. Mowiol® 4-88 (Polysciences, Inc.). To prepare Mowiol solution: Add 4.8 g of Mowiol 4-88 and 12 g of glycerol to a 100 mL beaker. Mix well using a stir bar. Add 12 mL of distilled water and continue stirring for several hours at room temperature. Add 24 mL of 0.2 M Tris–HCl (pH 8.5). Continue stirring. Heat occasionally to 50 °C in water bath for approximately 10 min. Continue stirring until Mowiol is dissolved. Once dissolved, centrifuge the solution at 500 × *g* for 15 min to clear the solution. Carefully remove the supernatant; store aliquots at ±20 °C.
9. Antifading mounting medium: Dissolve 2.5 g of 1,4-diazabicyclooctane in 50 mL Mowiol solution prepared above. Store aliquots at −20 °C. Warm up to room temperature before use.
10. Humidified chamber.

2.5. Immunogold Labeling of Thin Sections

1. A mixture of primary antibodies of known specificity in PBS containing 2 % fish gelatin-15 mM NaN_3. The concentration of the antibody is dependent on the specificity used. The most optimum concentration is determined experimentally or according to manufacturer's instruction.
2. Rabbit secondary antibodies. The antibody is diluted with PBS containing 2 % fish gelatin-15 mM NaN_3. The concentration should be determined experimentally or according to manufacturer's instruction.
3. Protein A-gold conjugate 15 nm (British BioCell International, Ltd.) (see Note 8).
4. Preparation of 0.05 M maleate-NaOH buffer, pH 5.5: Add 50 mL of 0.2 M solution of acid sodium maleate (8 g of NaOH and 23.2 g of maleic acid in 1,000 mL of distilled water) to 100 mL of distilled water. Adjust the pH of the solution with 0.2 M NaOH to 5.5 and adjust the volume of the solution to 200 mL. Uranyl acetate in maleate buffer is very stable and will not precipitate.
5. 2 % Uranyl acetate in 0.05 M maleate-NaOH buffer: Dissolve 2 g of uranyl acetate in 100 mL of 0.05 M maleate-NaOH buffer, pH 5.2.
6. 40 mM lead citrate: add 1.33 g of $Pb(NO_3)_2$, 1.76 g of $Na_3(C_6H_5O_7)\cdot 2H_2O$, and 30 mL of distilled water in 50 mL conical flask and stir on a magnetic stirrer plate for 30 min. Add 8 mL of 1 N NaOH to the solution and adjust the total volume to 50 mL. Aspirate 25 mL lead citrate solution into two 30 mL plastic syringes with needles, seal the needle with silicone rubber, and store at 4 °C (see Note 9).
7. High precision anticapillary and antimagnetic tweezers (5TTH. SA, ideal-tek s.a.).

2.6. Contrasting and Carbon Coating of Sections

1. Plastic dishes (5 cm in diameter).
2. PARAFILM "M"® (PECHINEY Plastic Packaging).
3. Carbon coater (EMS 150T ES, Electron Microscopy Sciences).

3. Methods

3.1. Fixation of Plant Tissues

1. Cut leaf, stem, rhizome, and root of *Panax ginseng*, and pumpkin cotyledons into 5–7 mm long sections.
2. Attach a cut piece of tissue to a sample stage of a Vibratome with instant adhesive and cover the tissue with low gelling temperature agarose.

3. Cool the stage on ice and set on a Vibratome.
4. Place cooled fixative into the sample stage and cut the tissue piece into slices of 100–200 μm thickness.
5. Transfer the slices to sample vials containing fixative solution and continue to fix at 4 °C for 60 min (see Note 10).

3.2. Dehydration and Embedding of Tissues (See Note 11)

1. Wash ten pieces of slices in vials with 3 changes of PBS for 5 min each.
2. Pipette cooled 50 % ethanol into the vials, keep at 4 °C, incubate for 15 min.
3. Dehydrate the sections in ascending ethanol series (70, 80, 90, 95, and 100 %) for 15 min in each solution at −20 °C. Repeat the 100 % ethanol dehydration step twice (15 min each step).
4. Aspirate ethanol using Pasteur pipette, add 0.5 mL of 100 % ethanol, and 0.5 mL of LR White resin mixture to each of the vials. Mix well and allow to stand at −20 °C for 60 min.
5. Add another 0.5 mL of LR White resin mixture to each of the vials containing 1 mL ethanol–resin mixture and mix well. This will bring the total volume of ethanol–LR White resin to 1.5 mL in each tube. Keep the vials containing tissue slices at −20 °C for 60 min.
6. Replace the ethanol–LR White mixture with 0.5 mL of pure LR White resin mixture and allow to stand overnight at −20 °C.
7. Warm the vials containing tissue slices in LR White resin mixture to room temperature. Transfer tissue pieces into polyethylene capsules standing on a capsule rack (see Note 12).
8. Fill the capsules completely with the resin until the level reaches the edge of the capsule (see Note 13). Pipette the resin at the level of the edge of capsule. Close the capsule with caps carefully to avoid making bubbles (see Note 14).
9. Transfer the capsules to the UV rays polymerization device, cool at −20 °C, and polymerize overnight at −20 °C.
10. Continue the polymerization at room temperature for 12 h. Store the sample blocks in a desiccator.

3.3. Preparation of Semithin and Ultrathin Sections

1. Mount the polymerized sample block in a specimen holder.
2. Under a stereomicroscope, mark a trimming area with a needle connected to 1 mL syringe.
3. Set the block holder in a vice bench and file off the end into a four-sided pyramid with walls at 40–50° angle and 1.5–3 mm^2 top surface.
4. Transfer the block holder to the arm of an ultramicrotome and shape the end into a four-sided pyramid with walls at a 45° angle and a 1–1.5 mm^2 top surface by hand motion mode as shown in Fig. 1 (see Note 15).

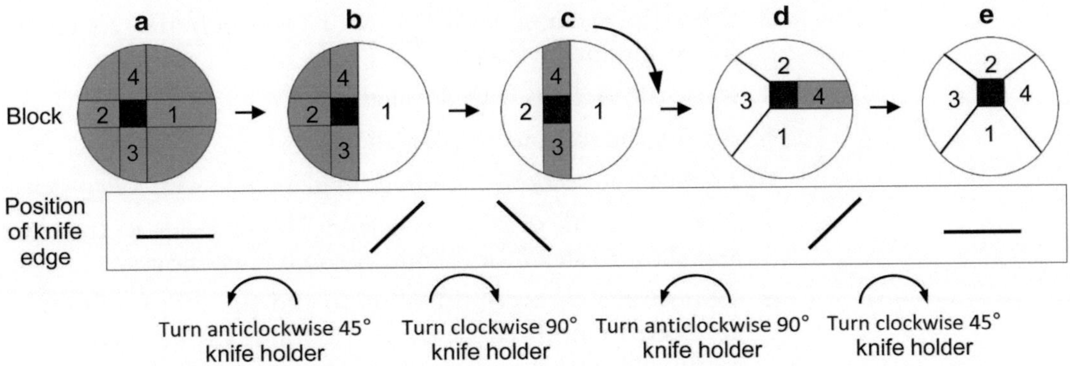

Fig. 1. Trimming block of LR White-embedded sample. Steps **a-e** show shapes of the block at each step of trimming. Numbers 1–4 show each face of the four-sided pyramid before (*shaded*) and after (*clear*) trimming. Bars inside the rectangle indicate position of knife edge by sectioning. Knife holder with the trimming knife is turned anticlockwise between step **a** and **b**, clockwise between step **b** and **c**, anticlockwise between step **c** and **d**, and clockwise between step **d** and **e**, respectively. Sample holder is turned clockwise after step **c**. At step **e** four-sided pyramid is completed and trimming of the block tip is started to carve out the tissue.

5. Cut carefully the top of the block with a trimming knife until tissue appears on the surface of the block (see Note 16).
6. Replace the trimming knife with a histodiamond knife.
7. Fill the trough of a histodiamond knife with distilled water at the level of cutting edge (see Note 17).
8. For immunofluorescence staining, cut semithin sections with thickness of 250 nm.
9. Transfer the sections to distilled water droplets placed on silicon-coated glass slide using a glass stick (see Note 18).
10. Transfer the glass slide to a hot plate heated at 90 °C, evaporate excess water, and store the glass slide with sections under clean condition.
11. For immunogold staining, cut dark yellow ultrathin sections approximately 90–100 nm thick (see Note 19).
12. Take out nickel grids from acetone and dry them on a filter paper.
13. Pinch the grid with a pair tweezers, carry it under the sections floating on water surface, and dip the sections with the grid.
14. Remove water from the grid using Whattman #1 filter paper.
15. Place the grid section side up on a clean filter paper inside the cover of a 50 mm plastic dish and allow it to dry overnight (see Note 20).

3.4. Immunofluorescence Staining (See Note 21)

1. Treat 250 nm thick sections mounted on glass slide with three changes of 0.05 % borohydride for 4 min each.
2. Rinse the section with PBS once.

3. Treat the sections with 0.1 M Tris–HCl buffer containing 0.1 M glycine for 20 min.
4. Rinse the sections with distilled water.
5. Air-dry the sections on glass slide.
6. Encircle the sections mounted on glass slide with a Pap-pen.
7. Pipette 0.1 mL of 2 % fish gelatin-PBS to inside of the circle and allow to stand for 30 min at room temperature.
8. Remove the excess of the solution and add 40 μL of primary antibody to inside of the circle. Continue the incubation with antibody in a humidity chamber overnight at 4 °C.
9. Wash the sections with PBS, pipette 40 μL diluted solution of Alexa Fluor®*546 FluoroNanogold-conjugated secondary antibody onto the sections, and continue to incubate for 60 min at room temperature.
10. Wash with PBS, add 10 μL of antifading mounting medium to the sections, and cover with the cover glass. Store the preparations at 4 °C.

3.5. Immunogold Labeling of Thin Sections

Throughout this procedure the sections should not be removed from the grids.

1. Immerse 90–100 nm thick sections on grids into 0.05 % solution of sodium borohydride, incubate 4 min, repeat the treatment three times.
2. Wash the sections with distilled water.
3. Incubate the sections in drops of 0.1 M Tris–HCl buffer containing 0.1 M glycine for 20 min (see Note 22).
4. Transfer the section into droplets of 2 % fish gelatin-PBS on parafilm for 20 min at room temperature.
5. Pick up the section, remove excess solution from the grids with use of the edge of a newly cut filter paper, and immerse the section on grids into ~10 μL droplets of primary antibody on parafilm placing in a moist chamber. The incubation with the antibody is continued overnight at 4 °C (see Note 23).
6. For control, incubate the section with antibody solution absorbed by antigen, preimmune serum, or unrelated antibody.
7. Wash the section with jet of PBS and put them in droplets of 2 % gelatin-PBS on parafilm.
8. Transfer the section to droplets of secondary probe and incubate with it for 30 min at room temperature (see Note 24).
9. Wash the section under the flow of distilled water, remove excess water with filter paper as mentioned earlier, and put them section side up on clean filter paper.
10. Dry the section on grids for 30 min at room temperature.

3.6. Contrasting and Carbon Coating of Sections

Throughout this procedure the sections should not be removed from the grids.

1. Float the section on grids section side down on the droplets of 20 μL 2 % uranyl acetate solution for 10 min.
2. Pick the grid with section carefully using tweezers, wash it with jet of water, remove excess water held between tips of tweezers, and put the grid on the filter paper.
3. Dry the section on grids for 10 min.
4. Float the sections on grids section side down on droplets of 20 μL lead citrate solution for 30 s.
5. Repeat washing step 2.
6. Put the sections on grids section side up on parafilm placed in a 50 mm plastic dish. Set the plastic dish with the section on grids in the chamber of a carbon coater and evaporate carbon for 5–7 s (see Note 25).
7. Put the sections in a grid case and store the case in a desiccator.
8. Examine sections in the electron microscope. Figure 2 shows localization of peroxisomal enzymes in pumpkin cotyledons by single and double labeling, and of ginsenoside-Rb1 in parenchymal cells of leaf and root of *Panax ginseng* by single labeling.

4. Notes

1. Fixation of animal tissues won't require the use of high concentration of glutaraldehyde.
2. Formalin gas is toxic; all procedures should be carried out in a hood. Paraformaldehyde is not soluble at room temperature and at neutral pH, so one should heat the solution to approximately 80 °C with constant stirring but never boil. When the solution becomes semitransparent, add a small amount of 0.1 M KOH until the solution becomes clear. High concentration of glutaraldehyde (5 %) is used for the localization of small soluble substances and low concentration (0.2 %) for staining proteins. The antigenicity of proteins is lost quickly at high glutaraldehyde concentrations (1).
3. Mixing LR White resin and accelerator should be performed gently, avoiding making bubbles. The resin will not polymerize in the presence of oxygen. If not perfectly mixed, resin may polymerize in part in areas with higher concentrations of the accelerator. Five microliters of the accelerator is sufficient for 40 mL of the resin. If the larger volume is added to 40 mL of

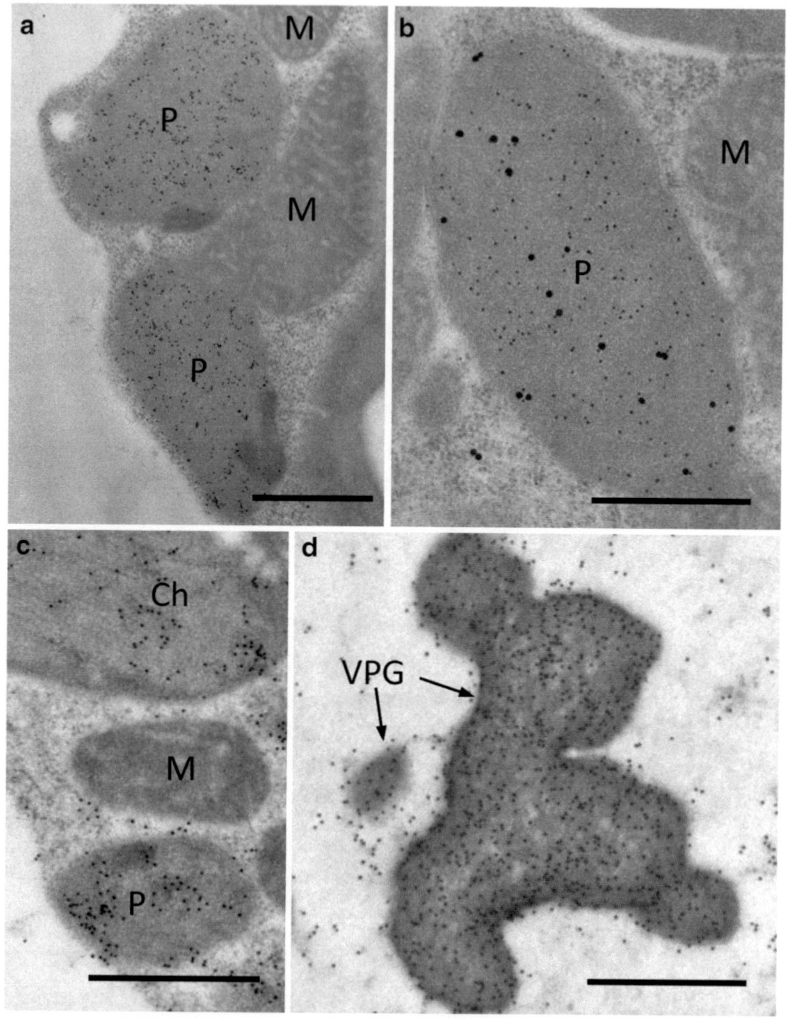

Fig. 2. Immunogold labeling of plant tissues. (**a**) Pumpkin cotyledon labeled for glycolate oxidase. Gold particles are observed in peroxisomes (P) but not in mitochondria (M). (**b**) Double labeling of citrate synthase (large particles) and hydroxyl pyruvate reductase (small particles) in pumpkin cotyledon. Note that large and small gold particles are present in peroxisome (P) but not in mitochondria. (**c**) Immunogold localization of ginsenoside-RB1 in leaf of *Panax ginseng*. Gold particles are seen in peroxisome (P), chloroplast (Ch) but not in mitochondrion (M). (**d**) Vacuolar protein granules (VPG) in vacuole of root parenchymal cell of *Panax ginseng*. VPG is strongly labeled for ginsenoside-RB1. Bar = 1 μm.

the resin, the resin may polymerize quickly during the embedding procedures without any UV light.

4. It is possible to make UV polymerization device, e.g., from a 20 W UV lamp used for sterilizing hoods. Make a suitable capsule holder from metal wire netting and position it 20 cm underneath the UV source.
5. Heat and pull apart a glass stick (3 mm in diameter) until it reaches approximately 10 cm. Break the stick to make a tip.

Heat the tip to round up sharp corners. Frost the tip with sandpaper. As a result of this treatment, sections don't attach on the tip.

6. Cleaning of grids is critical. If the grids are contaminated with oil or similar oily substances, thin sections may detach from the grids during labeling procedures.
7. A hair (~10 mm long) is cut from a brush and attached with glue to a pointed thin stick (10 cm long and 3 mm in diameter).
8. Preparation of protein A–colloidal gold conjugate is described previously (7).
9. At the beginning, crystals of trisodium citrate are visible in white milky mixture of lead nitrate and trisodium citrate but 30 min later the crystals disappear. If the crystals are still visible, add a small amount of distilled water and continue to stir. When the crystals disappear, add 8 mL of 1 N NaOH to the solution. The solution is immediately clear. The pH of the lead citrate solution prepared is approximately 12.0. It is not necessary to neutralize because the staining capacity is lost at the neutral pH.
10. For the staining of small soluble molecules, use fixative containing 5 % glutaraldehyde. Use 0.2 % glutaraldehyde if working with proteins. Fixation time affects the antigenicity of proteins, 60 min long fixation is sufficient to preserve the ultrastructure of cell but retain the antigenic properties of proteins. If tissue slices are floating on the surface of fixative, sink them by gentle repeated evacuation of the air in a vacuum desiccator.
11. The dehydration steps are performed at −20 °C, except for the 50 % ethanol step. During procedures, the vials should be capped and gently shaken on a rotator.
12. Cut the caps of capsules to prevent them from opening because of elastic force of the connection. Note the orientation of the sample (the tissue to be embedded in resin) in the capsule before polymerization of resin. It is difficult to determine the direction of tissue after polymerization.
13. This helps to avoid bubbles.
14. LR White resin is not polymerized if in contact with air. Hence bubbles between under the cap will prevent polymerization of the resin. To avoid bubbling, don't exert pressure on the cap (i.e., do not press with finger), because when released, the negative pressure will form within the capsule which may draw in the air.
15. Polymerized LR White block is very fragile, and it may be easily cracked with a razor. We recommend using microtome for

trimming as shown in Fig. 1. Turn the knife holder with a trimming knife anticlockwise 45° and cut the edge of sample block along the line marked at step 2. Next turn the knife holder clockwise 90° and cut similarly that of the block. Then, turn the block holder clockwise 90°, cut the block, turn the knife holder anticlockwise 90°, cut the block. By following these procedures, a four-sided pyramid will be obtained with walls at a 45° angle and 1–1.5 mm^2 top surfaces.

16. The block should be trimmed very carefully using thickness setting of less than 2 μm. When trimmed at 3 μm or more, deep cracks may appear on the surface and inside the sections. It has been previously shown that if the acryl resin-embedded material is cut along the biological structures at the front of knife edge, the antigenic sites exposed on the surface of sections might increase more than Epon section (8, 9). Therefore, generally, immunolabeling on acryl resin sections is stronger than that on Epon sections.
17. LR White-embedded materials are easily cut into large sections with histodiamond knife. LR White sections are fragile, compared with Epon sections, so that the sections are broken in part during immunoelectron microscopic procedures. If relatively larger sections are cut, unbroken parts are remained after immunoelectron microscopic procedures.
18. Bring the tip of a glass stick beneath the section floating on water surface, pick up the section on the tip, and transfer the tip into a water droplet on a glass slide.
19. It is better to use such thick sections because LR White sections are very weak as mentioned earlier.
20. Sections may be damaged during treatment so we recommend making ten or more mounted sections. The sections should be dried overnight; otherwise they escape from the grids during immunoelectron microscopic procedures. They can be stored several months in a desiccator without any loss of antigenicity.
21. It is recommended to stain sections by immunofluorescence technique. Image information at light microscopic level is very useful to look down upon the localization of antigens in wide area of plant tissue and helps to identify regions for IEM.
22. When tissue slices are fixed with low concentration of glutaraldehyde (below 0.2 %), steps 1 and 2 can be omitted, and one can start from step 3. When tissue slices are fixed with high concentration of glutaraldehyde, start from step 1. The borohydride treatment is necessary for the samples fixed with high concentration of glutaraldehyde to prevent nonspecific binding of antibody to the free aldehyde radicals of glutaraldehyde (10). Furthermore, nonspecific binding of antibody is blocked

by treatments of section with solution of 0.1 M Tris–HCl + 0.1 M glycine and with 2 % fish gelatin.

23. The binding of antibodies to antigenic sites on the surface of sections is normally completed after 8 h after incubation (1). For double labeling of different antigens in identical section, the section on grids is washed with distilled water after step 2 and dried on filter paper. Afterward, the treatments steps 3–9 are performed for one antigen on the one side of section by floating section on droplets. Next, the same treatments are carried out for the other antigen using other side of the section.
24. Use protein A-gold probe for staining rabbit and guinea pig primary antibodies. Use relevant secondary IgG for staining primary antibodies from other species. Optical density of diluted protein A-gold probe at 530 nm is approximately 0.1.
25. Carbon coating of the section on grids is carried out indirectly. This can be done by placement of a metal plate between the section and source of carbon evaporation. When directly coated, the surface of the section is damaged by heat and irregular flecks may appear.

Acknowledgements

The work was supported by the Nagasaki International University research fund and in part by a grant-in-aid (17570158) from the Ministry of Education, Science, Culture and Sport and by the Science Research Promotion Fund from the Promotion and Mutual Aid Corporation for Private Schools of Japan.

References

1. Yokota S, Fahimi HD (1981) Immunocytochemical localization of catalase in rat liver. J Histochem Cytochem 29:805–812
2. Yokota S, Fahimi HD (1981) Immunocytochemical localization of albumin in the secretory apparatus of rat liver parenchymal cells. Proc Natl Acad Sci U S A 78: 4970–4974
3. Burry RW, Vandré DD, Hayes DM (1992) Silver enhancement of gold antibody probes in pre-embedding electron microscopic immunocytochemistry. J Histochem Cytochem 40: 1849–1856
4. Roth J, Bendayan M, Carlemalm E, Villiger W, Garavito M (1981) Enhancement of structural preservation and immunocytochemical staining in low temperature embedded pancreatic tissue. J Histochem Cytochem 29:663–671
5. Yokota S, Okada Y (1997) Quantitative evaluation of preparation procedures affecting immunogold staining in post embedding immunocytochemistry. Acta Histochem Cytochem 30:497–504
6. Takizawa T, Robinson JM (2000) FluoroNanogold is a bifunctional immunoprobe for correlative fluorescence and electron microscopy. J Histochem Cytochem 48: 481–485
7. Yokota S (2010) Preparation of colloidal gold particles and conjugation to protein A, IgG, $F(ab')_2$, and streptavidin. In: Schwartzbach SD, Osafune T (eds) Immunoelectron microscopy. Methods and protocols. Springer, New York, pp 109–119
8. Acertain J-D, Carlemalm E, Kellenberger E, Villiger W (1987) Correlation of some mechan-

ical properties of embedding resins with their behavior in microtomy. J Electron Microsc Tech 6:63–79

9. Yamamoto A, Tashiro Y (1994) Visualization by an atomic force microscope of the surface of ultra-thin sections of rat kidney and liver cells embedded in LR White. J Histochem Cytochem 42:1463–1470

10. Osborn M, Webster RE, Weber K (1978) Individual microtubules viewed by immunofluorescence and electron microscopy in the same PtK2 cell. J Cell Biol 77:R27–R34

Chapter 6

RNA Quantification Using Noble Metal Nanoprobes: Simultaneous Identification of Several Different mRNA Targets Using Color Multiplexing and Application to Cancer Diagnostics

João Conde, Gonçalo Doria, Jesus M. de la Fuente, and Pedro Viana Baptista

Abstract

Nanotechnology provides new tools for gene expression analysis that allow for sensitive and specific characterization of prognostic signatures related to cancer. Cancer is a multigenic complex disease where multiple gene *loci* contribute to the phenotype. The ability to simultaneously monitor differential expression originating from each *locus* allows for a more accurate indication of degree of cancerous activity than either *locus* alone. Metal nanoparticles have been widely used as labels for in vitro identification and quantification of target sequences.

Here we describe the synthesis of nanoparticles with different noble metal compositions in an alloy format that are then functionalized with thiol-modified ssDNA (nanoprobes). We also show how to use such nanoprobes in a non-cross-linking colorimetric method for the direct detection and quantification of specific mRNA targets, without the need for enzymatic amplification or reverse transcription steps. The different metals in the alloy provide for distinct absorption spectra due to their characteristic plasmon resonance peaks. The color multiplexing allows for simultaneous identification of several different mRNA targets involved in cancer development. Comparison of the absorption spectra of the nanoprobes mixtures taken before and after induced aggregation of metal nanoparticles allows to both identify and quantify each mRNA target. We describe the use of gold and gold:silver-alloy nanoprobes for the development of the non-cross-linking method to detect a specific *BCR–ABL* fusion gene (e.g., e1a2 and e14a2) mRNA target associated with chronic myeloid leukemia (CML) using 10 ng μL^{-1} of unamplified total human RNA. This simple methodology takes less than 50 min to complete after total RNA extraction with comparable specificity and sensitivity to the more commonly used methods.

Key words: Gold nanoparticles, Gold:silver alloy nanoparticles, Nanoprobes, Colorimetric method, mRNA detection, Gene expression, Cancer

Mikhail Soloviev (ed.), *Nanoparticles in Biology and Medicine: Methods and Protocols*, Methods in Molecular Biology, vol. 906, DOI 10.1007/978-1-61779-953-2_6, © Springer Science+Business Media, LLC 2012

1. Introduction

Gene expression detection can provide powerful insights into the chemistry and physiology of biological systems. Cancer cells exhibit deregulation of the cell cycle resulting in uncontrolled growth and they are resistant to programmed death as a result of abnormalities in one or more proteins that inhibit apoptosis. Moreover, alternatively and abnormally expressed mRNAs (e.g., alternatively spliced mRNAs, fusion gene products) are considered as triggers of tumor development. Better understanding of the molecular mechanisms underlying biological processes can be achieved by comparing gene expression between cells in different states or between cells from different tissues, using techniques such as reverse transcription-polymerase chain reaction (RT-PCR) (1–3). However, methods based on reverse transcription are actually based on detection of amplified cDNAs instead of the RNA itself. Direct detection of RNA, especially mRNA, has remained a challenge.

The National Cancer Institute envisions that over the next years, nanotechnology will result in significant advances in early detection, molecular imaging, targeted and multifunctional therapeutics, prevention and control of cancer (4). The use of nanotechnology (materials, devices, or systems) for diagnostics purposes, i.e., nanodiagnostics, has delivered improved techniques for clinical diagnostics with increased sensitivity at lower costs. Due to their optical properties, gold nanoparticles (AuNPs) have been used for DNA/RNA screening approaches, namely via functionalization with thiolated oligonucleotides (Au-nanoprobes), capable of specifically hybridizing with a complementary oligonucleotide sequence (5–12). These optical properties derive from the characteristic surface plasmon resonance (SPR) band that can be easily tailored through the synthesis of NPs with different metal composition, either in an alloy or core–shell structure, e.g., different gold:silver ratios (13, 14).

Baptista and co-workers (15) developed a colorimetric non-cross-linking method where Au-nanoprobes are used to detect the presence of specific DNA and/or mRNA target sequence, which was successfully applied to tuberculosis diagnostics, gene expression studies, and cancer diagnostics (16–20). The method consists in a visual and/or spectrophotometric comparison of solutions before and after salt induced Au-nanoprobe aggregation and the method is outlined in Fig. 1. The presence of a complementary target prevents aggregation and the solution remains red with a strong surface plasmon absorbance peak at ±520 nm. Noncomplementary/mismatched targets do not prevent Au-nanoprobe aggregation, resulting in a visible change of color from red to blue characterized by a concomitant shift in the surface plasmon absorbance from 520 to 600–650 nm. More recently, the same group has developed an

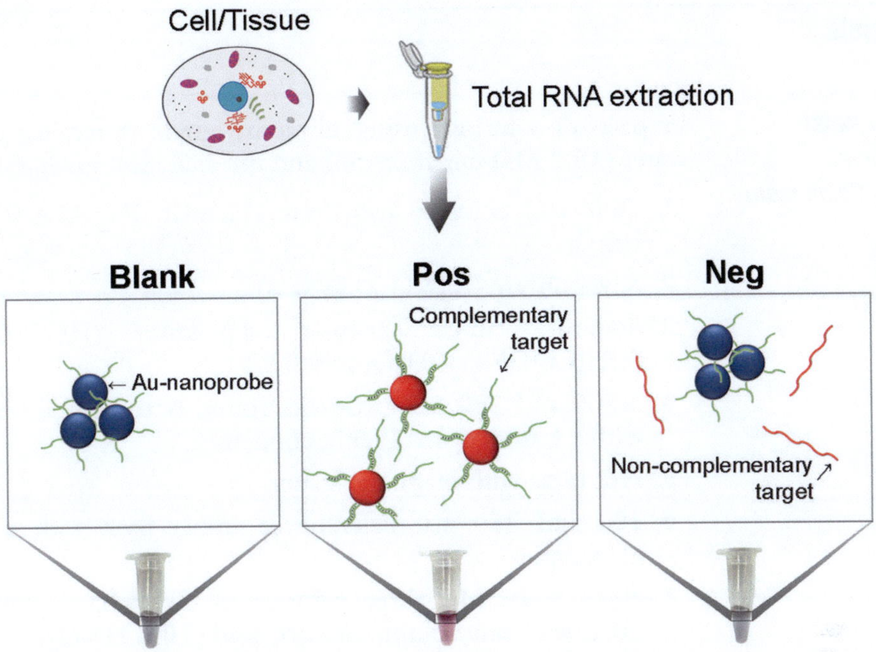

Fig. 1. Schematic of the non-cross-linking assay for RNA quantification. The assay relies on visual comparison of test solutions before and after salt-induced nanoprobe aggregation. "Blank" denotes nanoprobe alone; "Pos" denotes a Positive sample containing complementary mRNA target; "Neg" denotes a Negative sample, where a noncomplementary mRNA is added to the nanoprobe.

one-pot multiplex detection assay by using gold:silver-alloy nanoprobes in conjugation with the Au-nanoprobes (21).

Chronic myeloid leukemia (CML) is a clonal neoplastic disease of the hematopoietic stem cell, and the hallmark molecular event of CML is the genetic t(9;22)(q34;q11) translocation known as the Philadelphia chromosome that is present in >90 % of CML patients (22, 23). This translocation of ABL (chromosome 9) and BCR (chromosome 22) sequences form a BCR–ABL fusion gene, leading to the expression of a chimeric BCR–ABL protein with tyrosine-kinase activity (24). The BCR–ABL abnormal kinase activity is characterized by deregulation of cellular proliferation, decreased adherence of leukemic cells to the bone marrow stroma, and reduced apoptotic activity (25).

Here, we report on the use of gold and gold:silver-alloy nanoprobes for the development of the non-cross-linking method to detect a specific *BCR–ABL* fusion gene (e.g., e1a2 and e14a2) mRNA target associated with CML using 10 ng μL^{-1} of unamplified total human RNA. This simple methodology takes less than 50 min to complete after total RNA extraction with comparable specificity and sensitivity to the more commonly used methods.

2. Materials

2.1. Noble Metal Nanoparticle Synthesis Components

Prepare all solutions using ultrapure grade water, e.g., Milli-Q water (18.2 MΩ-cm at 25 °C) and use analytical grade reagents.

1. Gold(III) chloride trihydrate ($HAuCl_4 \cdot 3H_2O$, ≥49.0 % Au basis).
2. Silver nitrate ($AgNO_3$, ≥99.0 %).
3. Sodium citrate tribasic dihydrate ($HOC(COONa)(CH_2COONa)_2 \cdot 2H_2O$, ≥99.0 %).
4. 4,500 and 250 mL two-neck round bottom flasks equipped with a water-cooled Liebig condenser.
5. Hot plate with magnetic stirrer.
6. 100 and 250 mL Erlenmeyer amber flask with a ground glass cap.
7. *Aqua regia*: Mix three volumes of hydrochloric acid (37 % HCl) with one volume of nitric acid (70 % HNO_3).

2.2. Noble Metal Nanoprobes Synthesis Components

1. 1 M DL-Dithiothreitol (DTT) solution, molecular biology grade. Store at 4 °C until use.
2. Thiol-modified oligonucleotides (5′-thiol-$(CH_2)_6$-ssDNA oligo) containing a complementary sequence to the target(s) of interest (see Note 1 for sequence design tips). The probes given in Table 1 are used to exemplify the specific detection of BCR–ABL fusion gene (e14a2) mRNA, BCR–ABL fusion gene (e1a2) mRNA, and normal ABL gene mRNA. Resuspend the lyophilized thiol-modified oligonucleotides in 100 μL of 1 M DTT and incubate at room temperature for 1 h. Add 900 μL of ultrapure sterile water and mix gently. Store at −20 °C until use.
3. 10 mM phosphate buffer (pH8): 9.32 mM Na_2HPO_4, 0.68 mM NaH_2PO_4. Sterilize by autoclaving and store at 4 °C until use.

Table 1
Oligonucleotide probes

Probes	Sequence (5′-3′)	Gene-Bank acc. number
BCR–ABL (e14a2)	HS-$(CH_2)_6$-CGCTGAAGGGCTTTTGAACT	AJ131466.1
BCR–ABL (e1a2)	HS-$(CH_2)_6$-CGCTGAAGGGCTTCTGCGTC	AF113911.1
ABL (a1a2)	HS-$(CH_2)_6$-CGCTGAAGGGCTTCTTCCAG	NM 005157.3

4. AGEI solution: 2 % (w/v) SDS, 10 mM phosphate buffer (pH 8). Sterilize by filtration (0.22 μm) and store at 4 °C until use. Warm up to 25 °C before use.
5. AGEII solution: 1.5 M NaCl, 0.01 % (w/v) SDS, 10 mM phosphate buffer (pH 8). Sterilize by filtration (0.22 μm) and store at 4 °C until use. Warm up to 25 °C before use.
6. PBS solution: 0.1 M NaCl, 10 mM phosphate buffer (pH 8).
7. Ethyl acetate ($CH_3COOC_2H_5$, ≥99.5 %).
8. NAP-5 columns (GE Healthcare, Sweden).
9. Ultrasound bath S10H (Elma, Germany) or equivalent sonicator (ultrasonic effective power: 30 W; Ultrasonic frequency: 37 kHz).

2.3. Cell Culture Components

1. DMEM: Dulbecco's modified eagle medium.
2. Fetal bovine serum: Heat inactivated. Store at −20 °C until use.
3. CO_2 incubator MCO-96 (Sanyo, UK).

2.4. Total RNA Isolation Components

1. TRIsure® (Bioline, UK).
2. DEPC-treated water: Add 0.1 mL Diethyl Pyrocarbonate (DEPC) for each 100 mL of water to be treated (i.e., 0.1 % (v/v) DEPC) and shake vigorously to bring the DEPC into solution. Incubate the solution at least 1 hour at 37 °C. Autoclave at least 15 min to inactivate any traces of DEPC. Store at 4 °C until use.
3. 75 % ethanol: Add 75 volumes of absolute ethanol to 25 volumes of DEPC-treated water (see Note 2). Store at −20 °C until use.
4. Chloroform.
5. Isopropyl alcohol.

2.5. Non-Cross-Linking Assay Components

1. $MgCl_2$ solution: 0.3 M $MgCl_2$ in deionized water. Store at room temperature.
2. 2.10 mM phosphate buffer (pH 8): as described in Subheading 2.2, step 3.
3. Ultra-Micro quartz cells (Hëllma, Germany) or 384 well small volume, LoBase Polystyrene microplates, black (Greiner Bio-One, Germany).
4. UV–Vis Spectrophotometer UV2 (UNICAM, UK) or Microplate reader Infinite M200 with Absorbance module (Tecan, Switzerland) or equivalent instrument capable of scanning wavelength range 400–800 nm.

Table 2
Control targets

Targets	Sequence (5′-3′)	Gene-Bank acc. number
BCR–ABL (e14a2) gene fusion	TGGATTTAAGCAGAGTTCAAAAGCCCTTCA GCGGCCAGTA	AJ131466.1
BCR–ABL (e1a2) gene fusion	TCCATGGAGACGCAGAAGCCCTTCAGCGGC CAGTAGCATC	AF113911.1
ABL gene	CTCCAGCTGTTATCTGGAAGAAGCCCTTCAG CGGCCAGTA	NM 005157.3
BCR gene	TGGATTTAAGCAGAGTTCAAATCTGTACTGC ACCCTGGAG	NM021574.2
Unrelated	GGCCGCTGCGGCGGGGCTCAGGGCACAAATT GGAACGTTC	n.a.

2.6. RNA and Control Samples (See Note 3)

1. K562 erythroleukemic cells (BCR–ABL positive cell line derived from CML patients in blast crisis, expressing the BCR–ABL gene fusion e1a2 and e14a2 transcripts).
2. 2.HL-60cell line, a human leukemic promyelocytic cell line (BCR–ABL negative).
3. Control oligonucleotide samples—see Table 2. “BCR–ABL (e14a2) gene fusion” and “BCR–ABL (e1a2) gene fusion” oligonucleotide control targets are derived from the reference sequences deposited in Gene Bank with accession no. AJ131466.1 and AF113911.1, respectively. “ABL gene” and “BCR gene” oligonucleotide control targets are derived from the reference sequences Gene Bank accession no. NM 005157.3 and NM021574.2, respectively. The “Unrelated” oligonucleotide control target was randomly generated and checked to present no complementarity to the nanoprobes. The BCR–ABL probes span over the region of fusion, i.e., breakpoint that is present in more than 90 % of CML patients.

3. Methods

3.1. Noble Metal Nanoparticle Synthesis

Treat all glass materials with freshly prepared *aqua regia* (see Note 4) by immersion for at least 1 h and wash vigorously afterwards with ultrapure water. Cover all metal materials used during synthesis (e.g., metallic spatulas) with Teflon to avoid metal corrosion and contamination.

3.1.1. 13 nm Gold Nanoparticle Synthesis

1. In a 500 mL round bottom flask, boil 225 mL of 1 mM $HAuCl_4$ (88.61 mg) until reflux with vigorous stirring.
2. While in reflux, quickly add 25 mL of 38.8 mM sodium citrate (285 mg) and keep reflux for 15 min with vigorous stirring (see Note 5).
3. Stop reflux maintaining the stirring and leave the colloidal solution to cool down to room temperature.
4. Transfer the colloidal solution to a 250 mL Erlenmeyer amber flask with a ground glass cap and store in the dark at room temperature until further use (see Note 6).
5. The colloidal gold concentration should be determined by measuring the absorption (see Note 7) assuming a molar absorptivity for the plasmon resonance band maximum (520 nm) of 2.33×10^8 $M^{-1} cm^{-1}$. The expected typical yield (concentration) should be ~14 nM AuNPs (see Note 8).

3.1.2. Alloy Gold:Silver Nanoparticles Synthesis

1. In a 250 mL round bottom flask, boil 95 mL of 0.126 mM $HAuCl_4$ (4.71 mg) and 0.127 mM $AgNO_3$ (2.04 mg) (see Note 9) until reflux with vigorous stirring.
2. While in reflux, quickly add 5 mL of 34 mM sodium citrate (50 mg) and keep reflux for 15 min with vigorous stirring.
3. Stop reflux maintaining the stirring and leave the colloidal solution to cool down to room temperature.
4. Centrifuge the colloidal solution at $233 \times g$ for 20 min and decant the supernatant to a 100 mL Erlenmeyer amber flask with a ground glass cap, to remove any silver precipitate that may have formed.
5. Store in the dark at room temperature until further use.
6. The colloidal alloy nanoparticle concentration can be determined by measuring the absorption (see Note 7) assuming a molar absorptivity 1.19×10^{10} $M^{-1} cm^{-1}$ for the typical plasmon resonance band maximum at 460 nm (see Note 9). The expected typical yield (concentration) should be ~0.1 nM alloy nanoparticles (50 % Ag: 50 % Au) (see Note 10).

3.2. Noble Metal Nanoprobe Synthesis

1. Extract one volume (100–500 μL) of thiol-modified oligonucleotide resuspended in 0.1 M DTT with 2 volumes of ethyl acetate. Mix thoroughly.
2. Centrifuge for 5 min at $21{,}460 \times g$ and discard the organic phase (i.e., upper phase).
3. Repeat the extraction of the aqueous phase 2 more times.
4. Purify the remaining aqueous phase using a desalting NAP-5 column, following the manufacturer's instructions and using 10 mM phosphate buffer (pH 8) as eluent. Typically, after column equilibration with the eluent, 500 μL of the aqueous

phase is added and let to enter the column. Afterwards, 1,000 μL of purified thiol-modified oligonucleotide can be collected by adding 1,000 μL of the eluent to the column.

5. Quantify the purified thiol-modified oligonucleotide by UV/Vis spectroscopy using the extinction coefficient at 260 nm provided by the oligonucleotide manufacturer. An online Oligonucleotide Properties Calculator (e.g., http://www.basic.northwestern.edu/biotools/oligocalc.html) can be used if the extinction values are not known.
6. Mix the purified thiol-modified oligonucleotide with the colloidal solution in a 1:200 [AuNPs]: oligonucleotide or 1:62,000 [AuAg-alloy-NPs]:oligonucleotide ratio for the alloy nanoparticles.
7. Add AGE I solution to achieve a final concentration of 10 mM phosphate buffer (pH 8), 0.01 % (w/v) SDS. Typically, 15.1 μL of AGE I solution is added to a volume of 3 mL of the solution prepared in previous step.
8. Sonicate the solution for 10 s using an ultrasound bath and incubate at room temperature for 20 min.
9. Afterwards, sequentially increase the ionic strength of the solution by adding the respective volume of AGE II solution up to a final concentration of 10 mM phosphate buffer (pH 8), 0.3 M NaCl, 0.01 % (w/v) SDS (see Note 11). Typically, 104.1, 111.6, 119.8, 129.1, 139.4, and 151 μL of AGE II solution is added sequentially to the 3,015.1 μL of the solution prepared in step 7 to achieve 50 mM incremental increase in NaCl concentration. After each increment, sonicate the solution for 10 s and incubate at room temperature for 20 min before the next increment.
10. Incubate the solution overnight at room temperature.
11. Distribute the functionalized nanoparticles in 1.5 mL microcentrifuge tube and centrifuge for 20 min at 21,460 × *g*, discard the supernatant.
12. Wash the resulting oily pellet in microcentrifuge tubes by adding 1 mL of 10 mM phosphate buffer (pH 8), centrifuge for 20 min at 21,460 × *g*, discard the supernatant.
13. Repeat the above washing step (two washes in total).
14. Wash the pellet in microcentrifuge tubes by adding 1 mL of PBS solution and centrifuge for 20 min at 21,460 × *g*, discard the supernatant.
15. Resuspend the pellet in 500 μL of PBS solution. Transfer the resulting solutions of each microcentrifuge tube into a single polypropylene or glass vial with a conical skirted base.
16. Determine the nanoprobe concentration via Lambert–Beer equation, using the absorbance (see Note 7) and the molar

absorptivity of the respective nanoparticles. Prepare aliquots of 15 or 0.3 nM of Au- or AuAg-alloy-nanoprobes, respectively, using PBS as eluent. Normally, aliquots of 1,300 μL should be sufficient.

17. Store the nanoprobe stock solutions in the dark at 4 °C until further use.

3.3. Cell Culture

1. Grow K562 and HL-60 cells in Dulbecco's modified Eagle's medium (DMEM) with 10 % heat inactivated fetal bovine serum and maintain at 37 °C in 5 % CO_2.
2. After cells reach >80 % confluence level, collect 10^6 cells by centrifugation (1,500 rpm, 5 min at 4 °C).

3.4. Total RNA Isolation

Any suitable method can be used for total RNA extraction and purification. We have tried several commercially available purification kits with equivalent results. Currently, we are using a protocol based on the single step method developed by Chomczynski and Sacchi (26, 27).

1. Isolate Total RNA from K562 and HL-60 cell lines using TRIsure® (use 1 mL of TRIsure per 5×10^6 cells).
2. Lyse cells with 1 mL of TRIsure® and pass the lysate several times through a pipette tip and vortex.
3. Incubate samples for 5 min at room temperature.
4. Add 0.2 mL of chloroform per 1 mL of TRIsure® used. Cap tubes securely and shake vigorously by hand for 15 s.
5. Incubate samples for 2–3 min at room temperature. Centrifuge samples at $12{,}000 \times g$ for 15 min at 4 °C. The sample will separate into a pale green, phenol–chloroform phase, an interphase, and a colorless upper aqueous phase that contains the RNA.
6. Transfer the aqueous phase to another tube. Precipitate the RNA by mixing with 0.5 mL of isopropyl alcohol per 1 mL of TRIsure® used.
7. Incubate samples for 10 min at room temperature then centrifuge at $12{,}000 \times g$ for 10 min at 4 °C.
8. Remove the supernatant and wash the pellet once with 75 % ethanol, adding at least 1 mL of ethanol per 1 mL of TRIsure® used. Vortex samples and centrifuge at $7{,}500 \times g$ for 5 min at 4 °C.
9. Air-dry the pellet for 10 min and dissolve in 25–50 μL of DEPC-treated water (use 25 μL per 10^6 cells) by pipetting the solution up and down, and incubating for 10 min at 55 °C.
10. Store RNA at −80 °C.

3.5. Colorimetric Non-Cross-Linking Assay

3.5.1. Noble Metal Nanoprobes Characterization

1. Prepare six solutions in 200 μL polypropylene thermal-resistant reaction tubes by only mixing the nanoprobe stock solution with the 10 mM phosphate buffer, see Table 3. Do not add $MgCl_2$ at this point!
2. Incubate the solutions for 10 min at 95 °C.
3. Allow the solutions to cool down at room temperature for 30 min.
4. Add the 0.3 M $MgCl_2$ according to Table 3, mix well and spin down the solutions.
5. Incubate the solutions for 15 min at room temperature and register their absorption spectra (350–800 nm) using a UV–visible spectrophotometer or microplate reader.
6. Plot Abs_{peak}/Abs_{600nm} vs. $[MgCl_2]_{final}$ to determine the minimum salt concentration needed for a complete nanoprobe aggregation, where Abs_{peak} is the absorbance peak of the initial dispersed nanoprobe (see Note 12). For Abs_{peak}/Abs_{600nm} ratios below 1, the nanoprobe is considered to be fully aggregated (see Note 13).

3.5.2. mRNA Detection and Quantification

1. In 200 μL polypropylene thermal-resistant reaction tubes, prepare at least four assay solutions (Blank, Positive and Negative control, and sample) by mixing the total RNA sample (final concentration 10–60 ng μL^{-1}) with the BCR–ABL (e14a2) Au-nanoprobe solution (final concentration of 2.5 nM) and an appropriate volume of 10 mM phosphate buffer (pH 8) to make up the total volume (total volume of 60 μL, considering the volume of salt that will be added in step 5). If more than one sample is to be analyzed, more assay solutions can be prepared accordingly.

Table 3
Setup for characterization of nanoprobe aggregation

Tube	Nanoprobe[a] (μL)	10 mM phosphate buffer (pH 8) (μL)	300 mM $MgCl_2$ (μL)	$MgCl_2$ concentration (mM)
1	10	50	–	0
2	10	48	2	10
3	10	46	4	20
4	10	44	6	30
5	10	42	8	40
6	10	40	10	50

[a]Nanoprobe stock solutions: 15 nM Au-nanoprobe or 0.3 nM alloy-nanoprobe (50 % Ag: 50 % Au)

2. As a Positive control use the complementary oligonucleotide BCR–ABL (e14a2) and as a Negative control use the BCR or ABL oligonucleotides (final concentration of 100 fmol μL^{-1})—see Table 1 for sequences. Additionally, prepare the blank solution by replacing the total RNA with an equivalent volume of 10 mM phosphate buffer (pH 8).

   For mRNA quantification, prepare five reference standard solutions using the complementary target ranging in concentration from 0 to 100 fmol μL^{-1}.

3. Incubate the solutions for 10 min at 95 °C.
4. Allow the solutions to cool down for 30 min at room temperature.
5. Add a predetermined volume of the concentrated $MgCl_2$ solution to attain the final concentration needed for aggregation of the nanoprobe as determined in Subheading 3.5.1, step 6, Mix well.
6. Spin down and incubate the solutions for 15 min at room temperature.
7. Register the absorption spectra (350–800 nm) using a UV–visible spectrophotometer or a microplate reader.
8. Determine the Abs_{peak}/Abs_{600nm} ratio (for the described 13 nm gold nanoparticles, Abs_{peak} is typically measured at 520 nm). Compare the ratios of the blank, positive and negative control with the ratios of the sample solutions. Typically, the blank sample, negative control, and negative samples yield ratios below 1, while positive control and positive samples yield ratios above 1 (see Fig. 2).

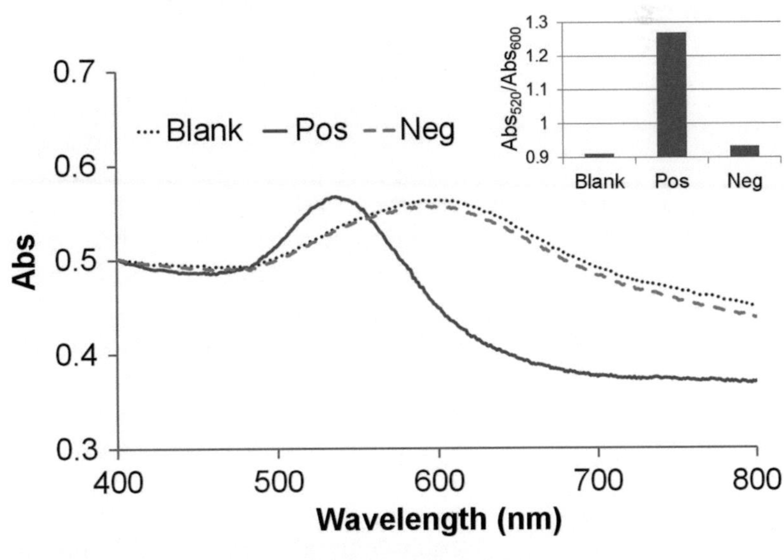

Fig. 2. Characteristic detection assay results. UV/visible spectra of Au-nanoprobe alone (Blank); in the presence of a complementary target (Pos); and in the presence of a non-complementary target (Neg), taken after 15 min incubation with salt. Inlet represents the Abs_{520nm}/Abs_{600nm} ratios of the presented UV/visible spectra.

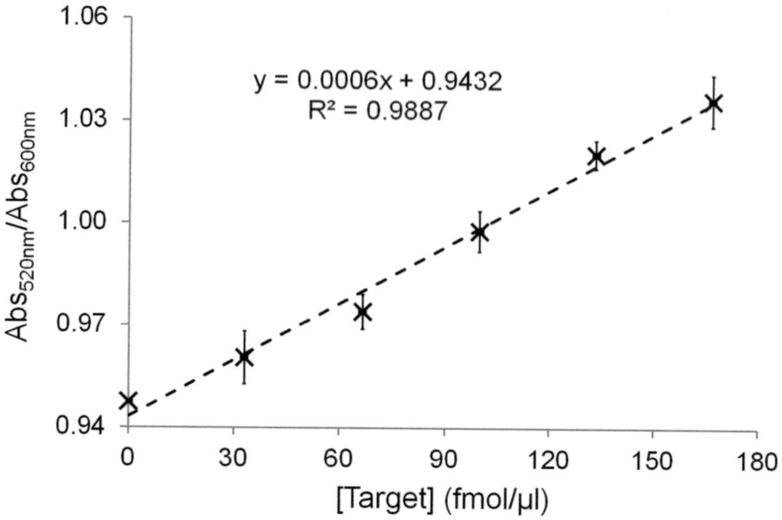

Fig. 3. Characteristic quantification assay results. Typical Abs_{520nm}/Abs_{600nm} ratios obtained for a range of complementary target concentrations. Dashed line represents a linear fit of the experimental data points and the error bars represent the standard deviation ($n = 3$).

9. For quantification, plot the Abs_{520nm}/Abs_{600nm} vs. mRNA concentration of the reference standard solutions and compare with the samples' Abs_{520nm}/Abs_{600nm} ratio to determine their mRNA concentration (see Fig. 3) (see Note 14).

3.5.3. One-Pot Multiplex Detection

1. In 200 μL polypropylene thermal-resistant reaction tubes, prepare at least six assay solutions (blank, three positive and one negative control, and sample) by mixing the total RNA sample (final concentration 10–60 ng μl^{-1}) with the ABL Au-nanoprobe solution (final concentration of 2.5 nM), the BCR–ABL (e1a2) AuAg-alloy-nanoprobe solution (final concentration of 0.05 nM), and an appropriate volume of 10 mM phosphate buffer (pH 8) to make up the total volume (total volume of 60 μL, considering the volume of salt that will be added in step 5). If more than one sample is to be analyzed, more assay solutions can be prepared accordingly.
2. Use the complementary oligonucleotides as Positive controls for each nanoprobe ABL (ABL Au-nanoprobe) and BCR–ABL (e1a2) (BCR–ABL (e1a2) AuAg-alloy-nanoprobe), both separately and mixed. Use the unrelated oligonucleotide sequence as a negative control (final concentration of oligonucleotides at 100 fmol μL^{-1})—see Table 2 for sequences. Additionally, prepare the blank solution by replacing the total RNA with equivalent volume of 10 mM phosphate buffer (pH 8).
3. Incubate all solutions for 10 min at 95 °C.

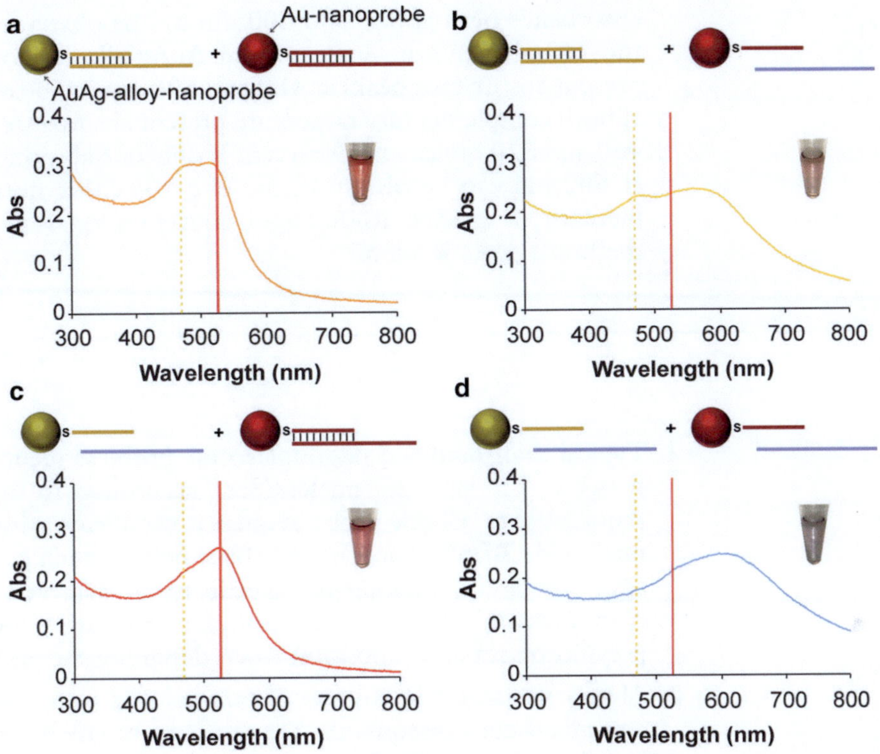

Fig. 4. Representative one-pot assay results. UV/visible spectra and digital photography of AuAg-alloy- and Au-nanoprobe mix in the presence of a complementary targets to both the AuAg-alloy- and Au-nanoprobe – the solution retains the initial orange color (**a**), a complementary target to the AuAg-alloy-nanoprobe – the solution changes from orange to yellow (**b**), a complementary target to the Au-nanoprobe – the solution changes from orange to red (**c**) or a noncomplementary target to both AuAg-alloy- and Au-nanoprobes – the solution changes from orange to blue (**d**). Vertical dashed and solid lines represent the position of absorption peak of the AuAg-alloy-nanoprobes (460 nm) and Au-nanoprobe (520 nm), respectively, when dispersed in solution. Schematics represent each nanoprobe and their complementary or noncomplementary targets. Reproduced from (16) with permission from IOP.

4. Allow the solutions to cool down for 30 min at room temperature.
5. Add a predetermined volume of the concentrated $MgCl_2$ solution to attain the final concentration needed for aggregation of the nanoprobes as determined in Subheading 3.5.1, step 6 (see Note 15).
6. Mix well, spin down, and incubate the solutions for 15 min at room temperature.
7. Measure the absorption spectra (350–800 nm) using a UV–visible spectrophotometer or a microplate reader.
8. Compare the spectra of the blank, positive and negative controls with the spectra of the sample(s)—as an example of typical multiplex results, see Fig. 4. Typically, if only the complementary target to the Au-nanoprobe is present an

absorbance peak at 520 and 600 nm can be observed. If only the complementary target to the AuAg-alloy-nanoprobe is present absorbance peaks at 460 and 600 nm can be observed. If both complementary targets are present absorbance peaks at 460 and 520 nm can be observed. If only one absorbance peak at 600 nm can be observed, the target is either noncomplementary or specific RNA target concentration is below the method's detection limit.

4. Notes

1. Typical thiol-modified oligonucleotide probe sequences range between 15 and 25 nucleotides, according to specificity requirements. Check probe sequence specificity using NCBI nucleotide BLAST tool (http://blast.ncbi.nlm.nih.gov/Blast.cgi). Increased nanoprobe specificity is achieved at the oligonucleotide's 3′ end, so consider any single nucleotide polymorphism at this position when designing the probe (28).
2. Always measure the volume of ethanol and water separately and mix them subsequently. Do not prepare ethanol dilutions based on volumetric flasks, as the total volume of ethanol/water will drop after mixing, leading to an erroneous dilution if the volumetric flask mark is used as reference (29).
3. Thiol-modified ssDNA, complementary to the fusion region of the BCR–ABL mRNA is used to functionalize the gold nanoparticles and produce specific Au-nanoprobes to detect the fusion gene. These nanoprobes are assessed in terms of specificity by means of total RNA mixtures spiked in with synthetic oligonucleotides (complementary target) harboring the fusion site BCR–ABL e14a2. It should be noted that, in reality, patients may only harbor one copy of the fusion gene and the remaining copies of normal ABL and BCR should be still functional, thus expressing the normal mRNA sequence. Two oligonucleotides harboring the normal sequence of the BCR or of the ABL genes are used to evaluate the probe's capability to discriminate from similar sequences.
4. Warning: *aqua regia* is potentially hazardous and highly corrosive. Always wear appropriate personal protective equipment (i.e., safety goggles, gloves, and laboratory coat) and work in a clean, well-ventilated chemical fume hood. Never store *aqua regia* in a sealed container, as pressure may escalate and the container may burst or explode. Dispose *aqua regia* adequately by dilution and neutralization.

5. After adding the citrate solution, the initial pale yellow color of the Au(III) solution should become instantly colorless and then gradually change to deep red due to the nanoparticle formation. The reduction process usually takes a few minutes to occur. During this process, a precursor called acetone dicarboxylic acid is formed as a result of the citrate oxidation. This precursor plays the roles of precursor, reducing and nucleating agents. The Au(III) ions are then reduced to Au(I) and when the solution becomes saturated of Au(I) atoms, they start to precipitate in the form of nanoparticles. The citrate acts as capping and stabilizing agent that sticks to the nanoparticle surface avoiding the aggregation of the nanoparticles.
6. The prepared nanoparticles are stable for months when stored in a clean container (glass or plastic) at room temperature. Do not freeze the nanoparticles as this may cause aggregation.
7. The Lambert–Beer law states that the absorbance of a homogeneous substance becomes linear with its concentration according to formula $A = \varepsilon \times l \times C$, where A is the substance absorbance typically at its wavelength peak, ε is the molar absorptivity for the wavelength of A, l is the optical path length, and C is the substance concentration. Care should be taken not to exceed an absorbance of 2 so as to avoid deviations to the Lambert–Beer law. In case the measured absorbance exceeds this value, dilute the sample and consider the dilution factor when calculating the original stock concentration.
8. Optionally, morphological characterization of the AuNPs can be performed by transmission electron microscopy (TEM) and dynamic light scattering (DLS).
9. To obtain different AuAg alloy nanoparticles with different plasmon resonance peaks, the ratio of $HAuCl_4$ and $AgNO_3$ can be varied (e.g., the molar concentrations presented in the protocol refers to a 50%Au:50%Ag ratio, with a typical SRP band peak at 460 nm and a molar absorptivity of 1.19×10^{10} M^{-1} cm^{-1}). See Fig. 5 for typical SPR peak variation as function of the Au:Ag metal ratio. Nanoparticle size may also vary with the Au:Ag metal ratio.
10. Optionally, morphological characterization of the alloy nanoparticles can be performed by TEM and DLS; the Au:Ag ratio can be determined by inductively coupled plasma (ICP) and/or Energy-dispersive X-ray spectroscopy (EDX).
11. Different degrees of nanoprobe functionalization (i.e., oligonucleotide density) are attained by varying the final NaCl concentration during the aging process (e.g., 0.1, 0.5, 0.7 M). Be aware that nanoprobe hybridization efficiency may decrease with increasing density due to steric hindrance. For further information, consult Doria et al. (28).

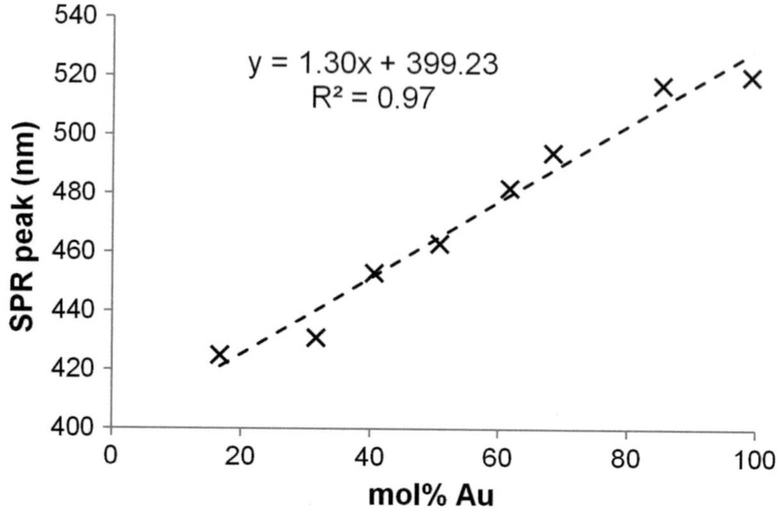

Fig. 5. SPR absorption band peak of alloy-nanoparticles as function of the gold:silver ratio. *Dashed line* represents a linear fit of the experimental data points.

12. Abs_{peak} is the absorbance maximum determined earlier for the particular nanoprobe peak of the initial dispersed nanoprobe. Typical values are 520 nm for Au-nanoprobes or 460 nm for alloy-nanoprobes: (50 % Ag: 50 % Au).
13. If the nanoprobes do not completely aggregate and change color within the suggested concentrations of $MgCl_2$, use higher concentrations of $MgCl_2$. If the nanoprobe changes color and fully aggregates at a final $MgCl_2$ concentration of 10 mM, then derivatization of the nanoparticles with the thiol-modified oligonucleotides may have not been effective. If so happens, repeat nanoprobe synthesis.
14. To discriminate between two significantly different aggregation levels, as for example in a YES/NO identification of a given target, the ratio between the peaks at 520 and 600 nm is usually used for Au-nanoprobes. However, for identifying small differences in aggregation levels between two quantities for the same target, there is a need to decrease the background noise in the spectra. Such background noise may be strongly reduced and even solved if an integral of the signal is used, i.e., the area under the curve (AUC).
15. In order for the one-pot assay to work, both nanoprobes must aggregate at similar minimum salt concentration. If the minimum salt concentration for aggregation varies slightly between nanoprobes, use the highest determined concentration of salt for the one-pot assay.

References

1. Bustin SA (2000) Absolute quantification of mRNA using real-time reverse transcription polymerase chain reaction assays. J Mol Endocrinol 25:169–193
2. Weis JH et al (1992) Detection of rare mRNAs via quantitative RT-PCR. Trends Genet 8:263–264
3. Freeman WM, Walker SJ, Vrana KE (1999) Quantitative RT-PCR: pitfalls and potential. Biotechniques 26:112–115
4. http://nano.cancer.gov/about/plan/ Accessed 11 May 2012
5. Baptista P et al (2008) Gold nanoparticles for the development of clinical diagnosis methods. Anal Bioanal Chem 391:943–950
6. Sato K, Hosokawa K, Maeda M (2005) Non-cross-linking gold nanoparticle aggregation as a detection method for single-base substitutions. Nucleic Acids Res 33:e4
7. Mirkin CA et al (1996) A DNA-based method for rationally assembling nanoparticles into macroscopic materials. Nature 382:607–609
8. Taton TA, Mirkin CA, Letsinger RL (2000) Scanometric DNA array detection with nanoparticle probes. Science 289:1757–1760
9. Qin WJ, Yung LY (2007) Nanoparticle-based detection and quantification of DNA with single nucleotide polymorphism (SNP) discrimination selectivity. Nucleic Acids Res 35:e111
10. Elghanian R et al (1997) Selective colorimetric detection of polynucleotides based on the distance-dependent optical properties of gold nanoparticles. Science 277:1078–1081
11. Sato K, Hosokawa K, Maeda M (2003) Rapid aggregation of gold nanoparticles induced by non-cross-linking DNA hybridization. J Am Chem Soc 125:8102–8103
12. Storhoff JJ et al (2004) Homogeneous detection of unamplified genomic DNA sequences based on colorimetric scatter of gold nanoparticle probes. Nat Biotechnol 22:883–887
13. Wilcoxon J (2009) Optical absorption properties of dispersed gold and silver alloy nanoparticles. J Phys Chem B 113:2647–2656
14. Liz-Marzan LM (2006) Tailoring surface plasmons through the morphology and assembly of metal nanoparticles. Langmuir 22:32–41
15. Baptista P et al (2005) Colorimetric detection of eukaryotic gene expression with DNA-derivatized gold nanoparticles. J Biotechnol 119:111–117
16. Doria G, Franco R, Baptista P (2007) Nanodiagnostics: fast colorimetric method for single nucleotide polymorphism/mutation detection. IET Nanobiotechnol 1:53–57
17. Conde J, de la Fuente JM, Baptista PV (2010) RNA quantification using gold nanoprobes—application to cancer diagnostics. J Nanobiotechnology 8:5
18. Veigas B et al (2010) Au-nanoprobes for detection of SNPs associated with antibiotic resistance in *Mycobacterium tuberculosis*. Nanotechnology 21:415101
19. Baptista PV et al (2006) Gold-nanoparticle-probe-based assay for rapid and direct detection of *Mycobacterium tuberculosis* DNA in clinical samples. Clin Chem 52:1433–1434
20. Costa P et al (2010) Gold nanoprobes assay for identification of mycobacteria from the *Mycobacterium tuberculosis* complex. Clin Microbiol Infect 16:1464–1469
21. Doria G et al (2010) Gold-silver-alloy nanoprobes for one-pot multiplex DNA detection. Nanotechnology 21:255101
22. Hehlmann R, Hochhaus A, Baccarani M (2007) Chronic myeloid leukaemia. Lancet 370:342–350
23. Shet AS, Jahagirdar BN, Verfaillie CM (2002) Chronic myelogenous leukemia: mechanisms underlying disease progression. Leukemia 16: 1402–1411
24. Ren R (2005) Mechanisms of BCR-ABL in the pathogenesis of chronic myelogenous leukaemia. Nat Rev Cancer 5:172–183
25. Melo JV, Hughes TP, Apperley JF (2003) Chronic myeloid leukemia. Hematology Am Soc Hematol Educ Program 132–152
26. Chomczynski P, Sacchi N (1987) Single step method of RNA isolation by acid guanidinum thiocyanate-phenol-chloroform extraction. Anal Biochem 162:156–159
27. Chomczynski P (1993) A reagent for the single-step simultaneous isolation of RNA, DNA and proteins from cell and tissue samples. Biotechniques 15:532–537
28. Doria G et al (2010) Optimizing Au-nanoprobes for specific sequence discrimination. Colloids Surf B 77:122–124
29. Parsons PW, Estrada FJ (1942) Changes in volume on mixing solutions. Ind Eng Chem 34:949–952

Chapter 7

Effect of Gold Nanoparticle Conjugation on the Activity and Stability of Functional Proteins

Julian Bailes, Sara Gazi, Rositsa Ivanova, and Mikhail Soloviev

Abstract

Immobilization of functional proteins such as enzymes on solid surfaces produces a variety of effects ranging from the reversal and strong inhibition to the enhancement of protein stability and function. Such effects are protein-dependent and are affected by the physical and chemical properties of the surfaces. Functional consequences of protein immobilization on the surface of gold nanoparticles (AuNPs) are protein-dependent and require thorough investigation using suitable functional tests. However, traditional approaches to making control samples, i.e., immobilized protein vs. protein in solution in absence of any nanoparticles do not provide sufficiently identical reaction conditions and complicate interpretation of the results. This report provides advice and methods for preparing AuNP-conjugated preparations generally suitable for studying the effects of immobilization on the activity and stability of different functional proteins. We use bovine catalase to illustrate our approach, but the methods are easily adaptable to any other enzyme or protein. The AuNP-immobilized enzyme showed increased stability at elevated temperatures compared to the same enzyme in solution.

Key words: Gold nanoparticles, Protein-binding capacity, Protein function, Protein stability, Thermal stability, Experimental design, Protein immobilization, Catalase

1. Introduction

Protein conjugation to gold nanoparticles (AuNPs) has been widely used in electron microscopy for visualizing immunostained preparations (1) or in light microscopy (2). AuNPs possess extinction coefficients far greater than that of organic dyes (3) and are therefore used widely for colorimetric detection in a variety of immunoassays (4–9), in biosensors (10–18), for drug delivery (19, 20), and in various nanobiotechnology applications (21–24). Furthermore, AuNPs are relatively easy to manufacture and a great

Mikhail Soloviev (ed.), *Nanoparticles in Biology and Medicine: Methods and Protocols*, Methods in Molecular Biology, vol. 906, DOI 10.1007/978-1-61779-953-2_7,

variety of chemical and biochemical methods have been reported (25–33). In most cases, AuNPs are conjugated directly to the proteins due to the combination of electrostatic and hydrophobic interactions. The presence of sulfhydryl groups, such as cysteine side chains of proteins form strong bonds with the gold. Many proteins readily adhere to the AuNPs surface especially at pH just above their pI values. The presence of cysteine residues, such as immunoglubulins, BSA, and many other proteins, results in very tight practically irreversible binding of the proteins to AuNPs. Tighter binding may affect protein structure and therefore function. It has been widely documented that immunoglobulins (large 150 kDa proteins) retain their antigen-binding activity upon conjugation to AuNPs. However, smaller proteins may be affected to a greater degree and would be in danger of losing their function. While some enzymes retain their activity (34, 35), others have been reported to be strongly affected (35, 36). In some cases, AuNP conjugation might improve enzyme activity and stability. For example, conjugation to AuNPs has been shown to increase the efficiency of trypsin by ~12 % (37), and improve specificity (38) and yield in PCR (39). The stabilizing and enhancement effects are not limited to AuNPs alone and have also been reported for other nanomaterials (40). A few opposing effects can significantly affect enzymatic rates and reaction yields. On one hand, the limited accessibility of an enzyme following immobilization (especially passive sorption) on AuNPs is likely to affect thermodynamics, kinetics, and lower the reaction rates (41, 42). On the other hand, immobilization on NP surfaces may stabilize a protein and increase its tolerance of thermal and pH gradients and reduce its susceptibility to chemical or biochemical degradation and denaturing (37, 43). The former effect will depend on the enzyme used and may be reduced by controlled orientation during immobilization or with the use of molecular spacers between solid surface and the protein to position it further away form the surface into the solution. That will however negate any positive effects on the increased stability following direct surface sorption. It is clear that any functional effect of protein conjugation to nanomaterials is affected by the physical and chemical properties of the AuNPs, is protein-dependent (44, 45), is widely applicable, and is not exclusive to AuNPs (46).

The variety of functional effects of gold and other nanomaterials on proteins justifies the need to understand and control the interactions of NPs with proteins. Stability and folding of AuNP-conjugated enzymes has been tested before using a variety of approaches. Physical methods such as UV, IR, or fluorescence spectroscopy for measuring changes in the extinction spectra, amide I/II shifts or fluorescence of aromatic residues, circular dichroism (35, 47–51) for estimating a degree of structural and hence functional changes due to conjugation to nanoparticles, are universally applicable but are indirect and although may indicate change in protein folding, may not yield information of whether enzymatic function was affected by the conjugation. Functional protein testing could provide direct means of testing stability of AuNP-conjugated proteins and their function, but the choice of the functional assay will depend on the protein being tested.

For example, conjugation of trypsin to AuNPs has been reported to reduce biocatalytic activity and affect chemical specificity of trypsin because of changes in the enzyme's conformation (36). Many of the differences identified between AuNP-conjugated trypsin and free trypsin in absence of the NPs were due to the sorption of proteolytic peptides on the NPs prior to separation by capillary electrophoresis (CE). And although the use of SDS helped to release some peptides, the report concluded that the presence of nanoparticles could have affected the outcome. Functional consequences of immobilization on solid surfaces have been reported in some cases to completely reverse the function of an enzyme, e.g., proteases could lose their ability to hydrolyze peptide bonds and instead catalyze peptide formation (52, 53). Such an example emphasizes the importance of a thorough functional analysis of functional proteins following their immobilization on solid surfaces such as AuNPs and highlights the need for better experimental design. An ideal test set should have two chemically and physically identical reaction mixtures (i.e., both containing the same buffer components, the same concentration of nanoparticles, the same amount of protein/enzyme, and the same amount of blocking reagents) but the protein under investigation should either be immobilized on the AuNP (the "unknown" sample) or be completely free in solution (control). Traditional approaches to making control samples, i.e., immobilized protein vs. protein in solution in absence of the nanoparticles (37) do not provide sufficiently identical reaction conditions and complicate interpretation of the results. This report provides advice and methods for preparing AuNP-conjugated preparations generally suitable for studying the effects of immobilization on the activity and stability of different functional proteins. We use bovine catalase to illustrate our approach, but the methods are easily adaptable to any other enzyme or protein.

2. Materials

1. Tetrachloroauric acid hydrate ($HAuCl_4$): make 0.02 % w/w solution of $HAuCl_4$ in deionized water (equivalent to 0.01 % w/w gold). Prepare 40 mL immediately before the experiment (see Note 1).

2. 1 % Sodium citrate solution in deionized water (see Note 1).
3. 10 % (w/v) NaCl solution in deionized water.
4. Bovine serum albumin (BSA) stock solution: 10 % (w/v) solution in water. Store stock solution frozen at −20 °C. Make working solution by diluting the stock 10x with deionized water on the day of experiment (final concentration 1 % or 10 mg/mL). These dilutions can be stored at 4 °C for a few days.
5. Assay buffer: make 50 mM potassium phosphate solution in deionized water using 11.4 mg of KH_2PO_4 per milliliter of water, adjust pH to 7.0 with 1 M KOH.
6. Bovine liver catalase: make 10 mg/mL stock solution in assay buffer. Store in aliquots at −20 °C until usage (see Notes 2 and 3).
7. Helios Beta UV-Vis spectrophotometer or other alternative spectrometer capable of measuring absorbance at 240 nm (see Note 4) and narrow 0.5 mL or smaller UV cuvettes.
8. Prepare 0.036 % (w/w) hydrogen peroxide (H_2O_2) solution in assay buffer for use in functional catalase assays (continuous spectrophotometric rate reduction assay) (see Note 5).
9. Eppendorf Mastercycler programmable thermal cycler for use in generating temperature gradients to allow multiple parallel incubations of the enzyme at different temperatures.

3. Methods

3.1. Making and Characterizing AuNPs

1. To make AuNPs, heat 40 mL of 0.02 % w/w solution of $HAuCl_4$ to boiling point under constant stirring and add 4 mL of 1 % sodium citrate. Continue to stir for another 15 min. Cool to room temperature and store at +4 °C.
2. To determine the protein-binding capacity of AuNPs, make a series of BSA dilutions in deionized water ranging from 1 μg/mL to 10 mg/mL (see Note 6). Use deionized water as negative (no protein) control.
3. Make eight identical 20 μL aliquots of AuNPs in small 0.5 mL microcentrifuge tubes.
4. Add 10 μL of BSA dilutions to 6 tubes and 10 μL of water to the remaining two tubes. Incubate at room temperature for 30 min.
5. Add 20 μL of 10 % NaCl to all tubes except one, where 20 μL of water is added instead (no aggregation control). Incubate at room temperature for 10 min.

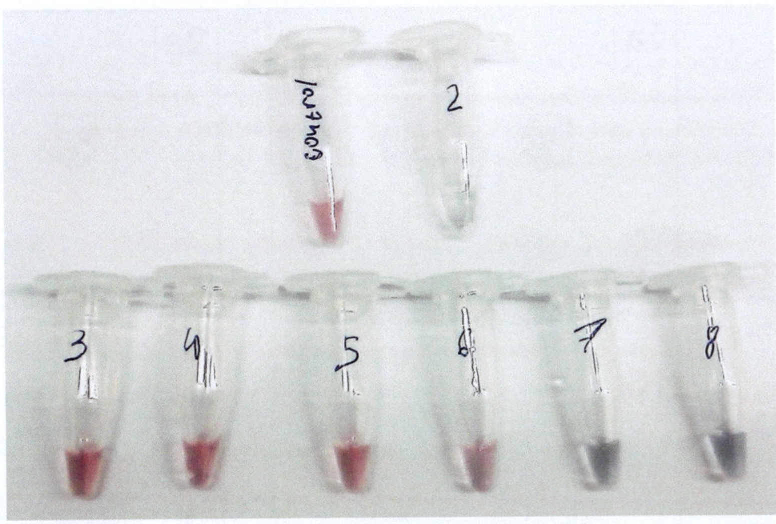

Fig. 1. Aggregation of AuNPs in the presence of salt. Tube 1 ("Control")—contains no salt (no protein, no salt control). Tube 2 is negative control (no protein, 4 % NaCl). Each sample had 20 μL of AuNPs preincubated with variable amounts of protein prior to the addition of salt (5, 2.5, 2, 1, 0.5, and 0.25 μg in tubes 3–8 respectively; final 4 % NaCl in each tube). The binding 2 μg per 20 μL of AuNP (0.1 μg BSA per 1 μL AuNPs) results in blocking the surface and prevents aggregation of nanoparticles.

6. Visual control is sufficient to determine the minimum protein-binding capacity of the AuNPs (see Fig. 1). In our case, the binding capacity of the AuNPs was estimated at ~0.1 μg BSA per 1 μL AuNPs. This value will vary depending on the nanoparticle used.

3.2. Enzyme Immobilization and Functional Assays

Thorough functional analysis of functional proteins following their immobilization on AuNPs is impractical or impossible without having properly designed negative controls. Traditional approaches to making control samples, i.e., immobilized protein vs. protein in solution in absence of the nanoparticles (37) do not provide sufficiently identical reaction conditions and complicate interpretation of the results. An ideal experiment should compare a set of two fully matching, chemically and physically identical samples (i.e., both should contain the same buffer components, the same concentration of nanoparticles, the same amounts of enzyme, and the same amount of blocking reagents). The only difference should be in the state of protein under investigation—it should either be immobilized on the NP (the unknown experimental sample) or be completely free in solution (negative control). We achieve this by adding the same amounts of enzyme and any blocking reagents

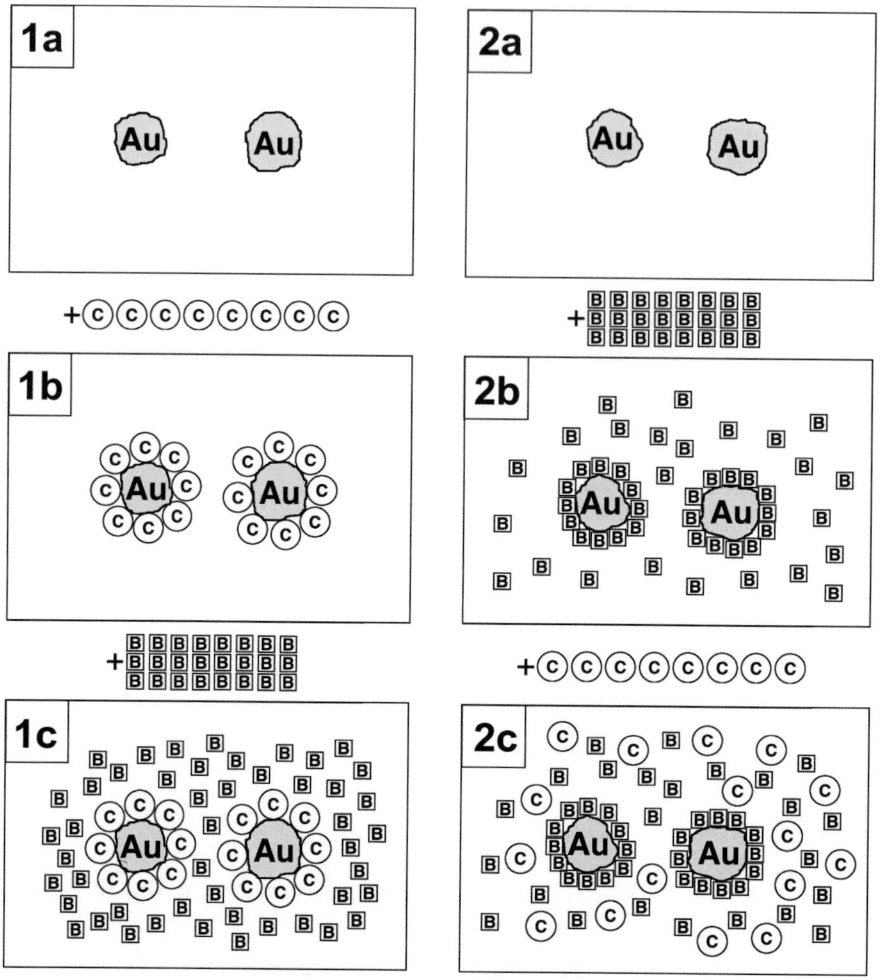

Fig. 2. An outline of the procedure. Panels (1a–1c) correspond to the immobilization of the specific protein, in our case the catalase (indicated as "C" in the figure). Panels 2a–2c refer to making blocked nanoparticles. The method generates a matched pair of samples which have identical chemical composition and the amount of AuNPs. The only significant difference is in the state of immobilized enzyme ("C"), which is either fully immobilized (as in panel 1c) or remains in solution (as in panel 2c), thus providing an ideal negative control for the experimental sample (1c).

but adding them in a different order. It is also important to use a large excess of blocking protein to ensure that its final concentration in solution is not significantly different whether it was applied to bare or derivatized AuNPs. Our procedure, as outlined in Fig. 2, produces perfectly matched sets of protein-modified AuNPs suitable for virtually any functional assay.

1. Immobilize sufficient amounts of enzyme to AuNPs. The determined protein-binding capacity (in our case this was ~0.1 μg BSA per 1 μL AuNP) may need to be adjusted for

proteins of different size (see Note 7). Use 4 μg of catalase per 20 μL AuNPs. (see Note 8). To make negative control add 40 μL of 0.1 % BSA to 20 μL of AuNPs. Incubate at room temperature for 25–30 min.

2. To complete preparation of the immobilized enzyme, add 40 μL of 0.1 % BSA to the sample of AuNP-catalase to block any remaining AuNP surface and to increase total protein concentration in the sample. Incubate at room temperature for 25–30 min. Store the prepared enzyme at +4 °C and use as soon as possible. Do not freeze.

3. To complete preparation of the negative control (in solution enzyme), add 4 μg of catalase to the sample of the blocked nanoparticles (AuNP-BSA). Incubate at room temperature for 25–30 min (see Note 9). Store at +4 °C and use as soon as possible. Do not freeze.

4. Repeat or scale-up the above procedures (see steps 1–3) for making sufficient amount of immobilized enzyme for stability studies.

5. No further sample purification is required, except dilution of both preparations to ~100 units of active catalase enzyme (whether free or AuNP immobilized) per milliliter of solution. Other functional proteins may require different dilutions.

6. Functional enzyme assays will vary depending on the enzyme used. For catalase, measure the rate of hydrogen peroxide degradation using a continuous spectrophotometric rate reduction assay measured at 240 nm. A typical assay would have 0.95 mL of 0.036 % (w/w) H_2O_2 solution (for 1 mL cuvettes) or have 0.45 mL of 0.036 % (w/w) H_2O_2 solution (for 0.5 mL cuvettes). Add 50 μL of the enzyme preparation to start the reaction and measure absorbance changes over time (for 5 min). Calculate the rate of absorbance change per minute (the initial rate).

7. To study the effect of temperature on the catalyse conjugates, incubate AuNP-conjugated and unconjugated enzyme at different temperatures prior to carrying out the functional assay. We incubated catalase samples at 25, 37, 45, 55, 60, and 70 °C for 30 min. Although total activity of the AuNP-conjugated enzyme is slightly lower at 37 °C (the optimal temperature), it shows much increased stability at a wider range of temperatures compared to unconjugated catalase (see Fig. 3). Other conditions such as pH, the presence of denaturants, salt, etc. may be easily compared for AuNP-immobilized and free enzyme.

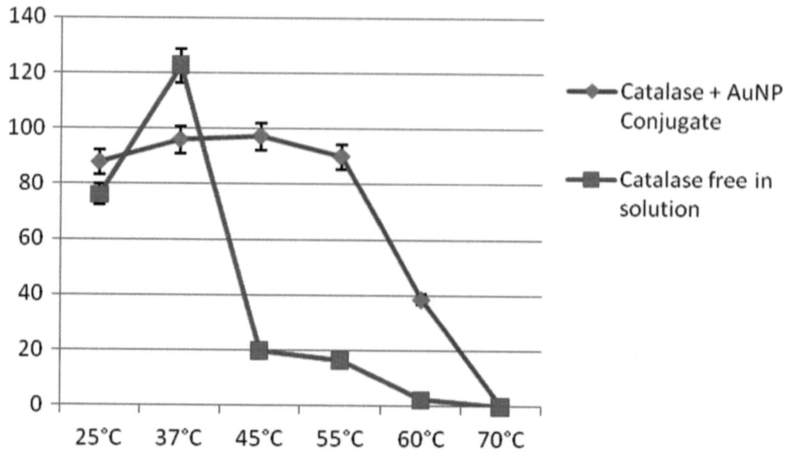

Fig. 3. The effect of temperature on catalase stability. Free or AuNP-conjugated catalase was incubated for 30 min at the temperatures shown prior to the enzymatic assay. All the reactions conditions were identical in all samples, including the presence of AuNPs, blocking protein (BSA), and buffer. The vertical axis shows relative enzyme activity (arbitrary values).

4. Notes

1. Alternatively, use ready-made citrate ion stabilized AuNPs, which can be easily obtained commercially.
2. Catalase is an antioxidant enzyme ubiquitously present in mammalian and nonmammalian aerobic cells containing a cytochrome system which catalyzes the decomposition of hydrogen peroxide (H_2O_2) to water and oxygen (54). Although we illustrate the method using bovine catalase, the approach can be easily used with virtually any other functional protein.
3. If enzyme does not dissolve completely, incubate the sample at 37 °C for one hour to obtain complete dissolution. High concentration of enzyme stock is necessary for AuNP derivatization, but the final concentration of the immobilized or free enzyme should not exceed ~1,000 units of active enzyme per milliliter of solution. One unit is defined as the amount of enzyme needed to decompose 1.0 μmole of H_2O_2 per minute at pH 7.0 at 25 °C. If the final catalase concentration (free or immobilized) is greater than 100 Units/mL in terms of activity units, further dilutions may be necessary.
4. This is needed to test the activity of catalase. Where alternative functional testing methods are needed, other instruments would be needed.
5. Check that absorbance of the solution measured at 240 nm is within the 0.52–0.55 range. Adjust the absorbance by adding more peroxide or more buffer, as necessary.

6. Following an initial test a narrower range of BSA concentration may be needed to make more accurate measurements.
7. Catalase is a 240 kDa enzyme 4× times bigger than 60 kDa BSA by mass. We therefore used an increased amount of catalase compared to the ratio obtained with BSA to allow for such mass difference.
8. Exceeding the protein capacity at this stage while not critical, is highly undesirable because it will result in the presence of increasingly large concentrations of free enzyme in solution thus reducing any functional differences between samples and their detection.
9. No enzyme is expected to bind to AuNPs at this stage, because the surface of the AuNPs is saturated by approximately a tenfold excess (by mass) of BSA. However, the final total amounts of all the proteins (catalase, BSA) and all other components (buffer, AuNPs) will be identical between the two preparations. Further controls may include enzyme preparation with citrate buffer alone without AuNP, e.g., for checking possible inhibitory effects of citrate ions on the enzyme.

References

1. Faulk WP, Taylor GM (1971) An immunocolloid method for the electron microscope. Immunochemistry 8(11):1081–1083
2. Neagu C, van der Werf KO, Putman CAJ et al (1994) Analysis of immunolabeled cells by atomic force microscopy, optical microscopy, and flow cytometry. J Struct Biol 112(1): 32–40
3. Zhao W, Chiuman W, Brook MA et al (2007) Simple and rapid colorimetric biosensors based on DNA aptamer and noncrosslinking gold nanoparticle aggregation. Chembiochem 8(7):727–731
4. Medley CD, Smith JE, Tang Z et al (2008) Gold nanoparticle-based colorimetric assay for the direct detection of cancerous cells. Anal Chem 80(4):1067–1072
5. Reynolds RA III, Mirkin CA, Letsinger RL (2000) A gold nanoparticle/latex microsphere-based colorimetric oligonucleotide detection method. Pure Appl Chem 72(1–2):229–235
6. Choi DH, Lee SK, Oh YK et al (2010) A dual gold nanoparticle conjugate-based lateral flow assay (LFA) method for the analysis of troponin I. Biosens Bioelectron 25(8):1999–2002
7. Rong-Hwa S, Shiao-Shek T, Der-Jiang C et al (2010) Gold nanoparticle-based lateral flow assay for detection of staphylococcal enterotoxin B. Food Chem 188(2):462–466
8. Girotti S, Eremin S, Montoya A et al (2010) Development of a chemiluminescent ELISA and a colloidal gold-based LFIA for TNT detection. Anal Bioanal Chem 396:687–695
9. Yu CY, Ang GY, Chua AL et al (2011) Dry-reagent gold nanoparticle-based lateral flow biosensor for the simultaneous detection of Vibrio cholerae serogroups O1 and O139. J Microbiol Methods 86(3):277–282
10. Schultz DA (2003) Plasmon resonant particles for biological detection. Curr Opin Biotechnol 14(1):13–22
11. Tanaka M, Matsuo K, Enomoto M et al (2004) A sol particle homogeneous immunoassay for measuring serum cystatin C. Clin Biochem 37(1):27–35
12. Thanh NTK, Rosenzweig Z (2002) Development of an aggregation-based immunoassay for anti-protein A using gold nanoparticles. Anal Chem 74(7):1624–1628
13. Thanh NTK, Rees JH, Rosenzweig Z (2002) Laser-based double beam absorption detection for aggregation immunoassays using gold nanoparticles. Anal Bioanal Chem 374(7–8): 1174–1178
14. Zhang CX, Zhang Y, Wang X et al (2003) Hyper-Rayleigh scattering of protein-modified gold nanoparticles. Anal Biochem 320(1): 136–140

15. Otsuka H, Akiyama Y, Nagasaki Y et al (2002) Quantitative and reversible lectin-induced association of gold nanoparticles modified with alpha-lactosyl-omega-mercapto-poly(ethylene glycol). J Am Chem Soc 123(34):8226–8230
16. Torrecilla JS, Mena LM, Yanez SP et al (2008) A neural network approach based on gold-nanoparticle enzyme biosensor. J Chemometr 22(1):46–53
17. Pingarro JM, Yanez SP, Gonzalez-Cortes A (2008) Gold nanoparticle-based electrochemical biosensors. Electrochim Acta 53(19): 5848–5866
18. Li J, Song S, Liu X et al (2008) Enzyme-based multi-component optical nanoprobes for sequence-specific detection of DNA hybridization. Adv Mater 20(3):497–500
19. Giulio FP, Kingston DGI, Tamarkin L (2006) Colloidal gold nanoparticles: A novel nanoparticle platform for developing multifunctional tumor-targeted drug delivery vectors. Drug Devel Res 67(1):47–54
20. Jun HK, Lee TR (2006) Discrete thermally responsive hydrogel-coated gold nanoparticles for use as drug-delivery vehicles. Drug Devel Res 67(1):61–69
21. Connolly S, Fitzmaurice D (1999) Programmed assembly of gold nanocrystals in aqueous solution. Adv Mater 11(14):1202–1205
22. Park SJ, Lazarides AA, Mirkin CA et al (2001) Directed assembly of periodic materials from protein and oligonucleotide-modified nanoparticle building blocks. Angew Chem Int Ed Engl 40(15):2909–2912
23. Niemeyer CM (2001) Nanoparticles, proteins, and nucleic acids: biotechnology meets materials science. Angew Chem Int Ed Engl 40(22):4128–4158
24. Mastroianni AJ, Claridge SA, Alivisatos AP (2009) Pyramidal and chiral groupings of gold nanocrystals assembled using DNA scaffolds. J Am Chem Soc 131(24):8455–8459
25. Ferrari E, Darios F, Zhang F et al (2010) Binary polypeptide system for permanent and oriented protein immobilization. J Nanobiotechnology 8:9
26. Brust M, Walker M, Bethell D et al (1994) Synthesis of thiol-derivatised gold nanoparticles in a two-phase liquid–liquid system. J Chem Soc Chem Commun 7(7):801–802
27. Perrault SD, Chan WC (2009) Synthesis and surface modification of highly monodispersed, spherical gold nanoparticles of 50-200 nm. J Am Chem Soc 131(47):17042–17043
28. Martin MN, Basham JI, Chando P et al (2010) Charged gold nanoparticles in non-polar solvents: 10-min synthesis and 2D self-assembly. Langmuir 26(10):7410–7417
29. Zhao W, Lin L, Hsing I (2009) Rapid synthesis of DNA-functionalized gold nanoparticles in salt solution using mononucleotide-mediated conjugation. Biconjug Chem 20(6): 1218–1222
30. Wangoo N, Bhasin KK, Mehta SK et al (2008) Synthesis and capping of water-dispersed gold nanoparticles by an amino acid: bioconjugation and binding studies. J Colloid Interface Sci 323(2):247–254
31. Satyavani K, Gurudeeban S, Ramanathan T et al (2011) Biomedical potential of silver nanoparticles synthesized from calli cells of *Citrullus colocynthis* (L.) Schrad. J Nanobiotechnol 9:43
32. Walter JG, Petersen S, Stahl F et al (2010) Laser ablation-based one-step generation and bio-functionalization of gold nanoparticles conjugated with aptamers. J Nanobiotechnol 8:21
33. Conde J, de la Fuente JM, Baptista PV (2010) RNA quantification using gold nanoprobes—application to cancer diagnostics. J Nanobiotechnol 8:5
34. Crumbliss AL, Stonehuerner J, Henkens RW et al (1994) The use of inorganic materials to control or maintain immobilized enzyme activity. New J Chem 18:327–339
35. Gole A, Dash C, Ramakrishnan V et al (2001) Pepsin–gold colloid conjugates: preparation, characterization, and enzymatic activity. Langmuir 17(5):1674–1679
36. Huang F, Huang CC, Chang HT (2003) Exploring the activity and specificity of gold nanoparticle-bound trypsin by capillary electrophoresis with laser-induced fluorescence detection. Langmuir 19(18):7498–7502
37. Lv M, Zhu E, Su Y et al (2009) Trypsin-gold nanoparticle conjugates: binding, enzymatic activity, and stability. Prep Biochem Biotechnol 39(4):429–438
38. Li H, Huang J, Lv J et al (2005) Nanoparticle PCR: Nanogold-assisted PCR with enhanced specificity. Angew Chem Int Ed Engl 44(32):5100–5103
39. Li M, Lin YC, Wu CC et al (2005) Enhancing the efficiency of a PCR using gold nanoparticles. Nucl Acids Res 33(21):184
40. Asuri P, Karajanagi SS, Yang H et al (2006) Increasing protein stability through control of the nanoscale environment. Langmuir 22(13):5833–5836
41. Halling PJ, Ulijn RV, Flitsch SL (2005) Understanding enzyme action on immobilised substrates. Curr Opin Biotechnol 16(4): 385–392
42. Basso A, Braiuca P, Ebert C et al (2006) Properties and applications of supports for

enzyme-mediated transformations in solid phase synthesis. J Chem Technol Biotechnol 81(10):1626–1640
43. Cortez J, Vorobieva E, Gralheira D et al (2011) Bionanoconjugates of tyrosinase and peptide-derivatised gold nanoparticles for biosensing of phenolic compounds. J Nanopart Res 13:1101–1113
44. Roach P, Farrar D, Perry CC (2006) Surface tailoring for controlled protein adsorption: effect of topography at the nanometer scale and chemistry. J Am Chem Soc 128(12):3939–3945
45. Hong R, Fischer NO, Verma A et al (2004) Control of protein structure and function through surface recognition by tailored nanoparticle scaffolds. J Am Chem Soc 126(3):739–743
46. Karajanagi SS, Vertegel AA, Kane RS et al (2004) Structure and function of enzymes adsorbed onto single-walled carbon nanotubes. Langmuir 20(26):11594–11599
47. Gole A, Dash C, Soman C et al (2001) On the preparation, characterization, and enzymatic activity of fungal protease-gold colloid bioconjugates. Bioconjug Chem 12(5):684–690
48. Keating CD, Kovaleski KM, Natan MJJ (1998) Protein:colloid conjugates for surface-enhanced Raman scattering: stability and control of protein orientation. Phys Chem B 102(47):9404–9413
49. Dykman LA, Bogatyrev VA, Khlebtsov BN et al (2005) A protein assay based on colloidal gold conjugates with trypsin. Anal Biochem 341(1):16–21
50. Laera S, Ceccone G, Rossi F et al (2011) Measuring protein structure and stability of protein–nanoparticle systems with synchrotron radiation circular dichroism. Nano Lett 11(10):4480–4484
51. Aubin-Tam ME, Hamad-Schifferli K (2005) Gold nanoparticle - cytochrome C complexes: the effect of nanoparticle ligand charge on protein structure. Langmuir 21(26):12080–12084
52. Kolobanova SV, Filippova IY, Lysogorskaya EN (2001) The enzymatic segment condensation of peptides on a solid phase in organic medium. Bioorg Khim 27(5):347–351
53. Doeze RHP, Maltman BA, Egan CL et al (2004) Profiling primary protease specificity by peptide synthesis on a solid support. Angew Chem Int Ed Engl 43(24):3138–3141
54. Zamocky M, Koller F (1999) Understanding the structure and function of catalases: clues from molecular evolution and in vitro mutagenesis. Prog Biophys Mol Biol 72(1):19–66

Part II

Semiconductor Nanoparticles: Synthesis and Applications

Chapter 8

Synthesis of Semiconductor Nanoparticles

Xianfeng Chen and Peter J. Dobson

Abstract

Here, we describe typical methods and provide detailed experimental protocols for synthesizing and processing various semiconductor nanoparticles which have potential application in biology and medicine. These include synthesis of binary semiconductor nanoparticles; core@shell nanoparticles and alloyed nanoparticles; size-selective precipitation to obtain monodisperse nanoparticles; and strategies for phase transfer of nanoparticles from organic solution to aqueous media.

Key words: Semiconductor, Nanoparticles, Quantum dots, Phase transfer

1. Introduction

Semiconductor nanoparticles (NPs) are crystalline clusters of a few hundred to a few thousand atoms. The shape is somewhat spherical and the size is in the range of 1–10 nm in diameter (shown in Fig. 1) (1). Because the NP size range is of the same order of magnitude as the de Broglie wavelength of electrons and holes at room temperature, the movement of the electrons and holes in the NPs is confined in all three dimensions and thus quantum confinement effect can be observed. Therefore, semiconductor NPs are often called quantum dots (QDs). Because of the quantum confinement effect, the bandgap of the semiconductor NPs can be controlled simply by changing their sizes. For example, the emission of CdSe NPs can be easily tuned to cover from red (centered at 650 nm) to blue (centered at 450 nm) as the size of NPs decreases from 7 to 2 nm (2, 3). This is one of the most intriguing features of semiconductor NPs.

During the past 3 decades, there has been extensive interest in making colloidal semiconductor NPs for different applications in optoelectronic devices, lasers, high density memory, medical

Mikhail Soloviev (ed.), *Nanoparticles in Biology and Medicine: Methods and Protocols*, Methods in Molecular Biology, vol. 906, DOI 10.1007/978-1-61779-953-2_8,

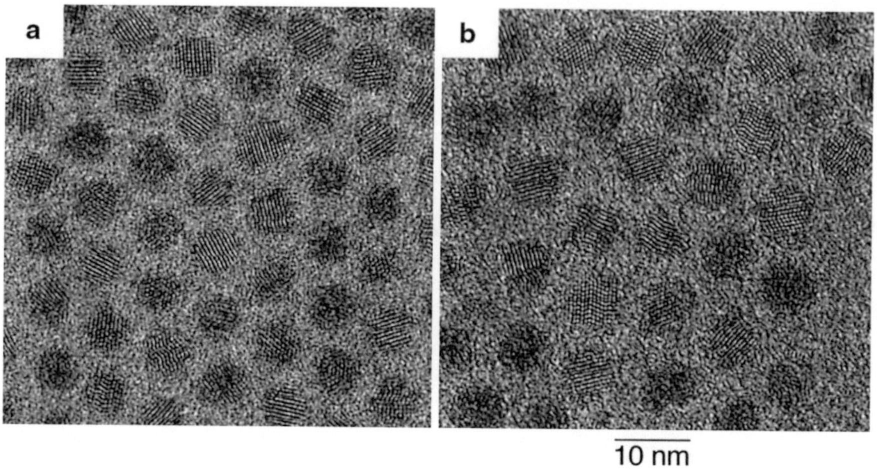

Fig. 1. High resolution TEM images of NPs. (**a**) CdSe nanoparticles and (**b**) CdSe@ZnS core@shell nanoparticles (Reproduced from Talapin et al. (1), with permission from American Chemical Society).

imaging and drug delivery (4, 5). In this chapter, the focus will be given to the synthesis and process of the semiconductor NPs which have potential applications in biology and medicine. The most successful synthesis method has been using organometallic compounds to form semiconductor NPs in coordinating solvents (e.g., TOPO) at high temperature in air-free environment. This approach was originally introduced by Murray et al. in 1993 (6). In the synthesis, organometallic reagents were rapidly injected into a hot coordinating solvent to produce homogeneous nucleation. The growth was then carried out at a certain temperature to fabricate nearly monodisperse NPs. Initially, dimethylcadmium was used as the cadmium precursor. Trioctylphosphine (TOP) and trioctylphosphine oxide (TOPO) were coordinating agents. Because dimethylcadmium is highly toxic, explosive and expensive, it was later replaced by other cadmium compounds such as cadmium oxide and cadmium acetate. Tributylphosphine (TBP) was also used to replace TOP to coordinate to selenium (2). It is worth noting that technical grade TOPO should be used for the NP fabrication in this method. If pure TOPO is used, $Cd(CH_3)_2$ will immediately decompose to form a metallic precipitate. It was later found that the impurity, probably phosphonic acid, was affecting the synthesis in a positive way (2, 7). However, variable impurity in different batches of TOPO led to fluctuations in the synthesis of NPs. Therefore, co-solvents (e.g., dodecylphosphonic acid (DPA), hexadecylamine (HDA)) were later used to supplement the impurity in TOPO, slow the nucleation process and provide resistance to Ostwald ripening (3). With all of the modifications made, high-quality NPs can be reproducibly made. However, these NPs are not soluble in aqueous solution and do not have the appropriate moieties for conjugation to biomolecules. To make the NPs

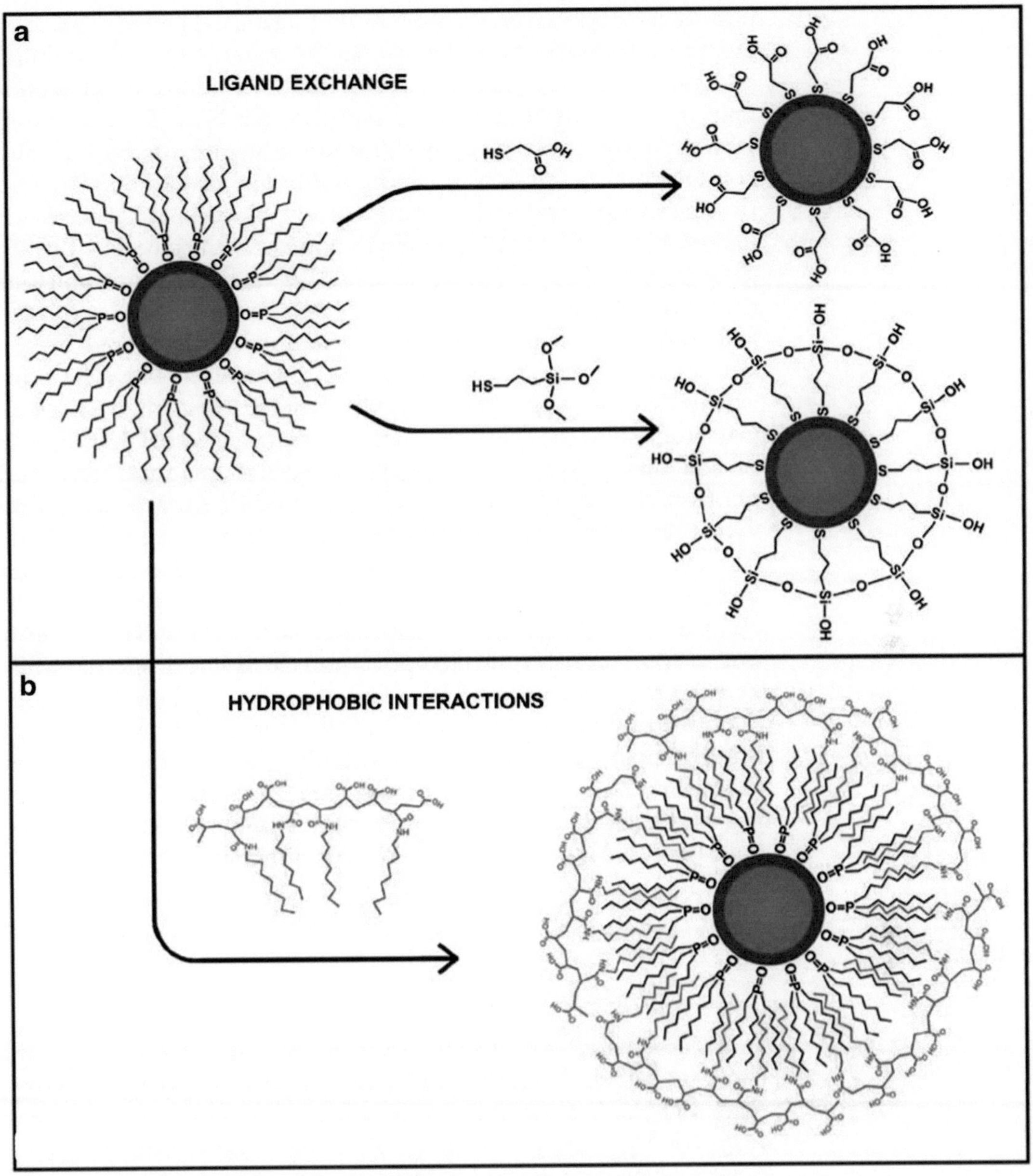

Fig. 2. Two general strategies to transfer TOPO-CdSe nanoparticles to aqueous solution. (**a**) TOPO ligands may be replaced by heterobifunctional molecules such as mecapto acids and silanes. (**b**) Attach an amphiphilic polymer on the existing TOPO ligands (Reproduced from Bailey et al. (8), with permission from Elsevier).

biocompatible, the surface TOPO ligands must be replaced or modified with surfactants allowing the NPs to dissolve in water or linked to biomolecules. Among the various techniques for surface modification, two general approaches are described in this chapter (shown in Fig. 2) (8). One is to replace the TOPO ligands with heterobifunctional ligands such as mecapto acids and silanes. The other is to attach an amphiphilic polymer like octylamine-modified polyacrylic acid on the existing TOPO ligands. The latter method has the advantage over the former in maintaining the stability and photoluminescence quantum yield. In addition, phospholipid micelles can be used to encapsulate the NPs (9). Compared with the above two surface modification strategies, the NP-micelles complex is very stable so it can be used in in vivo studies without the risk of any aggregation. Alternatively, biocompatible semiconductor NPs can be readily made in aqueous solution (10, 11). In general, the monodispersity of the NPs made through aqueous routes is not as good as that of the NPs fabricated from organometallic routes. To reduce the size distribution of as-prepared NPs, a size selective precipitation process can be applied (11). The principle is that the largest particles will precipitate first when a nonsolvent is added to the solution. For the NPs made from either organic or aqueous routes, their photoluminescence quantum yield and photostability can be dramatically improved by surface passivation with a second semiconductor layer, resulting in a core/shell structure (1, 12). The above mentioned NPs provide the opportunity to change their sizes to tune the corresponding properties. In comparison, another type of alloyed semiconductor NP can allow one to continuously tune the optical and electronic properties while keeping the size constant (13). This new class of alloyed semiconductor NPs further broadens the application of NPs in developing fluorescent probes for in vivo molecular imaging and biomarker detection. Herein, we describe the typical methods for synthesizing and processing various semiconductor nanoparticles which have potential application in biology and medicine.

2. Materials

In general, nanoparticle synthesis is performed in a round bottom three-neck-flask heated with a heating mantle (e.g., Thermo Scientific heating mantles coated with PTFE Resin) under active magnetic bar stirring. One neck of the flask is connected to a condenser which is connected to inert gas atmosphere. The other necks are sealed with a thermometer bearing caps and a rubber septum for injection of precursors or collecting nanoparticle samples. Deionized water is used in the synthesis unless stated otherwise.

2.1. Organometallic Approach to the synthesis of CdSe Nanoparticles in Coordinating Solvents

1. Trioctylphosphine oxide (TOPO) (technical grade, 90 % purity).
2. Tributylphosphine (TBP) (97 %).
3. Hexadecylamine (HDA) (technical grade, 90 %).
4. Octadecylamine (ODA) (97 %).
5. Stearic acid (95 %).

6. Cadmium oxide (CdO) (99.99 %).
7. Se powder (99.999 %).
8. Methanol (HPLC grade).
9. Toluene (HPLC grade).
10. Dodecylamine (99 %).
11. Chloroform (HPLC grade).
12. Acetone (HPLC grade).
13. Selenium precursor: the Se solution containing 0.079 g (1 mmol) of Se dissolved in 0.238 g (1.18 mmol) of TBP and 1.681 g of dioctylamine is prepared in a nitrogen glove box.
14. Argon gas.
15. Centrifuge capable of spinning 50 mL tubes at 5,000 × *g*.

2.2. Organometallic Approach to the synthesis of CdSe Nanoparticles in Non-Coordinating Solvents

1. Cadmium oxide (CdO) (99.99 %).
2. Oleic acid (OA) (technical grade, 90 %).
3. 1-Octadecene (ODE) (technical grade, 90 %).
4. Chloroform ($CHCl_3$) (HPLC grade).
5. Methanol (HPLC grade).
6. Sulfur (99.98 %, powder).
7. Sulfur injection solution (0.04 M): dissolve 0.0064 g of sulfur in 5 mL of ODE at 200 °C.

2.3. Organic Route to Produce Water-Soluble CdTe Nanoparticles

1. Poly(acrylic) acid (M_W ~1,800 D).
2. Dodecylamine (98 %).
3. Dimethylformamide (DMF) (HPLC grade).
4. Dicyclohexylcarbodiimide (DCC) (99 %).
5. Polyethylene glycol (PEG, M_W ~350).
6. Cadmium oxide (CdO) (99.99 %).
7. Tellurium (Te) (99.999 %).
8. Tributylphosphine (97 %).
9. Tellurium precursor: dissolve 0.05 mmol of Te in 25 μL of TBP and 1 mL of PEG at 200 °C in Argon gas atmosphere.
10. 1 M Hydrochloric acid.
11. 0.1 mM Sodium hydroxide (pH ~10).
12. Argon gas.
13. Magnetic stirrer plate.
14. Vacuum filtration apparatus.

2.4. Aqueous Routes to Produce Nanoparticles

1. 0.02 M of sodium selenosulfate: add 0.5 g sodium sulfite and 0.05 g selenium powder to 33 mL of water, incubate at 80 °C

under magnetic bar stirring for 5 h. The prepared colorless sodium selenosulfate solution is stored under dry nitrogen for later use.

2. Cadmium acetate ($Cd(CH_3COO)_2 \cdot 2H_2O$) (99.99 %).
3. 3-Mercaptopropionic acid (3-MPA) (99 %).
4. Mercaptosuccinic acid (2-MSA) (99 %).
5. Nitrogen gas.
6. 1 M NaOH (98 %).
7. Lambda UV/Vis spectrophotometer.

2.5. Core@shell Semiconductor Nanoparticles

1. Tri-*n*-octylphosphine oxide (TOPO) (technical grade, 90 %).
2. Hexadecylamine (HDA) (technical grade, 90 %).
3. Tri-*n*-octylphosphine (TOP) (technical grade, 90 %).
4. Se powder (99.999 %).
5. Diethylzinc (95 %).
6. Bis(trimethylsilyl)sulfide ($((CH_3)_3Si)_2S$) (98 %).
7. Dimethylcadmium ($Cd(CH3)_2$) (99.99 %).
8. 1 M TOPSe stock solution: dissolve 7.896 g of selenium in 100 mL of TOP in a glass flask in a nitrogen glove box.
9. Zn:S stock solution: add 0.4 mmol of diethylzinc and 0.51 mmol of bis(trimethylsilyl)sulfide to 3 mL of TOP.

2.6. Alloyed Semiconductor Nanoparticles

1. Cadmium oxide (CdO) (99.99 %).
2. Selenium stock solution (0.4 M): dissolve 0.79 g of Selenium shot (Se, 99.999 %) in 25 mL of TOP. The solution should be colorless.
3. Tellurium stock solution (0.4 M): dissolve 1.28 g of tellurium powder in 25 mL of TOP. The solution should become yellow. If required, a gentle heating can be used to dissolve the remaining tellurium powder.
4. Premixed Se and Te solutions (0.4 M total concentration): prepare Se:Te mixtures from individual stock solutions (0.4 M each) and molar ratios 100:0, 75:25, 50:50, 25:75, or 0:100.
5. Trioctylphosphine (TOP) (technical grade, 90 %)
6. Trioctylphosphine oxide (TOPO) (technical grade, 90 %)
7. Hexadecylamine (HDA) (technical grade, 90 %).
8. Tellurium powder (Te, 99.999 %).
9. Argon gas.
10. Lambda UV/Vis spectrophotometer.

2.7. Size-selective precipitation of Nanoparticles

1. 2-Propanol (HPLC grade).
2. Rotary evaporator.
3. Centrifuge capable of spinning 50 mL tubes at 5,000 × *g*.

2.8. Surface Capping Exchange (Thiol-Coated)

1. Tetramethylammonium hydroxide pentahydrate (97 %).
2. 11-Mercaptoundecanoic acid (MUA, 95 %).
3. Anhydrous ethyl ether (HPLC grade).
4. Ethyl acetate (HPLC grade).
5. Methanol (HPLC grade).
6. Argon gas.
7. Vacuum oven.
8. Centrifuge capable of 5,000 × *g*.

2.9. Surface Capping Exchange (Silica-Coated)

1. Trioctylphosphine oxide (TOPO) (technical grade, 90 %).
2. (3-Mercaptopropyl)trimethoxysilane (3-MPS) (95 %)
3. Tetramethylammonium hydroxide in methanol (TMAH) (25 wt.% solution is available from Sigma-Aldrich).
4. 120 mL of anhydrous methanol, adjust pH to ~10 by adding 750 μL of TMAH.
5. Anhydrous methanol (HPLC grade).
6. Tetramethylammonium hydroxide pentahydrate (97 %).
7. 2 mL of chlorotrimethylsilane, adjust pH to by adding ~3 g of solid TMAH pentahydrate.
8. 3-(Trihydroxysilyl)propyl methylphosphonate solution in water (42 wt%, solution is available from Sigma-Aldrich).
9. Chlorotrimethylsilane (99 %).
10. 10 mM phospahte buffer pH 7.4.
11. Dialysis tubing (M_W cut off 10,000).
12. Acrodisc syringe filters with 0.45 mm pore size nylon membrane (Pall Corporation).
13. Centricon devices (YM-100 M_W cut off 100,000, Centiplus, Millipore).
14. NAP columns (#17-0852-01, Amersham Pharmacia).
15. Dialysis cassettes (Slide-A-Lyzer, from Pierce) are used according to the instructions.
16. 0.22 μm pore size acetate filter (GE Water Systems).
17. Centrifuge capable of 20,000 × *g*.

2.10. Additional layer on TOPO Ligands

1. Poly(acrylic) acid (PAA, M_W = 1,800 D).
2. 1-Ethyl-3-(3-dimethylamino propyl)carbodiimide (EDC).

3. Octylamine (99 %).
4. Tetramethylammonium hydroxide (97 %).
5. Chloroform ($CHCl_3$) (HPLC grade).
6. 50 mM Borate buffer, pH 8.0.
7. Centrifuge capable of spinning 50 mL tubes at 5,000 × *g*.
8. PVDF syringe filter (0.2 μm).
9. Ultrahydrogel® 1000 column.

2.11. Encapsulating Quantum Dots in Phospholipid Micelles

1. 1,2-dipalmitoyl-sn-glycero-3-phosphoethanolamine-*N*-(methoxy(polyethylene glycol)-2000) (mPEG-2000 PE) (Avanti Polar Lipids Inc.)
2. 1,2-Dipalmitoylglycero-3-phosphocholine (DPPC) (Avanti Polar Lipids Inc.).
3. Chloroform ($CHCl_3$) (HPLC grade).
4. Ultracentrifuge capable of achieving 500,000 × *g*.
5. Branson 1200 ultrasonic cleaner.

3. Methods (See Notes 1–7)

3.1. Organometallic Approach to the synthesis of CdSe Nanoparticles in Coordinating Solvents (See Notes 8–10)

This method was reported by Qu and Peng (3). It can reproducibly synthesize high-quality CdSe NPs with photoluminescence (PL) quantum yield up to 80 % and size distribution within 5–10 % at the condition of a large excess of the selenium precursor. The initial Cd:Se ratio of the precursors in the solution is equal to 1:10. The optical properties of the as-prepared CdSe NPs are stable in air for several months.

1. Add 0.0127 g of CdO (0.1 mmol) and 0.1140 g of stearic acid (0.4 mmol) into a 25 mL three-neck flask and heat to 150 °C under vigorous stirring under Argon flow.
2. After the CdO is completely dissolved, allow the mixture to cool to room temperature.
3. Add 1.94 g of TOPO and 1.94 g of hexadecylamine to the flask and heat the mixture to 320 °C under stirring under Argon flow to obtain an optically clear solution.
4. Dissolve 0.079 g of Se (1 mmol) in 0.238 g of TBP (1.18 mmol) and 1.681 g of dioctylamine and quickly inject the solution into the reaction flask.
5. Grow NPs at 290 °C. At various time intervals (around 1 min), remove aliquots of the reaction mixture using a needle piercing through the rubber septum in one neck during synthesis (without stopping synthesis) and dilute with chloroform.

If there is insoluble white solid in the mixture, the NPs can be isolated by centrifugation.

6. Once the desired size is reached, allow the reaction mixture to cool to 30–50 °C. Before the reaction mixture is completely solidified, add approximately 15 mL of chloroform into the mixture. The NPs can be precipitated by adding methanol or acetone and then isolated by centrifugation at 5,000 × *g* for 10 min.

3.2. Organometallic Approach to the synthesis of CdSe Nanoparticles in Non-Coordinating Solvents (See Notes 8, 9 and 11–13)

This approach to making semiconductor NPs was introduced by Yu and Peng (14). Compared with the approaches using coordinating solvents, this method is simple and inexpensive. The approach can yield monodisperse CdS NPs and their size can be adjusted within 1–6 nm range. The standard deviation of the size distribution is about 5–15 %. The UV-Vis absorption and photoluminescence spectra are characterized by very narrow peaks. This approach provides the flexibility for tuning the temporal course of the nucleation and growth of the semiconductor NPs by simply changing the concentration of the ligands in a non-coordinating solvent (e.g., ODE), which is impossible to achieve for the synthesis in coordinating solvents.

1. Mix 0.0128 g of CdO (0.10 mmol), 0.5649 g of oleic acid (2 mmol) and 3.4223 g of ODE in a round-bottom three-neck-flask. Heat the mixture to 300 °C with magnetic stirrer using a heating mantle.
2. Quickly inject 1.25 mL of the sulfur injection solution into this hot solution and cool the reaction mixture to 250 °C to allow CdS NPs to grow. The synthesis may be carried out under argon or in air.
3. Take aliquots at regular intervals (a few tens of seconds to a few minutes) by using a needle piercing through the rubber septum in one neck during synthesis (without stopping synthesis) to monitor the growth of the CdS NPs. At the end of the synthesis separate the unconsumed cadmium precursor from the NPs by repeated extraction of the reaction aliquots with a mixture of Chloroform/Methanol (1:1, v:v) (see Note 13).

3.3. Organic Route to Produce Water-Soluble CdTe Nanoparticles (See Notes 8, 9 and 14)

The one-pot approach to make water soluble NPs was introduced by Kairdolf et al. (15, 16). This method yields CdTe NPs with a narrow size distribution (approx. 10 % standard deviation). The photoluminescence quantum yield is about 30 %. The NPs are soluble in a wide range of hydrophilic and hydrophobic solvents such as water, DMF, acetone and chloroform.

The amphiphilic polymer (~3,500 D) (for use as a surfactant to stabilize NPs) is synthesized using standard carbodiimide chemistry. The mechanism is the well known reaction of alkylamine on carboxyl group in an aprotic solvent in the presence of dicyclohexylcarbodiimide.

1. Dissolve 518 mg of poly(acrylic) acid and 533 mg of dodecylamine in 10 mL of DMF.
2. Dissolve 609 mg of dicyclohexylcarbodiimide in a minimum amount of DMF.
3. Add dicyclohexylcarbodiimide (DCC) solution dropwise to the solution of poly(acrylic) acid.
4. Stir the mixture vigorously for 24 h with magnetic stirring bar. The reaction will yield an amphiphilic polymer with 40 % of the carboxylic acid functional groups modified with a 12-carbon aliphatic chains.
5. Cool the polymer solution in an ice bath to quickly precipitate the dicyclohexylurea (DCU) byproduct and then remove the DCU using vacuum filtration. This will leave a transparent and slightly yellow filtrate.
6. Add the polymer filtrate dropwise to a 1 L of aqueous sodium hydroxide solution (pH ~10) under vigorous stirring.
7. Use vacuum filtration to remove any insoluble contaminants (trace DCU, unreacted DCC and dodecylamine).
8. Add 1 M hydrochloric acid dropwise to the solution until the pH reduces to below 5 to precipitate the polymer. Use vacuum filtration to collect the precipitated polymer.
9. Wash the polymer with deionized water to remove water soluble contaminants.
10. The solid product is collected by vacuum filtration and lyophilized to yield a dry, white powder.
11. Add 170 mg of amphiphilic polymer (0.6 mmol COOH groups) to 1.5 mL of PEG in a three-neck-flask. Incubate at 100 °C under vacuum to remove water and dissolved gases. Add 12.84 mg of CdO (0.1 mmol) and heat the solution to 200 °C under argon to form the carboxylate precursor.
12. Add 2.8 mL of PEG to dilute the solution and heat the solution to 300 °C under vigorous stirring.
13. Immediately inject the prepared tellurium precursor solution into the three-neck-flask.
14. Take aliquots at regular intervals (a few tens of seconds to a few minutes) by using a needle piercing through the rubber septum in one neck during synthesis (without stopping synthesis) to monitor the growth of the CdTe NPs.
15. Once the desired size is reached, cool the reaction mixture to room temperature to stop the NP growth.

3.4. Aqueous Routes to Produce Nanoparticles (See Notes 8 and 15–17)

This method was reported by Chen et al. (10). High quality CdSe NPs were produced at high cadmium to selenium precursor rations. The photoluminescence quantum yield of the as-prepared CdSe NPs is up to 20 %.

1. Add 0.064 g of $Cd(Ac)_2{\cdot}2H_2O$ (0.24 mmol) and 50.2 μL of 3-mercaptopropionic acid (or 0.086 g of mercaptosuccinic acid) (0.58 mmol) to 48 mL of deionized water in a three-neck-flask and adjust pH with 1 M NaOH to pH 9.3 (see Note 15). The solution should become clear and colorless. The solution is then degased by bubbling with nitrogen gas for 1 h.
2. Under vigorous stirring inject 1.2 mL of the stored oxygen-free 0.02 M sodium selenosulfate solution (Cd:Se precursor ratio 10:1) for 3-MPA stabilized NPs. Alternatively inject 0.4 mL of 0.02 M sodium selenosulfate solution (Cd:Se precursor ratio 30:1) for 2-MSA stabilized NPs.
3. Heat the reaction system to 100 °C and maintain the temperature for the duration of NP synthesis.
4. Remove 0.2 mL aliquots of the reaction mixture at regular intervals (every few hours). Monitor the growth of the CdSe NPs by measuring the first absorption peak using a UV-Vis spectrophotometer (wavelength ranges from 300 to 800 nm).
5. Once the desired size is reached, cool the reaction mixture to room temperature.

3.5. Core@shell Semiconductor Nanoparticles (See Notes 8, 10, and 18–21)

This approach was reported by Talapin et al (1). This method produces highly monodisperse NPs. The size distribution has a standard deviation usually below 4 % and quantum yield may be in the range of 10–25 %. After capping with a ZnS layer, the photoluminescence quantum yield increases to between 50 and 66 % for a wide range of the ZnS shell thickness and CdSe core sizes. The advantage of this method is that the CdSe NPs from the crude solution can be used directly for the growth of the ZnS shell. In addition, a rather high concentration of the CdSe NPs can be used as cores to fabricate core-shell NPs (12). The prepared CdSe and CdSe@ZnS NPs are shown in Fig. 1.

1. Mix 10 g of TOPO and 5 g of HDA (55:45, molar ratio) and heat to 300 °C.
2. Add 1 mL of 1 M TOPSe stock solution and 1.35 mmol of dimethylcadmium to 5 mL of TOP. Under vigorous stirring inject this mixture quickly into the TOPO-HDA mixture prepared earlier (see step 1).
3. The reaction system is maintained at 300 °C for NP growth. Once the desired size is reached, cool the reaction mixture to room temperature.

4. Take 2.5 mL of the crude solution of CdSe NPs prepared as described above and mix it with 5 g of TOPO and 2.5 g of HDA, heat to 220 °C.
5. The amount of Zn:S stock solution necessary to obtain the desired shell thickness is calculated from the ratio between the core and shell volumes using bulk lattice parameters of CdSe and ZnS. The required amount of Zn:S stock solution is then added dropwise to the vigorously stirred solution of CdSe NPs to form CdSe@ZnS core@shell NPs (see Notes 19 and 21).

3.6. Alloyed Semiconductor Nanoparticles (See Note 10)

This type of semiconductor NPs was firstly demonstrated by Bailey and Nie (13). With this technique, one can continuously tune the optical and electronic properties of alloyed semiconductor NPs while keeping the NP size constant. The alloyed NPs are crystalline and monodisperse (size distribution generally within 5–10 %). These NPs also exhibit excellent optical properties such as narrow photoluminescence spectra widths, high quantum yield (30–60 %) and high photostability.

1. Add 9 g of TOPO, 3 g of HDA and 16 mg of CdO (0.125 mmol) to a 125 mL round-bottom flask, heat to ~150 °C, degas under a vacuum of 20 Pa for 15 min.
2. Fill the flask with argon gas, heat the flask to 325 °C. After the precursor CdO is dissolved completely in the solvent, the temperature is lowered to 300 °C.
3. Once the temperature stabilizes at 300 °C, quickly inject 2.5 mL of the premixed Se and Te solution (less than 1 s) into the colorless TOPO/HDA/CdO solution. The NPs are prepared under cadmium limited-conditions (molar ratio of Cd:Se = 1:8).
4. Remove 0.2 mL aliquots of the reaction mixture at regular intervals (every few hours). Monitor the growth of the CdSe NPs by measuring the first absorption peak using a UV-Vis spectrophotometer (wavelength ranges from 400 to 1,000 nm).
5. Once the desired size is reached, cool the reaction mixture to room temperature.

3.7. Size-selective Precipitation of Nanoparticles

Normally, NPs produced using organometallic approaches yield narrower size distribution compared to the ones produced through aqueous routes. To reduce the size distribution, a post-preparative size-selective precipitation process is applied. The principle is that the largest particles will precipitate first when a nonsolvent is added to the solution. We describe a typical procedure using thiol-stabilized CdTe NPs as an example (based on ref. (11)).

1. Concentrate CdTe NP aqueous solution to ~0.05 M using a rotary evaporator.

2. Add 2-propanol into CdTe NP solution dropwise under stirring until the solution becomes slightly turbid.
3. Continue stirring the turbid dispersion for 15 min.
4. Separate the first fraction of CdTe NPs by centrifugation at 5,000 × *g* for 10 min.
5. Repeat steps 2–4 above for another ~10 times to obtain size-selected fractions of CdTe NPs.
6. Dissolve the precipitated CdTe NP fractions in water. Store in a dark environment (ideally in a refrigerator).

3.8. Surface Capping Exchange (Thiol-Coated) (See Notes 22 and 23)

To make the TOPO stabilized NPs soluble in water for subsequent application in medicine and biology, one way is to replace the TOPO molecules with heterobifunctional ligands such as mecapto acids and silanes. The typical procedures are described (based on ref. (17)).

1. Transfer 20 mg of MUA to a three-neck-flask.
2. Add 15 mL of methanol, adjust pH to >10 with tetramethylammonium hydroxide pentahydrate.
3. Protect the flask from light, under argon flow add 20 mg of TOPO-CdSe NPs to the mixture. Heat and incubate overnight at 65 °C under reflux (see Note 22).
4. Stop the reaction by cooling to ~28 °C.
5. Collect 2 mL of the reaction mixture and add ethyl acetate and ether dropwise until the NPs precipitate from the solution.
6. Centrifuge the mixture at 5,000 × *g* for 10 min and decant the supernatant.
7. Dissolve the NPs with 2 mL of methanol and precipitate the NPs with ethyl acetate.
8. Repeat step 6 and dry the washed NPs in a vacuum oven at 60 °C for 5 min.

3.9. Surface Capping Exchange (Silica-Coated)

Unlike the method described in Subheading 3.8, the approach described here produces very stable water soluble NPs. The illustration of this approach is shown in Fig. 3. In the process, a glass shell is formed around the NPs. However, this method is laborious and it is difficult to scale up (based on ref. (18)).

1. Suspend ~5 mg of TOPO stabilized CdSe/ZnS NPs in 50 μL of MPS in a microcentrifuge tube, vortex for 5 s. The solution should appear cloudy.
2. Add 5 μL of 25 % TMAH in methanol. The solution should become optically clear.
3. Dilute the mixture with 120 mL of anhydrous methanol (pH ~10). Transfer the solution to a 500 mL three-neck-flask and stirred for 60 min under nitrogen atmosphere (see Note 24).

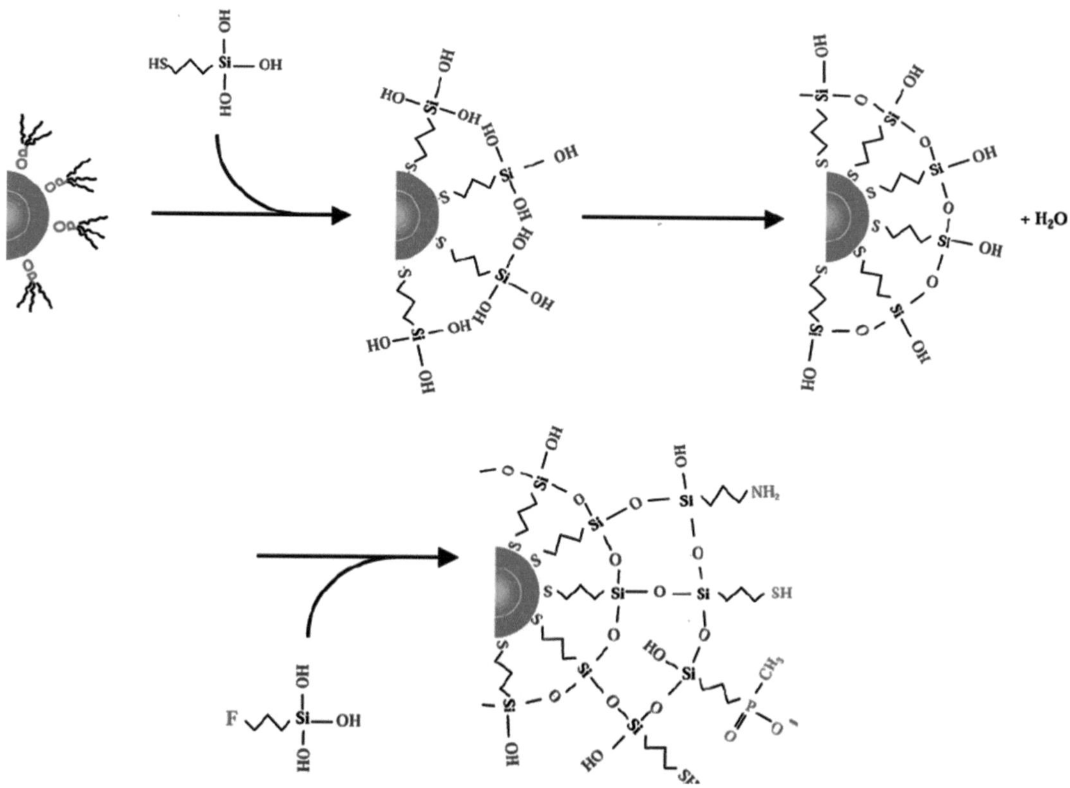

Fig. 3. The use of silane molecules to replace TOPO ligands (Reproduced from Gerion et al. (18), with permission from American Chemical Society).

4. Gently heat the solution to 65 °C, incubate for 30 min. Allow the solution to cool down to room temperature.
5. Add 90 mL of methanol, 10 mL of deionized water, 600 μL of (trihydroxysilyl)propyl methylphosphonate and 20 μL of MPS to the solution. Stir for ~2 h and then heat to 65 °C for about 5 min. Cool down to room temperature (see Note 24).
6. The remaining silanol groups are quenched with a mixture of 20 mL of methanol, and 2 mL of basified chlorotrimethylsilane. Stir for ~2 h (see Note 24).
7. Heat the solution to ~65 °C for ~30 min, and then leave at room temperature for 2–4 days under stirring under nitrogen atmosphere.
8. Evaporate the solution to reduce the volume by a factor of 2–5 in a rotary evaporator, and leave at room temperature for 24 h.
9. Dialyze the solution against methanol for 1 day. Filter the dialyzed solution through a 0.45 μm pore size nylon syringe filter.
10. Remove the excessive silane by concentrating the solution using centrifugal filter devices YM-100. After filtering, the volume of the solution reduces to about 2 mL. The solution of

silanized NPs in methanol should be left to stand in a Falcon tube for at least 12 h.

11. Then the 2 mL solution is passed through a NAP column. The movement of solution can be visualized by a UV lamp monitoring the fluorescence of the NPs. After the silanized solution entered the column, add 2 mL of phosphate buffer to the column. Collect fractions, discard the front and trail part of the elution, keep the fractions containing the NPs. Combine and filter the NP samples through a 0.22 μm pore size acetate filter.
12. Centrifuge the solution at 20,000 × *g* for 30 min and discard the precipitate. Store the supernatant in an Eppendorf tube.

3.10. Additional Layer on TOPO Ligands

This method was reported by Wu et al. (19) and Ku et al. (20). It makes use of the hydrophobic interaction between TOPO molecules and the hydrophobic part of the capping layer. The produced NPs are soluble in water.

1. Transfer 3.6 g of PAA and 2.3 g of EDC to a 100 mL round-bottom flask. Add 40 mL of DMF to dissolve the mixture.
2. Add 1.2 mL of octylamine dropwise into the flask.
3. Stir the mixture vigorously for 24 h with magnetic stirring bar. The reaction will yield an amphiphilic polymer.
4. Remove DMF with a rotary evaporator. Mix the residue with 10 mL of acetone and transfer into a centrifuge tube.
5. Add 40 mL of water. Separate the gummy precipitation at 5,000 × *g* and wash thoroughly in water (3 × 25 mL).
6. Dissolve the octylamine-modified PAA in 2 mL of chloroform. Add tetramethyl-ammonium hydroxide to the polymer solution till the pH is 10.
7. Transfer the polymer solution to a 20 mL round-bottom flask containing 5 mL of chloroform.
8. Stir for 1 min to mix the polymer solution well.
9. Under active stirring, add 0.033 mmol of CdSe/ZnS NPs to the polymer solution.
10. Stir the solution for 2 min to mix the solution well. Remove the chloroform with a rotary evaporator.
11. Add 5 mL of water to the flask to disperse the dry film. Stir the dispersion overnight at room temperature.
12. Centrifuge the dispersion at 4,000 × *g* to remove aggregates. Filter the solution through a 0.2 μm pore size PVDF syringe filter.
13. Purify the solution from excess polymer by gel filtration.

3.11. Encapsulating Quantum Dots in Phospholipid Micelles

This method was introduced by Dubertret et al. (9). This method produces highly stable NP-micelle complexes. The size is between 10 and 15 nm. The aqueous suspension can be stable for months (no aggregation even in 1 M salt). This high stability ensures that these NP-micelles are able to be used in in vivo studies. Furthermore, the NP-micelles provide efficient fluorescence, are resistant to photobleaching and have low nonspecific adsorption.

1. Disperse 5.5×10^{-6} moles of phospholipids containing 40 % of mPEG-2000 PE and 60 % of DPPC in 1 mL of chloroform in a round bottom flask (see Note 25).
2. Sonicate 20 min at 37 °C.
3. Add the solution to 17 mg of dried TOPO stabilized CdSe@ZnS NPs and mix well.
4. Completely dry the chloroform using a rotary evaporator at room temperature. The last trace of the solvent can be removed by putting the flask under a high vacuum overnight.
5. Heat the residue to at 80 °C using a heating mantle and add 1 mL of water and mix well. The suspension should be optically clear. Except micelles containing the nanoparticles, there are also empty micelles in the solution.
6. Cool down the solution and remove the empty micelles by ultracentrifugation at $500{,}000 \times g$ for 2 h. The micelles containing the nanoparticles are precipitated and the empty micelles stay in suspension.
7. Decant the supernatant and resuspend the water soluble complexes (micelles containing the nanoparticles) in water and store at room temperature.

4. Notes

1. There are various semiconductor NP systems. One can choose the suitable one depending on the requirement of the properties of NPs (e.g., photoluminescence quantum yield, stability, potential side effect) for a certain application and synthesis techniques (e.g., simplicity, cost).
2. In general, the procedures described in this chapter are suitable for a range of semiconductor NPs. However, one needs to choose the relevant materials for each specific type of NPs. For instance, the procedures described in Subheading 3.2 can also be used to fabricate CdSe NPs by injecting a solution of selenium/tributylphosphane (1:1.1) in ODE as precursor (14).
3. When measuring the photoluminescence quantum yield of NPs, the solvent and concentration of the "optical standard" may affect their photoluminescence quantum yield. For example, for the commonly used standard, Rhodamine 6G has different photoluminescence quantum yields in different solvents

(e.g., 95 % in ethanol vs. 90 % in H_2O) and at different concentrations (low photoluminescence quantum yield at high concentration) (21, 22).

4. In NP synthesis, if the growth temperature is chosen to be lower than the solvent temperature before injecting precursors, one can remove the three-neck flask from heating mantle, set the growth temperature, inject precursors and then put the flask back into the heating mantle.
5. A new rubber septum should be used for injecting precursors or removing samples in every experiment to ensure high quality seal.
6. To achieve homogeneous nucleation when make monodisperse binary semiconductor NPs, the first injection of precursors into the solution must be as quick as possible (<0.1 or 1 s). Do not inject the stock solution along the inside wall of the three-neck-flask.
7. If larger semiconductor NPs are sought (e.g., >3.9 nm), a second injection precursors (e.g., Se precursor) might be needed. This will improve the growth rate and should yield NPs with narrow size distribution (23).
8. In application, various sizes of NPs may be needed for different purposes. The growth of NPs in reaction systems can be monitored by collecting a small amount of samples (e.g., 0.2 mL) and measuring the first absorption peak which corresponds to the bandgap of the NPs (details refer to the chapter describing the characterization of NPs using UV-Vis spectroscopy techniques).
9. In addition to the typical methods described in Subheadings 3.1, 3.2, and 3.3, great efforts have been made to further simplify the synthesis of semiconductor NPs. Yang et al. reported one non-injection method in which all strongly air-sensitive reagents (e.g., tributylphosphine or other organophosphines) were removed from the synthesis (24). Pradhan et al. used a short aliphatic amine (e.g., dodecylamine) to replace long chain ones (e.g., octadecylamine) so the synthesis temperature was greatly reduced from generally 300 to about 150 °C while maintaining the high quality of as-prepared NPs (25). Moreover, single precursor approaches for synthesizing semiconductor NPs were also extensively studied (26). Despite these efforts, the TOPO/TOP–CdO method is still the classical approach to make semiconductor NPs. The most successful system for organometallic approaches is CdSe NPs.
10. In traditional organometallic approaches for making semiconductor NPs, technical grade TOPO must be used. If pure TOPO is used, $Cd(CH_3)_2$ will immediately decompose to form a metallic precipitate (2).

11. In step 1 of Subheading 3.2, any amount of oleic acid (with the range between 0.30 and 21.2 mmol) may be used in the experiments.
12. No air-sensitive reagents are used in the synthesis of semiconductor NPs, therefore, the synthesis can be carried out in air. However, if one wants to avoid the oxidation of NPs at high temperature in the synthesis, argon can be used.
13. After adding Chloroform/Methanol into the reaction mixture, because the density of $CHCl_3/CH_3OH$ is greater than that of OA/ODE, two separate liquid layers will be formed. The top layer is CdS NP solution while the bottom is the mixture of impurity and $CHCl_3/CH_3OH$.
14. A shell of semiconductor layer is able to be directly deposited on the core CdTe NPs by injecting the corresponding precursors into the reaction mixture (16). In other words, it is not necessary to isolate the CdTe cores from the original reaction solution. This is also a method for preparing core@shell NPs.
15. In aqueous routes for making semiconductor NPs, the pH value of the mixture of cadmium precursor and thiols is often fixed at 11.2–11.8 in literature (11, 27). However, at this high pH, the photoluminescence quantum yield is usually low (this might be because of leaching of sodium ions from the glassware and also the effect of the pH). A low pH of 9 can be used to enhance the photoluminescence quantum yield of the NPs while maintaining their stability.
16. The diffraction peaks of the NPs are intermediate between the values of the cubic CdSe and the cubic CdS phases. The reason is the partial hydrolysis of mercapto acids caused an incorporation of the sulfur from the mercapto acids molecules into the growing NPs under prolonged refluxing. Therefore, the NPs may have CdSe(S) structure (10, 27). To speed up the NP growth and improve the NP quality in aqueous solution, microwave irradiation can be used to heat the reaction solution (28).
17. High quality CdTe NPs can be prepared through aqueous routes, at low cadmium to tellurium ratios (e.g., 2:1) (11).
18. The ratio between TOPO and HDA strongly affects the growth kinetics of NPs. The general tendency is that, with increasing HDA content, the initial particle size and the growth rate decreases. If there is more than about 80 mol % of HDA, CdSe NPs will precipitate when temperature is above 200 °C. To prepare large CdSe NPs (>4.5–5.0 nm), additional injections of selenium precursor solution are required (1).
19. For core-shell structure synthesis, the injection of the shell material precursors must be slow enough to avoid new

nucleation of NPs. For example, for the CdSe@ZnS NP growth in Subheading 3.5, the injection of the Zn:S stock solution needs to be as slow as one drop per 30 s (1).

20. Other typical synthesis methods using successive ion layer adhesion and reaction (SILAR) to prepare core-shell NPs can be referred to (12, 29).
21. To grow ZnS shells with desired thickness on CdSe core NPs, the amounts of Zn and S precursors should be determined as follows. First, measure the average radius of the CdSe core NPs using high-resolution transmission electron microscopy (HRTEM) images. Then calculate the number of ZnS molecules on each CdSe core NP by dividing the volume of ZnS shell by the volume of ZnS molecules ($V_{ZnS} = a^3/4$, $a = 0.541$ nm Lattice constant of Zinc blende ZnS). Then calculate the amount of Zn and S precursor through the molar amount of CdSe core NPs. An example of such calculations is available from the supporting information in (30).
22. In the surfactant exchange process described in Subheading 3.8, the stability of thiol-coated NPs is often related to the original TOPO-coated NPs. In general, freshly prepared TOPO-coated NPs yield more stable thiol-coated NPs than those dried or aged ones (17).
23. For surfactant exchange, thiols exhibit strong affinity to CdSe NPs but their instability toward oxidation is one major drawback. This problem can be circumvented by using carbodithioate ligands. A versatile strategy for NP ligand exchange using dithiocarbamate has also been developed to overcome this problem (31).
24. The addition of methanol hydrolizes the methoxysilane groups (Si–OCH3) into silanol groups (Si–OH), which form a primary polymerization layer. The addition of the fresh silane precursors are incorporated into the shell. Thus, the NP surface functionality is tailored for different applications. In other words, different functional groups (e.g., –NH2, –SH) are introduced to the surface of the NPs in this step. In step 6, the remaining free hydroxyl groups are converted into methyl groups. This blocks further silica growth.
25. Alternatively 100 % mPEG-2000 PE can be used for encapsulation (without DPPC). If amino PEG-PE is used in the encapsulation to replace up to 50 % of the PEG-PE, a primary amine will be formed on the outer surface of micelles. Then thiol-modified DNA can be covalently coupled to the amines with a hetero-bifunctional coupler (9).

Acknowledgments

The authors would like to thank Dr. Zhen Li in the Australian Institute for Bioengineering and Nanotechnology at the University of Queensland for his advice. The authors also acknowledge Prof. Shuming Nie, Prof. Wolfgang Parak, Dr. Brad Kairdolf, Dr. Marcel Bruchez and Dr. Daniele Gerion for providing the details of the experiments.

References

1. Talapin DV, Rogach AL, Kornowski A et al (2001) Highly luminescent monodisperse Cdse and Cdse/Zns nanocrystals synthesized in a hexadecylamine-trioctylphosphine oxide-trioctylphospine mixture. Nano Lett 1:207–211
2. Peng XG, Thessing J (2005) Semiconductor nanocrystals and silicate nanoparticles. In: Peng X, Minogos DMP (eds) Structure and bonding, vol 118. Springer, Berlin, pp 79–119
3. Qu LH, Peng XG (2002) Control of photoluminescence properties of Cdse nanocrystals in growth. J Am Chem Soc 124:2049–2055
4. Erwin SC, Zu LJ, Haftel MI et al (2005) Doping semiconductor nanocrystals. Nature 436:91–94
5. Yin Y, Alivisatos AP (2005) Colloidal nanocrystals synthesis and the organic-inorganic interface. Nature 437:664–670
6. Murray CB, Norris DJ, Bawendi MG (1993) Synthesis and characterization of nearly monodisperse Cde (E = S, Se, Te) semiconductor nanocrystallites. J Am Chem Soc 115: 8706–8715
7. Peng ZA, Peng XG (2001) Mechanisms of the shape evolution of CdSe nanocrystals. J Am Chem Soc 123:1389–1395
8. Bailey RE, Smith AM, Nie SM (2004) Quantum dots in biology and medicine. Physica E 25:1–12
9. Dubertret B, Skourides P, Norris DJ et al (2002) In vivo imaging of quantum dots encapsulated in phospholipid micelles. Science 298:1759–1762
10. Chen XF, Hutchison JL, Dobson PJ et al (2009) Highly luminescent monodisperse CdSe nanoparticles synthesized in aqueous solution. J Mater Sci 44:285–292
11. Gaponik N, Talapin DV, Rogach AL et al (2002) Thiol-capping of Cdte nanocrystals: an alternative to organometallic synthetic routes. J Phys Chem B 106:7177–7185
12. Li JJ, Wang YA, Guo WZ et al (2003) Large-scale synthesis of nearly monodisperse Cdse/Cds core/shell nanocrystals using air-stable reagents via successive ion layer adsorption and reaction. J Am Chem Soc 125:12567–12575
13. Bailey RE, Nie SM (2003) Alloyed semiconductor quantum dots: tuning the optical properties without changing the particle size. J Am Chem Soc 125:7100–7106
14. Yu WW, Peng XG (2002) Formation of high-quality Cds and other Ii-Vi semiconductor nanocrystals in noncoordinating solvents: tunable reactivity of monomers. Angew Chem Int Ed Engl 41:2368–2371
15. Kairdolf BA, Nie SM (2011) Multidentate-protected colloidal gold nanocrystals: pH control of cooperative precipitation and surface layer shedding. J Am Chem Soc 133:7268–7271
16. Kairdolf BA, Smith AM, Nie S (2008) One-pot synthesis, encapsulation, and solubilization of size-tuned quantum dots with amphiphilic multidentate ligands. J Am Chem Soc 130:12866–12867
17. Aldana J, Wang YA, Peng XG (2001) Photochemical instability of Cdse nanocrystals coated by hydrophilic thiols. J Am Chem Soc 123:8844–8850
18. Gerion D, Pinaud F, Williams SC et al (2001) Synthesis and properties of biocompatible water-soluble silica-coated Cdse/Zns semiconductor quantum dots. J Phys Chem B 105:8861–8871
19. Wu XY, Liu HJ, Liu JQ et al (2003) Immunofluorescent labeling of cancer marker Her2 and other cellular targets with semiconductor quantum dots. Nat Biotechnol 21:41–46
20. Ku MJ, Dossin FM, Choi Y et al (2011) Quantum dots: a new tool for anti-malarial drug. Assays Malaria J 10(118)

21. Magde D, Wong R, Seybold PG (2002) Fluorescence quantum yields and their relation to lifetimes of Rhodamine 6G and fluorescein in nine solvents: improved absolute standards for quantum yields. Photochem Photobiol 75:327–334
22. Penzkofer A, Leupacher W (1987) Fluorescence behaviour of highly concentrated Rhodamine 6G solutions. J Lumin 37:61–72
23. Peng XG, Wickham J, Alivisatos AP (1998) Kinetics of Ii-Vi and Iii-V colloidal semiconductor nanocrystal growth: "focusing" of size distributions. J Am Chem Soc 120: 5343–5344
24. Yang YA, Wu HM, Williams KR et al (2005) Synthesis of Cdse and Cdte nanocrystals without precursor injection. Angew Chem Int Ed Engl 44:6712–6715
25. Pradhan N, Reifsnyder D, Xie RG et al (2007) Surface ligand dynamics in growth of nanocrystals. J Am Chem Soc 129:9500–9509
26. Malik MA, Revaprasadu N, O'Brien P (2001) Air-stable single-source precursors for the synthesis of chalcogenide semiconductor nanoparticles. Chem Mater 13:913–920
27. Rogach AL, Kornowski A, Gao MY et al (1999) Synthesis and characterization of a size series of extremely small thiol-stabilized Cdse nanocrystals. J Phys Chem B 103:3065–3069
28. Li L, Qian HF, Ren JC (2005) Rapid synthesis of highly luminescent Cdte nanocrystals in the aqueous phase by microwave irradiation with controllable temperature. Chem Commun 4:528–530
29. Xie RG, Kolb U, Li JX et al (2005) Synthesis and characterization of highly luminescent Cdse-Core Cds/Zn0.5cd0.5s/Zns multishell nanocrystals. J Am Chem Soc 127:7480–7488
30. Dong BH, Cao LX, Su G et al (2010) Facile synthesis of highly luminescent UV-blue emitting Znse/Zns core/shell quantum dots by a two-step method. Chem Commun 46: 7331–7333
31. Dubois F, Mahler B, Dubertret B et al (2007) A versatile strategy for quantum dot ligand exchange. J Am Chem Soc 129:482–483

Chapter 9

Size-Selective Synthesis of Ultrasmall Hydrophilic CdSe Nanoparticles in Aqueous Solution at Room Temperature

Yeon-Su Park, Yukihiro Okamoto, Noritada Kaji, Manabu Tokeshi, and Yoshinobu Baba

Abstract

Hydrophilic semiconductor nanoparticles are very attractive for various biological applications, such as in optical sensing, tracing, and imaging of biological molecules-of-interest, because of their broad excitation wavelength, tunable emission wavelength, strong photoluminescence, and relatively high stability against photobleaching and chemicals. Compared to organic phase synthesis and subsequent surface modification, aqueous phase synthesis approaches provide multiple advantages for obtaining hydrophilic semiconductor nanoparticles. Here, we describe methods for the size-selective growth and stabilization of ultrasmall hydrophilic CdSe nanoparticles in aqueous solution at room temperature by using amino acid cysteine or one of its derivatives as a surface capping agent.

Key words: Aqueous synthesis, CdSe, Cysteine, Hydrophilic, Nanoparticle, Selective growth, Semiconductor, Stability, Water-soluble

1. Introduction

Semiconductor nanoparticles (NPs) have attracted great attention because of their broad excitation wavelength, size-dependent emission wavelength, strong photoluminescence (PL), and relatively high stability against photobleaching and chemicals (1–3). Among them, CdSe NPs have been most extensively investigated because of the abundance of well-established methods for producing high-quality hydrophobic CdSe NPs (4–8) as well as their strong visible PL that is highly beneficial to various fluorescence-based biological applications such as optical sensing, tracing, and imaging of biological molecules-of-interest (1, 3, 9–12). For most biological applications, water-solubility of CdSe NPs is essential.

Mikhail Soloviev (ed.), *Nanoparticles in Biology and Medicine: Methods and Protocols*, Methods in Molecular Biology, vol. 906, DOI 10.1007/978-1-61779-953-2_9, © Springer Science+Business Media, LLC 2012

Aqueous phase synthesis routes (13–18) can provide a simpler, safer, more convenient, more reliable, more economical, and more eco-friendly way of producing water-soluble CdSe NPs, unlike organic phase synthesis of hydrophobic CdSe NPs followed by NPs surface modification by ligand exchange (10, 19–22), silica shell formation (9, 23), or amphiphilic copolymer coating (12, 24). However, despite many advantages of the aqueous phase synthesis routes, there have been relatively few publications reporting aqueous phase synthesis of hydrophilic CdSe NPs (13–18). Furthermore, the physical and optical properties of the water-soluble CdSe NPs synthesized directly in aqueous solution (13–18) are often inferior to those of the water-soluble CdSe NPs fabricated by the surface modification of organic phase-synthesized hydrophobic CdSe NPs (9, 10, 19–24). Recently, we introduced simple and convenient methods for synthesizing very stable hydrophilic CdSe NPs of various sizes directly in aqueous solutions at room temperature (25). Use of cysteine (Cys) or one of its derivatives as a capping ligand resulted in ultrasmall hydrophilic CdSe NPs ($d = 1.2$–2.1 nm). These NPs show an excellent aqueous solubility due to their relatively strong negative surface charge originated from the free-carboxylate moieties of their capping ligand. The aqueous solubility and free-carboxylate moieties of the NPs allow their easy and convenient surface modifications, in aqueous media, with various biological molecules for their applications in labeling, signaling, assay, and detection. The surface of the NPs could be modified with oligonucleotides, nucleic acids, receptors, peptides, antibodies, antigens, or other proteins using 1-ethyl-3-(3-dimethylaminopropyl) carbodiimide hydrochloride (EDC) as a coupling agent allowing formation of amide bonds between the carboxylate moieties on the NP surface and the primary amines of those biological molecules. Here, we report detailed procedures for the size-selective synthesis of ultrasmall hydrophilic CdSe NPs. The size-selective synthesis is achieved by means of sequential addition of the required amounts of the components (NaOH, a capping agent such as Cys or one of its derivatives, Cd precursor solution, and Se precursor solution) into glass vials filled with predetermined volumes of water. In addition, as an example of surface modification for biological applications, this chapter describes detailed procedures for the purification and enrichment of the NPs and their surface modification with streptavidin.

2. Materials

Store all chemical reagents (analytical grade) as specified by manufacturers and use them as received without further purification. Prepare all aqueous solutions using deionized water (i.e., Milli-Q water with $R > 18.2$ MΩ cm) and store all at room temperature.

Use deionized water (for simplicity, referred to as "water") for preparation of CdSe NPs and their purification. Strictly follow all waste disposal regulations and guidelines when disposing waste materials.

2.1. Preparation of Se Precursor Solution

1. Sodium sulfite (Na_2SO_3).
2. Selenium (Se) powder.
3. A magnetic hot plate stirrer.
4. A glass vial (20 mL capacity or larger).

2.2. Synthesis of CdSe NPs

1. 1 M Sodium hydroxide (NaOH).
2. L-Cysteine (Cys).
3. Cd precursor solution: 0.15 M $CdSO_4 \cdot 8/3\ H_2O$ in water.
4. A magnetic hot plate stirrer.
5. A glass vial (20 mL capacity or larger).
6. Aluminum foil.

2.3. Sample Purification

1. Membrane filters: Amicon Ultra-15 membrane filters (Millipore), sample capacity = 15 mL, MWCO = 10,000.
2. A centrifuge with a swing bucket rotor which can hold Amicon Ultra-15 membrane filters and capable of achieving rotation speed of 4,000 rpm (2,580 × *g*), e.g., model 3740 (KUBOTA).
3. Aluminum foil.

2.4. Sample enrichment

1. Membrane filters: Amicon Ultra-4 membrane filters (Millipore), sample capacity = 4 mL, MWCO = 10,000.
2. A centrifuge with a swing bucket rotor which can hold Amicon Ultra-4 membrane filters and capable of achieving rotation speed of 4,000 rpm (2,580 × *g*), e.g., model 3740 (KUBOTA).

2.5. Characterization

1. A UV-visible spectrophotometer UV-3600 UV-VIS-NIR (Shimadzu) or equivalent, wavelength range 250–700 nm or wider.
2. A pair of two quartz cuvettes (light path length = 1 cm, two transparent sides).
3. Spectrofluorometer FP-6500 (JASCO) or equivalent: excitation wavelength range = 250–500 nm or wider, detection wavelength range = 250–700 nm or wider.
4. One quartz cuvette (light path length = 1 cm, four transparent sides).
5. X-ray diffraction (XRD) spectrophotometer RINT2000 (Rigaku) or equivalent.
6. XRD sample holders (Si or glass).

7. Transmission electron microscope system H7650 (Hitachi) or equivalent.
8. Carbon-coated Cu-mesh grids for transmission electron microscopy (TEM).
9. NanoScope IIIa (Digital Instruments) or equivalent scanning probe microscope system.
10. Atomic force microscopy (AFM) tips.
11. Mica substrates.
12. 3-Aminopropyltriethoxysilane (APS).

2.6. Surface Modification for Biological Application

1. 10 mM borate buffer: 0.3092 g of H_3BO_3 in 500 mL of water (pH adjusted to 7.4 with NaOH).
2. 50 mM borate buffer: 1.5458 g of H_3BO_3 in 500 mL of water (pH adjusted to 8.3 with NaOH).
3. Streptavidin solution (10 mg/mL): 10 mg of streptavidin in 1 mL of 10 mM borate buffer.
4. EDC solution: 10 mg of EDC in 1 mL of water. Prepare fresh each time.
5. Membrane filters: Amicon Ultra-4 membrane filters (Millipore), sample capacity = 4 mL, MWCO = 10,000.
6. A centrifuge with a swing bucket rotor which can hold *Amicon Ultra-4 membrane filters* and capable of achieving rotation speed of 4,000 rpm (2,580 × *g*), e.g., model 3740 (KUBOTA).
7. A glass vial (20 mL capacity or larger).
8. Acrodisc® syringe filters with hydrophilic membranes: pore size ~0.2 μm.

3. Methods

All procedures are carried out at room temperature unless specified otherwise. Unlike many similar procedures, which require the use of inert gas to prepare aqueous solution free of dissolved oxygen species, deionized water is suitable for our methods and no further degassing is required. Wear personal protective equipment (gloves, goggles, a lab coat).

3.1. Preparation of Se Precursor Solution

1. Place a glass vial (20 mL) in the center of a magnetic hot plate stirrer. Put a magnetic stirring bar into the vial.
2. Weigh 0.02 g of Se powder and transfer it into the vial (see Note 1). Weigh 0.1 g of Na_2SO_3 and transfer it into the vial. After that, add 5 mL of water into the vial using a pipette (see Note 2).

3. Tightly cap the vial immediately after the water addition and then cover the capped vial completely with aluminum foil for more efficient heat delivery and heat keeping (see Note 3).
4. Set magnetic stirring speed at 800 rpm and then turn on the magnetic stirrer (see Note 4). Set hot plate temperature at 95 °C and then turn on the hot plate (see Note 5).
5. Check the vial visually 2–3 h after starting heating and then shake it gently several times to move down unreacted Se powder (grey color) attached on its inner-wall onto its bottom (see Note 6).
6. Continue heating the vial for 12 h or longer (see Note 7).
7. Just before starting nanoparticle synthesis, turn off both the hot plate and the magnetic stirrer, unwrap the aluminum foil from the vial, and then place the vial on bench at room temperature to slowly cool down the resulting Se precursor solution (0.05 M Na_2SeSO_3) to room temperature (see Note 8).

3.2. Synthesis of CdSe NPs

All procedures are carried out in a fume hood unless otherwise specified. Size-selective synthesis of CdSe NPs is achieved by the sequential addition of desired amounts of 1 M NaOH, Cys capping agent, 0.15 M Cd precursor solution, and 0.05 M Se precursor solution (see Subheading 3.1) into glass vials filled with predetermined volumes of water. Table 1 specifies the quantities and corresponding concentrations of the reactants used for the synthesis of CdSe NPs stabilized with Cys. Procedures for synthesizing CdSe NPs stabilized with one of the Cys-derivatives are the same as those for synthesizing CdSe NPs stabilized with Cys except for the choice of different Cys-derivatives instead of Cys (used at the same molar ratios) (see Note 9). Table 2 lists the size, size distribution, and stability of the manufactured CdSe NPs depending on the capping agent used.

1. Place a glass vial (20 mL) in the center of a magnetic stirrer. Put a magnetic stirring bar into the vial.
2. Add 9.45 mL of water into the vial using a pipette. Set the stirring speed at 1,500 rpm and then turn on the stirrer. After that, add 0.375 mL of 1 M NaOH into the vial using a pipette.
3. Weigh 0.016 g of Cys, immediately transfer it into the vial, and then allow 5 min for thorough mixing.
4. Add 0.1 mL of 0.15 M Cd precursor solution into the vial and then allow 10 min or slightly longer for complete mixing (see Note 10). After that, transfer 0.075 mL of 0.05 M Se precursor solution at room temperature into the vial.
5. Tightly cap the vial immediately after the Se precursor addition. Wrap the vial thoroughly with 2–3 layers of aluminum foil (see Note 11).

Table 1
Quantities and concentrations of the reactants for synthesizing CdSe NPs stabilized with Cys

	Water	1 M NaOH	Cys	0.15 M Cd^{2+}	0.05 M Se^{2-}
Quantity	9.45 mL	375 μL	16 mg	100 μL	75 μL
C/mM		37.5	13.2	1.5	0.375

Table 2
Capping agent dependence of the size, size distribution, and stability of CdSe NPs. (Reproduced with permission from Park et al. (25), Copyright 2010 American Chemical Society)

			Stability[a]	
Capping agent	Size (nm)	Size distribution[a]	Short-term	Long-term
Cys	1.7	E	E	E
CysME	1.7	E	E	P
	2.1	P	G	G
CysEE	1.7	E	G	P
	2.1	P	G	G
N-ACys	1.7	P	P	P
	1.5	E	P	P
	1.2	E	G	P
HomoCys	1.7	P	P	P
	1.5	G	E	E

[a] *E* Excellent; *G* Good; *P* Poor

6. Keep stirring the reaction solution at room temperature in the dark for 7 days (see Note 12). Turn off the stirrer and store the CdSe NP solution at room temperature in the dark (see Note 13).

3.3. Sample Purification

Purification of CdSe NPs is done at room temperature in ambient atmosphere. This section describes sample purification procedures which will yield 5 mL of CdSe NP solution ("sample solution") (see Note 14).

1. Place two membrane filters into the tube racks on an experimental bench and uncap them (see Note 15).
2. Transfer 5 mL of the sample solution onto the membrane unit of one of the membrane filters using a pipette. Add 9 mL of

water onto the membrane unit filled with the sample solution using a pipette. Mix the solution in the membrane unit by pipetting it up and down 4 times.

3. Cap the membrane filter tightly and then weigh the capped membrane filter containing the sample solution ("sample filter").
4. Fill the membrane unit of the other membrane filter with 14 mL of water ("balancing filter") and cap the filter. Weigh the capped balancing filter and then adjust its total weight to be the same as that of the capped sample filter by adding water onto or removing water from the membrane unit of the balancing filter (see Note 16).
5. Centrifuge at 2,000 rpm (640×*g*) for 10 min at 20 °C (see Notes 17 and 18).
6. Pull out the sample filter from the bucket immediately after the centrifugation and discard the filtered waste solution in the container unit. After that, add water onto the membrane unit of the sample filter using a pipette until the total solution volume reaches to ~14 mL. Mix the solution thoroughly using a pipette and cap the sample filter tightly. Weigh the capped sample filter and then place it into the previous sample bucket in the centrifuge.
7. Take out the balancing filter from the bucket in the centrifuge and then adjust total weight of the capped balancing filter to be the same as the weight of the capped sample filter by removing water from the filter or by adding water onto the filter (see Note 16).
8. Insert the capped balancing filter into the previous balancing bucket and then perform centrifugation at the same conditions described in step 5.
9. Repeat a series of purification steps (see steps 6–9) 4 times (5 in total).
10. Take out the sample filter from the bucket immediately after the centrifugation and then check the solution volume in its membrane unit. Perform centrifugation further at the same rotation speed until the solution volume decreases to 4–4.5 mL if the volume is larger than 4.5 mL. After that, place both filters into the tube racks on the bench.
11. Uncap the sample filter and homogenize the resulting water-purified sample solution in the membrane unit by pipetting it up and down 4 times.
12. Transfer all of the solution in the membrane unit into an empty glass vial using a pipette and then adjust the total solution volume in the vial to 5 mL by adding water using a pipette (see

Note 19). Mix the solution well by vigorous pipetting and then cap the vial to finish preparation of water-purified CdSe NPs.

13. Wrap the vial completely with 2–3 layers of aluminum foil and then store it in a refrigerator at 4 °C.

3.4. Sample Enrichment

Enrichment of water-purified CdSe NP solution is carried out at room temperature in ambient atmosphere. This section describes sample enrichment procedures which will yield 0.5 mL of 5 times concentrated water-purified CdSe NP solution (see Note 14).

1. Uncap two membrane filters and place them into the tube racks on the bench.
2. Using a pipette transfer 2.5 mL of the water-purified CdSe NP solution in the vial onto the membrane unit of one of the membrane filters ("sample filter"). Take out the membrane filter containing the sample solution, cap it tightly, and then weigh the capped sample filter.
3. Fill the membrane unit of the other membrane filter ("balancing filter") with 2.5 mL of water and then cap it. Weigh the capped balancing filter and adjust its weight to be the same as that of the capped sample filter by adding water onto or removing water from the membrane unit of the balancing filter (see Note 16).
4. Centrifuge the filter units at 2,000 rpm (640 × *g*) for 15 min at 20 °C (see Note 18).
5. Take out the sample filter from the bucket immediately after the centrifugation and then check the solution volume in its membrane unit. Perform centrifugation further at the same rotation speed until the solution volume decreases to 0.25–0.30 mL if the volume is larger than 0.30 mL.
6. Place both filters into a tube rack on bench. Uncap the sample filter and then homogenize the resulting concentrated, water-purified sample solution in the membrane unit using a pipette (see Note 20).
7. Transfer all of the remaining solution in the membrane unit into a 1.5 mL microcentrifuge tube using a pipette and adjust the total solution volume in the centrifuge tube to 0.5 mL by adding water using a pipette. Mix the solution well by gently shaking the centrifuge tube several times after tight capping to finish preparation of concentrated water-purified CdSe NPs.
8. Completely wrap the tube containing the concentrated sample solution with 2–3 layers of aluminum foil and then store it in a refrigerator at 4 °C.

3.5. Characterization

1. Take absorption spectra of as-prepared CdSe NP solution, using a UV-visible spectrophotometer and a pair of two

sides-transparent quartz cuvettes, to trace the progress and completion of the NP growth (see Figs. 1 and 2 and Note 21) as well as to obtain information about the size and size distribution of the NPs (see Note 22).

2. Take emission spectra of as-prepared CdSe NP solution, using a spectrofluorometer and an all sides-transparent quartz cuvette, to obtain PL characteristics of the NPs (see Fig. 3).
3. Take XRD spectra of concentrated, water-purified CdSe NPs dried on sample holders, using an XRD spectrophotometer, to obtain information about chemical composition and crystalline structure of the NPs (see Fig. 3).
4. Take TEM images of dried, water-purified CdSe NPs on carbon-coated Cu-mesh grids, using a transmission electron microscope system, to obtain information about morphology (i.e., size and shape) of the NPs (see Fig. 4).
5. Take AFM images of CdSe NPs immobilized on fresh mica substrates by APS linker molecules, in a fluid tapping mode

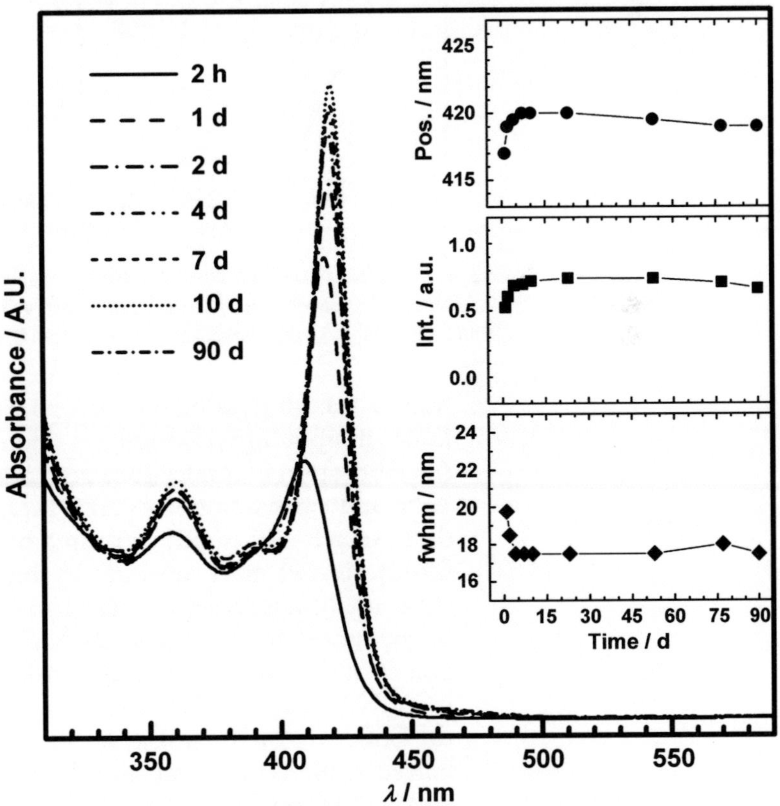

Fig. 1. UV-visible absorption spectra showing growth and stabilization of Cys-capped CdSe NPs. The *insets* show time-dependent changes in the position (*top*), intensity (*middle*), and full width at half-maximum (*bottom*) of their first absorption peak (Reproduced from Park et al. (25), with permission from American Chemical Society).

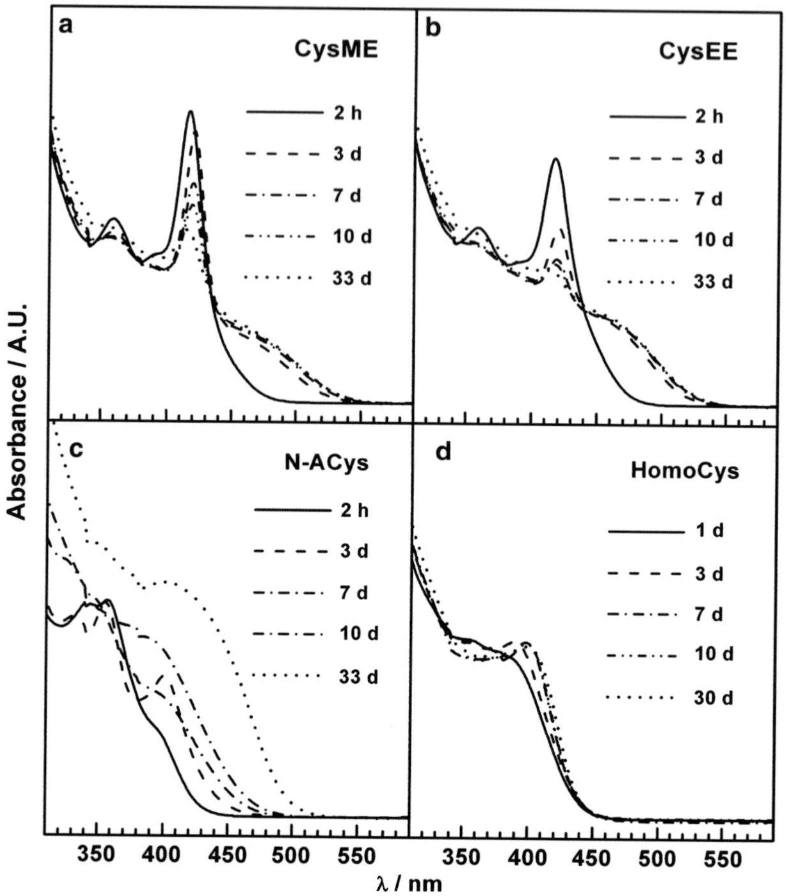

Fig. 2. UV-visible absorption spectra showing capping agent-dependent growth and stabilization of CdSe NPs: (**a**) CysME, (**b**) CysEE, (**c**) N-ACys, and (**d**) HomoCys (Reproduced from Park et al. (25), with permission from American Chemical Society).

using a scanning probe microscope system, to obtain close-to-intact morphology information about the NPs (Fig. 4). For the fabrication of a mica substrate coated with a monolayer of APS, add small amount of APS (i.e., 20–50 μL) into a Petri dish (keep 5–10° tilted) to form a bottom pool of APS, put a freshly cleaved mica substrate in the center of the dish, fully cover the dish with a cover glass, and incubate for 1 h at room temperature. Fix the APS-coated mica substrate in the sample holder of the microscope and transfer aliquots of the prepared CdSe NPs onto the APS-coated mica substrate to immobilize the CdSe NPs on the substrate. The mica substrate with immobilized CdSe NPs is immersed in the aliquot solution during AFM operation.

3.6. Surface Modification for Biological Application

1. This section describes procedures for conjugating streptavidin to the water-purified CdSe NPs for targeting biotinylated biological molecules. The concentration of the water-purified

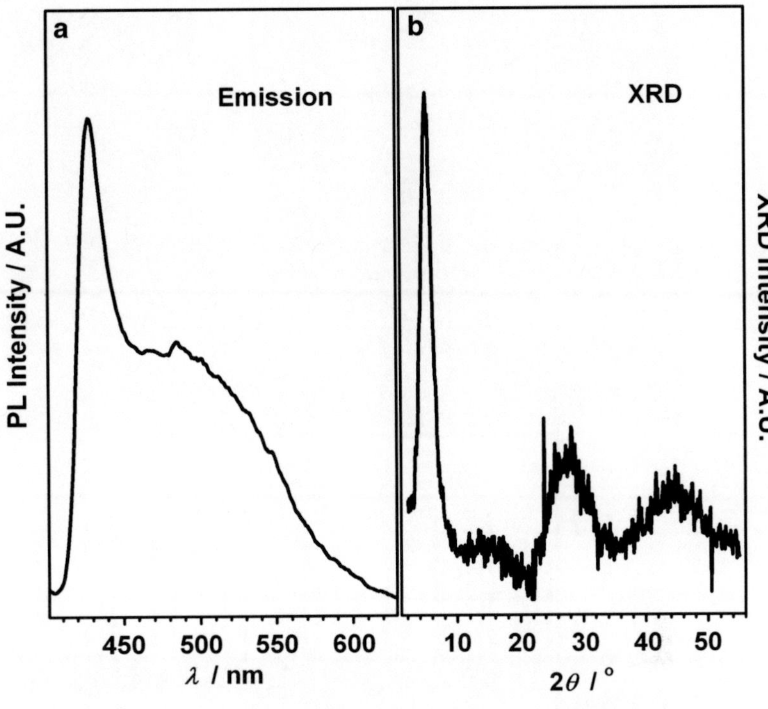

Fig. 3. Spectra showing optical emission and X-ray diffraction properties of Cys-capped CdSe NPs. (**a**) PL spectrum of the as-prepared CdSe NP sample. (**b**) XRD profile of water-purified CdSe NP sample (Reproduced from Park et al. (18) with permission from American Chemical Society).

CdSe NPs stabilized with Cys from step 3.3 is ~20 μM (see Note 23).

2. Place a glass vial (20 mL capacity) in the center of a magnetic stirrer. Add 250 μL of purified 20 μM CdSe NPs followed by 1.750 mL of 10 mM borate buffer. Start stirring at 800 rpm at room temperature and continue it for 5 min to ensure complete mixing of the solution.
3. Transfer 240 μL of the streptavidin (10 mg/mL) into the vial with CdSe NPs. Keep stirring for additional 10 min.
4. Transfer 57 μL of the freshly-prepared EDC solution (10 mg/mL) to the vial containing CdSe NPs and streptavidin (see Note 24). Keep stirring for 1–2 h for completing the conjugation.
5. Filter the solution through an Acrodisc® syringe filter to remove large agglomerates and transfer the filtered solution to the membrane unit of an Amicon Ultra-4 membrane filter ("sample filter"). Cap the membrane filter tightly.
6. Weigh the capped sample filter.
7. Fill the membrane unit of another membrane filter with water ("balancing filter") so that the weight of the capped filter is the same as that of the capped sample filter (see Note 16).

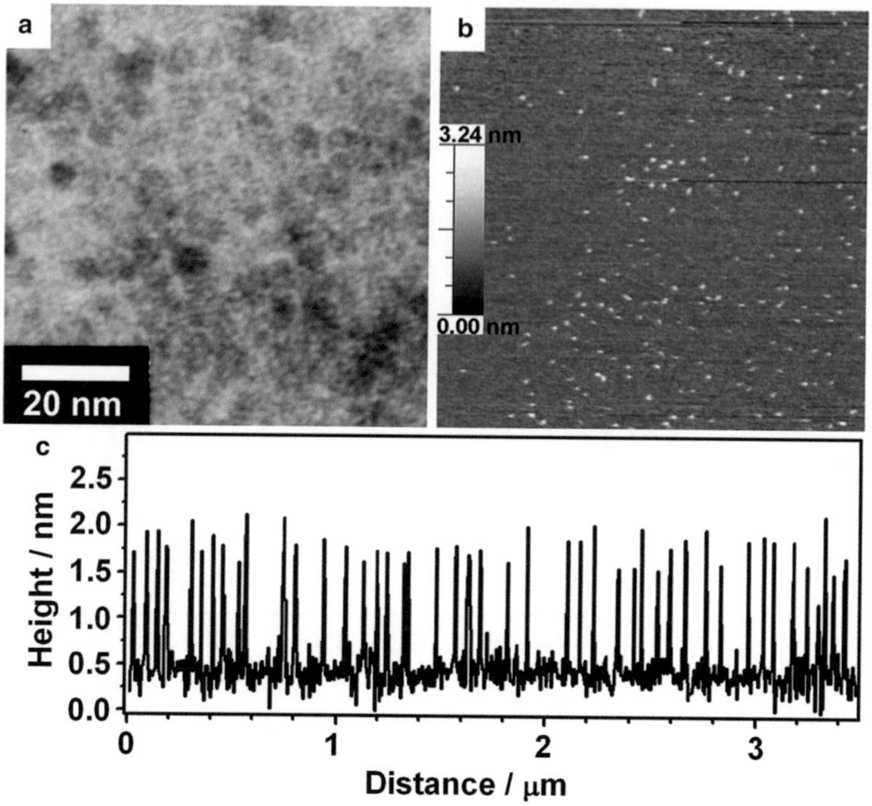

Fig. 4. Micrographs showing morphology of Cys-capped CdSe NPs. (**a**) TEM micrograph of the water-purified CdSe NP sample dried on a carbon-coated Cu-mesh grid. (**b**) AFM micrograph of the CdSe NPs immobilized on a mica substrate. (**c**) AFM height profile from the AFM micrograph in (**b**) (Reproduced from Park et al. (18) with permission from American Chemical Society).

8. Centrifuge for 15 min at 2,000 rpm (640 × *g*) at room temperature (see Note 25).
9. Immediately after the centrifugation, take out the sample filter from the bucket. Uncap the sample filter and discard the filtered solution in the container unit.
10. Fill the membrane unit of the sample filter with 50 mM borate buffer to make the solution volume to be ~2.5 mL and then cap the filter. After that, redisperse the conjugate solution by shaking the capped filter vigorously.
11. Repeat steps 5–9 four times to remove any unbound EDC and streptavidin molecules.
12. After last centrifugation add 50 mM borate buffer into the membrane unit of the sample filter to make the solution volume to be slightly less than 2 mL.
13. Cap the sample filter and then redisperse the conjugate solution by shaking the filter vigorously.

14. Filter the conjugate solution in 50 mM borate buffer through an Acrodisc® syringe filter.
15. Transfer the purified conjugate solution into a fresh 2 mL microcentrifuge tube and adjust the total solution volume to 2 mL by adding 50 mM borate buffer. Wrap in aluminum foil and store the purified conjugate solution tightly closed at 4 °C.

4. Notes

1. Put Se powder into the vial before adding water. At room temperature Se powder will not mix with water owing to its hydrophobic nature. A large portion of the powder spreads along the wet area of the inner-wall of the vial if the inner-wall is wet with water before Se powder addition.
2. Add water gradually using a 1 mL-volume pipette. Fast delivery results in severe spreading of Se powder along the inner-wall of the vial.
3. The simplest way is to cover the vial with an aluminum foil cap made of 2–3 layers of aluminum foil. Use an additional cover structure, such as a small empty iron can or a small empty paper box, above the aluminum foil cap to achieve more homogeneous temperature distribution in the synthesis solution as well as to minimize heat loss to atmosphere.
4. Magnetic stirring should not be vigorous (use 800 rpm or lower). Vigorous magnetic stirring often leads to severe spreading of Se powder onto the inner-wall of the vial.
5. Use slightly lower temperature (80–90 °C) if observing any shape change of the cap. Lower reaction temperatures lead to the same results, but requires slightly longer reaction times.
6. The vial is hot. Wear additional heat-protect gloves and then handle the vial. Make sure that the vial is capped well. The shaking process moves undissolved Se powder on the inner-wall onto the bottom of the vial. Repeat the shaking process until almost no powder is observed on the inner-wall. A small amount of Se powder remaining on the inner-wall is acceptable because it will be dissolved completely in couple of hours.
7. Check the solution visually. Well-prepared Se precursor solution is transparent without showing any particulates. Typically, it takes about half a day to obtain a transparent Se precursor solution. Total reaction time should not exceed 4 days. Longer reaction time may result in oxidation of the Se precursor. Do not use any Se precursor solution showing brown precipitates which indicate the presence of selenium oxides.

8. Use the Se precursor solution as soon as it is cooled down to room temperature. Prolonged storage at room temperature will lead to the formation of grey precipitate. Discard the solution if observing gray precipitates.
9. To prepare CdSe NPs stabilized with one of the Cys-derivatives, use corresponding Cys-derivative, instead of Cys, as follows: 0.0227 g of L-Cysteine methylester hydrochloride (CysME), 0.0245 g of L-Cysteine ethylester hydrochloride (CysEE), 0.0215 g of *N*-Acetyl-L-Cysteine (N-ACys), or 0.0178 g of DL-Homocysteine (HomoCys).
10. Formation of white species is observed along the trajectory of the injected Cd precursor solution immediately after its injection into the solution in the vial. The white species are poorly soluble Cd-Cys complexes, which will eventually be solubilized upon continuous mixing. It is very important to continue magnetic stirring of the solution for 10 minutes or longer, before moving to the next step, to ensure complete dispersion of Cd-Cys complexes.
11. CdSe NPs, like the other semiconductor NPs, are vulnerable to photobleaching upon long exposure to light. The simplest way to solve this problem is to wrap the vial containing CdSe NPs with 2–3 layers of aluminum foil. Use of an additional cover structure, such as a small empty iron can or a small empty paper box, above the aluminum foil cap, is helpful to achieve better protection from light.
12. Total reaction time required for preparing CdSe NPs depends strongly on the capping agents used. Preparation of CdSe NPs stabilized with Cys takes 6–7 days. Completion of the preparation is confirmed by the analysis of their UV-visible absorption spectra (see Figs. 1 and 2). Spectral properties will stabilize and won't change any longer on the completion of NP synthesis.
13. Store the CdSe NP solution in the dark, for example in a drawer, to protect it from light. Do not unwrap the aluminum foil. Cap tightly.
14. It might be convenient to use a centrifuge with a swing bucket rotor which can hold six buckets (a pair of large buckets located across each other and a pair of two small bucket sets located across each other). Each large bucket may accommodate a 15 mL-capacity membrane filter, whereas each small bucket may accommodate a 4 mL-capacity membrane filter.
15. Membrane filter units with larger membrane area are preferred because they permit to handle larger volume of the solution or to reduce number of purification cycles.
16. Make sure of weighing both filters after capping. Weight difference between the two capped filters should be 0.001 g or smaller. Much larger weight difference may cause large irregular

vibration and high level of noise, which may eventually lead to serious damage to the centrifuge.

17. Balance the centrifuge well. The sample and balancing filters should be located across each other to maintain balance during centrifugation. An improper location of the filters results in large irregular vibration of the equipment as well as high level of noise. Those may cause serious damage to the centrifuge.

18. Adjust the rotation speed and time appropriately to achieve the yield of ~5 mL. Use higher rotation speeds and/or longer time when the resulting solution volume is much larger than 5 mL (i.e., >5.5 mL) and vice versa. Be sure that the resulting solution volume is not much less than 5 mL (i.e., <4.5 mL). Smaller volume of the concentrated solution means unnecessarily strong centrifugation force that typically leads to the large amount of precipitated NPs inside the filter membrane. It may be difficult to redisperse these NPs into solution again.

19. First, transfer the water-purified, well-dispersed sample solution into an empty vial as completely as possible using a pipette ("the first transfer"). Remember or write down the volume of the solution transferred. Then add small amount (~0.5 mL) of water into the membrane unit, mix the solution by using a pipette, and then transfer a predetermined volume of the solution into the vial filled with water-purified sample solution ("the second transfer"). The solution volume transferred during the second transfer is determined by subtracting the volume transferred at the first transfer from 5 mL (the initial sample volume). The total volume transferred should therefore always be 5 mL. This will improve reproducibility of results.

20. After this final centrifugation, a relatively large amount of the concentrated, water-purified CdSe NPs may be located at the bottom of the membrane unit as well as inside the top layer of the membrane owing to the relatively strong centrifugation conditions used. Set the pipette volume to 0.20–0.30 mL and mix the solution in the membrane unit by pipetting it up and down for no less than 10 times. Wash out CdSe NPs from all areas of the membrane.

21. Spectral properties will stabilize and won't change any longer on the completion of NP synthesis.

22. The absorption spectra provide means to estimate the size and size distribution of the NPs. The size (*d* in nm) of CdSe NPs can be correlated with their first absorption peak position (λ in nm) as follows (26):

$$d = (1.6122\times10^{-9})\lambda^4 - (2.6575\times10^{-6})\lambda^3 + (1.6242\times10^{-3})\lambda^2 - (0.4277)\lambda + 41.57$$

The sharpness of their first absorption peak decreases with widening their size distribution.

23. Molar concentration of the water-purified CdSe NPs can be determined from their UV-visible absorption spectrum using Lambert–Beer's law:

$$C = A / \varepsilon L$$

where C is the molar concentration (in mol/L) of the CdSe NPs, A is the absorbance at their first absorption peak, ε is their extinction coefficient (in cm^{-1} M^{-1}), and L is the light path length (1 cm). The extinction coefficient is calculated using an empirical fitting function,

$$\varepsilon = 5857 \times (d)^{2.65}$$

where d is the CdSe NP size in nm (26). The determined concentration of the water-purified CdSe NPs stabilized with Cys is ~20 μM.

24. EDC looses its activity fast in water with time.
25. The solution volume after the centrifugation should be between 0.25 and 0.5 mL. If not, adjust the centrifugation conditions so that the resulting solution volume is in the range.

Acknowledgments

This work was supported by the Japan Society for the Promotion of Science through its "Funding Program for World-Leading Innovative R&D on Science and Technology" (FIRST Program), Japan.

References

1. Biju V et al (2008) Semiconductor quantum dots and metal nanoparticles: syntheses, optical properties, and biological applications. Anal Bioanal Chem 391:2469–2495
2. Wood V, Bulović V (2010) Colloidal quantum dot light-emitting devices. Nano Reviews 1:5202–5209
3. Biju V, Itoh T, Ishikawa M (2010) Delivering quantum dots to cells: bioconjugated quantum dots for targeted and nonspecific extracellular and intracellular Imaging. Chem Soc Rev 39:3031–3056
4. Murray CM, Norris DJ, Bawendi MG (1993) Synthesis and characterization of nearly monodisperse CdE (E=S, Se, Te) semiconductor nanocrystallites. J Am Chem Soc 115: 8706–9715
5. Ptatschek V et al (1998) Quantized aggregation phenomena in II-VI-semiconductor colloids. Ber Bunsenges Phys Chem 102:85–95
6. Kasuya A et al (2004) Ultra-stable nanoparticles of CdSe revealed from mass spectrometry. Nat Mater 3:99–102
7. Kudera S et al (2007) Sequential growth of magic-size CdSe nanocrystals. Adv Mater 19:548–552
8. Ouyang J et al (2008) Multiple families of magic-sized CdSe nanocrystals with strong

bandgap photoluminescence via noninjection one-pot syntheses. J Phys Chem C 112: 13805–13811
9. Bruchez M Jr (1998) Semiconductor nanocrystals as fluorescent biological labels. Science 281:2013–2016
10. Chan WCW, Nie S (1998) Quantum dot bioconjugates for ultrasensitive nonisotopic detection. Science 281:2016–2018
11. Mattoussi H et al (2000) Self-assembly of CdSe-ZnS quantum dot bioconjugates using an engineered recombinant protein. J Am Chem Soc 122:12142–12150
12. Gao X et al (2004) In vivo cancer targeting and imaging with semiconductor quantum dots. Nat Biotechnol 22:969–976
13. Rogach AL et al (1999) Synthesis and characterization of a size series of extremely small thiol-stabilized CdSe nanocrystals. J Phys Chem B 103:3065–3069
14. Zhou X et al (2005) Strong luminescing CdSe nanoparticles by surface modification with cadmium (II) hydrous oxide. Int J Mod Phys B 19:2835–2840
15. Yang YJ, Xiang BJ (2005) Wet synthesis of nearly monodisperse CdSe nanoparticles at room temperature. J Cryst Growth 284: 453–458
16. Park Y-S et al (2007) Aqueous-phase synthesis of ultra-stable small CdSe nanoparticles. J Nanosci Nanotechnol 7:3750–3753
17. Xia YS, Zhu CQ (2008) Aqueous synthesis of luminescent magic sized CdSe nanoclusters. Mater Lett 62:2103–2105
18. Park Y-S et al (2010) Aqueous phase synthesized CdSe nanoparticles with well-defined numbers of constituent atoms. J Phys Chem C 114:18834–18840
19. Wuister SF, de Mello Donegá C, Meijerink A (2004) Influence of thiol capping on the exciton luminescence and decay kinetics of CdTe and CdSe quantum dots. J Phys Chem B 108:17393–17397
20. Kloepfer JA, Bradforth SE, Nadeau JL (2005) Photophysical properties of biologically compatible CdSe quantum dot structures. J Phys Chem B 109:9996–10003
21. Algar WR, Krull UJ (2007) Luminescence and stability of aqueous thioalkyl acid capped CdSe/ZnS quantum dots correlated to ligand Ionization. Chem Phys Chem 8:561–568
22. Liu W et al (2007) Compact cysteine-coated CdSe(ZnCdS) quantum dots for in vivo applications. J Am Chem Soc 129:14530–14531
23. Selvan ST, Tan TT, Ying JY (2005) Robust, non-cytotoxic, silica-coated CdSe quantum dots with efficient photoluminescence. Adv Mater 17:1620–1625
24. Nann T (2005) Phase-transfer of CdSe@ZnS quantum dots using amphiphilic hyperbranched polyethylenimine. Chem Commun 13:1735–1736
25. Park Y-S et al (2010) Size-selective growth and stabilization of small CdSe nanoparticles in aqueous solution. ACS Nano 4:121–128
26. Yu WW et al (2003) Determination of the extinction coefficient of CdTe, CdSe, and CdS nanocrystals. Chem Mater 15:2854–2860

Chapter 10

Synthesis and Functionalization of CdSe/ZnS QDs Using the Successive Ion Layer Adsorption Reaction and Mercaptopropionic Acid Phase Transfer Methods

Ana Sofia Miguel, Christopher Maycock, and Abel Oliva

Abstract

Detailed protocols for the preparation of CdSe nanocrystals coated with a multishell structure of ZnS and their surface modification based on the successive ion layer adsorption and reaction method are described. The first phase of the synthesis produces a hydrophobic surface consisting of a TOPO/HDA ligand mixture. This is followed by a surface modification of the quantum dots (QDs) with 3-mercaptopropionic acid by the phase transfer method. The modified QDs become soluble in aqueous systems such as water or buffers for further biological applications.

Key words: Quantum dots, 3-Mercaptopropionic acid, Water-soluble, Phase transfer method, Hydrodynamic diameter, ξ-Potential, Bio-conjugation

1. Introduction

Quantum dots (QDs) have emerged as a new class of fluorescent label with unique optical properties. The development of water-soluble and bio-conjugated QDs has made them very appealing for applications in bio-sensing and bio-imaging. QDs made of CdSe core and ZnS shells have been widely used for bio-conjugation (1). Several procedures for the synthesis of CdSe/ZnS QDs have been reported; including the method of successive ion layer adsorption and reaction (SILAR) (2–4). The core-shell CdSe/ZnS QDs synthesis reported here is based upon a modified SILAR technique. The obtained cores are monodisperse and after coating of the CdSe (TOPO/HDA) with at least five monolayers of ZnS show a high fluorescence, are stable in organic solutions and are suitable for the posterior functionalization reactions with different ligands to

Mikhail Soloviev (ed.), *Nanoparticles in Biology and Medicine: Methods and Protocols*, Methods in Molecular Biology, vol. 906, DOI 10.1007/978-1-61779-953-2_10, © Springer Science+Business Media, LLC 2012

obtain hydrophilic and biocompatible QDs. However, CdSe/ZnS core-shell QDs are synthesized in organic medium and have a surface layer of lipophilic ligands which makes them insoluble in aqueous solution. We describe here the generation of water-soluble CdSe/ZnS QDs by ligand exchange of TOPO/HDA with 3-mercaptopropionic acid (MPA). Ligand exchange by the phase transfer method leads to a change of the surface polarity from hydrophobic to hydrophilic and yields water-soluble nanoparticles. The reported technique is advantageous due to its practical simplicity and the potential for further derivatization through the introduced carboxylic acid groups (5, 6).

2. Materials

All of the reagents used for the synthesis of CdSe/ZnS and functionalization of MPA-coated QDs can be stored at room temperature, unless indicated otherwise. All solutions should be freshly prepared under an argon atmosphere. We recommend using Pt 100 probe and a digital thermometer for the temperature control. Place the probe inside the reaction flask. For aqueous solutions, ultrapure water is recommended (18 MΩ at 25 °C, Millipore Milli-Q). All common solvents should be p.a. ACS grade.

2.1. CdSe Core Nanocrystal Components

1. UV/Vis spectrophotometer Beckman DU-70 or similar instrument.
2. Dynamic light scattering (DLS) instrument, Zetasizer Nano ZS (Malvern Instruments).
3. Precision cells made of Quartz Supracil®, Type Nr 101-QS, 10 mm, Hellma.
4. Centrifuge for use with 50 mL centrifuge tubes and capable of 5,000 × *g*, e.g., Eppendorf 5804R.
5. Rotary evaporator, BUCHI rotavapor R-205 with vacuum controller V-800 and heating bath B-490.
6. Ultrasound bath, Branson 1200.
7. Vortex mixer.
8. pH meter.
9. High vacuum pump.
10. Heating mantle.
11. Magnetic stirrer.
12. Digital thermometer with Pt 100 sensor.
13. Glass Pasteur pipettes (230 mm).
14. Disposable syringe (1, 2, and 5 mL).

15. Polypropylene centrifuge tubes (50 mL).
16. Syringe needles (19-gauge stainless steel, 10 in., with Luer hub and deflecting tip).
17. 100 mL three-neck flask.
18. 10 mL two-neck flask.
19. Chloroform.
20. Methanol (puriss. p.a., Reag. ACS).
21. Cadmium oxide (99.99 %).
22. Trioctylphosphine oxide (TOPO, technical grade).
23. Oleic acid (Ph Helv).
24. 1-Hexadecylamine (HDA, tech. 90 %).
25. Selenium/trioctylphosphine (Se/TOP) solution: add 0.0632 g of selenium powder (99.5 %, powder, 100 mesh) into a 10 mL two-neck flask under an argon flow. Add 2.4 mL of trioctylphosphine (TOP, tech. 90 %) and sonicate the solution until a colorless optically clear solution is obtained.

2.2. CdSe/ZnS Core-Shell Nanocrystal Components

1. Centrifuge for use with 50 mL centrifuge tubes and capable of 8,000 × *g*, e.g., Eppendorf 5804R.
2. Rotary evaporator, BUCHI rotavapor R-205 with vacuum controller V-800 and heating bath B-490.
3. Ultrasound bath, Branson 1200.
4. High vacuum pump.
5. Heating mantle.
6. Magnetic stirrer.
7. Digital thermometer with Pt 100 sensor.
8. Glass Pasteur pipettes (230 mm).
9. Filter (PTFE membrane, 0.2 μm pore size, filter-∅: 15 mm).
10. Polypropylene centrifuge tubes (15 and 50 mL).
11. Single-use syringes (1, 2, and 5 mL).
12. Syringe needles (19-gauge stainless steel, 10 in., with Luer hub and deflecting tip).
13. 100 mL three-neck flask.
14. 10 mL two-neck flasks.
15. 250 mL glass funnel.
16. Chloroform.
17. Methanol (puriss. p.a., Reag. ACS).
18. Hexane.
19. Acetone.
20. TOPO (technical grade).

21. 1-Hexadecylamine (tech. 90 %).
22. Zinc oxide precursor solution. To prepare 0.1 M solution transfer 0.041 g of ZnO to a 10 mL two-neck flask, under argon flow and add 1.4 mL of oleic acid (Ph Helv). Heat the mixture at 300 °C until the solution is completely clear. Add 3.6 mL of 1-octadecene (tech. 90 %) and keep the final solution at 80 °C until the growth shell procedure is complete.
23. Elemental sulfur solution: The sulfur precursor solution, 0.1 M, is prepared by dissolving sulfur in 1-octadecene (tech. 90 %) at 180 °C. In a 10 mL two-neck flask, under argon, containing 0.016 g of elemental sulfur (99.98 %, powder), add 5 mL of 1-octadecene. Heat the solution until a clear solution is obtained, indicating that all sulfur is completely dissolved in the noncoordinating solvent. Maintain this solution at 80 °C until the shell growth procedure is complete.

2.3. MPA-Coated QDS Components

1. UV/Vis spectrophotometer, Beckman DU-70.
2. Centrifuge for use with 15 mL centrifuge tubes and capable of 7,500 × *g*, e.g., Eppendorf 5804R.
3. pH meter.
4. Glass Pasteur pipettes (230 mm).
5. Centrifugal filter devices (Vivapsin 6, MWCO 10 kDa PES membrane).
6. Polypropylene centrifuge tubes (15 mL).
7. 10 mL volumetric flask.
8. Glass vials with polypropylene snap cap (5 and 10 mL).
9. Polypropylene micro tubes (2 mL).
10. 250 mL Duran Shott Bottle.
11. Chloroform.
12. Sodium hydroxide (p.a. ACS grade).
13. Stock solution of MPA: In a 10 mL graduated volumetric flask, add 400 μL of MPA (99 %) (4.59 mmol) to 10 mL of MeOH followed by the addition of 500 mg of KOH.
14. 10 mM NaOH in deionized water (pH 12).

2.4. Optical Characterization Components

1. UV/Vis spectrophotometer, Beckman DU-70.
2. Fluorescence spectrophotometer SPEX Fluorolog or Fluoromax-4 (Horiba Jobin Yvon).
3. DLS instrument, Zetasizer Nano ZS (Malvern Instruments).
4. Precision cells made of Quartz Supracil®, Type Nr 101-QS, 10 mm.
5. Universal "Dip" cell (ZEN 1002, Malvern Instruments) (see Note 1).

6. Filters (PTFE membrane, 0.2 μm pore size, filter-Ø: 15 mm).
7. Rhodamine 6G in EtOH solution. Adjust concentration to achieve the optical density at the peak of approximately 0.1 or less.
8. 10 mM NaOH in deionized water (pH 12).
9. Single-use syringes (1 and 2 mL).
10. Disposable needles (Ø0.80 × 50 mm).

3. Methods

3.1. Synthesis of CdSe Core Nanocrystals

1. In a 100 mL three-neck flask, under an argon atmosphere, place 0.042 g of cadmium oxide (CdO) and 0.6 mL of oleic acid. Heat the mixture under stirring to 100 °C to obtain a yellow optically clear solution.
2. Then, remove the flask from the heat source and add 2 g of 1-hexadecylamine (HDA) and 2 g of TOPO. Heat the solution again, but this time to 280–300 °C. When all is completely dissolved, the temperature is stabilized at 280 °C and the mixture should be optically clear.
3. While keeping the mixture at 280 °C, in one single injection, add 2.4 mL of Se/TOP solution prepared previously (see Note 2).
4. After 2 min at 280 °C remove the heating mantle. When the temperature reaches 85 °C inject 20 mL of MeOH, to precipitate the nanoparticles. Leave the reaction mixture to cool to room temperature.
5. Transfer the nanoparticle suspension to five 50 mL polypropylene centrifuge tubes (10 mL in each tube). Add 20 mL of chloroform to each tube to force the precipitation of nanocrystals and shake the tubes by hand to mix the solutions. Leave the tubes to stand for 90 min at room temperature.
6. Centrifuge the tubes at 5,000 × *g* at room temperature for a period of 10 min and discard the cloudy supernatant.
7. Wash the concentrated pellet at the bottom of the tubes at least three times with 20 mL of MeOH, centrifuging at 5,000 × *g* for a period of 10 min each time.
8. Dry the final red-brownish pellet under vacuum and the nanocrystals can then be dispersed in chloroform or hexane for further processing. Otherwise, they can be stored as a powder in the dark at 4 °C for several months.
9. This procedure typically yields CdSe nanocrystals with the first absorption peak at 568 nm with a diameter of about 3.5 nm.

3.2. Synthesis of CdSe/ZnS Core-Shell Nanocrystals, Based on Successive Ion Layer Adsorption Reaction Method

For shell growth, the SILAR method was found to be the best approach. This technique is based on alternating injections of precursor's solutions containing metal oxides such as ZnO and elemental sulfur into the solution containing the CdSe-core nanocrystals. Using this method, layers of ZnS can be grown successively around the core. The noncoordinating solvent, 1-octadecene is the most appropriate solvent for the growth of high-quality nanocrystals in general. It has a low melting point, high boiling point, low cost, low toxicity, low reactivity to precursors and excellent solvation power for many compounds at elevated temperatures (7, 8). 1-Hexadecylamine (HDA) is used as ligand for the core-shell nanocrystals because alkylamines have proven to be very good ligands for highly luminescent plain CdSe/ZnS core-shell nanocrystals (9).

1. Add 3 mL of 1-octadecene and 1 g of 1-hexadecylamine to a 100 mL three-neck flask. Heat the flask under vacuum at 70 °C during 1 h to remove any traces of water and then cool to room temperature.
2. Disperse ~30 mg of CdSe nanocrystals (from Subheading 3.1), in a polypropylene tube in 15 mL of chloroform and filter using a 0.2 μm PTFE membrane.
3. Place the chloroform suspension containing the CdSe nanocrystals in the flask and remove the chloroform and other volatiles by heating under vacuum at 70 °C for 30 min (see Note 3).
4. Remove the vacuum and heat the system under an argon atmosphere to 245 °C. Maintain the reaction at this temperature for the duration of the shell growth procedure.
5. To calculate the amount of Zn- and S-precursor for the growth of the first monolayer, assume that the surface of the CdSe-cores consists equally of Se and Cd atoms. For a solution containing CdSe nanocrystals with 3.5 nm (1.01×10^{-4} mmol), 0.47 mL of Zn and S precursor solutions should be added for the first layer; 0.64 mL of each injection solution for the second layer; 0.84 mL of each injection solution for the third layer; 1.064 mL of each injection solution for the fourth layer, and 1.32 mL of each injection solution for the fifth layer (see Note 4).
6. After the final injection, incubate the reaction mixture for 30 min at 260 °C.
7. Cool the reaction to room temperature and add 30 mL of hexane. Transfer the solution to a 250 mL glass extraction funnel and remove the unreacted compounds and possible byproducts by successive MeOH extractions (20 mL, 6–8 washes) (see Note 5).

8. Transfer the hexane/ODE phase containing the nanoparticles to a 50 mL centrifuge tube and wash with acetone. Centrifuge at 8,000 × *g* at room temperature for 10 min (see Note 6).
9. Suspend the final pellet in about 15 mL of hexane and precipitate the nanoparticles with the addition of the same volume of MeOH. Centrifuge at 8,000 × *g* at room temperature for 10 min.
10. Discard the colorless supernatant and dry the final red-brownish pellet under high vacuum for 1 h. The final QDs can be dispersed in chloroform or hexane for further processing. Otherwise, it can be stored as a powder in the dark at 4 °C for several months.

3.3. Water-Soluble QDs: QDs Capped with 3-Mercaptopropionic Acid

The MPA-coated QDs are obtained by the phase transfer method (4, 10). When reacted with ZnS-capped CdSe QDs in chloroform, the mercapto-group binds to the Zn atom and the polar carboxylic acid group renders the QDs water-soluble. The free carboxyl group is also available for covalent coupling to various biomolecules (such as proteins, peptides, and nucleic acids) by cross-linking to reactive amine groups (6).

1. Add 10 mg of CdSe/ZnS (TOPO/HDA) QDs to a 5 mL glass vial and disperse in about 1.5 mL of chloroform to give a high concentration solution (optical density, OD = 1.5). Then, add 200 μL of a fresh stock solution containing MPA, close the vial with a snap cap and manually shake for 2 min (see Note 7).
2. Add 1.5 mL of NaOH solution 10 mM (pH 12) to the flocculate solution and manually shake it again. Two phases appear which correspond to the transfer of QDs from the organic to the aqueous phase.
3. For the complete phase separation, transfer the two-phase solution to a 15 mL polypropylene tube and centrifuge at 5,500 × *g* at room temperature for 5 min. Recover the aqueous phase (colored phase). Repeat steps 2 and 3 to ensure the total collection of water-soluble QDs.
4. Purify and concentrate the aqueous solution containing the MPA-capped CdSe/ZnS QDs using a Vivapsin 6 tube (cutoff 10 kDa) at 7,500 × *g*, at room temperature for 15 min and wash (three times) with NaOH solution 10 mM (pH 12).
5. Disperse the concentrated sample in 1 mL of NaOH solution 10 mM (pH 12) affording an orange optically clear solution. The sample can be stored in the dark at 4 °C (see Note 8).

3.4. Characterization of CdSe/ZnS and Corresponding MPA-QDs

3.4.1. Optical Characterization

Record UV–Vis absorption spectra using a spectrometer and photoluminescence (PL) spectra with a spectrofluorometer. Quantum yields (QY) can be measured relative to Rhodamine 6G in ethanol solution with excitation at 530 nm. Solutions of QDs in chloroform (CdSe/ZnS QDs) or water (MPA-QDs) and dye in ethanol are optically matched at the excitation wavelength ($\lambda = 530$ nm). Fluorescence spectra of QDs and dye are measured under identical spectrometer conditions. The optical density at the peak should be maintained below 0.1, and the integrated intensities of the emission spectra, corrected for differences in index of refraction and concentration, should be used to calculate QY using the expression (see Note 9):

$$QY_{QD} = QY_{Dye} \times (abs_{Dye}/abs_{QD}) \times (\text{peak area}_{QD}/\text{peak area}_{Dye}) \times (\eta_{QD\,solvent})^2/(\eta_{Dye\,solvent})^2$$

where $QY_{Dye} = 0.95$ (the quantum yield of the solution containing Rhodamine 6G dye in EtOH), "abs" is absorbance of the solutions, "peak area" = corrected areas under the emission peaks, and η is average of refractive indexes of QDots and the dye solvents.

1. Using standard 10 mm path length Quartz cuvettes, record the UV–Vis absorbance spectra of the CdSe@ZnS QDots in chloroform, MPA-QDots in water and Rhodamine 6G (dye) in ethanol. The optical density for both samples and dye at excitation wavelength should be below 0.1.
2. Record the emission spectra at 530 nm of the solutions prepared in step 1 in the 10 mm fluorescence cuvettes (Fig. 1).
3. Correct the fluorescence spectra obtained in step 2 for both samples and dye and calculate the integrated fluorescence intensity, that is, the area of the fluorescence spectra, using the appropriate spectrometer software.
4. Calculate the Quantum Yield using integrated intensities of the emission spectra, corrected for differences in index of refraction and concentration. Use the formula shown above (see Note 9).

3.4.2. Concentration Determination of QD Solutions

The determination of QD concentration in solution, water, or a given medium is not straightforward (see Note 10). Measurements of optical density can in principle be used to find molar concentrations if there are established extinction coefficients for these materials. However, the published extinction coefficients for quantum dots vary by more than an order of magnitude. These discrepancies may arise in part from the sensitivity of absorption coefficients to both quantum dot diameter and composition. According to some literature, the absorption spectrum method seems to be the most practical and convenient way to determine the particle concentrations. Thus, the optical density in water is measured and an

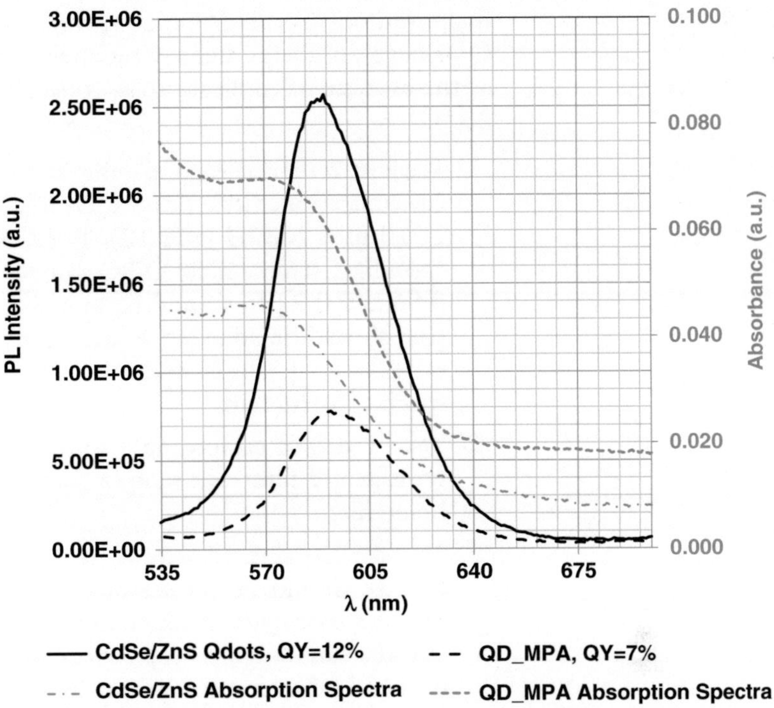

Fig. 1. Absorbance and fluorescence spectra. Absorbance spectrum of CdSe/ZnS core-shell (QD585) in $CHCl_3$ and MPA-coated QDs in water (*gray lines*). Fluorescence spectra with absorption normalized at the excitation of QD585 ($\lambda = 530$ nm) in $CHCl_3$ (QY = 12 %) and in water after ligand exchange with MPA (QY = 7 %) (*black lines*).

absorption coefficient reported in the literature (see Note 11) for CdSe core QDs is applied (to calculate the concentration using the Lambert–Beer's Law, $A = \varepsilon bC$, where A is the optical density, b is the light path length, C is the core/shell QD concentration, and ε is the extinction coefficient) (see Note 12).

1. Resuspend ~10 mg of CdSe core QDs in 1.5 mL of chloroform.
2. Using 10 mm standard path length Quartz cuvettes, record the UV–Vis absorbance spectra of the solution prepared in step 1.
3. From the UV–Vis spectra obtained in the previous step, take the absorbance value of the highest absorption wavelength (absorption maximum) between 400 and 700 nm and determine NPs concentration (see Note 12).

3.4.3. Dynamic Light Scattering Measurements of Hydrophobic and Hydrophilic CdSe/ZnS QDs

Light scattering analysis is performed using a dynamic light scatterer. With this equipment, it is possible to analyze the hydrodynamic diameter (HD) and zeta potential (ξ) of synthetic lipophilic CdSe/ZnS QDs and their corresponding hydrophilic variant. HD is obtained from number-weighted size distribution analysis and reported as the mean of triplicate measurements. ξ-Potential is measured only for hydrophilic QDs, because synthetic CdSe/ZnS QDs have no charge density on their surface. Values are reported as the average of triplicate runs consisting of 20 measurements at 25 °C.

1. Prepare 2 mL of each sample in the appropriate solvent and concentration: CdSe/ZnS QDs in chloroform and MPA-QDs in 10 mM NaOH (pH 12). For CdSe/ZnS and their corresponding hydrophilic QDs, sample concentrations between 0.06 and 0.3 μM should be prepared. The samples concentration should be adjusted to accommodate the scattering properties of your sample and/or the optical requirements of your specific instrument.
2. Take 1 mL of each sample solution using a 1 mL disposable syringe and filter the sample through a 0.2 μm PTFE membrane filter.
3. Transfer the filtered sample into the cuvette for analysis. The cuvette should be filled slowly to avoid the formation of air bubbles. For hydrodynamic diameter measurements, close the cuvette with supplied cap before analysis to avoid contamination and solvent evaporation. In zeta potential measurements, the cap is replaced by the universal "dip" cell which has to be

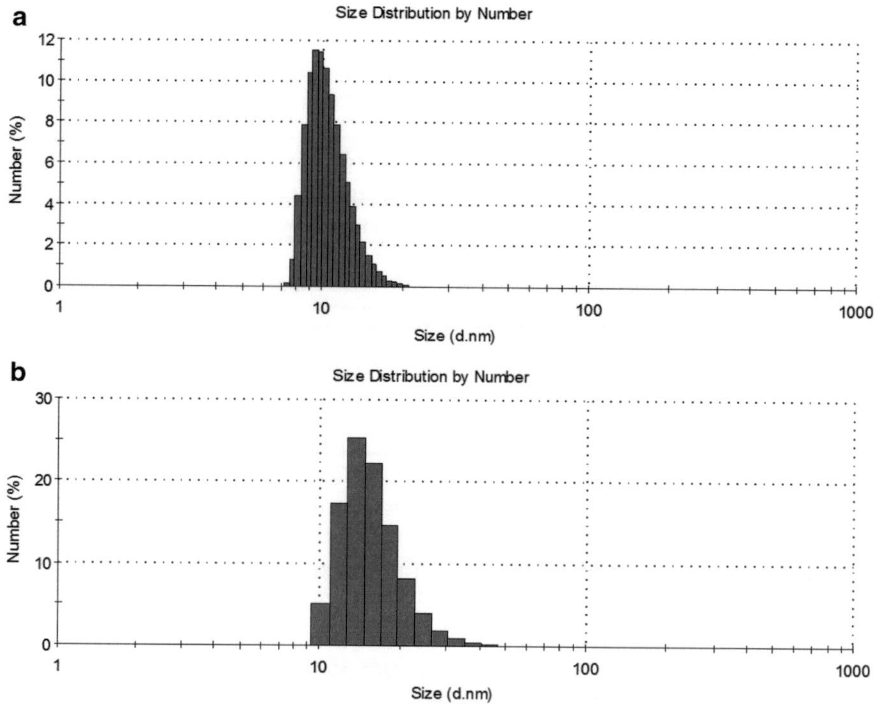

Fig. 2. Dynamic light scattering measurements. Representative dynamic light scattering histogram of (**a**) CdSe/ZnS QDs in $CHCl_3$ and (**b**) after ligand exchange with MPA, giving HD = 9.3 and 13.5 nm respectively.

inserted inside the cuvette according to the manufacturer's instructions (see Note 13).

4. Place the cuvette correctly in the sample holder (quartz windows should be facing the incident beam and detector).
5. Perform three measurements per sample at 25 °C (±1 min) (during 1 min/20 runs for size and zeta potential measurements respectively) with an equilibration time of 2 min.
6. To analyze the obtained data, use the instrument's software to evaluate parameters such as autocorrelation data, size distribution, count rate, and zeta potential values (Fig. 2).

4. Notes

1. 10 mm precision cells made of quartz should be used to ensure that the optimum signal is achieved in size and zeta potential measurements and due to their high resistance to aqueous and nonaqueous solutions. Therefore, an additional universal "dip" cell is required (according to instrument manufacturer's specifications) to measure zeta potential of the aqueous samples.
2. After the addition of the Se/TOP solution, there is always a decrease in the temperature of the mixture inside the flask and could reach 255–265 °C. In this case, you should wait until the temperature stabilizes again to 280 °C to control the crystal growth time.
3. As the system is under high vacuum, the temperature should not exceed 70 °C in order to avoid possible evaporation/ loss of the mixture of solvent/ligand 1-octadecene/1-hexadecylamine from the reaction vessel.
4. With fast stirring in place, there is no need for dropwise injections. A period of 10 min between each addition is sufficient for the reaction to be completed. The whole procedure for covering CdSe cores with five monolayers takes about 1 h 30 min.
5. In the extraction procedure, the undesired products are soluble in MeOH. Therefore this phase should be discarded, keeping the hexane/ODE phase which contains the nanocrystals.
6. This step is important for the elimination of excess ODE. This noncoordinating solvent can be retained by the nanoparticles during the MeOH washes. If it is not removed completely, it may appear on the walls of the tube in the form of a pale yellow powder. Washing with acetone helps to eliminate ODE, which will be discarded in the yellow cloudy supernatant. Repeat this wash four times for a complete removal of ODE.

7. With the addition of MPA stock solution, the chloroform solution may flocculate and became turbid. The solution is not optically clear at a pH = 5–7, possibly due to inter-particle hydrogen bonding between the carboxyl functions of the ligands.
8. With time, the functionalized MPA-QDs can form aggregates. To remove them, centrifuge the sample in 2 mL polypropylene tube at 2,000 × *g* at room temperature for 2 min. Recover the supernatant and store again in the dark at 4 °C for further use. In this case, the new concentration has to be determined.
9. This formula is from ref. (11).
10. The published extinction coefficients for QDs can vary by more than an order of magnitude. Some authors (12) measure the optical density of the core-shell-structured QDs in water applying an absorption coefficient that is the average of three reported in the literature (13–15). In this case, a large error on the extinction coefficient could exist and the optical density can at best give concentrations that are only accurate to 50 %.
11. Accurate determination of the extinction coefficient for QDs is described in ref. (13). The results show that the extinction coefficient of nanocrystals in the strong quantum confinement size regime increases as their size increases approximately as a square to a cubic function. Furthermore, the integrated ε of the first absorption peak seemed to be independent of temperature, nature of the surface capping groups, the refractive index of the solvents, the photoluminescence, and the methods employed for the synthesis of the QDs.
12. For example, for CdSe core nanocrystals with the first excitonic absorption peak at 568 nm, according to a sizing curve from Peng and colleagues (13) the typical size is around 3.5 nm and consequently the size-dependent extinction coefficient of these nanocrystals is about 1.5×10^5 M/cm. If the solution measured has $A = 1.5$ in a 1 cm path length,

$$c = A / \varepsilon = 1.5 / 1.5 \times 10^5 = 1 \times 10^{-5}\ \text{M} = 10\ \mu\text{M}$$

13. For both measurements, inspect the cuvette to ensure that no air bubbles are clinging to the optical window area or interfering with the metal electrodes of "dip" cell in zeta potential analysis.

Acknowledgements

This work was supported by a grant to A.S.M., the grant #PEST-OE/EQB/LA004/2011 and the project NTec/SQA/0131/2007 from the Fundação para a Ciência e Tecnologia.

References

1. Liu W, Howarth M, Greytak AB, Zheng Y, Nocera DG, Ting AY, Bawendi MG (2008) Compact biocompatible quantum dots functionalized for cellular imaging. J Am Chem Soc 130:1274–1284
2. Aldana J, Wang YA, Peng X (2001) Photochemical instability of CdSe nanocrystals coated by hydrophilic thiols. J Am Chem Soc 123:8844–8850
3. Li JJ, Wang A, Guo W, Keay JC, Mishima TD, Johnson MB, Peng X (2003) Large-scale synthesis of nearly monodisperse CdSe/CdS core/shell nanocrystals using air-stable reagents via successive ion layer adsorption and reaction. J Am Chem Soc 125: 12567–12575
4. Xie R, Kolb U, Li J, Bashé T, Mews A (2005) Synthesis and characterization of highly luminescent CdSe-core CdS/$Zn_{0.5}Cd_{0.5}S$/ZnS multishell nanocrystals. J Am Chem Soc 127: 7480–7488
5. Wang Q, Xu Y, Zhao X, Chang Y, Liu Y, Jiang L, Sharma J, Seo D, Yan H (2007) A facile one-step in situ functionalization of quantum dots with preserved photoluminescence for bioconjugation. J Am Chem Soc 129: 6380–6381
6. Chan WCW, Nie S (1998) Quantum dot bioconjugates for ultrasensitive nonisotopic detection. Science 281:2016–2018
7. Yu WW, Peng X (2002) Formation of high-quality and other II–VI semiconductor nanocrystals in noncoordinating solvents: tunable reactivity of monomers. Angew Chem Int Ed 41:2368–2371
8. Battaglia D, Peng X (2002) Formation of high quality InP and InAs nanocrystals in a noncoordinating solvent. Nano Lett 2:1027–1030
9. Peng X, Schlamp MC, Kadavanich AV, Alivisatos AP (1997) Epitaxial growth of highly luminescent CdSe/CdS core/shell nanocrystals with photostability and electronic accessibility. J Am Chem Soc 119:7019–7029
10. Gill R, Freeman R, Xu J, Willner I, Winograd S, Shweky I, Banin U (2006) Probing biocatalytic transformations with CdSe-ZnS QDs. J Am Chem Soc 128:15376–15377
11. Williams ATR, Winfield SA, Miller JN (1983) Relative fluorescence quantum yields using a computer-controlled luminescence spectrometer. Analyst 108:1067–1071
12. Yu WW, Chang E, Falkner JC, Zhang J, Al-Somali AM, Sayes CM, Johns J, Drezek R, Colvin VL (2007) Forming biocompatible nonaggregated nanocrystals in water using amphiphilic polymers. J Am Chem Soc 129:2871–2879
13. Yu WW, Qu L, Guo W, Peng X (2003) Experimental determination of the extinction coefficient of CdTe, CdSe and CdS nanocrystals. Chem Mater 15:2854–2860
14. Leatherdale CA, Woo WK, Mikulec FV, Bawendi MG (2002) On the absorption cross section of CdSe nanocrystal quantum dots. J Phys Chem B 106:7619–7622
15. Striolo A, Ward J, Prausnitz JM, Parak WJ, Zanchet D, Gerion D, Milliron D, Alivisatos AP (2002) Molecular weight, osmotic second virial coefficient and extinction coefficient of colloidal CdSe nanocrystals. J Phys Chem B 106:5500–5505

Chapter 11

In Vivo, Site-Specific, Covalent Conjugation of Quantum Dots to Proteins via Split-Intein Splicing

Anna Charalambous, Maria Andreou, Ioanna Antoniades, Neophytos Christodoulou, and Paris A. Skourides

Abstract

The ability to target proteins with nanostructures and/or nanodevices in vivo is important for understanding and controlling their biological function. Quantum dots (QDs) serve as an ideal model nanostructure due to their superior optical properties that permit visual confirmation of in vivo targeting and localization and due to their potential as a bio-imaging tool. Here, we describe the site-specific covalent conjugation of quantum dots to target proteins in vivo using an intein-based method. Experimental procedure includes the following steps: (1) fusion of Pleckstrin-homology (PH) domains with the N-terminus half of a split intein (I_N); (2) conjugation of the C-terminal (I_C) intein-derived peptide to streptavidin-coated QDs in vitro; and (3) in vivo expression of PH–I_N following microinjection of PH–I_N RNA and I_C–QDs into *Xenopus* embryos. Intein-splicing results in covalent conjugation of QDs with the C-terminus of the PH domain without interfering with protein localization or function. Such produced QD–PH conjugates could be monitored in real time within live embryos.

The use of near infrared-emitting QDs allows monitoring of QD conjugates within the embryo at depths where EGFP is undetectable demonstrating the advantages of QDs for this type of experiment. The reported approach therefore allows the covalent conjugation of QDs or other similar nanostructures to proteins in vivo and the targeting of such nanomaterial to any intracellular compartment or signaling complex within the cells of the developing embryo.

Key words: Quantum dots, Inteins, Protein trans-splicing, Site-specific conjugation, In vivo targeting

1. Introduction

Characterization of proteins in the context of a living organism is of crucial importance for a complete understanding of their function. Although organic dyes and fluorescent proteins have been used extensively for single molecule imaging, they generally

Mikhail Soloviev (ed.), *Nanoparticles in Biology and Medicine: Methods and Protocols*, Methods in Molecular Biology, vol. 906, DOI 10.1007/978-1-61779-953-2_11,

emit weak signals that typically bleach in <10 s. Nanometer-scale semiconductor crystallites or quantum dots (QDs) are an alternative probe since, in contrast to traditional fluorophores, QDs act as robust, broadly tunable nanoemitters that can be excited by a single light source, offer extremely high fluorescence intensity, wide excitation spectra, narrow and tunable emission spectra, large stokes shift and resistance to photobleaching (1). Moreover, there is currently a limited number of fluorescent proteins with emission in the NIR region. Despite claims of improved optical properties they are still far from optimal in terms of brightness and photostability, in comparison to NIR-QDs (2–4). The NIR region of the spectrum (700–950 nm) is ideal for imaging in tissues because light scattering diminishes with increasing wavelength, and hemoglobin electronic and water vibrational overtone absorptions approach their minimum over this spectral domain. Furthermore living tissue autofluorescence also reaches a minimum at this range and the fluorescent signal can, even in the case of organic fluorophores, be detected in vivo at subnanomolar quantities and at depths sufficient for experimental or clinical imaging (5). The full potential of QDs is yet to be realized however because of limitations related to their relatively large size (typically 20–30 nm for biocompatible red-emitting QDs) (1), multivalency, and the inability to genetically encode them. The first two issues have been resolved to a large extent with the synthesis of new types of QDs with much smaller hydrodynamic radii (6) and monovalent nanocrystals (7). The third issue remains elusive and therefore addressed in this work using a simple intein-based method that allows the site-specific conjugation of QDs to protein targets in vivo, effectively overcoming the requirement to genetically encode QDs for tagging target proteins. In addition, this approach can be used to conjugate other nanostructures or nanodevices to target proteins and as a result to any intracellular compartment or protein signaling complex within the cell.

Existing methods of QD–protein conjugation generally use either random chemical coupling with reactive amino acids (e.g., $-NH_2$, $-COOH$, $-SH$) on the protein surface or noncovalent complexation mediated by electrostatic interactions and ligand recognition. A survey of site-specific bioconjugation methods led us to the intein-mediated ligation system. Inteins are polypeptide sequences that are able to self-excise, rejoining the two flanking extein sequences by a native peptide bond (8–10). Inteins catalyze the splicing reaction through formation of an active thioester intermediate and have been widely used for in vitro protein semisynthesis (9), segmental isotopic labeling (11), and in vivo protein cyclization (12). This is the first time however that this approach has been used successfully in a vertebrate embryo to label proteins with QDs (13).

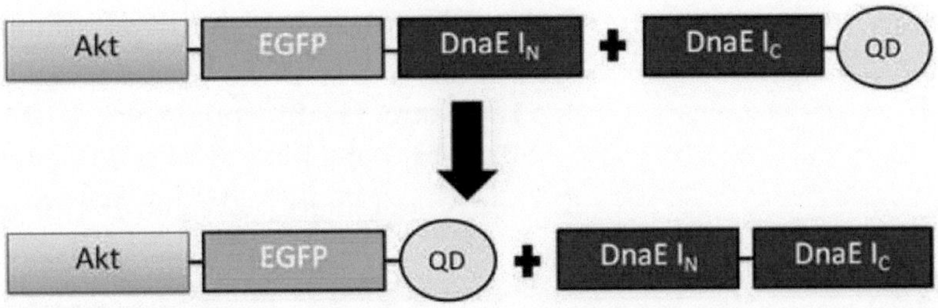

Fig. 1. In vivo conjugation of QDs to Akt–PH–EGFP via intein-mediated protein splicing. Schematic representation of site-specific intein-mediated conjugation of QDs to target protein.

We selected to illustrate our methods by tagging the PH domains of two proteins Akt and Btk with the N-terminus half of a split intein (I_N). These were chosen due to their ability to translocate to the cell membrane upon PIP_3 production by PI3K (14) and would thus provide a clear visual confirmation of the conjugation in the intact embryo. A synthetic peptide encoding the complementary C-terminus half of the intein (I_C) is biotinylated and conjugated in vitro to streptavidin-coated QDs. Following their co-injection together with RNAs encoding EGFP tagged Akt–PH–I_N or Btk–PH–I_N into *Xenopus* embryos, the in vivo association of the intein halves in the cytosol triggers protein trans-splicing, resulting in the ligation of the QD to the target protein through a peptide bond (Fig. 1).

We show in situ labeling of the PH domain of Akt with QDs using the above described intein-mediated ligation system. Our approach offers significant advantages over traditional labeling using organic fluorophores. We also show that localization of the PH–QD conjugates can be monitored in real time in the developing *Xenopus* embryo and that the QD tags do not affect the primary function of PH domains which is to recognize PIP_3; their ability to translocate from the cytosol to the plasma membrane is not compromised.

2. Materials

Prepare all solutions using analytical/molecular biology grade reagents. Prepare and store all reagents at 4 °C unless indicated otherwise. Diligently follow all waste disposal regulations when disposing waste materials.

2.1. I_c Peptide–QD Conjugates

1. Biotinylated Intein (I_C) peptide (I_C-Biotin) (HOOC)-MVKVI GRRSLGVQRIFDIGLPQDHNFLLANGAIAANCFDYKD DDDK(Ahx-Biotin)G-NH_2 (see Note 1).
2. Dimethyl sufoxide (DMSO): Molecular Biology Grade.

3. 1× PBS (phosphate buffered saline): 10 mM sodium phosphate, pH 7.4, 0.9 % NaCl.
4. QDot–streptavidin conjugates: 1 μM solution in 1 M betaine, 50 mM borate, pH 8.3 with 0.05 % sodium azide (Invitrogen).
5. Centrifugal filters (MW cutoff 100 kD).

2.2. Generation of mRNA Encoding N-Terminus of Intein Fused to the Akt–PH–EGFP

1. PCR machine.
2. Primers: IGpr61, IGpr63, Apr1, Apr3 (see Table 1).
3. Plasmids (used as template DNA for PCR reactions):
   - pJJDuet30 bearing DNA sequence encoding DnaE Intein (Uniprot: P74750).
   - pAkt–PH–EGFP–N1 bearing DNA sequence corresponding to Akt–PH–EGFP (Plasmid 18836 from Addgene).
4. 10× PCR Mix, which contains 22 U/mL recombinant DNA polymerase from *Thermococcus* species strain KOD, 66 mM Tris-SO_4 (pH 8.4), 30.8 mM $(NH_4)_2SO_4$, 11 mM KCl, 1.1 mM $MgSO_4$, 330 μM dNTPs.
5. T4 DNA ligase.
6. The pCS2++ vector bearing an Sp6 promoter.
7. Chemically competent bacteria, e.g., DH5a (Invitrogen).
8. Mini-prep kit for DNA extraction, e.g., NucleoSpin (Macherey-Nagel).
9. In vitro transcription: All plasmids are transcribed into capped RNA using Sp6 promoter kit (mMESSAGE mMACHINE® Sp6 Kit, Ambion). After the reaction, 1 μL of TURBO DNase (2 U/μL) treatment removes the template DNA. The mRNAs are purified using a purification kit.
10. UltraPure 10× TAE running buffer: A pH-adjusted blend of ultra pure grade Tris-acetate buffer and EDTA which, when diluted to a 1× working concentration, contains 40 mM Tris-acetate and 1 mM Na_2EDTA.

Table 1
Primer sequences

Primers	Sequence[a]	Restriction site
IGpr61-forward	AAG**GAATTC**AAGTTTGCGGAATATTGCCTCAGTTTTGG	EcoRI
IGpr63-reverse	AAG**CTCGAG**TTATTTAATTGTCCCAGCG	XhoI
Apr1-forward	AAG**ATCGAT**ATGAGCGACGTGGCTATTG	ClaI
Apr3-reverse	AAG**GAATTC**CTTGTACAGCTCGTCCATGCCGAG	EcoRI

[a] Highlighted sequences correspond to restriction enzyme site

11. Agarose: Certified free of RNAses, DNAses, and inhibitors.
12. Ethidium bromide.
13. 100 bp or 1 kb DNA ladder.
14. Gel Documentation Imaging system.

2.3. Embryos and Microinjections

1. *Xenopus laevis* eggs collected from induced spawning of adult females (Nasco), in vitro fertilized using testes from adult males.
2. 10× MMR: Add 58.44 g NaCl (0.1 M), 1.5 g KCl (1.8 mM),2.2 g $CaCl_2$ (2 mM), 1.2 g $MgCl_2$ (1 mM), 12 g HEPES (5 mM), 0.37 g EDTA (0.1 mM) to distilled water up to 1,000 mL and adjust pH to 7.8 with NaOH. Add gentamycin gel to a final concentration of 0.05 mg/mL.
3. 2 % cysteine: 2 g L-cysteine in 100 mL 0.33× MMR, pH 7.8 with NaOH.
4. 4 % Ficoll in 0.33× MMR: 4 g Ficoll 400 in 100 mL 0.33× MMR.
5. Human chorionic gonadotropin (hCG) 800 U/mL in PBS (Intervet).
6. 0.05 % benzocaine.
7. Leibovitz L15 medium.
8. FCS (fetal calf serum): heat-inactivated.
9. Gentamycin (50 mg/mL).
10. Petri dishes.
11. The embryos are injected with 300 pg RNA and 4–6 nL Intein (I_C) peptide–QD conjugates at the two- and four-cell stage. After injections, the embryos are cultured in 4 % Ficoll in 0.33× MMR until stage 8 and then cultured in either 0.1× MMR or 400 nM Wortmannin (for some experiments) at room temperature.

2.4. Imaging of Injected Embryos

1. Foil tray.
2. Sylgard 184 silicone elastomer kit (Dow Corning).
3. Razor blade.
4. 26×76 mm slides (Menzel Glaser).
5. Sylgard-coated slides: arrange slides in a foil tray, tightly packed. Prepare the Sylgard silicone elastomer according to the manufacturer's instructions and pour over the slides. The volume of Sylgard poured over the slides can be adjusted to produce slides of different thickness (~1 mm for gastrula stage embryos). Cure slides overnight at room temperature or for 2 h at 60 °C. Cut each slide with a razor blade and then cut well(s) into the cured Sylgard layer.

6. Cover slips.
7. Widefield epi-fluorescence microscope: Injected embryos are visualized using an upright Zeiss Axioimager Z1 equipped with QD and GFP filter sets (see Note 2), cooled monochromatic Zeiss Axiocam MR3 and the Axiovision software 4.6 for in vivo monitoring of GFP–QD fluorescence. Additionally, the Zeiss apotome structured illumination system was used to obtain optical sections.
8. Forceps.
9. Kimwipes.
10. 400 nM Wortmannin in PBS.

3. Methods

3.1. Preparation of I_C Peptide–QD Probe

1. The lyophilized biotinylated I_C peptide is dissolved in PBS containing 60 % DMSO to an initial concentration of 1 mM (see Note 3). It is then diluted in sterile PBS (1:20) to a final working concentration of 50 μM.
2. Mix 15 μL of 50 μM biotinylated I_C peptide with 15 μL of streptavidin-coated QDs (1 μM) (see Note 4).
3. Incubate at room temperature for 1 h.
4. Equilibrate a centrifugal filter device (cutoff: 100 kD) by adding 400 μL of 1× PBS and spinning at 8,000 × *g*, at 4 °C.
5. Add 360 μL of 1× PBS to the I_C peptide–QD conjugation reaction and spin at 8,000 × *g*, at 4 °C (see Note 5). Repeat three times until the volume of the I_C peptide–QD conjugate is concentrated down to 15 μL. Place probes at 4 °C until ready to inject.

3.2. Preparation of mRNA Encoding the Fusion Protein Akt–PH–EGFP–I_N

3.2.1. Cloning of a DNA Fragment Encoding Akt–PH–EGFP into the pCS2++ Vector (See Fig. 2)

1. Set up a PCR reaction to amplify the DNA fragment encoding Akt–PH–EGFP. Use 10 ng of the pAkt–PH–EGFP–N1 plasmid as template, 22.5 μL of PCR mix and each of the following primers, Apr1 and Apr3 (see Table 1) at a final concentration of 0.1 μM. Use the following PCR program: 95 °C for 5 min, 35 cycles each of: 95 °C for 15 s, 62 °C for 30 s, 68 °C for 70 s followed by 68 °C for 10 min and 4 °C until reaction termination.
2. Analyze the PCR product on a 1 % agarose gel and extract the 450 bp band corresponding to the N-terminus of the intein from the gel using a DAN extraction kit like Nucleospin extract II.

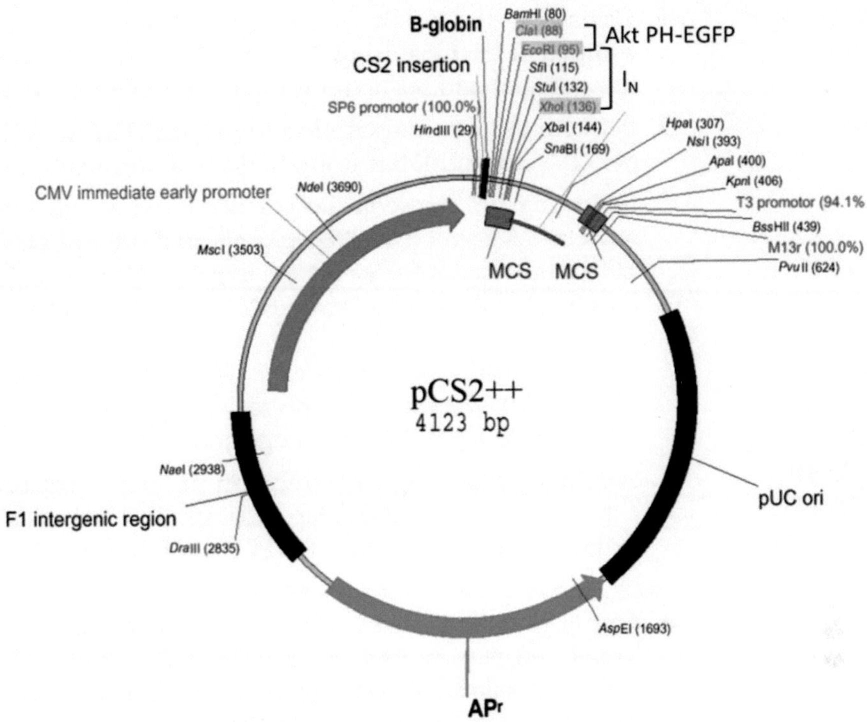

Fig. 2. Plasmid map of pCS2++ indicating restriction sites used to clone Akt–PH–EGFP and I_N. DNA fragments encoding Akt–PH–EGFP and I_N with five N-terminal extein residues (KFAEY) are inserted in frame into the pCS2++ plasmid bearing an Sp6 promoter. Reproduced from ref. (13).

3. Clone the PCR product into the multiple cloning site of pCS2++ plasmid using ClaI and EcoRI restriction enzymes.

3.2.2. Cloning of a DNA Fragment Encoding I_N into the pCS2++ Vector (See Fig. 2 and Note 6)

1. Set up a PCR reaction to amplify the DNA fragment encoding I_N with five N-terminal extein residues (KFAEY). Use 10 ng of the pJJDuet30 plasmid as template, 22.5 μL of PCR mix and each of the following primers, IGpr61 and IGpr63 (see Table 1) at a final concentration of 0.1 μM. Use the following PCR program: 95 °C for 5 min, 35 cycles each of: 95 °C for 15 s, 62 °C for 30 s, 68 °C for 40 s followed by 68 °C for 10 min and 4 °C until reaction termination.
2. Analyze the PCR product on a 1 % agarose gel and extract the 450 bp band corresponding to the N-terminus of the intein from the gel using a DAN extraction kit like Nucleospin extract II.
3. Clone the PCR product into the multiple cloning site of pCS2++ plasmid downstream of Akt–PH–EGFP using EcoRI and XhoI restriction enzymes.

3.2.3. Akt–PH–EGFP–I_N mRNA Preparation

1. Transform the newly constructed plasmid into chemically competent bacteria using 2 μL of the ligation reaction. Pick a few colonies and set up mini-preps for DNA extraction.
2. Perform in vitro transcription of capped RNA from linearized pCS2++ plasmid DNA using Sp6 RNA polymerase.
3. Following the transcription reaction, remove unincorporated nucleotides using a Nucleospin plasmid kit and resuspend in 100 μL distilled RNAse/DNase free water (see Note 7). Divide into 50 μL aliquots and store RNA at −80 °C until injection.

3.3. Embryo Isolation and Microinjection of mRNAs Encoding I_N Fused with Target Protein and I_C–QD Probe

1. Induce ovulation of female frogs (*X. laevis*) by injecting 800 U of hCG into the dorsal lymph sac. Keep female frogs at room temperature for 9–10 h until egg collection.
2. Sacrifice a male frog by immersion in 0.05 % benzocaine for 1 h. Isolate testes and keep in a sterile Petri dish in 90 % Leibovitz's L-15 medium, 10 % FCS and 50 μg/mL Gentamycin (at 4 °C).
3. The eggs collect in the sac near the cloaca so simultaneous lateral and vertical pressure should release them. Collect the eggs manually by massaging the belly with one thumb over a clean Petri filled with 0.33× MMR.
4. Fertilize eggs immediately after collection (see Note 8). Cut a small piece of testis with scissors and macerate it with forceps so it is evenly distributed throughout the media. Leave the embryos for about 20 min. The eggs should rotate within the vitelline membrane so the animal hemisphere faces upward.
5. About 2 h after fertilization, the first cleavage is visible. De-jelly the embryo for microinjection by adding 2 % cysteine in 0.33× MMR at pH 7.8 (with NaOH) for 3–5 min and then wash with 0.33× MMR. Keep the embryos in 0.33× MMR.
6. During injections, keep embryos in 4 % Ficoll in 0.33× MMR. Micro-inject both blastomeres of two-cell stage *Xenopus* embryos with 2 nL of the probe (I_C–$QDot_{585}$). Allow the embryos to develop to the four-cell stage and then micro-inject three out of four blastomeres with 200 pg RNA encoding the target protein (e.g., Akt–PH–EGFP–I_N) (see Note 9). Allow injected *Xenopus* embryos to develop to stage 10.

3.4. Embryo Imaging to Visualize In Vivo Conjugation of QDs to the PH Domain of Akt

1. Place stage 10 *Xenopus* embryos in Sylgard-coated slides in 0.1× MMR (see Note 10). In order to enable use of high power lenses with small working distance, the depth of the chamber should be close to 1 mm or less (the approximate diameter of the embryo).

2. Remove excess buffer so that the chamber is only slightly overfilled. Using forceps turn the embryos so that the injected area faces up.
3. Carefully place a coverslip over the embryos. Remove excess liquid using kimwipes, this will set the embryos and from this point on repositioning cannot be performed (see Note 11).
4. Image the embryos on an upright epi-fluorescence microscope equipped with a cooled monochromatic CCD camera and the suitable filter sets for QDs and GFP (see Notes 2 and 12). Use the lowest intensity excitation that will result in acceptable signal-to-noise ratio in order to minimize phototoxicity and photodamage. Use of the field diaphragm to limit excitation in the area of interest is also advisable if embryos will be imaged for long periods of time. In the case of Akt, embryos should be imaged before and 5 min after a 10 s exposure to UV (DAPI filter works well at excitation: 360/40), to demonstrate membrane translocation of QDs after PI3K activation and PIP3 production (see Note 13).
5. If optical sectioning is required, structured illumination can provide relatively good results when imaging superficial blastomeres in early embryos. In this case, elevated levels of excitation will be necessary in order to bring exposure times to reasonably small values since long exposure times using this method lead to poor image quality (see Note 14).
6. If using the PH domain of Akt as a target, specificity of membrane translocation can be demonstrated via Wortmannin inhibition of PI3K. Allow injected *Xenopus* embryos to develop to stage 10, incubate for 1 h in Wortmannin (final concentration 400 nM), place on a slide and image as above (see Note 15).

4. Notes

1. We have synthesized this 47 amino acid peptide, corresponding to C-terminal intein (I_C), on a 0.5 mmol scale on 4-methylbenzhydrylamine (MBHA) resin according to the in situ neutralization/HBTU activation protocol for Boc SPPS. In order to put a biotin at the C-terminus, it is necessary to add an extra amino acid, Lys, at the C-terminus. This Lys serves as a linking point for biotin as well as a spacer (Ahx 6 carbon inert linker) between the peptide and biotin. The peptide contains a cysteine protected with the NPyS group which

is added as the last amino acid in the synthesis. Following chain assembly, global de-protection and cleavage from the support is achieved by treatment with HF containing 4 % v/v pcresol, for 1 h at 0 °C. Following removal of the HF, the crude peptide product is precipitated and washed with anhydrous cold Et2O before being dissolved in aqueous acetonitrile (50 % B) and lyophilized. The crude peptide is purified by preparative HPLC using a linear gradient of 25–45 % B over 60 min. The purified peptide is characterized as the desired product by ESMS. Optionally, the peptide may be ordered from a variety of commercial peptide synthesis providers.

2. Commercially available bandpass sets optimized for the selected emission may be used with QDs. Such sets often have wide excitation filters centered around 400 nm, optimized for maximum excitation. However, it is preferable to use custom sets with suboptimal longer wavelength excitation to limit photodamage.
3. This peptide is relatively hydrophobic and thus difficult to dissolve in water alone. It should therefore be dissolved in a small amount of organic solvent such as DMSO. This should be added drop wise, followed by vortexing after every drop until the peptide dissolves. Note, if DMSO solvent is at all wet it will not freeze at −20 °C, so freeze at −30 °C. DMSO is hygroscopic and will absorb water if subjected to repeated freeze–thaw cycles. To minimize these issues, prepare multiple aliquots.
4. The peptide is used in ~50-fold excess relative to the streptavidin-coated QDs during conjugation. This is to ensure that all biotin binding sites on the streptavidin-coated QDs are bound with peptide. According to the manufacturers, commercially available QDs from Invitrogen are multivalent, bearing 4–10 streptavidin molecules/QD, giving 16–40 biotin binding sites/QD.
5. This step is performed to clear I_c peptide–QD conjugates from any unbound peptide. Given that the MW cutoff of the microcentrifugal device is 100 kD and given that the MW of the I_C peptide is insignificant compared to the MW of the streptavidin-coated QDs (250–550 kD depending on the number of streptavidin molecules bound), this method ensures that the unbound peptide is removed without any loss of I_c peptide–QD conjugate.
6. When cloning DNA fragment encoding I_N with five N-terminal extein residues (KFAEY) immediately downstream of the

Akt–PH–EGFP into the pCS2++ vector, one must ensure that all sequences are in frame.

7. Subsequent to in vitro transcription, the amount and quality of RNA are examined using a spectrophotometer and gel electrophoresis, respectively.
8. The excess 0.33 %× MMR has to be removed from the Petri dish containing the eggs when ready to fertilize.
9. The presence of EGFP on the PH domain allows the monitoring and comparison of the distribution of the QDs vs. the Akt–PH.
10. Sylgard-coated slides are very useful mounting chambers for imaging embryos on an upright microscope.
11. The coverslip should slightly depress the embryo creating a small flattened area. If the embryo requires repositioning after addition of the coverslip, this can be adjusted by sliding the coverslip.
12. Imaging NIR QDs requires that the IR blocking filter (in the case of Zeiss axiocam BG 40) is removed from the camera. Most cameras have such IR filters installed.
13. Previous studies have reported that UVB irradiation induces Akt activation and consequent translocation to the plasma membrane via a PI3K/PDK-dependent pathway as well as via Erks and p38 kinase through their downstream kinase, mitogen and stress-activated protein kinase Msk1 (15, 16). In agreement with these studies, upon exposure to UV light both the Akt–PH–EGFP and the QD conjugates translocated to the cell membrane within minutes, suggesting UV-induced activation of PI3K (Fig. 3a).
14. Using structured illumination is not advisable when imaging dynamic movement of proteins or fluorophores. This method requires acquisition of three frames per optical section. Acquisition times length is also compounded due to the presence of the grid at the position of the field diaphragm limiting the intensity of excitation. Any motion during the acquisition of the three frames will be lost and lead to poor image quality.
15. The translocation of Akt–PH–EGFP and the QD conjugates to the plasma membrane is completely eliminated by Wortmannin, a PI3K-specific inhibitor, suggesting that the observed translocation is PI3K-dependent (Fig. 3b).

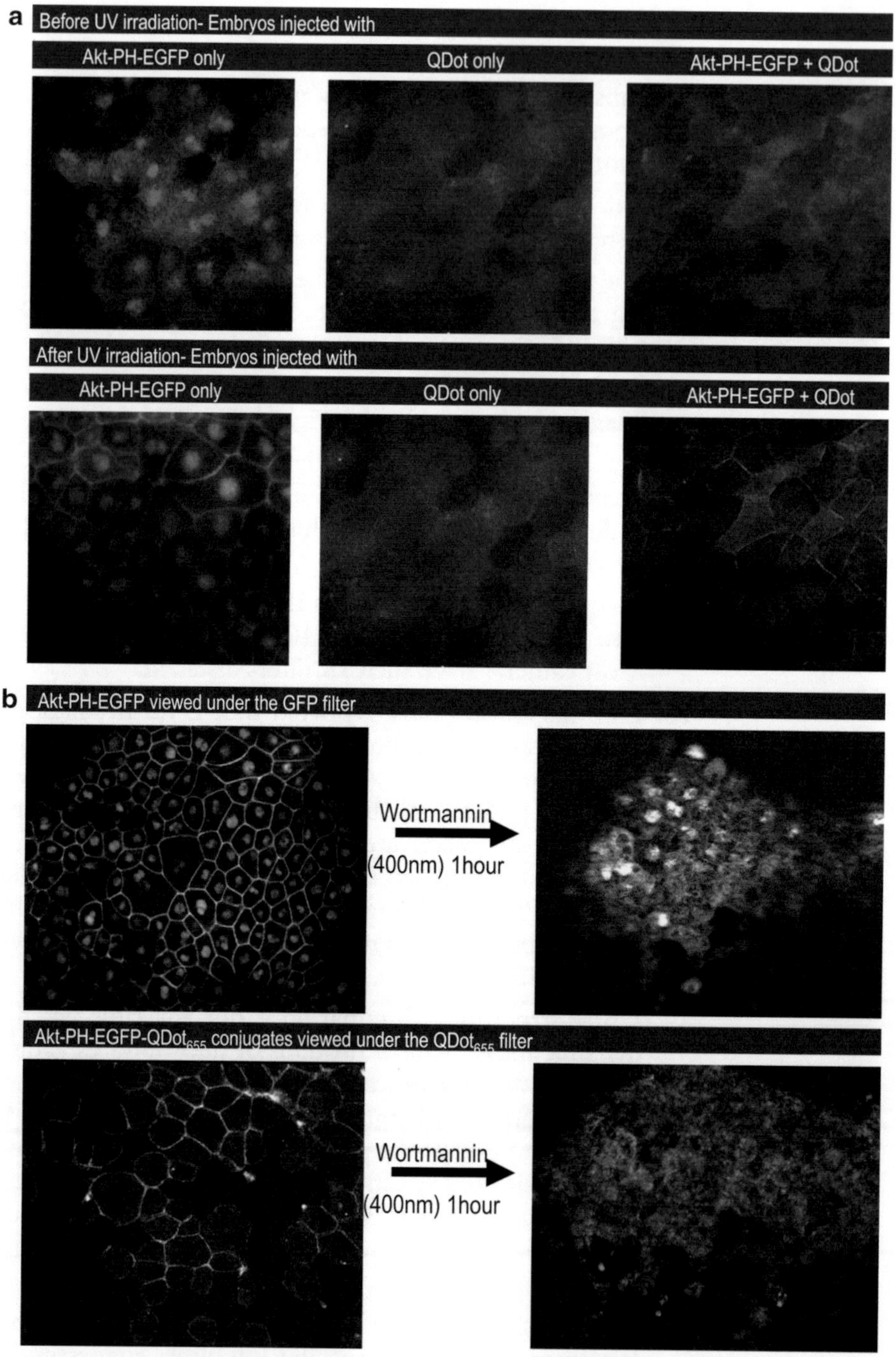

Fig. 3. UV-inducible and Wortmannin-sensitive translocation of QD–Akt–PH–EGFP conjugates to the membrane. (**a**) Akt–PH–EGFP protein fusions and Akt–PH–$QDot_{585}$ conjugates translocate to the cell membrane upon exposure of injected *Xenopus* embryos to UV radiation. Live *Xenopus* embryos injected as described were imaged on a Zeiss Axioimager to visualize the localization of Akt–PH–EGFP and Akt–PH–QD conjugate before and after exposure to UV radiation for 5 min. Both Akt–PH–EGFP and Akt–PH–QD conjugates translocate to the cell membrane following brief exposure to UV radiation. (**b**) Translocation of Akt–PH–QD conjugates ($QDot_{655}$) to the cell membrane is Wortmannin-sensitive. Live *Xenopus* embryos were imaged on a Zeiss Axioimager to visualize the localization of Akt–PH–QD conjugate before and after treatment with Wortmannin (400 nM), a PI3K inhibitor, for 1 h. The Akt–PH–QD conjugates become diffusely localized in the cytosol after treatment.

References

1. Michalet X, Pinaud FF, Bentolila LA, Tsay JM, Doose S, Li JJ, Sundaresan G, Wu AM, Gambhir SS, Weiss S (2005) Quantum dots for live cells, in vivo imaging, and diagnostics. Science 307:538–544
2. Shcherbo D, Merzlyak EM, Chepurnykh TV, Fradkov AF, Ermakova GV, Solovieva EA, Lukyanov KA, Bogdanova EA, Zaraisky AG, Lukyanov S, Chudakov DM (2007) Bright far-red fluorescent protein for whole-body imaging. Nat Methods 4:741–746
3. Stepanenko OV, Verkhusha VV, Kuznetsova IM, Uversky VN, Turoverov KK (2008) Fluorescent proteins as biomarkers and biosensors: throwing color lights on molecular and cellular processes. Curr Protein Pept Sci 9:338–369
4. Shu X, Royant A, Lin MZ, Aguilera TA, Lev-Ram V, Steinbach PA, Tsien RY (2009) Mammalian expression of infrared fluorescent proteins engineered from a bacterial phytochrome. Science 324:804–807
5. Rao J, Dragulescu-Andrasi A, Yao H (2007) Fluorescence imaging in vivo: recent advances. Curr Opin Biotechnol 18:17–25
6. Liu W, Choi HS, Zimmer JP, Tanaka E, Frangioni JV, Bawendi M (2007) Compact cysteine-coated CdSe(ZnCdS) quantum dots for in vivo applications. J Am Chem Soc 129: 14530–14531
7. Howarth M, Liu W, Puthenveetil S, Zheng Y, Marshall LF, Schmidt MM, Wittrup KD, Bawendi MG, Ting AY (2008) Monovalent, reduced-size quantum dots for imaging receptors on living cells. Nat Methods 5: 397–399
8. Evans TJT, Xu MQ (2002) Mechanistic and kinetic considerations of protein splicing. Chem Rev 102:4869–4884
9. Giriat I, Muir TW (2003) Protein semi-synthesis in living cells. J Am Chem Soc 125: 7180–7181
10. Muralidharan V, Muir TW (2006) Protein ligation: an enabling technology for the biophysical analysis of proteins. Nat Methods 3:429–438
11. Zuger S, Iwai H (2005) Intein-based biosynthetic incorporation of unlabeled protein tags into isotopically labeled proteins for NMR studies. Nat Biotechnol 23:736–740
12. Scott CP, Abel-Santos E, Wall M, Wahnon DC, Benkovic SJ (1999) Production of cyclic peptides and proteins in vivo. Proc Natl Acad Sci U S A 96:13638–13643
13. Charalambous A, Andreou M, Skourides PA (2009) Intein-mediated site-specific conjugation of quantum dots to proteins in vivo. J Nanobiotechnol 7:9
14. Varnai P, Bondeva T, Tamas P, Toth B, Buday L, Hunyady L, Balla T (2005) Selective cellular effects of overexpressed pleckstrin-homology domains that recognize PtdIns(3,4,5)P3 suggest their interaction with protein binding partners. J Cell Sci 118:4879–4888
15. Kabuyama Y, Hamaya M, Homma Y (1998) Wavelength specific activation of PI 3-kinase by UVB irradiation. FEBS Lett 441:297–301
16. Nomura M, Kaji A, Ma WY, Zhong S, Liu G, Bowden GT, Miyamoto KI, Dong Z (2001) Mitogen- and stress-activated protein kinase 1 mediates activation of Akt by ultraviolet B irradiation. J Biol Chem 276:25558–25567

Chapter 12

Evaluating the Potential of Quantum Dots for In Vitro Biological Studies: Effects on Gene Expression Using Microarray Analysis

Babu R. Prasad, Enda O'Connell, Terry J. Smith, Valérie A. Gérard, Yurii K. Gun'ko, and Yury Rochev

Abstract

Quantum dots have potential applications in the biomedical field and especially in bioimaging owing to their tunable fluorescent properties. Although many phenotypic studies have been carried out using QDs on different cell lines, only very few of them involved the analysis of the effect of QDs on gene expression. Here, we describe the application of microarray gene expression analysis for studying the differential expression of genes in the cells treated with QDs.

Key words: HPMEC cells, Quantum dots, RNA microarray, Gene expression

1. Introduction

Quantum dots (QDs) are nanocrystalline semiconducting fluorophores, typically having diameters of 2–10 nm, whose excitons are confined in all three dimensions spatially, a phenomenon known as quantum confinement (1). QDs in the size range of 2–6 nm are important for bioimaging and drug delivery as their physical dimensions are similar to those of biological macromolecules such as nucleic acids (DNA and RNA) and proteins (2). Quantum dots can be used as diagnostic and therapeutic tools, but they bear the significant drawback of a heavy metal composition, which has raised genuine concern of cytotoxicity (3–5). In an effort to combat this problem, much research has been conducted towards the identification of the mechanisms of QDs toxicity (6–12) and

Mikhail Soloviev (ed.), *Nanoparticles in Biology and Medicine: Methods and Protocols*, Methods in Molecular Biology, vol. 906, DOI 10.1007/978-1-61779-953-2_12, © Springer Science+Business Media, LLC 2012

into the ways of reducing it. One approach is the development of nontoxic coatings (13). QD size has also been proven to be a crucial factor affecting their uptake and adverse effects on cellular activity (14).

When studying the biological response and cytotoxicity of QDs, three major parameters should be taken into consideration: their size, QD surface modifications, and time of exposure to cells. Studying gene expression using microarrays provides a unique opportunity for assessing the complex issue of QD cytotoxicity by looking at the expression levels of a large number of genes, rather than investigating a small number of expressed markers by immunostaining. The procedure described below involves exposing human pulmonary microvascular endothelial cells (HPMEC-ST1.6R cells) (15) to gelatinated (13) and nongelatinated (16) QDs for extended periods of time, investigating QD's cellular activity using various assays like alamarBlue and PicoGreen and analyzing gene expression using microarrays. Because QDs are luminescent and may interfere with optical readouts, all cellular activity assays should include negative controls without cells (14). Microarray data analysis was carried out using Agilent GeneSpring GX v11 and subsequent functional analysis was performed using the internet-based analysis package DAVID (Database for Annotation, Visualization, and Integrated Discovery) (17–19).

2. Materials

2.1. Viability and Quantification of DNA from HPMEC-ST1.6R Cells Treated with QDs by AlamarBlue and PicoGreen Assays

1. HPMEC-ST1.6R cells (endothelial cell line derived by transfection of HPMEC with plasmids encoding the SV-40 large T-antigen (pSV3neo) and human telomerase (pC1.neo.hTERT)).
2. Gelatinated and nongelatinated red and orange quantum dots, with the following characteristics as shown in Table 1 (see Note 1). Store NP stock solutions at 4 °C to ensure the stability of QDs. For incubation with cells, dilute the stock solutions of QDs with nuclease-free water to 10^{-8} M final concentration. The diluted QDs may be stored up to 3–4 weeks at 4 °C.
3. Human pulmonary microvascular endothelial cell (HPMEC) medium: use medium 199 (M199) supplemented with 20 % fetal calf serum (FCS), 2 mM glutamax, 100 U/100 μg/mL penicillin/streptomycin, 50 μg/mL heparin, and 50 μg/mL endothelial cell growth supplement (ECGS).
4. Hanks balanced salt solution (HBSS).
5. Trypsin–EDTA (0.05 % Trypsin/0.53 mM EDTA in HBSS).
6. T-75 and T-175 tissue culture flasks (Sarstedt).

Table 1
Quantum dots

QD type	Surface	Absorbance peak (nm)	PL emission peak (nM)	Quantum yield (%)	Core size (nM)	Hydrodynamic diameter (nM)	Zeta potential (mV)
Red nongel	TGA	586	608	30	4.7	11.7	–30
Orange nongel	TGA	515	546	23	2.4	3.6	–27
Red gel	TGA-gelatine	579	610	34	4.5	14.3	–29
Orange gel	TGA-gelatine	522	550	29	2.6	5.3	–42

7. Nunc tissue culture-treated 48-well micro-plates
8. 96-Well flat tissue culture micro-plates.
9. AlamarBlue reagent (Biosource International): dilute alamarBlue with HBSS (1:10, v/v).
10. Quant-iT PicoGreen DNA assay kit (Molecular Probes, Invitrogen). All the solutions should be prepared according to the kit manual:
    - Solution 1 (1× TE buffer): add 8.075 mL of distilled water to the 425 μL 20× TE buffer.
    - Solution 2 (2 mg/mL DNA standard): add 588 μL of 1 × TE buffer to 12 μL of 100 mg/mL DNA.
    - Solution 3 (50 ng/mL DNA standard): add 586 μL 1 × TE buffer to 15 μL of 2 mg/mL DNA (from solution 2).
    - Solution 4 (Pico Green): add 5.174 mL 1× TE buffer (Solution 1) to 26 μL concentrated PicoGreen reagent. Make fresh just before use.
11. Perkin-Elmer Victor Multilabel Plate Reader.

2.2. Gene Expression in HPMEC-ST1.6R Cells Treated with QDs by Microarray Analysis

1. 1.5 mL nuclease-free tubes.
2. Nuclease-free water.
3. TRI reagent (Ambion).
4. RNeasy kit (Qiagen).
5. DNase set (Qiagen).
6. Chloroform.
7. Absolute ethanol.
8. RNA 6000 Nano Kit (Agilent Technologies).
9. Whole human genome microarray kit, 4x44K (Agilent Technologies). Five glass slides each formatted with four high-definition 44K arrays targeting 19,596 Entrez Gene RNAs.
10. Quick-Amp Labeling Kit (one-color) (Agilent Technologies)
11. Gene Expression Hybridization Kit (Agilent Technologies).
12. SureHyb Gasket Slides (Agilent Technologies).
13. Gene Expression Wash Pack (Agilent Technologies).
14. RNA spike in kit (Agilent Technologies) for labeling and hybridization quality control.
15. NanoDrop ND-1000 UV–VIS Spectrophotometer for measuring the total RNA concentration and cRNA concentration.
16. Agilent 2100 Bioanalyzer for measuring integrity of total RNA.
17. Agilent DNA Microarray Scanner for scanning the microarray slides.

18. GeneSpring GX v11 software (Agilent) for microarray data analysis.
19. DAVID for functional analysis of microarray gene expression data.

3. Methods

Cell viability and quantification of DNA of HPMEC-ST1.6R cells in response to QD treatment (see Note 2) is measured using alamarBlue and PicoGreen assays, respectively.

3.1. Viability and Quantification of DNA of HPMEC-ST1.6R Cells Treated with QDs by AlamarBlue and PicoGreen Assays

1. Seed HPMEC-ST1.6R cells at a density of 10^5 cells/cm^2 per well in 48-well micro-plates and incubate at 37 °C in 5 % CO_2 for 24 h. Prepare separate plates for alamarBlue and PicoGreen assays for three different time points (24, 48, and 72 h).
2. Add QDs to the cells (use red gelatinated, red nongelatinated, orange gelatinated, or orange nongelatinated) to 10^{-8} M final concentration (in our case this was 70 μL of QDs per 700 μL of medium in each well). Perform each treatment in triplicate (three separate wells). For positive control, use untreated cells. For negative control, use QDs suspended in cell culture medium in the absence of cells.
3. Incubate plates at 37 °C in 5 % CO_2 for 24, 48, and 72 h.
4. Perform alamarBlue and PicoGreen assays at 24, 48, and 72 h.

3.1.1. AlamarBlue Assay

1. After the cells have been incubated with QDs remove the medium and wash the wells with HBSS.
2. Add 200 μL of alamarBlue/HBSS solution to each well and incubate for 1 h.
3. Transfer 100 μL of reduced alamarBlue solution from each well into a fresh tissue culture 96-well micro-plate.
4. Measure absorbance at the lower wavelength of 550 nm and the higher wavelength of 595 nm with a measurement time of 5.0 s using Perkin Elmer Victor Multilabel Plate Reader.

3.1.2. PicoGreen Assay

1. After the cells have been incubated with QDs, remove the medium and wash the wells with HBSS.
2. Add 200 μL of deionized double-distilled water to each well and lyse the cells by performing three consecutive freezing/thawing cycles: freeze at −80 °C for 15 min and thaw at room temperature for 15 min.

Table 2
Calibration curve for PicoGreen assay

	DNA from Solution 2					DNA from Solution 3		
Constituents	**1,000**	**500**	**100**	**50**	**0**	**25**	**10**	**5**
DNA (μL)/ buffer (μL)	100/0	50/50	10/90	5/95	0/100	100/0	40/60	20/80

3. Construct the PicoGreen assay standard curve using final concentrations of the standards of 1,000, 500, 100, 50, 25, 10, 5, and 0 ng/μL in triplicates as shown in Table 2.
4. Transfer 100 μL of lysed cell solution from each well into a fresh tissue culture 96-well plate.
5. Add 100 μL of diluted PicoGreen reagent (solution 4) to each test well in the 96-well plate.
6. Analyze the plate by fluorescence using a Perkin Elmer Victor Multilabel Plate Reader (excitation at 485 nm and emission at 535 nm, with a read time of 1.0 s) and calculate the DNA levels in each sample using the standard curve.

3.2. Gene Expression of HPMEC-ST1.6R Cells Treated with QDs by Microarray Analysis

3.2.1. HPMEC-ST1.6R Cell Culture and Cell Harvesting

1. Transfer HPMEC-ST1.6R cells in a T75 flask, seeding 1×10^6 cells in 16 mL of HPMEC medium.
2. Culture the cells at 37 °C and 5 % CO_2 until fully confluent.
3. Detach and collect the cells with 1× EDTA Trypsin and split them into five T-75 flasks with 16 mL of HPMEC medium in each flask. Once cells have reached confluence, split again each of the five T-75 flasks into five T-175 flasks with 36 mL of HPMEC medium in each flask.
4. Culture the cells at 37 °C and 5 % CO_2 until fully confluent.
5. Detach and collect the cells with 1× EDTA Trypsin and split them into 25 T-75 flasks with 16 mL of HPMEC medium in each flask.
6. Divide the 25 T-75 flasks with HPMEC-ST1.6R cells with 16 mL of HPMEC medium in each flask into five sets of five biological replicates.
7. Culture the cells at 37 °C and 5 % CO_2 for 24 h.
8. Add QDs to a final concentration of 10^{-9} M to the different flasks sets as follows:

   Set 1 (five flasks): red gelatinated QDs.

   Set 2 (five flasks): red nongelatinated QDs.

   Set 3 (five flasks): orange gelatinated QDs.

Set 4 (five flasks): orange nongelatinated QDs.

Set 5 (five flasks): No QDs (control culture).

9. Incubate for 72 h at 37 °C and 5 % CO_2.

3.2.2. RNA Extraction

Extract total RNA using a combination of Trizol and Qiagen protocol for RNA extraction

1. Discard the cell culture medium and wash cells using HBSS medium.
2. Lyse the cells by adding 1 mL Trizol into the tissue culture flask with cells.
3. Using a cell scraper, lift all the homogenate and pipette the homogenized cells from the tissue culture flask.
4. Transfer the homogenate in 1.5 mL nuclease-free tubes.
5. Store the homogenate for 5 min at room temperature to allow complete dissociation of the nucleoprotein complexes
6. Heat at 37 °C for 10 min.
7. Vortex for 15 s.
8. Add 0.2 mL of chloroform per 1 mL of Trizol reagent.
9. Shake the tubes vigorously for 15 s.
10. Incubate for 15 min at room temperature.
11. Centrifuge at 12,000 × *g* for 15 min at 4 °C. This will separate the suspension into three phases—a lower red phenol–chloroform phase, an interphase, and an upper aqueous phase (colorless).
12. Remove the clear upper aqueous phase and mix with an equivalent volume (~650 μL) of 70 % ethanol in a 1.5 mL tube (see Note 3). To purify RNA follow the protocol provided by the kit manufacturer, e.g., we used RNeasy kit and associated DNase kit (see Note 4). This procedure will yield approximately 16 μg of total RNA.

3.2.3. RNA Concentration, RNA Purity, RNA Integrity

1. Determine RNA concentration using NanoDrop ND-1000 UV–VIS Spectrophotometer or similar instrument. Blank the baseline of the spectrometer with water. Measure the absorbance at 260 nm (1 unit of A_{260} = 40 ng/μL of RNA) and at 280 nm.
2. Calculate the ratio of absorbance at 260 and 280 nm, i.e., A_{260}/A_{280}. This provides an estimate of the purity of RNA with respect to protein contamination as shown in Table 3, which absorbs light at 280 nm (see Note 5).
3. Determine RNA integrity according to Protocol of Agilent RNA 6000 Nano Assay kit. Alternatively, RNA integrity can be determined using denaturing gel electrophoresis. Figure 1

Table 3
Correlation between A_{260}/A_{280} and RNA purity

A_{260}/A_{280}	RNA purity (%)
2.00	100
1.98	90
1.97	80
1.94	70
1.91	60
1.87	50
1.81	40
1.73	30
1.59	20
1.32	10

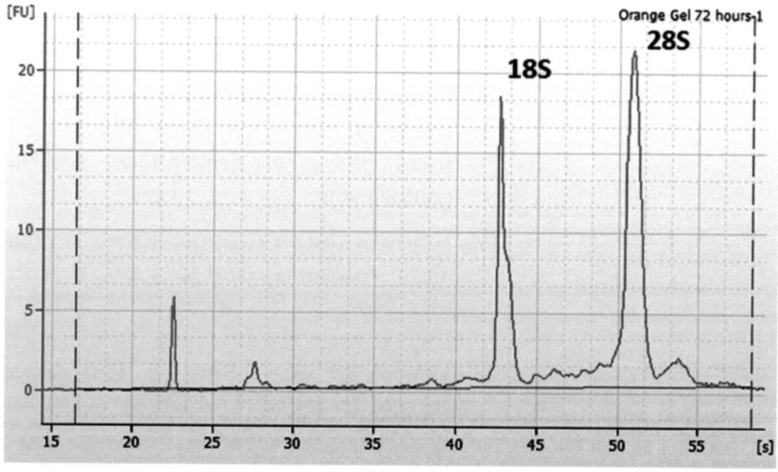

Fig. 1. The electropherogram of total RNA. Marker peak, Small interfering RNA (siRNA) and the two ribosomal peaks corresponding to 18S and 28S ribosomal RNA are shown.

shows a typical electropherogram of one of the RNA samples. According to the recommended microarray practices, the RNA integrity number (RIN) of the extracted total RNA should be above 7.5 on a maximum scale range of 10 (see Note 6) to be suitable for labeling and amplification for use with one color microarray kit.

3.2.4. Agilent One-Color Microarray-Based Gene Expression Analysis

Affymetrix and illumina kits can also be used for gene expression microarray analysis. In Agilent microarray gene expression analysis, the method uses T7 RNA polymerase, which simultaneously amplifies target material (mRNA) and incorporates cyanine 3-labeled CTP. Amplification is typically at least 100-fold from total RNA to cRNA with the use of this kit. Agilent's Quick Amp Labeling Kit generates fluorescent cRNA (complimentary RNA) with a sample input 200 ng of total RNA for one-color processing.

1. The amount of total RNA for use with Agilent's Quick Amp Labeling Kit can range from 200 to 1,000 ng.
2. Use Agilent one-color microarray gene expression analysis kit and the kit protocol to produce cRNA.
3. Quantify the generated cRNA using NanoDrop ND-1000 UV–VIS Spectrophotometer or a similar instrument and note the following values:
   - Cyanine 3 dye concentration (pmol/μL)
   - RNA absorbance ratio (260/280 nm)
   - cRNA concentration (ng/μL)
4. Determine the yield and specific activity (see Note 7). The yield and specific activity of all the samples of cRNA should be more than 1.65 μg and 9.0 pmol Cy3 per μg cRNA, respectively, in order to proceed with microarray hybridization (see Note 8).

3.2.5. Microarray Data Analysis

To effectively mine QD gene expression microarray data, a number of approaches can be taken in relation to experimental design. When working with QDs, physical characteristics such as size and surface modifications should also be taken into consideration, in addition to concentration and time of exposure. Therefore, the analysis of microarray data, when studying the effect of QDs on cells, is different to the analysis of a drug, in which only the concentration and time dependency are usually the main variables to be considered. Simple pairwise comparisons between untreated cells and cells treated with various types of QDs reveal the effect of the treatment on gene expression. The subsequent use of Venn diagrams to link two or more pairwise comparisons (see Fig. 2) can identify overlapping lists of genes affected by all types of QDs and genes uniquely differentially regulated by a particular QD treatment.

To examine the specific effect of QD size or gelatinization, data from cells treated with the appropriate QDs can be compared directly (e.g., red nongelatinated vs. orange nongelatinated). A more complex multifactor analysis is also possible (e.g., by using a two-way ANOVA), to assess the effect of QD size, gelatinization and the interaction between both parameters. Our study found genes involved with the Metallothionein pathway, the p53 signaling pathway and the ATM signaling pathway, among others, are

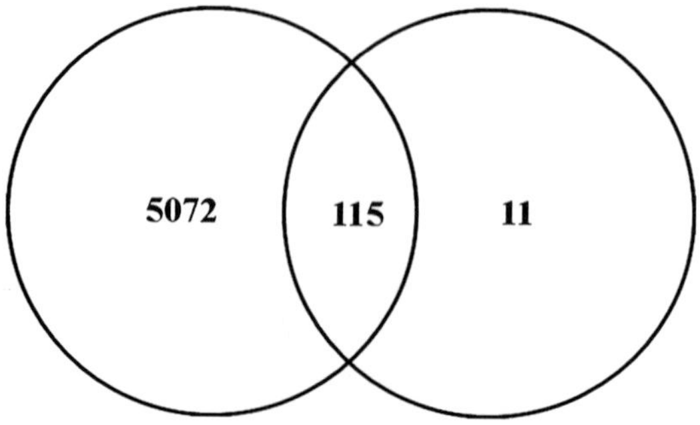

Fig. 2. Venn diagram showing the relationship between QD size and number of differentially expressed genes. Differentially expressed genes in the cells treated with red gel QDs (*left*) vs. differentially expressed genes in the cells treated with orange gel QDs (*right*).

affected by treatment with QDs, and are of particular interest for further analysis for pathway studies.

A number of commercially available (e.g., Agilent GeneSpring GX, DNASTAR ArrayStar) and free or open source (http://www.Bioconductor.org, http://www.TM4.org) data analysis software packages are available. While individual steps vary across platforms, and microarray manufacturers often recommend particular methodologies, typically the following steps are used:

1. *Data preprocessing*: Various algorithms may be used to merge data from replicate spots and/or subtract background noise or nonspecific hybridization (e.g., RMA, GCRMA, MAS5).
2. *Normalization*: Data are adjusted to account for technical variability and to easily compare genes expressed at different levels (e.g., Lowess, Per-chip normalization, median polishing).
3. *Sample quality control*: Samples may be removed from further analysis if identified as outliers based on signal intensity criteria (e.g., signal-to-noise ratio, principal components analysis) and/or labeling, hybridization and scanning controls (e.g., spiked-in controls, housekeeping genes, printed controls).
4. *Data filtering*: Signal intensity-based filters may be used to remove nonexpressed or lowly expressed genes while standard deviation filters may be applied to remove unreliable genes across replicate samples.
5. *Statistical analysis*: To identify significantly differentially expressed genes between differently sized QDs and/or gelatinated and nongelatinated QDs (shown in Table 4 (see Note 9)), statistical tests (*T*-Test, ANOVA, SAM) are applied along with a Multiple Testing Correction algorithm (e.g., Benjamini & Hochberg or Bonferroni multiple testing algorithms) to correct for the large number of genes, as well as a Fold Change Filter to apply a cutoff level.

Table 4
The number of differentially expressed genes compared for different *QD* treatments

QD treatments	Number of differentially expressed genes
Red gel vs. control	5,187
Orange gel vs. control	126
Red gel vs. orange gel	4,736
Red gel vs. red nongel	327
Orange gel vs. orange nongel	103
Red nongel vs. control	5,849
Orange nongel vs. control	404
Orange nongel vs. red nongel	5,731

6. *Cluster analysis*: unsupervised clustering algorithms identify patterns of expression within gene lists to find co-regulated and functionally related genes (Hierarchical, K-Means, Self-Organizing Maps).
7. *Functional analysis*: to identify gene families and functionally related genes, particularly from large sets of differentially expressed genes as in the case of cells treated with certain types of QDs, associated gene annotation terms are statistically analyzed for overrepresentation in gene lists (Gene Ontology, Gene Set Enrichment Analysis, DAVID).
8. *Pathway analysis*: analyzing gene lists for a significant overrepresentation in existing pathways, or to build new networks, involves software (Cytoscape, IPA) identifying pathway-related annotations and known interactions in curated databases (KEGG, Pathguide).

4. Notes

1. The gelatine shell size cannot be measured by electron microscopy. There is an increase in hydrodynamic diameter when comparing bare and gelatinated QDs with the same core size. This accounts for the presence of gelatine; however, it does not provide the actual thickness of the shell. The hydrodynamic diameter includes the layer of water around the particles in solution.
2. In our hands, cells treated with gelatinated QDs showed better viability (nearly equal to positive controls) than cells treated with nongelatinated QDs.

3. During this process, QDs along with cellular components (proteins, lipids, and carbohydrates) will be removed, so that there is no chance of any interference of QDs mixing with nucleic acids (DNA or RNA).
4. DNase enzyme removes contaminating DNA from downstream applications, e.g., Cy3 labeling. RNeasy columns selectively bind RNA, while any other remaining molecules (including QDs) are washed away.
5. Samples are considered 100 % protein-free when they exhibit a ratio of A_{260}/A_{280} greater than or equal 2.0 All the RNA samples measured in our experiment showed a ratio of A_{260}/A_{280} above 2.0, indicating 0 % protein contamination, see Table 2 for the correlation between A_{260}/A_{280} and RNA purity.
6. The RIN of all our samples tested measured above 9.0. The RIN of all the set of samples (treated with red gel and nongel QDs, orange gel and nongel QDs and untreated controls) measured does not show significant variation, indicating good quality, intact RNA was purified,
7. Use the concentrations of cRNA (ng/μL) and cyanine 3 (pmol/μL) to determine the specific activity as follows:

$$(\text{Concentration of Cy3}) / (\text{Concentration of cRNA}) \times 1{,}000$$
$$= \text{pmol Cy3 per } \mu\text{g cRNA}$$

8. In our experiments, the cRNA yield and specific activity was not affected by the treatment of the cells with QDs.
9. Cells treated with the larger red gel and red nongel QDs showed a very high number of differentially expressed genes relative to untreated cells, while cells treated with smaller orange gel and orange nongel QDs showed smaller numbers of differentially expressed genes relative to untreated cells, though the total RNA and cRNA concentration profiles are similar.

Acknowledgements

This work was funded by the Science Foundation Ireland (SFI) and National University of Ireland Galway Millennium Fund 2008. HPMEC-ST1.6R cells were kindly provided by Ronald E. Unger, Institute of Pathology, Johannes Gutenberg University, Mainz, Germany.

References

1. Drummen GP (2010) Quantum dots-from synthesis to applications in biomedicine and life sciences. Int J Mol Sci 11:154–163
2. Chan WC, Maxwell DJ, Gao X, Bailey RE, Han M, Nie S (2002) Luminescent quantum dots for multiplexed biological detection and imaging. Curr Opin Biotechnol 13:40–46
3. Kondoh M, Araragi S, Sato K, Higashimoto M, Takiguchi M, Sato M (2002) Cadmium induces apoptosis partly via caspase-9 activation in HL-60 cells. Toxicology 170:111–117
4. Rikans LE, Yamano T (2000) Mechanisms of cadmium-mediated acute hepatotoxicity. J Biochem Mol Toxicol 14:110–117
5. Nel A, Xia T, Madler L, Li N (2006) Toxic potential of materials at the nanolevel. Science 311:622–627
6. Hoshino A, Fujioka K, Oku T, Suga M, Sasaki YF, Ohta T, Yasuhara M, Suzuki K, Yamamoto K (2004) Physicochemical properties and cellular toxicity of nanocrystal quantum dots depend on their surface modification. Nano Lett 4:2163–2169
7. Derfus AM, Chan WCW, Bhatia SN (2004) Probing the cytotoxicity of semiconductor quantum dots. Nano Lett 4:11–18
8. Guo G, Liu W, Liang J, He Z, Xu H, Yang X (2007) Probing the cytotoxicity of CdSe quantum dots with surface modification. Mater Lett 61:1641–1644
9. Lovrić J, Cho SJ, Winnik FM, Maysinger D (2005) Unmodified cadmium telluride quantum dots induce reactive oxygen species formation leading to multiple organelle damage and cell death. Chem Biol 12:1227–1234
10. Kirchner C, Liedl T, Kudera S, Pellegrino T, MunozJavier A, Gaub HE, Stolzle S, Fertig N, Parak WJ (2005) Cytotoxicity of colloidal CdSe and CdSe/ZnS nanoparticles. Nano Lett 5:331–338
11. Chang E, Thekkek N, Yu WW, Colvin VL, Drezek R (2006) Evaluation of quantum dot cytotoxicity based on intracellular uptake. Small 2:1412–1417
12. Wang L, Nagesha D, Selvarasah S, Dokmeci M, Carrier R (2008) Toxicity of CdSe nanoparticles in Caco-2 cell cultures. J Nanobiotechnology 6:11
13. Byrne SJ, Williams Y, Davies A, Corr SA, Rakovich A, Gun'ko YK, Rakovich YP, Donegan JF, Volkov Y (2007) "Jelly dots": synthesis and cytotoxicity studies of CdTe quantum dot-gelatine nanocomposites. Small 3:1152–1156
14. Prasad BR, Nikolskaya N, Connolly D, Smith TJ, Byrne SJ, Gerard VA, Gun'ko YK, Rochev Y (2010) Long-term exposure of CdTe quantum dots on PC12 cellular activity and the determination of optimum non-toxic concentrations for biological use. J Nanobiotechnology 8:7
15. Krump-Konvalinkova V, Bittinger F, Unger RE, Peters K, Lehr HA, Kirkpatrick CJ (2001) Generation of human pulmonary microvascular endothelial cell lines. Lab Invest 81:1717–1727
16. Byrne SJ, Corr SA, Rakovich TY, Gun'ko YK, Rakovich YP, Donegan JF, Mitchell S, Volkov Y (2006) Optimisation of the synthesis and modification of CdTe quantum dots for enhanced live cell imaging. J Mater Chem 16:2896–2902
17. Huang DW, Sherman BT, Lempicki RA (2009) Systematic and integrative analysis of large gene lists using DAVID bioinformatics resources. Nat Protoc 4:44–57
18. Huang DW, Sherman BT, Lempicki RA (2009) Bioinformatics enrichment tools: paths toward the comprehensive functional analysis of large gene lists. Nucleic Acids Res 37:1–13
19. Barton G, Abbott J, Chiba N, Huang DW, Huang Y, Krznaric M, Mack-Smith J, Saleem A, Sherman BT, Tiwari B et al (2008) EMAAS: an extensible grid-based rich Internet application for microarray data analysis and management. BMC Bioinformatics 9:493

Chapter 13

The Use of Quantum Dots for Immunochemistry Applications

Helena Montón, Mónica Roldán, Arben Merkoçi,
Emma Rossinyol, Onofre Castell, and Carme Nogués

Abstract

Immunocytochemistry and histochemistry are two most valuable immunochemistry techniques routinely used in biological laboratories. These techniques rely on the use of antibodies to label epitopes of interest in cells. At present, there is a wide range of commercially available organic dyes for labeling antibodies. However, limited extinction coefficients of organic dyes often make it difficult to achieve the optimal exposure and therefore fluorescence detection limit. Quantum dots (QDs) are fluorescent semiconductor nanocrystals which are advantageous over the organic fluorescent dyes in many aspects, principally their long-term luminescence stability, high brightness, and multicolor detection. Here, we describe the use of QDs for immunocytochemistry applications. We used three different antibodies—anti-β-tubulin monoclonal antibody for visualizing the microtubule network, the GM130 antibody for staining the Golgi complex, and the EEA1 antibody for detecting the endosomal system. We use the anti-mouse IgG antibody directly conjugated to QD655 quantum dots or anti-mouse IgG conjugated to biotin for tertiary detection with streptavidin-conjugated QD655.

Key words: Immunocytochemistry, Fluorescence, Quantum dots, Cell culture, Antibody, Confocal laser scanning microscope

1. Introduction

There are several types of immunocytochemistry (IC) methods (1). Direct IC is a one-step method which utilizes only one labeled antibody that binds directly to the target. Another type is indirect IC, which requires at least two antibodies, the first (primary antibody) is not labeled and recognize the epitope of interest, while the second (secondary antibody) is labeled and reacts against the primary one. When the signal needs to be amplified, the secondary antibody is conjugated to biotin and a third labeled molecule (avidin, streptavidin, neutravidin) is necessary to bind to the biotin (1).

Mikhail Soloviev (ed.), *Nanoparticles in Biology and Medicine: Methods and Protocols*, Methods in Molecular Biology, vol. 906, DOI 10.1007/978-1-61779-953-2_13, © Springer Science+Business Media, LLC 2012

Organic dyes with diverse spectral properties are currently used to label the different antibodies and molecules involved in IC procedures. However, in some cases their fluorescence is not sufficient for obtaining a good detection limit due to the limited extinction coefficients and quantum yield of organic dyes.

In the last decades, nanotechnology has become relevant leading to the development of semiconductor nanocrystals (quantum dots (QDs)) with unique optical properties. QDs have intrinsic brightness, high photostability, high molar extinction coefficient, potentially narrow emission band, and they can be excited with a wide range of wavelengths (2). QDs can be modified with different kinds of molecules in order to make them water-soluble and conjugate to different biomolecules, such as antibodies (3–7).

Here we describe a simple protocol, suitable for a range of immunostaining applications using commercially available QDs (8). We exemplify our protocols by immuostaining microtubule network, Golgi complex and endosome system using anti-mouse secondary IgG antibody directly conjugated to QDs or biotinylated anti-mouse IgG and streptavidin conjugated to QDs. Image resolution is comparable to that achieved with traditional organic dyes, but unlike organic dyes, QDs staining is more photostable allowing longer-term luminescence detection.

2. Materials

2.1. Cell Culture

1. Minimum essential media (MEM) supplemented with 10 % fetal calf serum (FCS) (Invitrogen).
2. Hank's balanced salt solution (HBSS) Ca^{2+} and Mg^{2+}-free, 1× (Invitrogen) (see Note 1).
3. Cell dissociation Trypsin/EDTA solution (TE): 0.05 % Trypsin-EDTA (Invitrogen).
4. Cell culture flasks 25 cm^2 and 24-well plates (Nunc) (see Note 2).
5. Cover slips: wrap individually each cover slip with aluminum foil, place them into a glass container and sterilize by autoclaving (see Note 3).

2.2. Immunocytochemistry Labeling with QDs

1. Phosphate buffered saline (PBS) 0.01 M: 0.0027 M potassium chloride and 0.137 M sodium chloride, pH 7. Dissolve one tablet of PBS (Sigma) in 200 mL of deionized water (see Note 4).
2. 20 % Paraformaldehyde (PF): dissolve 20 g of paraformaldehyde (Electron Microscopy Sciences) in 80 mL of deionized water, add a magnetic stir bar, warm the mixture gently to 60 °C and mix the solution on a magnetic stirrer hotplate. Adjust the

pH to 7.0–7.4 by adding NaOH drop-wise until the solution clears. Fill up to 100 mL and store in 1 mL aliquots at −20 °C.

3. To prepare 4 % PF working solution: dilute 20 % PF in PBS to a final concentration of 4 % with PBS. Prepare fresh just before use.
4. Permeabilization solution: 0.25 % Triton X-100 in PBS.
5. Blocking buffer: 6 % bovine serum albumin (BSA) in PBS (see Note 5).
6. All antibodies are diluted from the stock solution in the dilution buffer: 6 % BSA in PBS.
7. Primary antibodies stock solutions: anti-β-tubulin monoclonal antibody (clone TUB 2.1) (2 mg/mL) (Sigma), anti-GM130 antibody (250 μg/mL) and anti-EEA1 antibody (250 μg/mL) (BD Biosciences) (see Note 6).
8. Secondary antibodies stock solutions: anti-mouse IgG antibody conjugated to QD655 (1 μM) (Invitrogen) (see Note 7); and anti-mouse IgG conjugated to biotin (1 mg/mL) (Boheringer Mannheim).
9. Streptavidin conjugated to QD655 stock solution (1 μM) (Invitrogen).
10. Nuclear stain: Hoechst 33342 solution stock (10 mg/mL) (Sigma).
11. Mounting medium: Fluoprep (bioMérieux® SA) (see Note 8).
12. Fluorescent microscope. We used a confocal laser scanning microscope (CLSM): Leica TCS-SP5 AOBS spectral, with a Plan-Apochromatic ×63 objective (NA 1.4, oil) (see Note 9).
13. Imaging processing software (Imaris v.6.1.0, Bitplane AG) for postanalysis and three-dimensional modeling.

3. Methods

3.1. Cell Seeding on Cover Slips

We exemplify our method with HeLa cell line, but the method can be used with other cell lines.

1. Maintain HeLa cells in cell culture flasks, in MEM supplemented with 10 % FCS at 37 °C and 5 % CO_2 in a humidified atmosphere.
2. Warm up MEM, HBSS, and TE in a water bath at 37 °C before use.
3. Aspirate culture medium and rinse cells twice in HBSS 1× for 5 min each.
4. To dissociate cells add enough 1× TE to cover the cells, incubate for 3 min at 37 °C. Neutralize the action of TE

by adding 1 mL MEM supplemented with 10 % FCS (see Note 10).

5. Unwrap sterile cover slips and place them in wells in 24-well plate. Seed 20,000 HeLa cells per well and fill up to 1 mL with MEM + 10 % FCS.
6. Maintain cell cultures at 37 °C and 5 % CO_2 in a humidified atmosphere up to 24 h to obtain a nonconfluent culture (see Note 11).

3.2. Immunocytochemistry Labeling Using QDs

1. Remove culture medium with aspiration and rinse cells twice in PBS for 5 min each.
2. Fix cells by adding 0.5 mL of freshly made 4 % PF in PBS for 15 min at room temperature (RT).
3. Remove PF and wash cells three times with PBS for 5 min each.
4. Permeabilize cells by adding 0.5 mL of permeabilization solution for 15 min at RT.
5. Wash cells twice in PBS for 5 min each.
6. Incubate cells in blocking buffer (0.5 mL) for 40 min at RT.
7. Remove blocking buffer and incubate each cell culture with the selected primary antibody (100 μL) diluted in BSA 6 % (β-tubulin at 4.0 μg/mL; GM130 at 10 μg/mL; and EEA1 at 2.5 μg/mL) for 1 h at 37 °C (see Notes 12 and 13).
8. Wash samples twice in blocking buffer, 2 min each washing step.
9. To perform secondary immunodetection, incubate cells with the corresponding secondary antibody (100 μL) diluted in BSA 6 % (anti-mouse IgG-QD655 at 30 nM or anti-mouse IgG biotin 1 μg/mL) for 1 h at 37 °C (see Notes 13 and 14).
10. Wash cells twice in blocking buffer, 2 min each washing step.
11. If nuclear staining is considered necessary go to step 16.
12. The indirect secondary IC is completed at this point. Allow samples to air-dry completely at RT and proceed to mount the sample (go to step 20).
13. For IC with unlabeled secondary antibodies incubate samples with streptavidin-conjugated QD655 (100 μL at 30 nM) for 1 h at 37 °C.
14. Wash cells twice in blocking buffer, 2 min each washing step.
15. If nuclear staining is not performed (if necessary go to step 16), allow the samples to air-dry completely at RT and proceed with the mounting (go to step 20).
16. Nuclear staining: remove blocking buffer and incubate cells with Hoechst 33342 (100 μL at 5 μg/mL) for 5 min at RT.
17. Rinse cells twice in PBS, 2 min each washing step.

18. Allow the samples to air-dry completely at RT.
19. Mount the cover slip in a drop of mounting medium onto a glass slide, aspirate off the excess from around the edges of the cover slip, and seal the edges with clear nail polish (see Note 15).
20. Store the slides in a dark box at 4 °C until microscope evaluation.
21. The slides are viewed under fluorescence or confocal microscope. Fluorescent microscope filters are selected according to the emission fluorescence. In case of confocal microscope, lasers are selected according to the required excitation wavelength. In our case, the laser line 561 nm was selected to excite QD655 (red emission) and blue diode 405 nm was used to detect the nuclei (blue emission).
22. Maximum intensity projections and three-dimensional models are obtained using the appropriate software (e.g., Imaris v.6.1.0) from CLSM image stacks.

As an example, a three-dimensional model of Golgi and endosomes are shown in Fig. 1. In both cases, the structures are well delimited with low noise and high contrast. However, microtubules show a poorly structured network with background and QD aggregates that had not selectively linked to β-tubulin (Fig. 2a). An example of the signal obtained with tertiary IC to detect β-tubulin is shown in Fig. 2b. Here, the staining against β-tubulin is more specific and tubulin network is clearly detected (Fig. 2b).

Image resolution is comparable to that achieved with traditional organic dyes, but unlike organic dyes, QDs staining is more photostable (see an example in Fig. 3), allowing longer-term luminescence detection.

4. Notes

1. HBSS Ca^{2+}- and Mg^{2+}-free medium is required to dissociate cells. These two ions are involved in cell adhesion. When using a buffer without Ca^{2+} and Mg^{2+}, the lack of these two ions weakens cell adhesion which facilitates the action of trypsin.
2. In this example we used 24-well plates to carry out the experiment, because it allows performing several immunocytochemistry at the same time. The same protocol can be done using 6-well plates.
3. Small round cover slips are convenient because they fit into each of 24-well plates.
4. Our preferred way is to use PBS tablets (e.g., from Sigma). One tablet of PBS is usually sufficient for making 200 mL of PBS. However, PBS can be prepared from individual components without affecting the results.

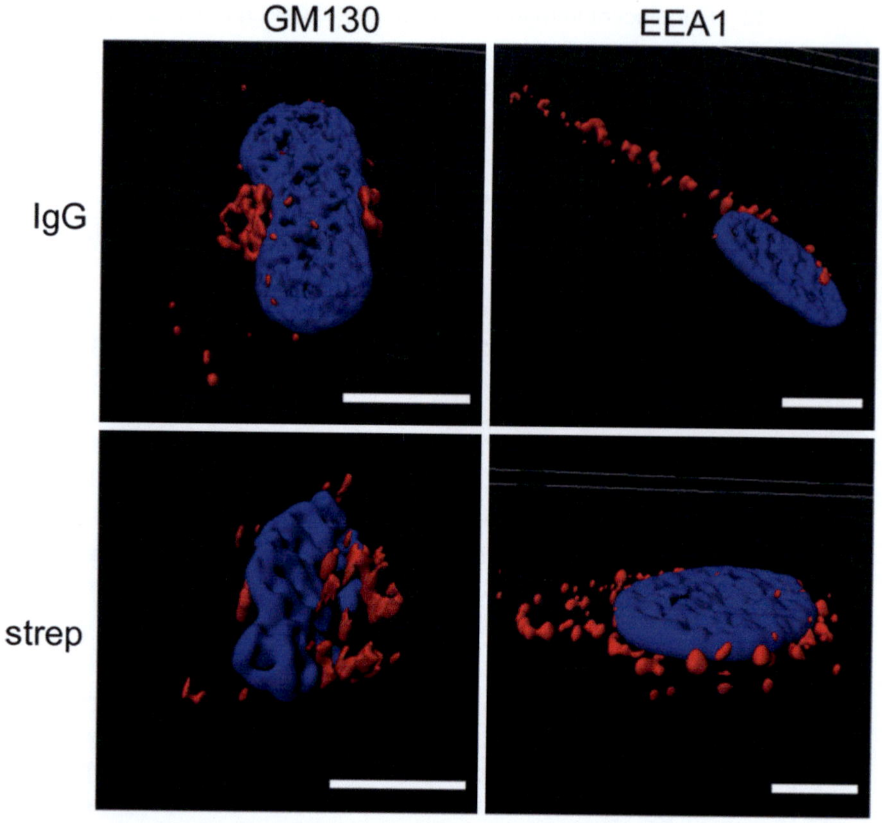

Fig. 1. Isosurface representations of Golgi and endosome systems in four different cells. Nuclei (*blue*) were labeled with Hoechst 33342, Golgi complex (left column: GM130 antibody) (*red*) and endosomal system (*right column*: EEA1 antibody) (*red*); both antibodies were labeled with QD655. Images acquired with a CLSM. Structures are well delimited with low noise and high contrast. QD655 was employed to label secondary antibody (*top*) or streptavidin (*bottom*). In this example, no differences between secondary and tertiary IC were detected. Scale bar = 10 μm.

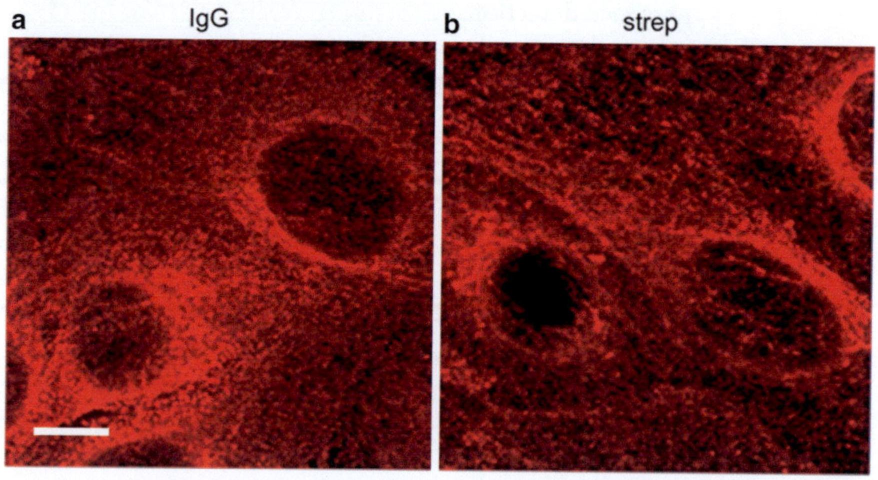

Fig. 2. CLSM specificity analysis of β-tubulin labeling. Maximum intensity projections of the distribution of the tubulin network labeled with QDs. When the same secondary antibody (anti-mouse IgG antibody conjugated QD655) against β-tubulin was used, the resolution of the tubulin network showed a poorly structured network with background and QD aggregates that have not been selectively linked to β-tubulin (**a**). But when streptavidin-conjugated QD655 were used to label biotin-conjugated secondary antibody, a better resolution was obtained and the tubulin network was clearly detected (**b**). Scale bar = 10 μm.

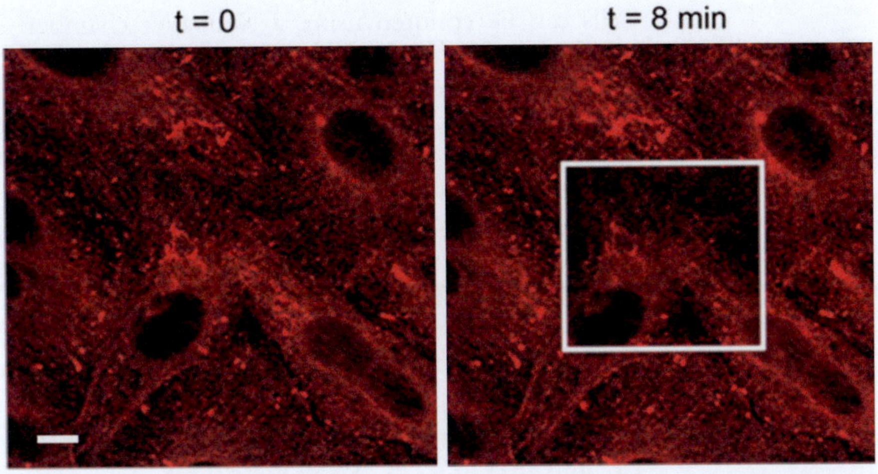

Fig. 3. Photostability analysis of QDs. Left-hand image corresponds to the emission signal of QDs conjugated to streptavidin before the irradiation period (*t*=0 min) with the 561 nm laser line at its maximum power. Right-hand image shows the emission signal of QDs at the end of the irradiation period (*t*=8 min). Note the loss of fluorescence intensity in the delimited area. A 10 % of fluorescence intensity reduction was observed for the first 90 s. After 8 min, the tubulin network was clearly visible and, the final intensity of the signal was between 10 and 40 % of the initial one. Scale bar=10 μm.

5. It is important to make sure that BSA is completely dissolved in order to obtain the correct concentration. First, mix gently manually and then mix vigorously with a vortex. When no traces of solid BSA are observed, let the solution rest at RT for about 30 min until the foam formed disappears.

6. The selection of primary antibodies is crucial for the detection of the protein of interest. As a simple rule and in the absence of other information consider using antibodies reported in the literature and proven to work.

7. Concentration of QDs-conjugated antibodies is critical to obtain good quality results. To optimize the protocol we tested several concentrations of QDs-conjugated antibodies. We observed that, when using the primary antibodies at the concentration routinely used in our laboratory to perform IC, concentration of QDs-conjugated antibodies below 30 nM gave too weak a signal. By contrast, higher concentrations caused the formation of aggregates which difficult the observation of the signal.

8. Fluoprep was used due to the better results obtained in our laboratory even in long-term observations. It should be taken into account that QDs fluorescence is susceptible to chemical quenching and that not all mounting media might be suitable.

9. Any fluorescent microscope equipped with a ×63 objective and with the appropriate fluorescent filters can be used to detect the signal. To obtain maximum projections or/and three-dimensional models, a confocal microscope is necessary.

10. The specified timing is for HeLa cells, but may need to be adjusted for other cell types.

11. Cells can be counted using a Neubauer chamber. The best results are obtained when cultures have not reached the confluence because cells are not in close contact and individual cells can be seen. Twenty-four hours after seeding, the 20,000 cells/well HeLa cells are well distributed across the surface of the cover slip and individual cells can be observed.
12. Here we used the anti-β-tubulin monoclonal antibody to detect the microtubule network, the GM130 antibody to detect the Golgi complex and the EEA1 antibody to detect the endosomal system (see Subheading 2 for concentrations used). Antibody stock solutions should be diluted in BSA 6 % prior to use.
13. Alternatively, the antibodies may be incubated for 2 h at RT. Some antibody may work better if incubated first at RT, for 1 h, and after at 37 °C for an extra hour.
14. Here, we used anti-mouse IgG antibody conjugated to QD655 or anti-mouse IgG conjugated to biotin (see Subheading 2 for concentrations used).
15. When mounting the samples, make sure that the side of the cover slip where the cells are attached is facing the glass slide and that mounting medium is between the cells and the slide. Gently place the cover slip onto the drop of mounting medium by using tweezers and again press gently to spread it throughout the sample, taking care that no bubbles are formed.

Acknowledgements

The authors would like to ackowledge the Servei de Microsòpia from the Universitat Autònoma de Barcelona.

References

1. Dodson A (2002) Review: modern methods for diagnostic immunocytochemistry. Curr Diagn Pathol 8:113–122
2. Resch-Genger U, Grabolle M, Cavaliere-Jaricot S et al (2008) Quantum dots versus organic dyes as fluorescent labels. Nat Methods 5:763–775
3. Mazumder S, Dey R, Mitra MK et al (2009) Review: biofunctionalized quantum dots in biology and medicine. J Nanomater. doi:10.1155/2009/815734
4. Chan WC, Maxwell DJ, Gao X et al (2002) Luminescent quantum dots for multiplexed biological detection and imaging. Curr Opin Biotechnol 13:40–46
5. Tiwary DK, Tanaka S-I, Inouye Y et al (2009) Synthesis and characterization of anti-HER2 antibody conjugated CdSe/CdZnS quantum dots for fluorescence imaging of breast cancer cells. Sensors 9:9332–9354
6. Shi C, Zhou G, Zhu Y et al (2008) Quantum dots-based multiplexed immunohistochemistry of protein expression in human prostate cancer cells. Eur J Histochem 52:127–134
7. Sweeney E, Ward TH, Gray N et al (2008) Quantitative multiplexed quantum dot immunohistochemistry. Biochem Bioph Res Co 374: 181–186
8. Montón H, Nogues C, Rossinyol E et al (2009) QDs versus Alexa: reality of promising tools for immunocytochemistry. J Nanobiotechnolgy. doi:10.1186/1477-3155-7-4

Chapter 14

Quantum Dots for Labeling Live Cells

Yun Chen

Abstract

Quantum dots (QDs) are semiconductor nanocrystals with unique photo-physical properties. QDs are brightly fluorescent and photostable, and therefore represent a novel class of fluorescence probes for bio-imaging and cell monitoring. There are different techniques for labeling cells with QDs. Moreover, there is evidence that optical properties of the QD can be changed as a result of interaction with live cells. This manuscript describes a simple method for labeling of rat endothelial progenitor cells using ODs and for characterization of their optical properties.

Key words: Stem cell tracking, Quantum dot, Confocal microscope, Endothelial progenitor cell

1. Introduction

Fluorescent quantum dots (QDs) are semiconductor nanocrystals that have attractive optical characteristics such as high quantum yield, narrow emission spectrum, high resistance to photobleaching, and large Stokes shift. The latter feature means that QDs can be excited at any wavelength lower than its emission wavelength. This makes the collected signal QD-specific and is one of the ways to mitigate the effects of autofluorescence. Moreover, the narrow emission spectrum and broad excitation spectrum of the QD enable spectral multiplexing and the simultaneous recording of multiple color images using single excitation wavelength.

Nano-sized QDs have been used successfully in many fluorescence-based optical imaging applications, including imaging of living cells (reviewed in (1, 2)). Other examples include their use for noninvasive imaging of dendritic cell migration into lymph nodes (3), for tracking stem cells (4), and for tracking leukocytes in atherosclerotic lesion (5).

A number of different approaches have been reported for QD labeling of different cell types. For example, transfection reagent

Mikhail Soloviev (ed.), *Nanoparticles in Biology and Medicine: Methods and Protocols*, Methods in Molecular Biology, vol. 906, DOI 10.1007/978-1-61779-953-2_14,

and peptide ligands have been used to increase cellular uptake of QDs (6, 7). But transfection agents such as lipofectamine cause some cell death and are not approved for clinical use, whilst peptide mediated delivery of QDs leads to non-uniform labeling (7). We and others have shown that carboxyl coated QD can be used to efficiently label cells (7–9). Furthermore, stem cells labeled with carboxyl coated QD had normal cell viability and function (7, 8), and can be used for long-term cell tracking in vivo (7) and for studying stem cell engraft and differentiation in the host environment. Another interesting phenomenon reported by us and others is the change in optical properties of the QD, i.e., blueshift of QD, as a result of interaction with live cells (10, 11). Understanding when and how the changes in optical properties of the QD occur is important for QD-based bio-imaging as well as potential QD-based in vivo sensors. Here I describe a protocol for labeling cells with carboxyl coated QD, using an easy adaptable labeling method without transfection reagents or the use of targeting molecules.

2. Materials

2.1. Cell Culture and Cell Labeling

1. Rat bone marrow-derived endothelial progenitor cells (EPCs) (see Note 1).
2. Complete endothelial growth medium-2 (EGM-2): Endothelial cell basal medium-2 (EBM-2) supplied with SingleQuots, 20 % fetal bovine serum (FBS), 100 U/mL penicillin, 100 μg/mL streptomycin, 29.2 μg/mL glutamine, and 0.25 μg/mL amphotericin B (EBM-2 and SingleQuots are from Lonza, Denmark; All the others are from Invitrogen, Sweden).
3. Solution of trypsin (0.05 %) and ethylenediamine tetraacetic acid (EDTA) (0.02 %).
4. Dulbecco's phosphate-buffered saline (PBS) with or without calcium and magnesium.
5. Forty-eight-well cell culture plate and 25 cm^2 cell culture flask (Nunclon™ Δ).
6. Nr. 1 round glass cover slips with diameter 10 mm.
7. Carboxyl coated QDs from Molecular Probe/Invitrogen (see Note 2).

2.2. Analysis by Flow Cytometry

1. FACS-Canto II flow cytometry equipped with Diva 6 software (BD Bioscience, Erembodegen, Belgium).
2. PBS with 2 % FBS.

2.3. Analysis by Confocal Microscopy

1. Leica TCS SP5 Spectral Confocal Microscope with the software Leica Application Suite 2.02 (see Note 3).

2. Mounting medium containing Prolong Gold Antifade Reagent and nuclear staining agent 4',6-diamidino-2-phenylindole (DAPI) (Invitrogen).
3. Paraformaldehyde: Prepare 4 % paraformaldehyde by heating 40 mL water to 65 °C in microwave and adding 2 g paraformaldehyde (working in a fume hood). Mix well and add 10 μL of 10 N NaOH. The solution should become clear. Add 5 mL of 10× PBS and bring volume to 50 mL with water.

3. Methods

The protocol described here was designed for labeling rat EPCs. Using this protocol we found that cells can be labeled efficiently by incubating with 8 nM of carboxyl coated QD705 (QD with emission peak at 705 nm) for 24 h (7). The signals of QD705 were located in the cytoplasm, mostly in endosomal/lysosomal vesicles. We also showed that the QD labeling did not change morphology, metabolic activity, or function of rat EPCs. We have also used our protocol for labeling other cells, such as mouse aortic endothelial cells, human umbilical vein endothelial cells, and human EPCs.

3.1. Cell Culture and Cell labeling

3.1.1. For analysis by Flow Cytometry

1. Prepare cell suspension in EGM-2 medium and add carboxyl coated QD to a final concentration of 8 nM. Seed 5×10^5 cells in a T-25 cell culture flask and incubate at 37 °C for 24 h (see Note 4).
2. Wash the cells with calcium- and magnesium-free PBS to remove all traces of serum. Dispense enough tripsin/EDTA solution to completely cover cell monolayer and place in 37 °C incubator until cells detach from the surface (see Note 5).
3. Add EGM-2 to the cell suspension. Gently pipette the cell suspension to break up the clumps. Pellet cells by centrifugation at $100 \times g$ for 5 min.
4. Resuspend cells in PBS with 2 % FBS and keep on ice until analysis.

3.1.2. For Analysis by Confocal Microscopy

1. Prepare cell suspension in EGM-2 medium and add carboxyl coated QD in a final concentration of 8 nM. Seed 1×10^4 cells per well in a 48-well plate with glass cover slip placed in the bottom of the well, and incubate at 37 °C for different time desired.
2. Wash the cells with PBS. Fix cells in 4 % PFA for 10 min. Repeat the PBS-washing step before mount the cells with DAPI mounting medium.

3.2. Analysis by Flow Cytometry

1. To determine QD labeling efficiency, a minimum of 50,000 viable cells are analyzed using flow cytometry.

2. Fluorescence is detected using the 488 nm laser and the LP655 nm and LP670 nm filters.
3. Unlabeled cells are used as negative control for determining the borderline of background fluorescence intensity. The QD labeling efficiency is calculated as percentage of QD-positive cells.

3.3. Analysis by Confocal Microscope

1. To determine intracellular location of QD: Configure image acquisition parameters by selecting acquisition mode XYZ, image size, and acquisition speed. Activate the UV laser line, the PMT detectors (one for QD signal, one for DAPI signal, and one for differential interference contrast (DIC) image), and choose the bandwidth according to QD used and for DAPI (see Note 6).
2. Acquire live images. Adjust laser output power and gain to maximize the QD signal and DAPI signal. Adjust offset to minimize background signal. Capture images.
3. To ensure the intracellular location of the QDs, three-dimensional confocal image can be generated using the *z*-axis cross-sectional scan mode and the software Leica Application Suite 2.02. With the help of the DIC cell image, choose start position of the Z-stack where the cell just appears; move the Z-stack to a position where the cell just disappears (end position of the Z-stack). Capture scanning images. A typical example of three-dimensional distribution and localization of QDs in rat EPCs is shown in Fig. 1.
4. To study optical properties of the QD: Select acquisition mode *xyλ* (*λ* for wavelength). Configure image acquisition parameters as described in step 1.
5. Select start and end positions of wavelength scanning and start scanning (see Note 7).

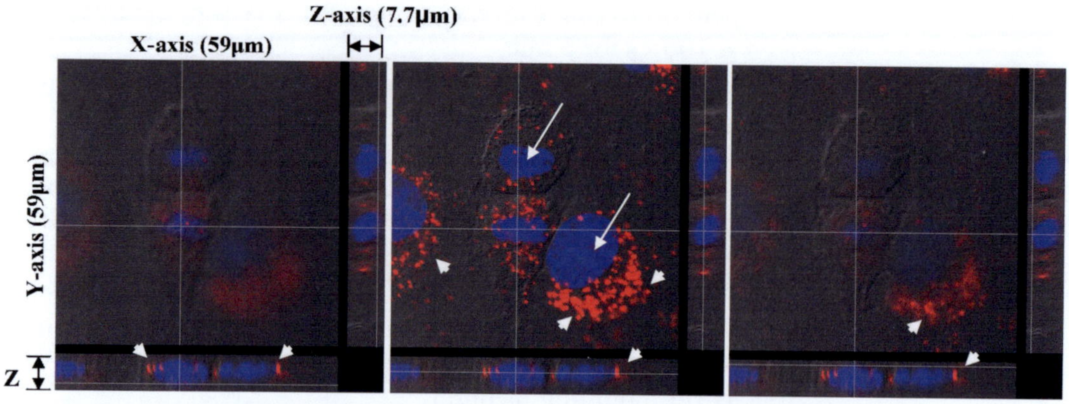

Fig. 1. Three-dimensional distribution and localization of QDs in rat endothelial progenitor cells. The sections shown are taken from three different *z*-positions as indicated by the *white lines*. *Arrows*: nuclear staining with DAPI; *Arrow heads*: QD fluorescence.

4. Notes

1. In this example, we used rat EPCs that are isolated and cultured according to previously described (7).
2. We tested both positively charged PEG-amine coated QDs and negatively charged carboxyl coated QDs. We found that very few cells internalized the positively charged QDs, whereas most cells were loaded with negatively charged QDs after 24-h incubation.
3. The Leica TCS SP5 Spectral Confocal Microscope has a spectrophotometer detection system containing Prisms and free adjustable barriers. This filter-free and tunable detection system allows acquiring complete emission spectra, a very convenient tool for characterizing spectral properties of the QD in a bio-environment.
4. Cells can also be plated on glass-bottom culture dishes and imaged in PBS. We observed no difference.
5. The time required for detachment is dependent on the cell type used. One should optimize it to keep it as short as possible.
6. One can use sequential acquisition to avoid "crosstalk" problems by acquiring each channel one after the other.
7. The start and end positions of wavelength scanning are set to cover the emission spectrum of the fluorophore used. This is a very useful tool. We always acquire emission spectrum to ensure that the fluorescence signals are specific for QDs. By selecting the region of interest, one can compare emission spectra of QDs within different locations.

Acknowledgments

This study was supported by VINNOVA (Project number P35914-1) and Wilhelm & Martina Lundgren Foundation.

References

1. Michalet X, Pinaud FF, Bentolila LA, Tsay JM, Doose S, Li JJ, Sundaresan G, Wu AM, Gambhir SS, Weiss S (2005) Quantum dots for live cells, in vivo imaging, and diagnostics. Science 307:538–544
2. Wang Y, Chen L (2011) Quantum dots, lighting up the research and development of nanomedicine. Nanomedicine. doi:10.1016/j.nano.2010.1012.1006
3. Noh YW, Lim YT, Chung BH (2008) Noninvasive imaging of dendritic cell migration into lymph nodes using near-infrared fluorescent semiconductor nanocrystals. FASEB J 22:3908–3918

4. Slotkin JR, Chakrabarti L, Dai HN, Carney RS, Hirata T, Bregman BS, Gallicano GI, Corbin JG, Haydar TF (2007) In vivo quantum dot labeling of mammalian stem and progenitor cells. Dev Dyn 236:3393–3401
5. Jayagopal A, Su YR, Blakemore JL, Linton MF, Fazio S, Haselton FR (2009) Quantum dot mediated imaging of atherosclerosis. Nanotechnology 20:165102
6. Yukawa H, Mizufune S, Mamori C, Kagami Y, Oishi K, Kaji N, Okamoto Y, Takeshi M, Noguchi H, Baba Y, Hamaguchi M, Hamajima N, Hayashi S (2009) Quantum dots for labeling adipose tissue-derived stem cells. Cell Transplant 18(5):591–599
7. Rosen AB, Kelly DJ, Schuldt AJ, Lu J, Potapova IA, Doronin SV, Robichaud KJ, Robinson RB, Rosen MR, Brink PR, Gaudette GR, Cohen IS (2007) Finding fluorescent needles in the cardiac haystack: tracking human mesenchymal stem cells labeled with quantum dots for quantitative in vivo three-dimensional fluorescence analysis. Stem Cells 25:2128–2138
8. Molnar M, Friberg P, Fu Y, Brisslert M, Adams M, Chen Y (2010) Effects of quantum dot labeling on endothelial progenitor cell function and viability. Cell Med B Cell Transplant 1:105–112
9. Zhang LW, Monteiro-Riviere NA (2009) Mechanisms of quantum dot nanoparticle cellular uptake. Toxicol Sci 110:138–155
10. Molnar M, Fu Y, Friberg P, Chen Y (2010) Optical characterization of colloidal CdSe quantum dots in endothelial progenitor cells. J Nanobiotechnology 8:2
11. Riegler J, Ditengou F, Palme K, Nann T (2008) Blue shift of CdSe/ZnS nanocrystal-labels upon DNA-hybridization. J Nanobiotechnology 6:7

Chapter 15

Labeling and Imaging Mesenchymal Stem Cells with Quantum Dots

Maria C. Collins, Philip R. Gunst, Wayne E. Cascio, Alan P. Kypson, and Barbara J. Muller-Borer

Abstract

Mesenchymal stem cells (MSCs) are multipotent cells with the potential to differentiate into bone, cartilage, adipose, and muscle cells. Adult derived MSCs are being actively investigated because of their potential to be utilized for therapeutic cell-based transplantation. Methods to track MSCs in vivo are limited, preventing long-term functional studies of transplanted cells. Quantum Dots (QDs) offer an alternative to organic dyes and fluorescent proteins to label and track cells in vitro and in vivo. Nanoparticles are resistant to chemical and metabolic degradation, demonstrating long-term photostability. Here, we describe the technique to label MSCs with QDs and demonstrate intracellular QD distribution in the labeled MSCs with laser scanning confocal fluorescent microscopy.

Key words: Quantum dot, Nanoparticle, Mesenchymal stem cell, In vivo/in vitro cell tracking, Fluorescence imaging, Confocal microscopy

1. Introduction

Quantum dots (QDs) are fluorescent semiconductor nanoparticles, with unique optical and chemical features which make them useful as fluorescent tags for long-term in vitro and in vivo cell imaging applications. QDs have narrow band emission spectra and broad excitation spectra are photostable and resistance to chemical and metabolic degradation (1–3). In addition, QDs demonstrate a direct relationship between size and emitted fluorescence. QDs emit light in the visible region of the spectra with the smallest QDs about the size of green fluorescent protein and largest about the size of red fluorescent protein (4). These properties make QDs amenable to multicolor imaging applications and the tracking of live cells in vitro and in vivo (5–7).

Mikhail Soloviev (ed.), *Nanoparticles in Biology and Medicine: Methods and Protocols*, Methods in Molecular Biology, vol. 906, DOI 10.1007/978-1-61779-953-2_15, © Springer Science+Business Media, LLC 2012

Here we describe a technique for labeling mesenchymal stem cells (MSCs) with QDs. Once labeled the MSCs can be grown with other cells and tracked in vitro or introduced and tracked in an in vivo environment using laser scanning confocal fluorescent microscopy. We have successfully tracked QD labeled MSCs in cardiomyocyte co-cultures for up to 7 days, while others have demonstrated in vivo tracking in rat hearts at 8 weeks post-transplantation (6, 7). For the purpose of evaluating QD location and distribution in MSCs we describe additional protocols using a second fluorescent label in fixed and live cell cultures. While additional fluorescent cell markers are not necessary for tracking QD labeled MSCs in vitro or in vivo, the technique of labeling cell membranes or organelles is effective in differentiating cell boundaries or for co-localization studies.

2. Materials

The materials are divided into several sections, including materials needed for labeling MSCs with QDs, preparing MSCs for labeling, fixing MSC cultures and secondary fluorescent labels for live cell imaging. Have the following materials available: disposable laboratory gloves, 37°C water bath or warmer, laboratory centrifuge, hot plate with magnetic stirrer, pH meter, plate rotator, 15 and 50 mL sterile conical tubes, 1.5 mL sterile microfuge tubes and sterile pipettes. Prepare and store reagents/buffers as directed. Perform all cell culture under a biological hood using sterile cell culture procedures.

2.1. Cell Culture Preparation

1. 10× Phosphate buffered saline (10× PBS): 0.1 M PBS, pH 7.2. Weigh 10.9 g Na_2HPO_4 (Sodium Phosphate, dibasic, anhydrous), 3.2 g NaH_2PO_4 (Sodium Phosphate, monobasic, anhydrous), and 90 g NaCl (Sodium Chloride) and transfer to a 1 L glass beaker. Add 800 mL of reagent grade deionized water (DH_2O). Stir with stir bar to dissolve. Transfer solution to a graduated cylinder. Bring final volume to 1 L by adding DH_2O. Transfer solution to a glass bottle, mix well and adjust pH to 7.2 using a pH meter. Date, cap, and store at room temperature.
2. 1× PBS, washing and blocking solution. In a glass bottle, add 80 mL of 10× PBS (step 1) to 720 mL DH_2O. Date, cap, and store at room temperature.
3. Round glass coverslips (#1.5) with a thickness of 170 nm, designed for high-resolution imaging. Place glass coverslip in 35 mm petri dish or multiwell plate. Alternatively coverglass bottom petri dishes or coverglass bottom multiwell plates can be used (see Note 1).

4. Poly-L-lysine coated coverslips: To prepare coverslip surface for MSCs add about 5 mL of poly-L-lysine to the petri dish until the coverslip is just covered. Incubate at room temperature for 5 min on a plate rotator at low speed to allow adequate coverage of the coverslip. Aspirate residual poly-L-lysine with pipette. Rinse the coverslip with 5 mL of 1× PBS and repeat rinse step with 1× PBS.
5. Sterilize coverslips for 1 h by placing under UV light.

2.2. Labeling MSCs with Quantum Dots

1. Q-Tracker® 605 Cell Labeling kit was purchased from Invitrogen (Invitrogen, Carlsbad, CA) (see Note 2).
2. Complete Media: To prepare cell culture media with 16.5 % Fetal Bovine Serum (FBS) add 100 mL FBS and 5 mL of Pen/Strep/Amphotericin-B to a 500 mL bottle of Dulbecco's Minimal Essential Medium (DMEM). Gently mix by swirling bottle. Date, label, and store media at 4°C. Media should be used within 1 month.

2.3. Preparing Fixed Quantum Dot Labeled MSCs for Imaging

1. 10 % Triton X-100 stock solution: Weigh 2.5 g of Triton X-100 and add to 22.5 mL 1× PBS in a 50 mL conical tube. Place conical tube in a beaker of H_2O and heat on hot plate at low heat until dissolved. Invert conical tube every few minutes to help Triton X-100 go into solution. Date, label, and store at 4 °C for 6 months.
2. 0.1 % Triton X-100 to permeabilize cell membranes: Use 1 % of the 10 % Triton X-100 stock solution. For example, in a 15 mL conical tube add 10 μL of 10 % Triton X-100 stock solution to 1 mL of 1× PBS for a 0.1 % Triton X-100 solution. Prepare enough to use in 1 day and discard after use.
3. 3 % Bovine Serum Albumin (BSA) for blocking and incubating primary antibody: For example, weigh 1.5 g of BSA, add to 50 mL 1× PBS and mix well. Prepare enough to use in 1 day and discard after use.
4. Purchase 4 % COLD Paraformaldehyde in PBS for fixing MSCs: Store at 4 °C.

2.4. Solutions for Secondary Labeling of Live Cell Cultures

1. FBS-FREE Complete Media: To prepare cell culture media with no FBS add 5 mL of Pen/Strep/Amphotericin-B to a 500 mL bottle of DMEM. Gently mix by swirling bottle. Date, label, and store at 4 °C. Media should be used within 1 month (see Note 3).
2. 1 mmol/L Calcein AM: Resuspend 50 μg Calcein AM in 50 μL of anhydrous dimethylsulfoxide (DMSO). Vortex vial to dissolve. Prepare just prior to labeling live cell cultures. Date unused, dissolved calcein AM, and store in desiccant, protected from direct light exposure, at 20 °C for up to 6 months (see Note 4).

3. 1 μmol/L Calcein AM in FBS-FREE Complete Media for intracellular labeling of live cell cultures: In a 15 mL conical tube or small sterile vial, add 2 μL of 1 mmol/L Calcein AM resuspended in DMSO to 2 mL of FBS-FREE Complete Media. This is enough volume to cover and label one 25 mm round glass coverslip in a 35 mm petri dish. Prepare media with fluoroprobe just prior to labeling live cell cultures. Protect from direct light exposure (see Note 5).
4. Phenol-Free Complete Media for imaging live cell cultures: Prepare Complete Media as in Subheading 2.2, item 2, using phenol-free DMEM. Store at 4 °C for 1 month (see Note 6).

3. Methods

The Methods describe protocols for preparing MSCs, labeling MSCs with QDs, fixing MSC cultures, and labeling fixed and live MSCs with secondary fluorescent markers. Labeling fixed or live cells with a secondary fluorescent marker is not necessary to track QD labeled MSCs in vitro or in vivo. These protocols are included to aid in imaging the MSC cultures when evaluating QD location and distribution.

3.1. Cell Culture Preparation

The following procedure applies to MSCs stored in DMSO in liquid nitrogen (see Note 7). Stem cells stored in liquid nitrogen need to be thawed quickly and diluted with media to reduce damage from DMSO. Most of the stem cells live through the thawing process but some do not. Counting the stem cells before plating will provide an indication of how many cells survived the freezing/thawing process. This procedure also removes the liquid DMSO that is used to freeze the cells. This procedure must be performed under a biological hood using sterile cell culture procedures.

1. Warm Complete Media (see Subheading 2.2, item 2) to 37 °C in water bath.
2. Remove cryotube with MSCs from liquid nitrogen and thaw quickly at 37 °C. Hold the cryotube with gloved hands in the water bath. Thaw MSCs until the pellet begins to move. This will take 30 s to 1 min.
3. Transfer the thawed MSCs to a 50 mL conical tube that contains 1 mL of prewarmed Complete Media and mix gently with pipette by gently aspirating and dispensing volume in conical tube three to four times.
4. Place the conical tube with cell suspension in a centrifuge and spin the cells to pellet at 1,500 rpm for 5 min.

5. Remove the media above the cell pellet and resuspend the cell pellet in 2 mL of Complete Media. Mix gently to remove any cell clumps by gently aspirating and dispensing volume in conical tube three to four times.
6. Count MSCs before plating (see Note 8).
7. Remove 1×10^6 cells and place in sterile 50 mL conical tube and centrifuge cells to pellet at 1,500 rpm for 5 min.

3.2. Labeling MSCs with Quantum Dots

1. Follow directions provided by manufacturer and premix 1 μL each of Qtracker Reagent A and B in a 1.5 mL microfuge tube. This is a 10 nmol/L concentration of QDs (see Note 9).
2. Incubate QDs in microfuge tube for 5 min at room temperature.
3. Add 0.2 mL of Complete Media to QDs and vortex for 30 s.
4. Add the 0.2 mL QD labeling solution to the cell pellet containing 1×10^6 MSCs (Subheading 3.1, step 7).
5. Incubate MSCs for 1 h at 37 °C, 5 % CO_2 in cell culture incubator.
6. Remove unincorporated QDs by washing MSCs with Complete Media. To wash MSCs add 1 mL of Complete Media to the MSCs. Centrifuge at 1,500 rpm for 5 min. Discard supernatant, i.e., the media above the cell pellet. Repeat wash sequence.
7. Resuspend washed MSCs in 1 mL of Complete Media.
8. Plate MSCs at a density of 15,000–20,000 cells/cm^2 on sterilized, poly-L-lysine coated glass coverslips (Subheading 2.1, item 4) and place in 37 °C, 5 % CO_2 cell culture incubator.
9. Live MSC cultures can be imaged within 1 h after plating (see Note 10).

3.3. Preparing Fixed Quantum Dot Labeled MSCs for Imaging

Here we describe a procedure for using a secondary fluorescent label, i.e., intracellular or membrane fluorescent marker, to aid in locating and imaging the fixed MSCs. The procedure outlined below is for MSCs plated on glass coverslips. Adapt technique if MSCs are plated in glass coverslip bottom petri dishes or multiwell plates. Have available DH_2O, microscope slides and slide mounting media.

1. Prepare all solutions for fixing cells prior to starting (Subheading 2.3).
2. Place coverslips, cell side up, in a 6 well plate, one coverslip per well. Add 2 mL of COLD 4 % Paraformaldehyde to each well for 10 min (see Note 11).
3. Wash MSCs with 2 mL of 1× PBS for 10 min. Repeat wash step twice.

4. To permeabilize MSCs remove 1× PBS and add 2 mL of 0.1 % Triton X-100 for 10 min. Wash with 2 mL 1× PBS for 10 min. Repeat wash step.
5. To block nonspecific binding of the antibodies remove 1× PBS and add 2 mL of 3 % BSA for 1 h at room temperature.
6. Wash with 2 mL 1× PBS for 10 min. Repeat wash step.
7. Label MSCs with an intracellular or membrane fluoroprobe. Follow laboratory protocol or vendor directions for incubation time for primary antibody (see Note 12).
8. At the end of the incubation step wash with 2 mL 1× PBS for 10 min. Repeat wash step twice.
9. Incubate with secondary antibody. Cover coverslips with foil to protect from light during incubation step (see Note 13).
10. Wash coverslips with 1× PBS for 10 min. Repeat wash step twice.
11. Dip coverslips in DH_2O twice to remove PBS salts.
12. Add one drop of mounting media to clean, dry microscope slide then place coverslip, cells side down, onto the mounting media. Let microscope slides dry flat overnight in closed slide holder protected from light. Seal with acetone free clear nail polish.

3.4. Preparing Live Quantum Dot Labeled MSCs for Imaging

Here we describe a procedure for using an intracellular fluorescent label to aid in locating and imaging the MSCs. The procedure outlined below is for MSCs plated on glass coverslips. Adapt technique if MSCs are plated in glass coverslip bottom petri dishes or multi-well plates. This intracellular labeling step is not necessary to track QD labeled MSCs in vitro or in vivo.

1. Prepare all solutions for labeling cells prior to starting (Subheading 2.4).
2. Wash MSCs plated to coverslip with FBS FREE Complete Media. To wash, use a pipette to aspirate the media from the MSCs, then add FBS-FREE Complete Media. Repeat wash sequence (see Note 3).
3. Remove media and add 1–2 mL of 1 μmol/L Calcein AM in FBS-FREE Complete Media to MSCs plated to coverslip (see Note 5).
4. Cover with aluminum-foil and allow to incubate at 37 °C, 5 % CO_2 in a cell culture incubator for 30 min (see Note 14).
5. Wash MSCs twice with Phenol-Free Complete Media. Use a pipette to aspirate the media from the MSCs, then add 2 mL Phenol-Free Complete Media. Repeat wash sequence twice.
6. Add 1–2 mL Phenol-Free Complete Media to MSCs and return to cell culture incubator for 10 min.

7. To prepare MSCs for imaging, place coverslip in a coverslip chamber with Phenol-Free Complete Media. If using glass bottom petri dishes or multiwell plates add Phenol-Free Complete Media to MSCs and place directly on the microscope stage (see Note 15).

3.5. Fluorescence Imaging of QD labeled MSCs

1. To image the intracellular distribution of QDs in MSCs place fixed MSCs (Subheading 3.3) or live MSCs (Subheading 3.4) on the stage of a laser scanning confocal fluorescence microscope (see Note 16 and 17).
2. To image QD labeled MSCs, select a high magnification objective with a numerical aperture close to 1 (see Note 18).
3. Use a 488 nm excitation wavelength to illuminate the QDs. The 488 nm excitation wavelength can be used to simultaneously collect emission wavelengths through a 580 nm long-pass filter (emission wavelength for QDs) and 515 ± 15 nm band pass filter (emission wavelength for calcein or fluoroprobes with green wavelength emission spectra) as shown in Figs. 1 and 2 (see Note 19).

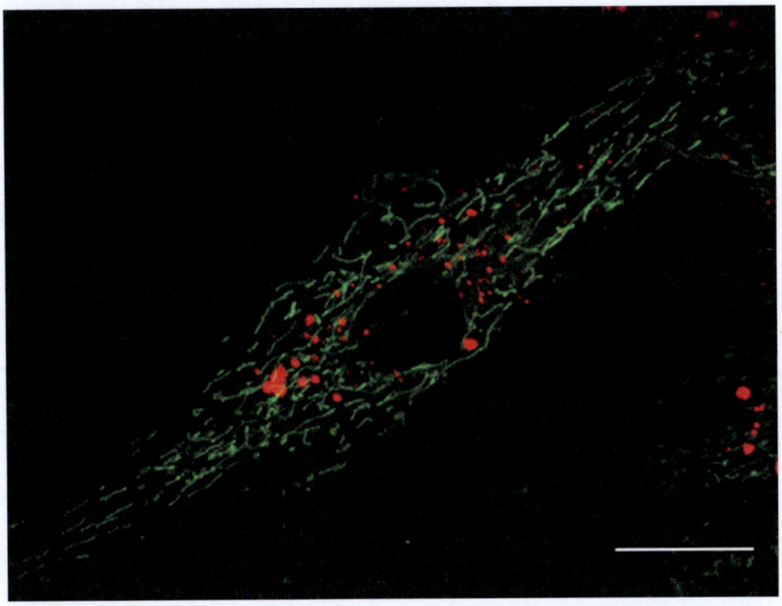

Fig. 1. QD distribution in fixed mesenchymal stem cells (MSCs) 24 h after QD labeling. This image illustrates fixed MSCs labeled with mito-tracker green (*green*) and QDs (*red*). Note the aggregates of the QD particles. This image was acquired using a Zeiss 510 laser scanning confocal fluorescent microscope equipped with a 63×/1.2 NA water immersion objective. An argon ion laser with an excitation wavelength of 488 nm was used and emission wavelengths were collected simultaneously through a 515 ± 15 nm band pass filter and 590 nm long-pass filter. Scale bar 20 μm.

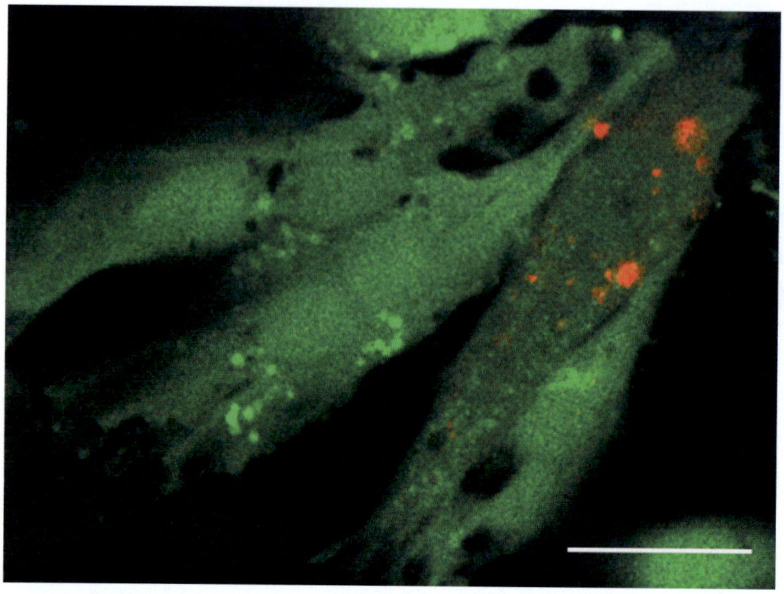

Fig. 2. QD labeled MSC in cardiac myocyte co-culture at 7 days. Images show a 1-μm optical section of a QD labeled MSC growing in culture with cardiac myocytes. The QD labeled MSC is shown adjacent to and surrounded by cardiac myocytes. All cells were labeled with the cytosolic fluoroprobe calcein AM. QDs are preferentially localized in the MSC. No QDs were found to be localized in the cardiac myocytes. This image was acquired on a Zeiss 510 laser scanning confocal fluorescent microscope with a 40×/1.3 NA oil immersion objective. Scale bar 20 μm.

4. Perform Z-stack imaging to acquire optical sections through the MSC cultures as shown in Fig. 3 (see Note 20).
5. Image processing software can be used to generate a 3D volume from the optical sections and evaluate the QD location and distribution in the cell culture as shown in Fig. 4 (see Note 20).

4. Notes

1. Select glass coverslips #1.5 with a thickness of 170 nm, designed for high-resolution imaging. The focal plane of the images will be shifted if the coverslip is too thin (<170 nm) or too thick (>170 nm), decreasing image resolution. Avoid using plastic coverslips. Plastic coverslips are autofluorescent and will prevent high-resolution imaging of the MSCs and QDs. Cells can be plated on 25 mm round #1.5 glass coverslips and placed in a petri dish. For live cell imaging place the coverslip in a coverslip chamber with media before imaging.

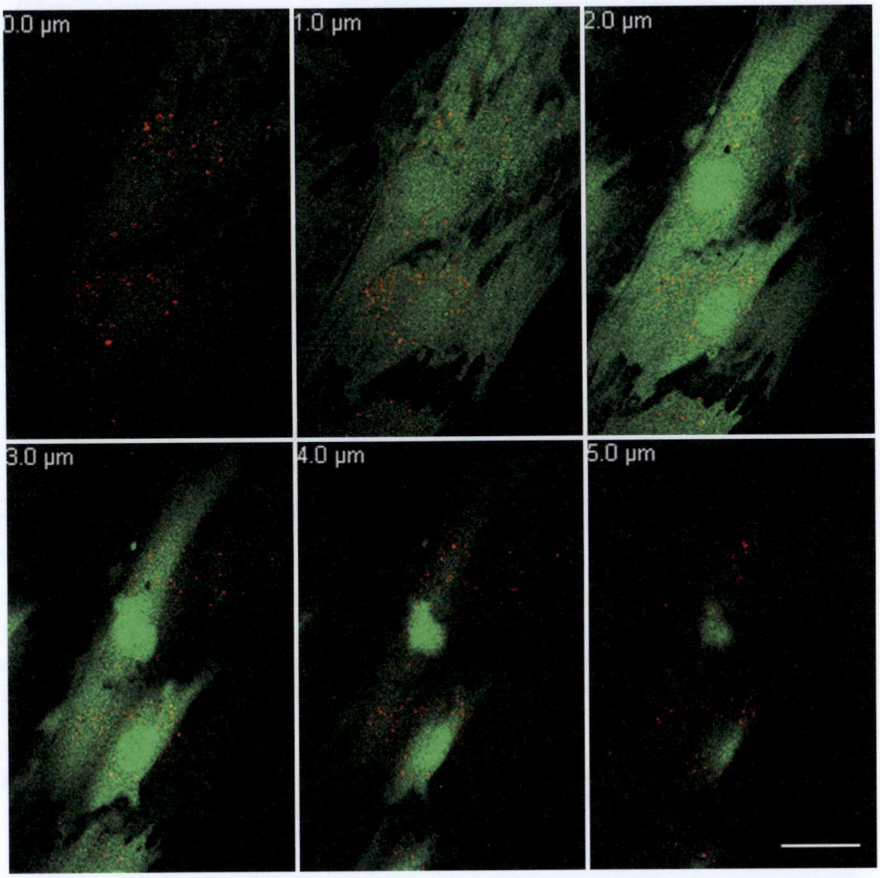

Fig. 3. QD location in labeled MSCs. Images show 1 μm optical sections acquired as a confocal Z stack with 1-μm spacing. The image labeled 0.0 μm shows the QD labeled MSC close to the coverslip. As acquired images advance into the cell culture (1.0–5.0 μm, away from coverslip) the QD aggregates appear to be located within and through the thickness of the MSC. To visualize cell boundaries MSCs were labeled with the cytosolic fluoroprobe calcein AM (*green*). All images were acquired on a Zeiss 510 laser scanning confocal fluorescent microscope with a water immersion 63×/1.2 NA objective. Scale bar 20 μm.

2. A Q-Tracker® 605 Cell Labeling kit was purchased with QDs composed of a cadmium selenium core and an inner zinc sulfide shell with a custom peptide bound to the QDs outer shell. The peptide allowed the QDs to be endocytosed into the MSC interior and exist in periplasmic vesicles (8, 9). QDs were approximately 10–15 nm in diameter with spectral characteristics of 488 nm wavelength excitation and 605 nm wavelength peak emission.
3. Prepare Complete Media without FBS to facilitate fluoroprobe intracellular labeling. It is important to wash the MSC cultures with the FBS free media to remove any residual FBS in the cell culture before adding the fluoroprobe. The FBS free media should be used only during incubation with the fluoroprobe.

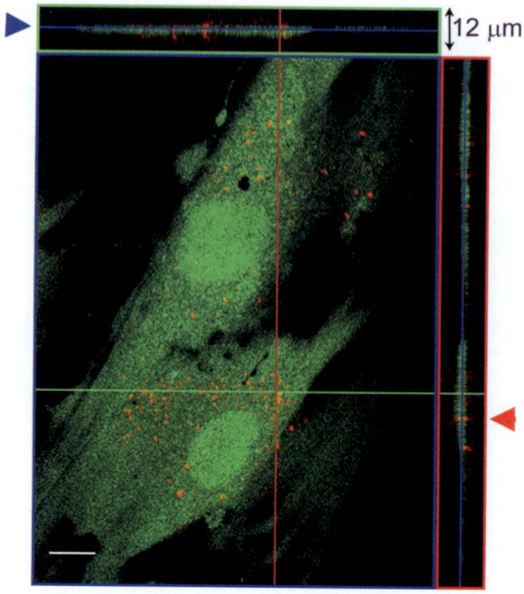

Fig. 4. 3D distribution and localization of QDs in MSCs. A cut view through 12 superimposed optical sections (12 μm) illustrating the 3D distribution of QDs in MSC cultures 24 h post-QD labeling. The image shows the intracellular space of the MSCs indicated by the *blue arrow*. The *red* (*vertical*) and *green* (*horizontal*) crosshairs are aligned near QD aggregates, *red arrow*. The QD aggregates are homogeneously distributed through the MSC cytosol and are clearly visualized in the intracellular space. Scale bar 20 μm. Reproduced from (6) with permission from Biomed Central.

4. If possible purchase and resuspend small aliquots of fluorescent dyes. Resuspended calcein AM can be thawed and refrozen several times without affecting cell labeling efficiency. Resuspend fluorescent dyes just prior to cell loading. Keep resuspended fluorescent dyes covered and protected from light exposure.
5. The intracellular fluoroprobe calcein is bright and photostable in MSCs for several hours. We have used calcein at concentrations ranging from 500 nmol/L to 1 μmol/L with success. Label MSCs only as necessary and preferably just before imaging. We have observed calcein being taken up by intracellular organelles and/or leaking out of the cells several hours after labeling.
6. Prepare phenol-free Complete Media to use for live MSCs imaging. The color of the phenol will introduce noise into the images acquired at wavelengths greater than 600 nm (i.e., images acquired in the red spectrum).
7. This procedure has been used with passage 2 and 3 MSCs. Results may differ with MSCs at later passages. This procedure was not evaluated on freshly isolated MSCs.

8. Count cells using a hemocytometer, remove 30 μL of cells in media and place in a 1.5 mL microfuge tube. Add 30 μL of trypan blue and mix well. Load the hemocytometer with 10 μL and count live cells only, excluding all blue cells from count.
9. In preliminary studies QD concentrations of 5 and 20 nmol/L were evaluated. MSCs labeled with the 20 nmol/L concentration showed a higher intracellular QD yield, but MSC proliferation was decreased. At all concentrations QD aggregates were present in the MSCs and easily identified.
10. For in vitro or in vivo applications QD labeled MSCs should be used within the first 24 h after QD labeling. In earlier studies our lab reported that detection of QD labeled MSCs decreased as MSCs proliferate in culture (6).
11. When adding buffer to the MSCs in the multiwell plates use a transfer pipette and let it run down the side of the wells. Turn the multiplate over onto an absorbent paper towel to empty buffer from wells. Coverslips will stay attached to the bottom of the multiplate.
12. Primary antibodies vary and this step may require an overnight incubation. Consult protocol supplied by vendor.
13. Incubation times for secondary antibodies will vary. Consult protocol supplied by vendor.
14. When working with live cells labeled with fluoroprobes keep light exposure to a minimum to reduce photobleaching.
15. When imaging live cell cultures an onstage incubator attachment with atmosphere and temperature control is beneficial to provide a stable environment for the cell cultures. If an onstage incubator system is not available replace the media covering the MSCs with fresh media about every 30 min. A heated stage set to 37 °C may also be used.
16. When using an inverted microscope place slide on microscope stage with coverslip facing down. When using an upright microscope place slide on microscope stage with the coverslip facing up.
17. Use an inverted microscope for live cell imaging. Gently clean the bottom of coverslip (cell free side) with lens paper and lens cleaning solution before placing on microscope stage.
18. For fixed MSCs a ×63/1.4 NA oil immersion objective should be used. For live MSCs a ×63/1.2 NA water immersion objective should be used. A ×40 objective also provides good images. We observed no enhancement of images with ×100 objectives.
19. The optimum wavelength to excite QDs is 405 nm but QDs can also be excited at 488 nm.

20. QD labeled MSCs were identified by punctate red fluorescent cellular inclusions. Acquiring 1 μm optical images through the cell culture will allow you to evaluate QD location within the cells. Image analysis software is available with the Zeiss microscopes. ImageJ software (http://rsb.info.nih.gov/ij) was used to evaluate QD location and distribution and count QD aggregates in MSCs.

Disclaimer

The chapter has been reviewed by the National Health and Environmental Effects Research Laboratory, U.S. Environmental Protection Agency, and approved for publication. The contents of this article should not be construed to represent Agency policy nor does mention of trade names or commercial products constitute endorsement or recommendation for use.

References

1. Jaiswal J, Mattoussi H, Mauro J, Simon S (2003) Long-term multiple color imaging of live cells using quantum dot bioconjugates. Nat Biotechnol 21:47–51
2. Chan W, Maxwell D, Gao X, Bailey R, Han M (2002) Luminescent quantum dots for multiplexed biological detection and imaging. Curr Opin Biotechnol 13:40–46
3. Alvisiatos A (2004) The use of nanocrystals in biological detection. Nat Biotechnol 22:47–52
4. Jaiswal JK, Simon SM (2004) Potentials and pitfalls of fluorescent quantum dots for biological imaging. Trends Cell Biol 14(9):497–504
5. Zheng J, Ghazani A, Song Q, Mardvani S, Chan W, Wang C (2006) Cellular imaging and surface marker labeling of hematopoietic cells using quantum dot bioconjugates. Lab Hematol 12:94–98
6. Muller-Borer BJ, Collins MC, Gunst PR, Cascio WE, Kypson AP (2007) Quantum dot labeling of mesenchymal stem cells. J Nanobiotechnolgy 5:1–9
7. Rosen A, Kelly D, Schuldt A, Potapova I, Doronin K, Robichaud R, Rosen M, Brink P, Gaudette G, Cohen I (2007) Finding fluorescent needles in the cardiac haystack: tracking human mesenchymal stem cells labeled with quantum dots for quantitative in vivo 3-D fluorescence analysis. Stem Cells 25:2128–2138
8. Jaiswal J, Goldman E, Mattoussi H, Simo S (2004) Use of quantum dots for live cell imaging. Nat Methods 1(1):73–78
9. Larson D, Zipfel W, Williams R, Clark S, Bruchez M, Wise F, Webb W (2004) Water-soluble quantum dots for multiphoton fluorescence imaging in vivo. Science 300:1434–1436

Chapter 16

The Application of Semiconductor Quantum Dots for Enhancing Peptide Desorption, Improving Peak Resolution and Sensitivity of Detection in Matrix-Assisted Laser Desorption/Ionization (MALDI) Mass Spectrometry

Julian Bailes and Mikhail Soloviev

Abstract

The interest in quantum dots (QD) and a number of reported life sciences applications increased dramatically over the last decade. The popularity of QDs stems from better photostability, higher extinction co-efficients, and unique optical properties such as superior light absorption. Here we report methods for improving matrix-assisted laser desorption/ionization desorption of crude tryptic protein digests by using CdSe/ZnS QDs. The addition of QDs to the matrix improves the signal-to-noise ratio, peak quality and increases the number of detected peptides and the overall sequence coverage.

Key words: Quantum dots, Mass spectrometry, MALDI, Nanoparticles, QD, Semiconductor, Desorption/ionization, Light absorption, Peptides

1. Introduction

Over the last few decades the use of semiconductor quantum dots has expanded from material science into other areas including the life sciences. The vast majority of applications involving quantum dots (QDs) in life sciences exploit their optical properties, namely their wide absorption spectra and tunable fluorescence emission spectra. The latter depend on the nanoparticle size and are therefore easily adjustable. Typical applications include the use of QDs as fluorescent biological labels instead of organic molecules (1), for in vivo imaging (2–5) and immunochemistry applications (6, 7), in fluorescence energy transfer (FRET) applications (8, 9), for the detection of DNA binding proteins (10) fluorescent labeling and

Mikhail Soloviev (ed.), *Nanoparticles in Biology and Medicine: Methods and Protocols*, Methods in Molecular Biology, vol. 906, DOI 10.1007/978-1-61779-953-2_16,

detection of DNA (11, 12) and protein arrays (13–15) and for multiplexed fluorescence coding and detection of molecules (16, 17). Other applications of QDs outside the life sciences include their use in QD-based light emitting diodes (QLEDs) (18–20) in a range of photovoltaics applications including solar cells (21–24), for downconversion of high energy to lower energy light (instead of conventional inorganic phosphors) (25, 26) and in image sensors in optoelectronics (27). In many of the aforementioned applications, the superior light absorption ability of QDs (over silicon based materials) provides the advantage, such as the reported twofold or higher increase in the incident photon-to-charge carrier generation efficiency for QD-sensitized solar cells for example (28, 29).

The development of matrix-assisted laser desorption/ionization mass spectrometry (MALDI MS) started with the use of fine metal powders suspended in glycerol (30, 31), and a massive array of matrix-free and matrix-assisted LDI techniques and applications followed, including ionization on silicon (DIOS) (32, 33), polymeric sol–gel assisted laser desorption/ionization (SGALDI) (34–38), surface-enhanced laserdesorption/ionization (SELDI) (39, 40) and many others. The reader might also be interested to further explore the effects of various additives such as glycerol (30, 41), paraffin (42), copper (43), and ammonium citrate (33). More recent reports have shown that noble metal nanoparticles and germanium semiconductor nanoparticles can also improve matrix-assisted desorption or, in some cases, substitute matrix altogether (44–48). In addition, we reported that QDs may also improve peptide detection in MALDI MS (49), with our findings subsequently confirmed independently by other groups (50, 51).

Here we describe a simple modification of the standard and widely practiced MALDI-MS procedure which includes the use of CdSe/ZnSQDs for improving peptide desorption and detection on MALDI-TOF-MS. The procedure is based on ref. (49) and results in an increase in the number of the peptides detected/observed, improved S/N ratios, and higher peptide peak resolution.

2. Materials

1. Alpha-cyano-4-hydroxycinnamic acid matrix solution in methanol, MALDI-QUALITY™ (Agilent Technologies).
2. Colloidal dispersions of CdSe/ZnS Core/Shell "EviDots" in toluene (Evident Technologies) ED-C11-TOL-0560 (566 nm) or ED-C11-TOL-0580 (579 nm) (see Notes 1 and 2). Alternatively other QDs may be used (see Note 3).

3. Acetone (99.5 %).
4. Acetic acid (99.8 %).
5. 20 mM potassium phosphate buffer (PBS), pH 9.
6. Trypsin: make 0.25 mg/mL working solution in 20 mM potassium phosphate, pH 9, immediately prior to the digestion (see Note 4).
7. Protein for the MALDI analysis. We use Bovine Serum Albumin (BSA) to illustrate the method (see Note 5).
8. Bruker Reflex III MALDI-TOF or similar MALDI-TOF mass spectrometer (see Note 6) and a suitable MALDI plate. Our instrument was equipped with a sealed nitrogen pulsed laser (337 nm, line width 0.1 nm, 4 ns pulse duration, 300 μJ rated pulse energy, average power 5 mW at 20 Hz).
9. A stainless steel MALDI plate.

3. Methods

1. Prepare proteolytic digestion using Trypsin:protein ratio of 1:10 in 20 mM PBS and incubate for 10 h at 37 °C (see Notes 5 and 7).
2. Prepare matrix solution by mixing 60 μL of the alpha-cyano-4-hydroxycinnamic acid matrix stock solution in methanol, 30 μL acetone and 1 μL 1 % acetic acid in a glass vessel (see Note 8).
3. Prepare QD-enhanced matrix solution by adding 5 μL of QD solution to 25 μL of the traditional matrix (from previous step) (see Note 9).
4. Mix the digested protein samples with the matrixes (±QDs) in 200 μL polypropylene tubes by adding 1 μL of protein to 10 μL of matrix. Resuspend the mixture using a hand pipette and immediately spot onto a stainless steel MALDI plate (see Note 8). The addition of QDs yields a noticeable improvement in the appearance of matrix crystals on the plate (see Fig. 1).
5. Follow the standard operating procedures for acquiring mass spectra (see Note 10).
6. To compare the matrices (with and without QDs), check the reported peak resolution values, the number of correctly identified (matched) peaks (see Note 11).

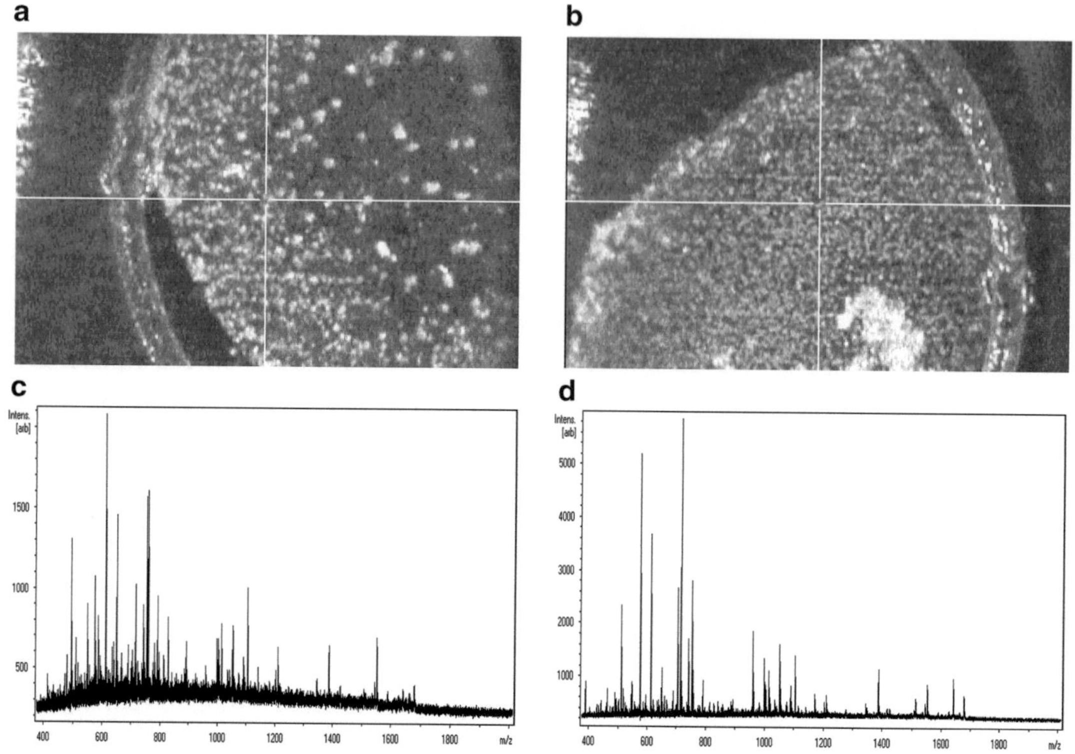

Fig. 1. Comparison of crystallized MALDI matrix (alpha-cyano-4-hydroxycinnamic acid) with and without addition of QDs. (**a**) Crude tryptic digest of BSA (no purification), crystallized MALDI matrix without QDs. (**b**) Same Tryptic digest of BSA as in (**a**) but containing a 50:50 % (v/v) mix "EviDots." (**c**) MALDI spectrum obtained from tryptic digests of BSA spotted with matrix only (no QDs). (**d**) MALDI spectrum obtained from tryptic digests of BSA spotted using alpha-cyano-4-hydroxycinnamic acid matrix mixed with QDs. The amount of protein and the matrix spotted are the same in (**c**) and (**d**). Spectra were acquired with 30 laser shots (at 10 Hz) using 50 % laser power setting in both cases (49).

4. Notes

1. We previously observed that QDs with emission around 575 nm (Ex = 337 nm) produced the best results (49).
2. The size of the QDs was estimated from TEM images.
3. The exact mechanism of the improved adsorption by QDs is not known. A number of possible explanations include the improvement of the matrix crystallization (with QDs forming nucleation centers for the growing matrix, see Fig. 1), improvement in the efficiency of light absorption during sample exposure to laser light pulse through higher extinction coefficients of QDs, or by means of increasing surface area, or because of spectral shifts introduced by QDs and the resulting increase in light absorption by matrix molecules, or due to the role of QDs as electron scavengers, or because of a combination of the

above (43, 49). It is therefore advisable to test any particular QD preparation by comparing desorption with and without QDs in the matrix with a sample of known proteolytically digested protein before using the procedure for unknown experimental samples.

4. Avoid storing Trypsin at neutral pH for longer periods of time to avoid autolysis.
5. Ensure that the final pH of the reaction mixture is within the suitable pH range for Trypsin (pH 8–9) or is ideally at the optimum pH of 8 (52).
6. We used the instrument in reflectron mode.
7. There is plenty of information and protocols in the literature, e.g., (53). It is generally advisable to follow the protocol suggested by the supplier.
8. It is advisable to rinse all the glassware and plasticware in methanol and air dry before use.
9. If QDs are dissolved in toluene, which is *immiscible* with water containing solutions (e.g., matrix with peptides), the mixture should be mixed by pipetting well and used immediately.
10. Spots containing the QDs in addition to the matrix are likely to require reduced laser power (approximately 50 % less than without the QDs (49)), and result in a better quality mass spectra (much improved S/N ratio, improved peak resolution, increased number of identified peaks) and therefore improved Mascot scores (49).
11. One way to objectively compare the number of identified peaks is to create a file with calibration masses using the list of expected proteolytic peptides, e.g., BSA peptides in this case, and then use the auto-calibration function in Bruker FlexControl or FlexAnalysis software to match the peaks. Alternatively, the automatically identified masses should be submitted for peptide mass fingerprinting and the matched masses counted. The latter approach will also yield Mascot scores, which can be useful in comparing the different adsorption procedures.

References

1. Bruchez M, Moronne M, Gin P et al (1998) Semiconductor nanocrystals as fluorescent biological labels. Science 281(5385): 2013–2016
2. JaiswalJK GER, Mattoussi H et al (2004) Use of quantum dots for live cell imaging. Nat Methods 1(1):73–78
3. Gao XH, Yang LL, Petros JA et al (2005) In vivo molecular and cellular imaging with quantum dots. Curr Opin Biotechnol 16(1):63–72
4. Michalet X, Pinaud FF, Bentolila LA et al (2005) Quantum dots for live cells, in vivo imaging, and diagnostics. Science 307(5709):538–544

5. Xing Y, Rao J (2008) Quantum dot bioconjugates for in vitro diagnostics & in vivo imaging. Cancer Biomark 4(6):307–319
6. Sukhanova A, Devy M, Venteo L et al (2004) Biocompatible fluorescent nanocrystals for immunolabeling of membrane proteins and cells. Anal Biochem 324(1):60–67
7. Xing Y, Chaudry Q, Shen C et al (2007) Bioconjugated quantum dots for multiplexed and quantitative immunohistochemistry. Nat Protoc 2(5):1152–1165
8. Medintz IL, Clapp AR, Mattoussi H et al (2003) Selfassembled nanoscale biosensors based on quantum dot FRET donors. Nat Mater 2(9):630–638
9. Zhang CY, Hu J (2010) Single quantum dot-based nanosensor for multiple DNA detection. Anal Chem 82(5):1921–1927
10. Ebenstein Y, Gassman N, Kim S et al (2009) Lighting up individual DNA binding proteins with quantum dots. Nano Lett 9(4):1598–1603
11. Giraud G, Schulze H, Bachmann TT et al (2009) Fluorescence lifetime imaging of quantum dot labeled DNA microarrays. Int J Mol Sci 10(4):1930–1941
12. Liang RQ, Li W, Li Y et al (2005) An oligonucleotide microarray for microRNA expression analysis based on labeling RNA with quantum dot and nanogold probe. Nucleic Acids Res 33(2):e17
13. Shingyoji M, Gerion D, Pinkel D et al (2005) Quantum dots-based reverse phase protein microarray. Talanta 67(3):472–478
14. Zajac A, Song D, Qian W et al (2007) Protein microarrays and quantum dot probes for early cancer detection. Colloids Surf B Biointerfaces 58(2):309–314
15. Gokarna A, Jin LH, Hwang JS et al (2008) Quantum dot-based protein micro- and nanoarrays for detection of prostate cancer biomarkers. Proteomics 8(9):1809–1818
16. Han MY, Gao XH, Su JZ et al (2001) Quantum-dot-tagged microbeads for multiplexed optical coding of biomolecules. Nat Biotechnol 19:631–635
17. Zhao Y, Zhao X, Tang B et al (2010) Quantum-dot-tagged bioresponsive hydrogel suspension array for multiplex label-free DNA detection. Adv Funct Mater 20(6):976–982
18. Bae WK, Kwak J, Lim J et al (2009) Deep blue light-emitting diodes based on Cd1-xZnxS@ZnS quantum dots. Nanotechnology 20(7):075202
19. Cho KS, Lee EK, Joo WJ et al (2009) High-performance crosslinked colloidal quantum-dot light-emitting diodes. Nat Photonics 3:341–345
20. Sun Q, Wang YA, Li LS et al (2007) Bright, multicoloured light-emitting diodes based on quantum dots. Nat Photonics 1:717–722
21. Beard MC, Ellingson RJ (2008) Multiple exciton generation in semiconductor nanocrystals: toward efficient solar energy conversion. Laser Photonics Rev 2(5):377–399
22. Sablon KA, Little JW, Mitin V et al (2011) Strong enhancement of solar cell efficiency due to quantum dots with built-in charge. Nano Lett 11(6):2311–2317
23. Ardalan P, Brennan TP, Lee HB et al (2011) Effects of self-assembled monolayers on solid-state CdS quantum dot sensitized solar cells. ACS Nano 5(2):1495–1504
24. Kamat PV (2008) Quantum dot solar cells. Semiconductor nanocrystals as light harvesters. J Phys Chem C 112(48):18737–18753
25. Nizamoglu S, Ozel T, Sari E et al (2007) White light generation using CdSe/ZnS core-shell nanocrystals hybridized with InGaN/GaN light emitting diodes. Nanotechnology 18:065709
26. Yeh DM, Huang CF, Lu YC et al (2008) White-light light-emitting device based on surface plasmon-enhanced CdSe/ZnS nanocrystal wavelength conversion on a blue/green twocolor light-emitting diode. Appl Phys Lett 92(9):09112
27. Noone KM, Ginger DS (2009) Doping for speed: colloidal nanoparticles for thin-film optoelectronics. ACS Nano 3(2):261–265
28. Lee HJ, Yum JH, Leventis HC et al (2008) CdSe quantum dot-sensitized solar cells exceeding efficiency 1 % at full-sun intensity. J Phys Chem B 112(30):11600–11608
29. Kongkanand A, Tvrdy K, Takechi K et al (2008) Quantum dot solar cells. Tuning photoresponse through size and shape control of $CdSe-TiO_2$ architecture. J Am Chem Soc 130(12):4007–4015
30. Tanaka K, Waki H, Ido Y et al (1988) Protein and polymer analyses up to m/z 100,000 by laser ionization time-of-flight mass spectrometry. Rapid Commun Mass Spectrom 2(8):151–153
31. Tanaka K (2003) The origin of macromolecule ionization by laser irradiation. Angew Chem Int Ed Engl 42(33):3860–3870
32. Wei J, Buriak JM, Siuzdak G (1999) Desorption-ionization mass spectrometry on porous silicon. Nature 399(6733):243–246
33. Thomas JJ, Shen ZX, Crowell JE et al (2001) Desorption/ionization on silicon (DIOS): a diverse mass spectrometry platform for protein characterization. Proc Natl Acad Sci U S A 98(9):4932–4937

34. Lin YS, Chen YC (2002) Laser desorption /ionization time-of-flight mass spectrometry on sol-gel derived 2,5-dihydroxybenzoic acid film. Anal Chem 74(22):5793–5798
35. Chen WY, Chen YC (2003) Reducing the alkali cation adductions of oligonucleotides using sol-gel-assisted laser desorption /ionization mass spectrometry. Anal Chem 75(16):4223–4228
36. Lin YS, Yang CH, Chen YC (2004) Glass-chip-based sample preparation and on-chip trypic digestion for matrix-assisted laser desorption/ ionization mass spectrometric analysis using a sol-gel/2,5-dihydroxybenzoic acid hybrid matrix. Rapid Commun Mass Spectrom 18(3):313–318
37. Chen CT, Chen YC (2004) Desorption/ionization mass spectrometry on nanocrystalline titania sol-gel-deposited films. Rapid Commun Mass Spectrom 18(17):1956–1964
38. Chen CT, Chen YC (2004) Molecularly imprinted TiO_2-matrix-assisted laser desorption/ionization mass spectrometry for selectively detecting alpha-cyclodextrin. Anal Chem 76(5):1453–1457
39. Tang N, Tornatore P, Weinberger SR (2004) Current developments in SELDI affinity technology. Mass Spectrom Rev 23(1):34–44
40. Seibert V, Wiesner A, Buschmann T et al (2004) Surface-enhanced laser desorption ionization time-of-flight mass spectrometry (SELDI TOF-MS) and ProteinChip technology in proteomics research. Pathol Res Pract 200(2):83–94
41. Berkenkamp S, Menzel C, Karas M et al (1997) Performance of infrared matrix-assisted laser desorption/ionization mass spectrometry with lasers emitting in the 3 um wavelength range. Rapid Commun Mass Spectrom 11(13):1399–1406
42. Kinumi T, Saisu T, Takayama M et al (2000) Matrix-assisted laser desorption/ionization time-of-flight mass spectrometry using an inorganic particle matrix for small molecule analysis. J Mass Spectrom 35(3):417–422
43. Okuno S, Nakano M, Matsubayashi G et al (2004) Reduction of organic dyes in matrix-assisted laser desorption/ionization and desorption/ionization on porous silicon. Rapid Commun Mass Spectrom 18(23):2811–2817
44. Shrivas K, Wu HF (2008) Modified silver nanoparticle as a hydrophobic affinity probe for analysis of peptides and proteins in biological samples by using liquid-liquid microextraction coupled to AP-MALDI-ion trap and MALDI-TOF mass spectrometry. Anal Chem 80(7):2583–2589
45. Shrivas K, Wu HF (2008) Applications of silver nanoparticles capped with different functional groups as the matrix and affinity probes in surface-assisted laser desorption/ionization time-of-flight and atmospheric pressure matrix-assisted laser desorption/ionization ion trap mass spectrometry for rapid analysis of sulfur drugs and biothiols in human urine. Rapid Commun Mass Spectrom 22(18): 2863–2872
46. Sudhir PR, Shrivas K, Zhou ZC et al (2008) Single drop microextraction using silver nanoparticles as electrostatic probes for peptide analysis in atmospheric pressure matrix-assisted laser desorption/ionization mass spectrometry and comparison with gold electrostatic probes and silver hydrophobic probes. Rapid Commun Mass Spectrom 22(19):3076–3086
47. Honda A, Sonobe H, Ogata A et al (2005) Improved method of the MALDI-TOF analysis of DNA with nanodot sample target plate. Chem Commun 42:5340–5342
48. Seino T, Sato H, Yamamoto A et al (2007) Matrix-free laser desorption/ionization-mass spectrometry using self-assembled germanium nanodots. Anal Chem 79(13):4827–4832
49. Bailes J, Vidal L, Ivanov DA et al (2009) Quantum dots improve peptide detection in MALDI MS in a size dependent manner. Nanobiotechnology 7:10
50. Shrivas K, Kailasa SK, Wu HF (2009) Quantum dots laser desorption/ionization MS: multifunctional CdSe quantum dots as the matrix, concentrating probes and acceleration for microwave enzymatic digestion for peptide analysis and high resolution detection of proteins in a linear MALDI-TOF MS. Proteomics 9(10):2656–2667
51. Liu CW, Chien MW, Chen GF et al (2011) Quantum dot enhancement of peptide detection by matrix-assisted laser desorption/ionization mass spectrometry. Anal Chem 83(17):6593–6600
52. Finehout EJ, Cantor JR, Le KH (2005) Kinetic characterization of sequencing grade modified trypsin. Proteomics 5(9):2319–2321
53. Rivers J, Simpson DM, Robertson DH et al (2007) Absolute multiplexed quantitative analysis of protein expression during muscle development using QconCAT. Mol Cell Proteomics 6(8):1416–1427

Part III

Magnetic Nanoparticles: Synthesis and Applications

Chapter 17

Near-Infrared Fluorescence Labeling of Iron Nanoparticles and Applications for Cell Labeling and In Vivo Imaging

Jinke Wang, Yingxun Liu, Yong Hou, Zhongpin Chen, and Ning Gu

Abstract

In recent years, the near-infrared fluorescence (NIRF) labeled iron nanoparticles were synthesized and applied to labeling human cells for monitoring the engraftment process, imaging tumors, testing intracellular molecular environment surrounding the nanoparticles, and tracing biodistribution of nanoparticles in vivo. These studies demonstrated that the NIRF-labeled iron nanoparticles provided an excellent method not only for cell labeling but also for in vivo monitoring and tracing of iron nanoparticles due to the excellent in vivo imaging performance of the NIR fluorophores. However, the availability of commercial iron nanoparticles labeled with suitable NIRF dyes is limited. Optimal wavelength for in vivo imaging is centered at 800 nm, where tissue autofluorescence is minimal. Here we describe the manufacture of 12-nm 3-dimercaptosuccinic acid-coated Fe_3O_4 magnetic nanoparticles, their labeling with a new near-infrared fluorophore, IRDye800CW (excitation/emission: 778/806 nm), and their applications for cell labeling and in vivo imaging.

Key words: Near-infrared fluorescence, Iron nanoparticles, Cell labeling, In vivo imaging

1. Introduction

The iron-based magnetic nanoparticles (MNPs) are well suited for use in cell labeling and in vivo imaging (1–8). At present, the in vivo imaging based on iron nanoparticles is achieved mainly by using magnetic resonance imaging (MRI) (9–14). However, MRI costs are prohibitively high and limit its use for routine biomedical research. Bifunctional iron nanoparticles suitable for both MRI and fluorescence imaging and detection have been developed to overcome these limitations (15, 16). For example, iron nanoparticles were labeled with fluorophores, such as FITC (17–19), rhodamine B (20, 21), and rhodamine 6G (14), and were used to investigate the intracellular distribution of MNPs in cells (13, 14, 22).

Mikhail Soloviev (ed.), *Nanoparticles in Biology and Medicine: Methods and Protocols*, Methods in Molecular Biology, vol. 906,
DOI 10.1007/978-1-61779-953-2_17,

However, nanoparticles labeled with conventional fluorophores having excitation wavelengths in the range 350–700 nm, cannot be used for in vivo imaging due to high autofluorescence of biological materials below 700 nm.

Near-infrared fluorescence (NIRF) imaging has been applied for monitoring biological functions in vitro and in vivo (23–25). The advantage of NIR fluorophores is that better signal to noise ratios can be achieved in the spectrum from 700 to 900 nm because of the low absorptivity of tissue chromophores (26). NIR fluorophores can also be imaged through several centimeters of tissue (27–29). NIRF has already been used to label nanoparticles and to study their biodistribution, clearance and biocompatibility for in vivo biomedical applications. Reported examples include silica nanoparticles labeled with DY776 (30) and calcium phosphate nanoparticles labeled with indocyanine green (ICG) (31).

The NIRF dyes have been used to label the iron nanoparticles. For example, the dextran-coated iron oxide nanoparticles (Feridex) were covalently labeled with Alexa Fluor 750 and used to label human hepatic stellate cell for monitoring the engraftment process (32). In another report iron oxide particles were labeled with VivoTag 680 and used to image tumors (33). Monocrystalline iron oxide nanoparticles with cross-linked dextran coating (cross-linked iron oxide or CLIO) nanoparticles, used as MRI contrast agent, were labeled with Cy5.5 (excitation/emission = 660/710 nm) and employed to sensor in vivo molecular environment surrounding the nanoparticles and tracing the in vivo biodistribution of CLIO in liver, spleen and kidneys (34). In other reports NIRF-labeled iron nanoparticles were shown to be suitable for labeling biomolecules, cells, and for in vivo imaging (35–38).

However, the availability of commercial iron nanoparticles labeled with suitable NIRF dyes is limited. Optimal wavelength for in vivo imaging is centered at 800 nm, where tissue autofluorescence is minimal. Here we describe the manufacture of 12-nm 3-dimercaptosuccinic acid (DMSA)-coated Fe_3O_4 MNPs, their labeling with a new near-infrared fluorophore, IRDye800CW (Excitation/emission: 778/806 nm), and their applications for cell labeling and in vivo imaging. The use of IRDye800CW provides several advantages. Firstly, IRDye800CW is a reactive dye (39), which can be easily conjugated to MNPs. Therefore the described methods can be easily adopted for the labeling of other DMSA-coated nanoparticles. Secondly, the excitation and emission maxima of IRDye800CW are in the spectral region where tissue autofluorescence, and scattering are minimal (800 nm), yielding high signal to background ratio (40, 41). Thirdly, this dye is highly water soluble and shows very low nonspecific binding to cellular components (39, 42). Furthermore, animal toxicity studies revealed that the single administration of IRDye800CW carboxylate intravenously at dose levels of 1, 5, and 20 mg/kg produced no pathological evidence of toxicity (39). A number of publications report successful use of IRDye800CW and its conjugates for in vivo imaging in small animals (42–47).

2. Materials

2.1. Preparation of IRDy800CW-MNPs

1. Solution 1 (Iron(III) chloride): dissolve 2.7 g of $FeCl_3{\cdot}6H_2O$ in 50 mL of methanol and then add 8.5 mL oleic acid. Make fresh just before use.
2. Solution 2 (Sodium hydroxide): dissolve 1.2 g of NaOH in 100 mL of methanol. Make fresh just before use.
3. 1-Octadecene (90 %).
4. Oleic acid (analytic reagent).
5. Ethanol (100 %).
6. Chloroform.
7. Triethylamine (analytic reagent).
8. Dimethyl sulfoxide (DMSO).
9. DMSA in DMSO mixture: dissolve 50 mg of DMSA in 10 mL of DMSO.
10. 1-Ethyl-3-[3-dimethylaminopropyl] carbodiimide hydrochloride (EDC) solution (0.55 mg/mL): dissolve 20 mg of EDC in 36 mL of water.
11. Sterile deionized water.
12. Streptavidin-IRDye 800CW (Li-Cor Biosciences).
13. Nitrogen gas.
14. Magnetic stirrer.
15. Ultrasonic cleaner (220 V, 100 W, 223 × 114 × 123 mm).
16. Centrifuge (Eppendorf centrifugator 5810R or similar).
17. Permanent magnet (0.5 × 0.5 × 0.2 cm; field strength 0.3 T).
18. Vacuum pump.
19. Vacuum drying oven.
20. Heat block with Programmable temperature controller. We used XMT-8000P controller (Shanghai Yatai Instrumentation Co), but any equivalent instrument would be suitable.
21. Rotator platform suitable for 15 mL tubes.

2.2. Characterization of IRDy800CW-MNPs

1. Transmission electron microscope (TEM) JEM-2100 (JEOL Ltd.).
2. Copper grid (Type: 300 mesh).
3. Odyssey Infrared Imaging System (Li-Cor Biosciences).

4. Fluorescence Spectrophotometer F-7000 (HITACHI).
5. Image Origin 6.1: a Scientific Graphing and Data Analysis Software developed from Origin Lab Corporation.
6. 96-Well microplate suitable for use with Fluorescence Spectrophotometer.

2.3. Cell Labeling with IRDy800CW-MNPs

1. Mouse leukemic monocyte macrophage cell line RAW 264.7.
2. Cell culture medium DMEM (Dulbecco's modified Eagle medium).
3. Complete DMEM cell culture medium: DMEM supplemented with 10 % fetal calf serum, penicillin (100 units/mL), streptomycin (100 μg/mL), and 10 mM 4-(2-hydroxyethyl)-1-piperazineethanesulfonic acid (HEPES).
4. Sodium chloride (NaCl) (99 %).
5. Potassium chloride (KCl) (99 %).
6. Sodium phosphate dibasic dehydrate ($Na_2HPO_4{\cdot}2H_2O$) (analytic reagent).
7. Potassium phosphate monobasic (KH_2PO_4) (99 %).
8. Paraformaldehyde (95 %).
9. Hydrochloric acid (HCl) (36.5 %).
10. Potassium ferrocyanide (99 %).
11. Aluminum sulfate [$Al_2(SO_4)_3$] (98 %).
12. Thymol (analytic reagent).
13. Phosphate Buffer Solution (PBS) (pH 7.4): 137 mM NaCl, 10 mM phosphate, 2.7 mM KCl, and a pH of 7.4. A 10 L stock of 10× PBS can be prepared by dissolving 800 g NaCl, 20 g KCl, 144 g $Na_2HPO_4{\cdot}2H_2O$ and 24 g KH_2PO_4 in 8 L of distilled water, and topping up to 10 L. The pH is ~6.8. 1× PBS can be prepared by diluting 10× PBS with distilled water, pH will be 7.4.
14. Perls reagent: Prepare 6 % (v/v) hydrochloric acid and 2 % (w/v) potassium ferrocyanide, respectively. Mix two solutions at 1:1 (v:v) immediately before use.
15. Nuclear fast red solution: 0.1 g nuclear fast red nuclear stain, 5 g aluminum sulfate, 100 mL Distilled water. Dissolve aluminum sulfate in water. Add nuclear fast red and slowly heat to boil and cool. Filter and add a grain of thymol as a preservative.
16. CO_2 incubator for cell culture.
17. Light microscope with Microscope Digital Camera (Olympus IX51-A12PH with DP71-SET) or a similar setup.

2.4. Measurement of the Relative Iron Loading of Cells

1. Mouse leukemic monocyte macrophage cell line RAW 264.7.
2. Complete DMEM cell culture medium.
3. PBS (pH 7.4), as in item 13 (Subheading 2.3).
4. Sodium dodecyl sulfate (SDS) (99 %).
5. Potassium peroxydisulfate ($K_2S_2O_8$) (99 %).
6. Potassium ferrocyanide (KSCN) (99 %).
7. Hydrochloric acid (HCl) (36.5 %).
8. CO_2 incubator for cell culture.
9. Odyssey Infrared Imaging System (Li-Cor Biosciences).
10. Digital camera (ordinary digital camera would be suitable).
11. White light transilluminator.
12. Microplate reader.

2.5. Measurement of Cell Viability

1. Mouse leukemic monocyte macrophage cell line RAW 264.7.
2. Complete DMEM cell culture medium.
3. PBS (pH 7.4), as in item 13 (Subheading 2.3).
4. DMSO (99.5 %).
5. 3-(4,5-Dimethyl-2-thiazolyl)-2,5-diphenyl-2H-tetrazolium bromide (MTT) solution: dissolve MTT (98 % purity) in PBS at the final concentration of 5 mg/mL, and filter through a 0.22 μm filter before use.
6. CO_2 incubator for cell culture.
7. Odyssey Infrared Imaging System (Li-Cor Biosciences).
8. Thermomixer (Eppendorf).
9. Microplate reader.

2.6. In Vivo Imaging with IRDy800CW-MNPs

1. IRDy800CW-MNPs solution (sterilized by filtration through 0.22 μm filter).
2. Naked mouse (CByJ-Cg-Foxn1nu/J).
3. Ether (99 %).
4. Anesthesia cylinder.
5. Pearl(TM) infrared fluorescence imager.(Li-Cor Biosciences).
6. Syringe (1 mL).

3. Methods

3.1. Preparation of IRDy800CW-MNPs

1. Under constant vigorous stirring, add the whole of solution 2 into solution 1. Continue vigorous stirring for another 2 h until brown precipitate forms.

2. Discard the supernatant and wash the brown precipitate containing iron nanoparticles 4–5 times with methanol. Dry the washed nanoparticles in vacuum drying oven at 60 °C overnight to remove all solvents.
3. The following day, dissolve the obtained waxy iron-oleate in 1-Octadecene at 70 °C; it may take about 30 min to disperse. Store at room temperature (this stock solution will be stable at room temperature for at least 1 year).
4. Mix 1 mL of the above solution containing nanoparticles with 4 mL of 1-Octadecene and 0.5 mL oleic acid.
5. Heat the mixture to 320 °C at a constant heating rate of 3.3 °C/min under nitrogen atmosphere in a heat block with Programmable temperature controller, and continue to incubate at this temperature for 30 min.
6. Cool the resulted solution to room temperature and precipitate the nanoparticles by adding 10 mL of ethanol. Centrifuge at 7,000 × *g* for 15 min.
7. Wash the precipitate containing Fe_3O_4 nanoparticles 4–5 times with 10 mL of ethanol. Centrifuge the nanoparticles after each wash at 7,000 × *g* for 10 min and finally dry it in a vacuum drying oven at 60 °C overnight to remove all solvents.
8. Take 100 mg of above Fe_3O_4 nanoparticles and disperse in 10 mL chloroform.
9. Add 50 μL of triethylamine and 10 mL of DMSA in DMSO mixture (see Note 1). Vortex at 60 °C for 12 h until black precipitate is formed.
10. Centrifuge the solution at 7,000 × *g* for 10 min to precipitate the nanoparticles.
11. Wash the precipitate 4–5 times with ethanol. Centrifuge the nanoparticles after each wash at 7,000 × *g* for 10 min. Dissolve the washed nanoparticles in 100 mL of ethanol for storage.
12. To transfer nanoparticles in water, precipitate all nanoparticles form the above solution using a permanent magnet and resuspend in 10 mL of water. The final concentration of nanoparticles in this solution will be of the order of 0.8 mg/mL Fe.
13. Transfer 3 mL of the water suspension of Fe_3O_4 nanoparticles into a 15-mL centrifuge tube.
14. Add 9 mL of EDC solution (see Note 2). Sonicate for 40 min at 4 °C with a general ultrasonic cleaner (100 W).
15. Centrifuge the solution at 18,000 × *g* for 10 min. Discard the supernatant and resuspend the precipitate in 12 mL sterile deionized water (see Note 3).
16. Add 15 μL of Streptavidin-IRDye800CW to the resuspended nanoparticles and keep nanoparticles solution at room temperature on a rotator overnight.

17. Centrifuge the nanoparticles solution at 18,000 × *g* for 10 min.
18. Wash the precipitate three times with deionized water. Centrifuge the solution at 18,000 × *g* for 10 min to precipitate the nanoparticles (see Note 4).
19. Resuspend the precipitated nanoparticles in 6 mL sterile deionized water (see Note 4). The final concentration of nanoparticles' solution should be ~0.4 mg/mL Fe.

3.2. Characterization of IRDy800CW-MNPs

1. Dilute 5 μL IRDye800CW-MNPs (0.4 mg/mL Fe) (Fig. 1a) with 61.7 μL water. The final concentration of this nanoparticles' solution is 30 μg/mL Fe. Add 15 μL diluted nanoparticles onto a copper grid and dry at room temperature.
2. Image the nanoparticles with TEM, and estimate the monodispersibility of nanoparticles (Fig. 1b).

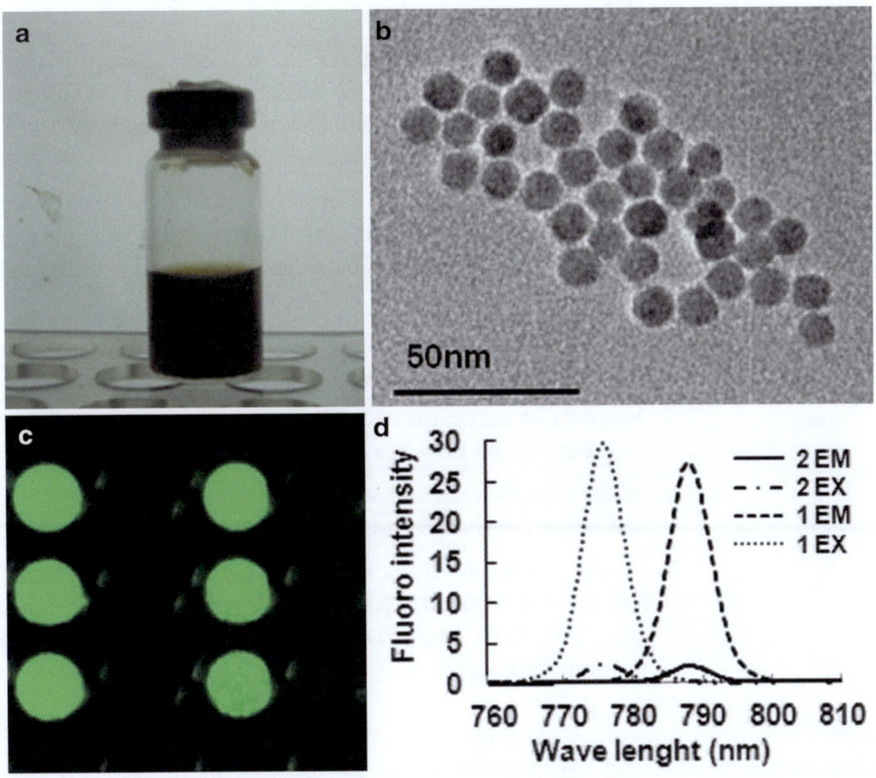

Fig. 1. Characterization of the IRDy800CW-MNPs. (**a**) The solution of IRDy800CW-MNPs. (**b**) TEM image of IRDy800CW-MNPs. (**c**) NIRF signal of IRDy800CW-MNPs. From *left* to *right*, Column 1: IRDy800CW-MNPs; Column 2: unlabeled MNPs. Column 3–4: The resuspended precipitate and supernatant of the IRDy800CW-MNPs solution after heating treatment and centrifugation. (**d**) Fluorescent spectrum of nanoparticles. 1: IRDy800CW-MNPs; 2: unlabeled MNPs. *Abs* absorbance wave length; *Em* emission wave length.

3. Measure the size of nanoparticles by analyzing TEM images with Origin 6.1. The size of IRDye800CW-MNPs is likely to average 11 nm in diameter, as exemplified in Fig. 1b.
4. Transfer 100 μL of IRDye800CW-MNPs solution to a 200-μL microcentrifuge tube and incubate solution at 60 °C for 15 min (see Note 5).
5. Centrifuge the solution at 18,000 × *g* for 5 min and transfer the supernatant into a well in a 96-well microplate (see Note 5).
6. Resuspend the precipitated nanoparticles in 100 μL water. Transfer the resuspended nanoparticles into another well of the same 96-well microplate, as in step 5 above.
7. Scan the 96-well microplate with Odyssey Infrared Imaging System (Fig. 1c) to confirm the covalent linking of IRDye800CW to MNPs (see Note 5).
8. Take the fluorescence spectrum of nanoparticles with Fluorescence Spectrophotometer (excitation wavelength: <780 nm; emission range: 780–810 nm) (Fig. 1d).

3.3. Cell Labeling with IRDy800CW-MNPs

1. Seed RAW264.7 cells into 48-well cell culture plates (15,000 cells per well) and culture in the complete DMEM cell culture medium in a humidified 5 % CO_2 atmosphere at 37 °C for 24 h.
2. Discard culture media from cells; add fresh media containing different concentrations (30–100 μg/mL) of unlabeled MNPs and IRDy800CW-MNPs (Fig. 2) to cells. Culture cells for 24 h.
3. Wash cells three times with PBS and fix with 4 % paraformaldehyde for 30 min at room temperature.
4. Wash cells three times with PBS and incubate cells with Perls reagent for 30 min at room temperature (see Note 6).
5. Wash cells three times with PBS and stain cells with nuclear fast red solution for 35 min at room temperature.
6. Wash cells three times with PBS. Observe the stained cells with light microscope and photograph with microscope digital camera. The iron nanoparticles will be stained blue with Perls reagent (Fig. 2).

3.4. Measurement of the Relative Iron Loading of Cells

1. Seed RAW264.7 cells into 48-well plate (15,000 cells per well) and culture for 24 h in the complete DMEM cell culture medium in a humidified 5 % CO_2 atmosphere at 37 °C.
2. Discard culture media from cells; add fresh media containing different concentrations of IRDy800CW-MNPs (Fig. 3) to cells. Culture cells for 24 h.
3. Wash cells three times with PBS, and scan microplate with Odyssey Infrared Imaging System (Fig. 3a).

a b

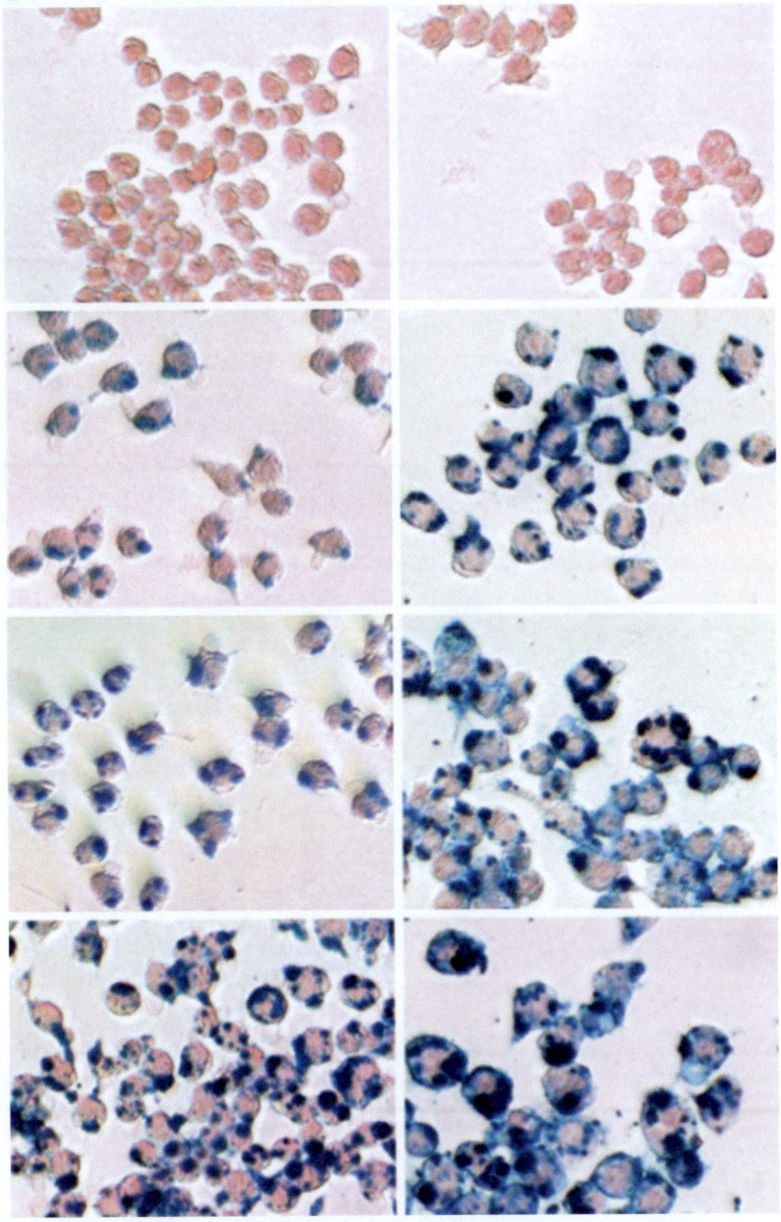

Fig. 2. Prussian blue staining of cells. The agglomerates of Fe_3O_4 nanoparticles are stained in *blue*. Cells are treated with unlabeled MNPs (**a**) and IRDy800CW-MNPs (**b**) at the different doses (nanoparticles' concentration: 0, 30, 40, and 50 μg/mL from *up* to *down*). Following the nanoparticles treatment, cells are stained with Prussian blue and observed with light microscope and photographed with microscope digital camera.

4. Divide cells into two groups, one will be used to detect relative iron loading by NIRF assay, and the other will be used to detect relative iron loading by the colorimetric assay.
5. To analyze iron loading of cells by NIRF assay, add 100 μL SDS lysis buffer (2 % SDS) to each well and keep microplate at

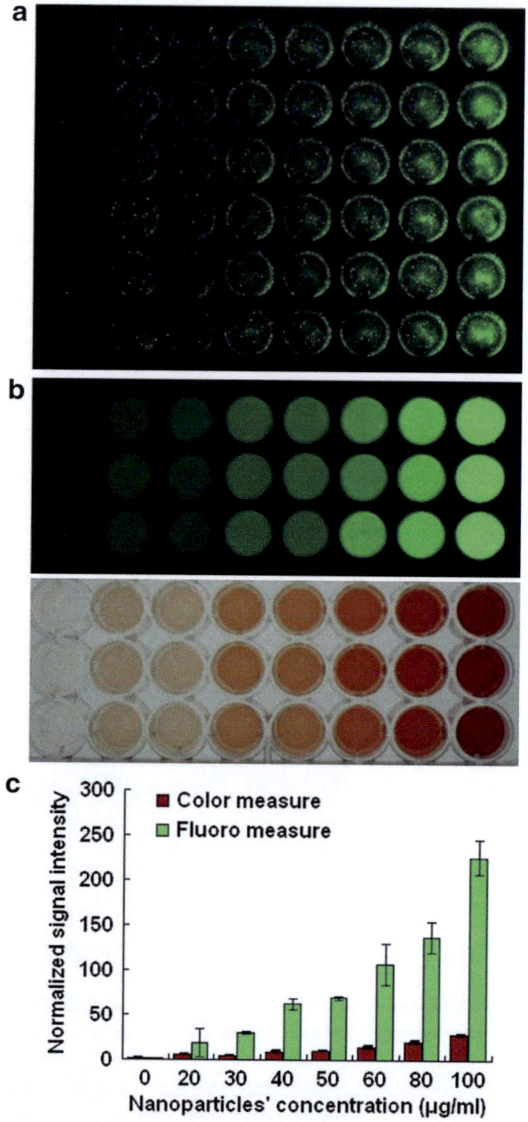

Fig. 3. Measurement of the relative iron loading of cells. (**a**) NIRF signal of cells labeled with the IRDy800CW-MNPs at the dose of 0, 20, 30, 40, 50, 60, 80, and 100 μg/mL (Column 1–8 from *left* to *right*). Each dose contained six repeats (Rows 1–6). Following the treatment of cells with IRDy800CW-MNPs, cells are washed three times with PBS and then image with Odyssey Infrared Imaging System. (**b**) Measurements of the relative iron loading of cells with the colorimetric and NIRF approaches. Row 1–3 (from *up* to *down*): NIRF detection; Row 4–6 (from *up* to *down*): Colorimetric assay. (**c**) The normalized colorimetric and NIRF signal intensities. The signals of the nanoparticle-labeled cells were normalized by that of the negative control cells.

room temperature for 10 min. Scan microplate and quantify NIRF signal intensity with Odyssey Infrared Imaging System (Fig. 3b, upper plate).

6. To analyze iron loading of cells by colorimetric assay, collect cells in each well into 1.5-mL tube. Precipitate cells by

centrifugation, and resuspend cells in 50 μL of hydrochloric acid and incubate at 60 °C for 4 h. At the end of incubation, prepare fresh detection reagent by dissolving $K_2S_2O_8$, KSCN, and HCl in water at the final concentration of 0.08 %, 8 %, and 3.6 %, respectively.

7. Precipitate cells by centrifugation, and transfer the supernatants into 96-well plates. Add 50 μL of detection reagent to each well, and keep plates at room temperature for 10 min.
8. Transfer the microplate onto a white light illuminator and take pictures of the microplate with a digital camera (Fig. 3b, lower plate). Measure the absorbance using a microplate reader at an excitation wavelength of 490 nm (see Note 7).
9. Normalize NIRF and colorimetric signal intensities of the nanoparticle-treated cells by subtracting the values obtained with the blank control cells. Plot the normalized NIRF and colorimetric signal intensities in a histogram for comparing two assays (Fig. 3c).

3.5. Measurement of Cell Viability

1. Seed RAW264.7 cells into 96-well plate (10,000 cells per well) and culture 24 h in the complete DMEM cell culture medium in a humidified 5 % CO_2 atmosphere at 37 °C.
2. Discard culture media from the cells; add fresh media containing different concentrations of unlabeled MNPs and IRDy800CW-MNPs (Fig. 4) to the cells. Culture cells for 24 h.
3. To evaluate the effect of nanoparticles on cell viability use MTT assay as described in following steps.
4. Scan microplate with Odyssey Infrared Imaging System (Fig. 4a). Remove media containing nanoparticles from cells, wash cells three times with PBS and scan microplate with Odyssey Infrared Imaging System again (Fig. 4b).
5. Add 180 μL of serum-free cell culture medium and 20 μL of MTT solution (5 mg/mL in PBS) to each well and incubate for 4 h. As a control, add 150 μL of PBS to cells of three wells.
6. Aspirate the supernatants of each well and add 150 μL of DMSO to each well to solubilize cells and MTT crystals. Shake microplate on a Thermomixer at 400 rpm at 37 °C for 10 min to dissolve all crystals.
7. Measure the absorbance using a microplate reader at an excitation wavelength of 570 nm (Fig. 4c).
8. Calculate the cell viability by comparing the absorption of nanoparticles-treated cells to those of blank control cells (see Note 8). Define the cell viability of blank control cells as 100 %. Nanoparticles are considered toxic if the difference between

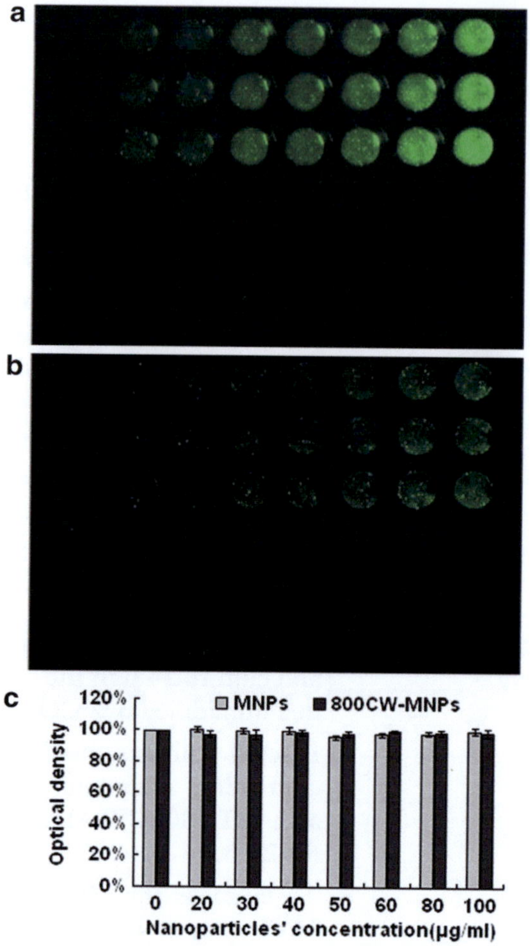

Fig. 4. Measurement of the cell viability. Cells are treated with IRDy800CW-MNPs (Rows 1–3) and unlabeled MNPs (Rows 4–6) and at the dose of 0, 20, 30, 40, 50, 60, 80, and 100 μg/mL (Column 1–8 from *left* to *right*) for 24 h. Cells are detected with Odyssey Infrared Imaging System before (**a**) and after (**b**) three washings of PBS. (**c**) Quantitative measurement of cell viability by MTT assay.

cell growth inhibition of control and exposed cells is statistically significant at the 5 % level, as determined by a *t*-test.

3.6. In Vivo Imaging with IRDy800CW-MNPs

1. Anesthetize mice (see Note 9) by inhaling ether in an anesthesia cylinder (see Note 10).
2. Image mice on Pearl Image (Li-Cor).
3. Inject IRDy800CW-MNPs to mice by tail vein at the dose of 2 or 5 mg Fe/kg body weight.
4. Breed mice with normal condition and image the animal at regular intervals with Pearl(TM) infrared fluorescence imager (see Fig. 5). Just before each imaging, anesthetize mice by using ether. Capture all images using the same setting for the excitation intensity (preset by manufacture and cannot be set by user).

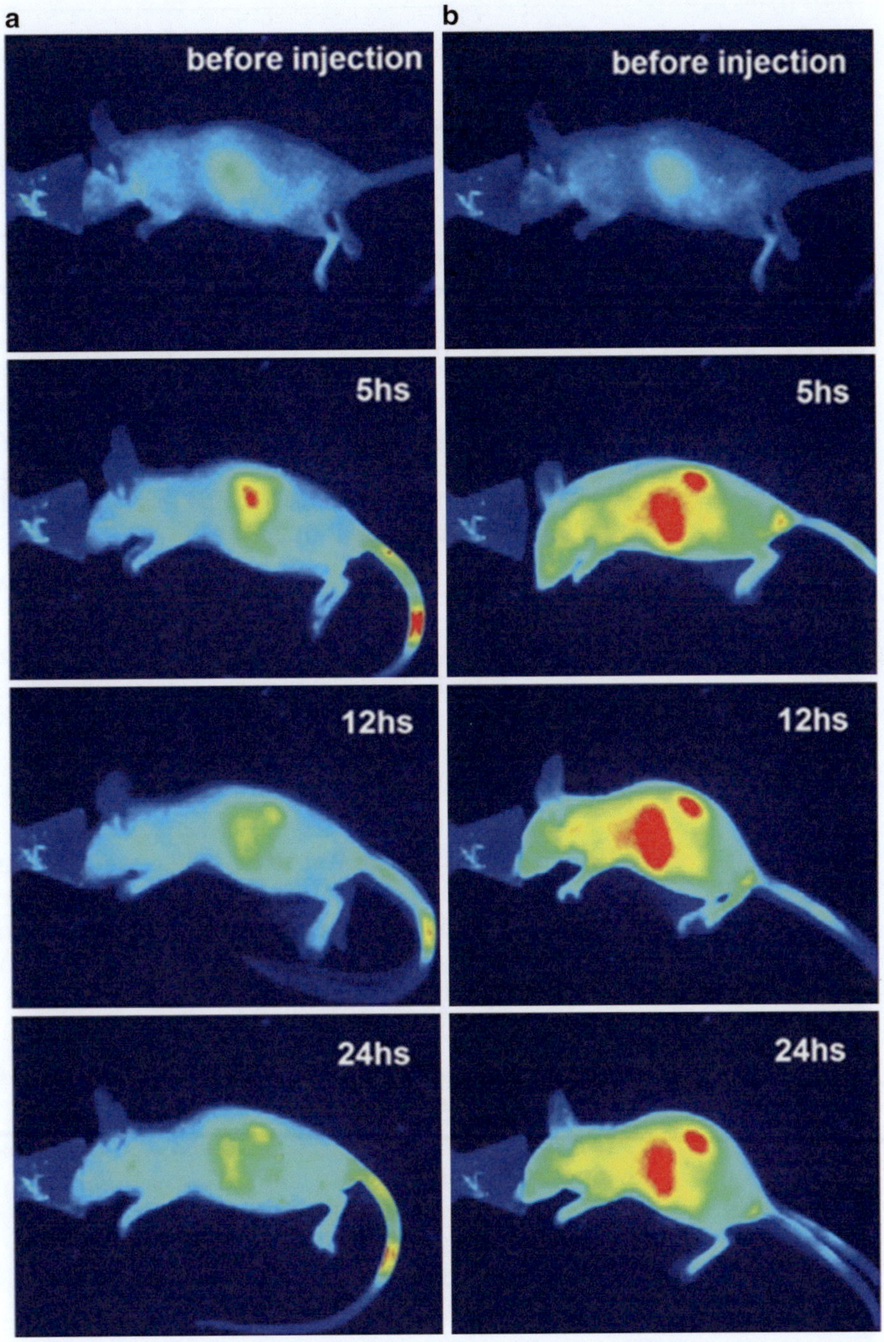

Fig. 5. NIRF imaging of the mice administered IRDye800CW-MNPs at the different dose. The mice are administered the IRDye800CW-MNPs at the dose of 2 mg/kg (**a**) and 5 mg/kg (**b**) body weight by intravascular injection (tail vein). The mice are imaged before and after injection of IRDy800CW-MNPs with NIRF imaging system, Pearl Image (Li-Cor). The images are displayed in pseudo color mode.

4. Notes

1. DMSA coating is important to the water solubility and NIRF labeling of the MNPs in this study. DMSA coating makes iron oxide nanoparticle hydrophilic and dispersible in water solution, improve the biocompatibility of the iron oxide nanoparticles. DMSA-coated iron nanoparticles have abundant carboxyl groups on their surface, which is helpful for covalent labeling of nanoparticles by fluorescent dyes.
2. The carboxyl groups on the surface of the DMSA-coated Fe_3O_4 nanoparticles are activated with EDC. EDC is a carboxyl and amine-reactive cross-linker, which reacts with a carboxyl and amino group and forms an amide bond. The streptavidin-IRDy800CW contains amino group, therefore can be covalently linked to the surface of MNPs.
3. Resuspend the nanoparticles by pipetting gently or sonicating at 4 °C.
4. Resuspend the nanoparticles by pipetting gently but not by sonicating. After sufficient pipetting, some aggregates may still exist; they can be filtered out by passing IRDy800CW-MNPs solution through 0.22 μm filter.
5. Heat treatment (step 4, Subheading 3.2) is needed when confirming the covalent linkage between IRDy800CW and the MNPs. The IRDy800CW-MNPs will retain the covalently bound dye and therefore NIRF signal after the heat treatment. However, any IRDy800CW nonspecifically absorbed to the surface of MNPs, will be desorbed by heat treatment and thus the NIRF signal may appear in the supernatant in Step 5 (Subheading 3.2).
6. To obtain high quality images, the Perls reagent solution should be prepared fresh.
7. The iron content can be determined by normalizing the obtained absorbance with a standard curve using $FeCl_3$ calibration standards that contained 0, 0.5, 1, 2, 4, 6, 10, 20, 40, 50, and 100 μg/mL of iron. The average iron content per cell can be calculated by dividing the total iron content of a cell sample with the number of cells in this sample.
8. To exclude nanoparticles' interference with the MTT assay, 180 μL of cell culture medium containing MNPs in different concentrations can be added to the blank wells (without cells). Following incubation under the same conditions as the cell groups, 20 μL of MTT solution should be added to each well and the absorbance of each well can be measured as described above. The average absorbance of this cell-free control is subtracted from the average absorbance of cell groups.

9. To obtain high quality images, the naked mouse should be used, for example, the mice of CByJ-Cg-Foxn1nu/J used in this study.
10. The mouse is sensitive to anesthesia with ether. Mice should receive sufficient exposure to remain still whilst undergoing imaging, but the exposure should be kept to a minimum to avoid the animal dying due to overexposure.

Acknowledgments

This study was partially funded by the National Important Science Research Program of China (2006CB933205; 2011CB933503), and US-China international S & T cooperation project (2009DFA31990).

References

1. Ito A, Shinkai M, Honda H, Kobayashi T (2001) Heat-inducible TNF-alpha gene therapy combined with hyperthermia using magnetic nanoparticles as a novel tumor-targeted therapy. Cancer Gene Ther 8:649–654
2. Hergt R, Dutz S (2007) Magnetic particle hyperthermia-biophysical limitations of a visionary tumour therapy. J Magn Magn Mater 311:187–192
3. Lin BL, Shen XD, Cui S (2007) Application of nanosized Fe3O4 in anticancer drug carriers with target-orientation and sustained-release properties. Biomed Mater 2:132–134
4. Shubayev VI, Pisanic TR II, Jin S (2009) Magnetic nanoparticles for theragnostics. Adv Drug Deliv Rev 61:467–477
5. Pankhurst QA, Connolly J, Jones SK, Dobson J (2003) Applications of magnetic nanoparticles in biomedicine. J Phys D: Appl Phys 36: R167–R181
6. Shaw SY, Westly EC, Pittet MJ, Subramanian A, Schreiber SL, Weissleder R (2008) Perturbational profiling of nanomaterial biologic activity. Proc Natl Acad Sci 105: 7387–7392
7. Bacon BR, Stark DD, Park CH, Saini S, Groman EV, Hahn PF, Compton CC, Ferrucci JT Jr (1987) Ferrite particles: a new magnetic resonance imaging contrast agent. Lack of acute or chronic hepatotoxicity after intravenous administration. J Lab Clin Med 110: 164–171
8. Weissleder R, Stark DD, Engelstad BL, Bacon BR, Compton CC, White DL, Jacobs P, Lewis J (1989) Superparamagnetic iron oxide: pharmacokinetics and toxicity. AJR Am J Roentgenol 152:167–173
9. Lacava LM, Lacava ZGM, Azevedo RB, Chaves SB, Garcia VAP, Silva O, Pelegrini F, Buske N, Gansau C, Da Silva MF, Morais PC (2002) Use of magnetic resonance to study biodistribution of dextran-coated magnetic fluid intravenously administered in mice. J Magn Magn Mater 252:367–369
10. Lacava LM, Garcia VAP, Kuckelhaus S, Azevedo RB, Lacava ZGM, Silva O, Pelegrini F, Gansau C, Buske N, Morais PC (2003) Magnetic resonance and light microscopy investigation of a dextran coated magnetic fluid. J Appl Phys 93:7563–7565
11. Briley-Saebo K, Bjornerud A, Grant D, Ahlstrom H, Berg T, Kindberg GM (2004) Hepatic cellular distribution and degradation of iron oxide nanoparticles following single intravenous injection in rats: implications for magnetic resonance imaging. Cell Tissue Res 316:315–323
12. Kalber TL, Smith CJ, Howe FA, Griffiths JR, Ryan AJ, Waterton JC, Robinson SP (2005) A longitudinal study of R2* and R2 magnetic resonance imaging relaxation rate measurements in murine liver after a single administration of 3 different iron oxide-based contrast agents. Invest Radiol 40:784–791
13. Kwon JT, Hwang SK, Jin H, Kim DS, Minai-Tehrani A, Yoon HJ, Choi M, Yoon TJ, Han DY, Kang YW, Yoon BI, Lee JK, Cho MH (2008) Body distribution of inhaled fluorescent magnetic nanoparticles in the mice. J Occup Health 50:1–6

14. Lee PW, Hsu SH, Wang JJ, Tsai JS, Lin KJ, Wey SP, Chen FR, Lai CH, Yen TC, Sung HW (2010) The characteristics, biodistribution, magnetic resonance imaging and biodegradability of superparamagnetic core-shell nanoparticles. Biomaterials 31:1316–1324
15. Laemmel E, Genet M, Le Goualher G, Perchant A, Le Gargasson JF, Vicaut E (2004) Fibered confocal fluorescence microscopy (Cell-viZio) facilitates extended imaging in the field of microcirculation. A comparison with intravital microscopy. J Vasc Res 41:400–411
16. Lim YT, Kim S, Nakayama A, Stott NE, Bawendi MG, Frangioni JV (2003) Selection of quantum dot wavelengths for biomedical assays and imaging. Mol Imaging 2:50–64
17. Hatanaka S, Matsushita N, Abe M, Nishimura K, Hasegawa M, Handa H (2003) Direct immobilization of fluorescent dyes onto ferrite nanoparticles during their synthesis from aqueous solution. J Appl Phys 93:7569–7570
18. Yoon TJ, Kim JS, Kim BG, Yu KN, Cho MH, Lee JK (2005) Multifunctional nanoparticles possessing a "magnetic motor effect" for drug or gene delivery. Angew Chem Int Ed Engl 44:1068–1071
19. Guo J, Yang W, Deng Y, Wang C, Fu S (2005) Organic-dye-coupled magnetic nanoparticles encaged inside thermoresponsive PNIPAM microcapsules. Small 1:737–743
20. Kwon JT, Kim DS, Minai-Tehrani A, Hwang SK, Chang SH, Lee ES, Xu CX, Lim HT, Kim JE, Yoon BI, An GH, Lee KH, Lee JK, Cho MH (2009) Inhaled fluorescent magnetic nanoparticles induced extramedullary hematopoiesis in the spleen of mice. J Occup Health 51:423–431
21. Bertorelle F, Wilhelm C, Roger J, Gazeau F, Menager C, Cabuil V (2006) Fluorescence-modified superparamagnetic nanoparticles: intracellular uptake and use in cellular imaging. Langmuir 22:5385–5391
22. Kim JS, Yoon TJ, Yu KN, Kim BG, Park SJ, Kim HW, Lee KH, Park SB, Lee JK, Cho MH (2006) Toxicity and tissue distribution of magnetic nanoparticles in mice. Toxicol Sci 89:338–347
23. Frangioni JV (2003) In vivo near-infrared fluorescence imaging. Curr Opin Chem Biol 7:626–634
24. Achilefu S (2004) Lighting up tumors with receptor-specific optical molecular probes. Technol Cancer Res Treat 3:393–409
25. Ntziachristos V, Bremer C, Weissleder R (2003) Fluorescence imaging with near-infrared light: new technological advances that enable in vivo molecular imaging. Eur Radiol 13:195–208
26. Weissleder R (2001) A clearer vision for in vivo imaging. Nat Biotechnol 19:316–317
27. Loo C, Lowery A, Halas N, West J, Drezek R (2005) Immunotargeted nanoshells for integrated cancer imaging and therapy. Nano Lett 5:709–711
28. Sevick-Muraca EM, Houston JP, Gurfinkel M (2002) Fluorescence-enhanced, near infrared diagnostic imaging with contrast agents. Curr Opin Chem Biol 6:642–650
29. Weissleder R, Ntziachristos V (2003) Shedding light onto live molecular targets. Nat Med 9:123–128
30. Kumar R, Roy I, Ohulchanskky TY, Vathy LA, Bergey EJ, Sajjad M, Prasad PN (2010) In vivo biodistribution and clearance studies using multimodal organically modified silica nanoparticles. ACS Nano 4:699–708
31. Altinoglu EI, Russin TJ, Kaiser JM, Barth BM, Eklund PC, Kester M, Adair JH (2008) Near-infrared emitting fluorophore-doped calcium phosphate nanoparticles for in vivo imaging of human breast cancer. ACS Nano 2:2075–2084
32. Maxwell DJ, Bonde J, Hess DA, Hohm SA, Lahey R, Zhou P, Creer MH, Piwnica-Worms D, Nolta JA (2008) Fluorophore-conjugated iron oxide nanoparticle labeling and analysis of engrafting human hematopoietic stein cells. Stem Cells 26:517–524
33. McCann CM, Waterman P, Figueiredo JL, Aikawa E, Weissleder R, Chen JW (2009) Combined magnetic resonance and fluorescence imaging of the living mouse brain reveals glioma response to chemotherapy. Neuroimage 45:360–369
34. Josephson L, Kircher MF, Mahmood U, Tang Y, Weissleder R (2002) Near-infrared fluorescent nanoparticles as combined MR/optical imaging probes. Bioconjug Chem 13:554–560
35. Jaffer FA, Nahrendorf M, Sosnovik D, Kelly KA, Aikawa E, Weissleder R (2006) Cellular imaging of inflammation in atherosclerosis using magnetofluorescent nanomaterials. Mol Imaging 5:85–92
36. Montet X, Ntziachristos V, Grimm J, Weissleder R (2005) Tomographic fluorescence mapping of tumor targets. Cancer Res 65:6330–6336
37. Jaffer FA, Sosnovik DE, Nahrendorf M, Weissleder R (2006) Molecular imaging of myocardial infarction. J Mol Cell Cardiol 41:921–933

38. Funovics M, Montet X, Reynolds F, Weissleder R, Josephson L (2005) Nanoparticles for the optical imaging of tumor E-selectin. Neoplasia 7:904–911
39. Marshall MV, Draney D, Sevick-Muraca EM, Olive DM (2010) Single-dose intravenous toxicity study of IRDye 800CW in Sprague-Dawley rats. Mol Imaging Biol 12:583–594
40. Blum G, von Degenfeld G, Merchant MJ, Blau HM, Bogyo M (2007) Noninvasive optical imaging of cysteine protease activity using fluorescently quenched activity-based probes. Nat Chem Biol 3:668–677
41. Adams KE, Ke S, Kwon S, Liang F, Fan Z, Lu Y, Hirschi K, Mawad ME, Barry MA, Sevick-Muraca EM (2007) Comparison of visible and near-infrared wavelength-excitable fluorescent dyes for molecular imaging of cancer. J Biomed Opt 12:024017
42. Tanaka E, Ohnishi S, Laurence RG, Choi HS, Humblet V, Frangioni JV (2007) Real-time intraoperative ureteral guidance using invisible near-infrared fluorescence. J Urol 178:2197–2202
43. Foster AE, Kwon S, Ke S, Lu A, Eldin K, Sevick-Muraca E, Rooney CM (2008) In vivo fluorescent optical imaging of cytotoxic T lymphocyte migration using IRDye800CW near-infrared dye. Appl Opt 47:5944–5952
44. Kovar JL, Volcheck W, Sevick-Muraca E, Simpson MA, Olive DM (2009) Characterization and performance of a near-infrared 2-deoxyglucose optical imaging agent for mouse cancer models. Anal Biochem 384:254–262
45. Wang GJ, Liu Y, Qin A, Shah SV, Deng ZB, Xiang X, Cheng Z, Liu C, Wang J, Zhang L, Grizzle WE, Zhang HG (2008) Thymus exosomes-like particles induce regulatory T cells. J Immunol 181:5242–5248
46. Duysen EG, Lockridge O (2008) Whole body and tissue imaging of the butyrylcholinesterase knockout mouse injected with near infrared dye labeled butyrylcholinesterase. Chem Biol Interact 175:119–124
47. Tanaka E, Choi HS, Fujii H, Bawendi MG, Frangioni JV (2006) Image-guided oncologic surgery using invisible light: completed pre-clinical development for sentinel lymph node mapping. Ann Surg Oncol 13:1671–1681

Chapter 18

Labeling Stem Cells with Superparamagnetic Iron Oxide Nanoparticles: Analysis of the Labeling Efficacy by Microscopy and Magnetic Resonance Imaging

Jasmin, Ana Luiza Machado Torres, Linda Jelicks, Antonio Carlos Campos de Carvalho, David C. Spray, and Rosalia Mendez-Otero

Abstract

Stem cell therapy has emerged as a potential therapeutic option for cell death-related heart diseases. Application of non-invasive cell tracking approaches is necessary to determine tissue distribution and lifetime of stem cells following their injection and will likely provide knowledge about poorly understood stem cells mechanisms of tissue repair. Magnetic resonance imaging (MRI) is a potentially excellent tool for high-resolution visualization of the fate of cells after transplantation and for evaluation of therapeutic strategies. The application of MRI for in vivo cell tracking requires contrast agents to achieve efficient cell labeling without causing any toxic cellular effects or eliciting any other side effects. For these reasons clinically approved contrast agents (e.g., ferumoxides) and incorporation facilitators (e.g., protamine) are currently the preferred materials for cell labeling and tracking. Here we describe how to use superparamagnetic iron oxide nanoparticles to label cells and to monitor cell fate in several disease models.

Key words: Stem cells, Cell labeling, Ferumoxides, Superparamagnetic iron oxide nanoparticles, Magnetic resonance imaging

1. Introduction

Stem cell therapy has emerged as a therapeutic option in several types of injury and disease. However, the lack of data concerning the distribution and engraftment of the cells presents a serious obstacle for the successful application of stem cell therapies in patients. Thus, the development of sensitive, non-invasive techniques for serially tracking cells will likely provide knowledge about

Mikhail Soloviev (ed.), *Nanoparticles in Biology and Medicine: Methods and Protocols*, Methods in Molecular Biology, vol. 906, DOI 10.1007/978-1-61779-953-2_18, © Springer Science+Business Media, LLC 2012

the poorly understood mechanisms responsible for the reported improvement in several lesion/disease models (1–3). Magnetic resonance imaging (MRI) offers an imaging modality with high-resolution to visualize the fate of cells after transplantation and to evaluate cell-based repair, replacement, and therapeutic strategies (4–8).

Several paramagnetic contrast agents have been successfully used for in vivo cell tracking and labeling of different mammalian cell types (1, 9–12). Ferumoxides, such as Feridex IV (Advanced Magnetics Inc., USA) and Endorem (Guerbet SA, France) are dextran-coated superparamagnetic iron oxide nanoparticles (SPIONs) clinically used to provide intravenous MRI contrast in order to analyze liver pathology. The particles are biodegradable and all the degradation products can be incorporated in normal metabolic processes (13). SPIONs, however, tend to aggregate, but this can be minimized by their coating with dextran or other polymers (14). Other types of human approved contrast agents have also been used to label and track transplanted cell, such as ferucarbotran (Resovist) and gadoliniun nanoparticles (15, 16). However, there have been reports that gadolinium based agents may adversely affect the labeled cells (17, 18).

The dextran-coated SPIONs do not show sufficient cellular uptake to enable tracking of nonphagocytic cells. However, uptake of SPIONs can be facilitated by incubation with cationic compounds such as poly-L-lysine (PLL) (19–21), protamine sulfate (21–24), FuGENE (3), Superfect or Lipofectamine (12). The primary advantage of the incubation labeling method is its simplicity. The primary disadvantage is the extended incubation time (4 h or more) required (25). Thus, other approaches to induce labeling of freshly isolated cells such as magnetoelectroporation (26, 27) and magnetosonoporation (28) offer alternative protocols and are preferable when the culture system must be avoided. Nevertheless, for adherent cells that can be maintained in culture, PLL and protamine are the most frequently used agents to facilitate the incorporation of SPIONs.

PLL is a synthetic cationic polymer commonly used to enhance cell adhesion to the surface of culture dishes. Its use has not yet been approved in humans and studies have shown that complexes formed by ferumoxides and PLL (FePLL) can be large. Such large complexes do not become incorporated into endosomes and, instead, remain adhered to the cell membrane (22, 29). These complexes can affect cell proliferation (21) and differentiation (30). Protamines are low-molecular-weight arginine-rich proteins (~4,000 Da), that are purified from the mature fish testis. Protamine is a clinically approved polycationic peptide primarily used as an antidote for heparin anticoagulation (31, 32).

Approval for clinical tracking of labeled stem cells by MRI depends on efficient cell labeling that does not cause toxic cellular effects and does not elicit side effects. Ferumoxide and protamine are the currently favored materials for cell labeling and tracking because of the following advantages: Efficiency of cell tracking by MRI, commercial availability, approval for human use, and the lack of toxic effects on biological properties of cells. Besides, recent studies have shown that the number of labeled cells or the amount of intracellular SPIONs in cells exposed to FeProt complexes for 4 h is the same as at 24 h (21) and 48 h (23).

Here we describe procedures for making FeProt complexes (derivatized SPION nanoparticles), techniques for labeling adherent cells with FeProt complexes and microscopy and MRI-based methods for analysis of incorporation efficiency of the SPIONs' incorporation efficiency.

2. Materials

All cell culture materials should be prepared and used under sterile conditions.

2.1. Labeling of Adherent Cells

1. Culture plates: the culture plate size should be chosen according to planned experiments. For detection of SPIONs (with Prussian Blue and/or immunostaining) use a 24-well plate and coverslips; for cell transplant in vivo and MRI detection (see Note 1) and/or for in vitro MRI use a 100 mm Petri dish.
2. Culture medium: use the appropriate culture medium for the cell requirements, e.g., for mesenchymal stem cells use DMEM high glucose supplemented with fetal bovine serum (FBS, see Note 2) and 100 U/mL penicillin and 100 μg/mL streptomycin, or with appropriate supplements used for the specific cell type.
3. Ferumoxide such as Endorem is commercially available (Guerbert SA) at 11.2 mg Fe/mL solution (particle size: 80–150 nm). The stock solution should be stored at 4 °C or at room temperature, but the solution should never be frozen. Until recently another type of ferumoxide was also available from Feridex IV (Advanced Magnetics Inc.).
4. Protamine sulfate (see Note 3): commercially available as 1 mL ampoules/injections containing 10 mg of protamine sulfate, 9 mg of sodium chloride in water. This solution should be maintained at room temperature.
5. Phosphate-buffered saline 1× (PBS): Phosphate-buffered saline 1× (PBS): 2.67 mM Potassium Chloride (KCl); 1.47 mM Potassium Phosphate monobasic (KH_2PO_4); 137.93 mM Sodium Chloride (NaCl) and 8.06 mM Sodium Phosphate dibasic ($Na_2HPO_4 \cdot 7H_2O$), pH 7.4.

2.2. Fixing Cells for Evaluation by Staining

1. A 24-well cell culture plate.
2. Round glass coverslips (diameter 13 mm).
3. Gelatin for coating coverslips (0.2 %): dilute 0.2 g of gelatin in 100 mL of deionized water, mix, and autoclave. This solution can be used for months, with sterile storage at 4 °C or at room temperature.
4. Formaldehyde for cell fixation (4 %): dilute 1 ampoule (10 mL) of 16 % formaldehyde solution in 30 mL of PBS and mix. After preparation, this solution can be stored in aliquots at −20 °C, if thawed do not refreeze.

2.3. Determining Labeling Efficiency (See Note 4)

Prepare all working solutions fresh before the staining experiment.

2.3.1. Prussian Blue Staining (See Note 5)

1. Deionized water.
2. Perl's Solution: Prepare 20 % potassium ferrocyanide by adding 0.2 g of potassium ferrocyanide to 1 mL of deionized water. To prepare 20 % hydrochloric acid, add 200 μL of 100 % hydrochloric acid to 800 μL of deionized water (see Note 6). Mix both solutions together, cover tube with aluminum foil to protect it from the light and shake until the reagents are completely dissolved. The final volume depends on the number of wells used, e.g. for each well of a 24-well plate prepare 300 μL. The solution should be prepared fresh.
3. Nuclear Fast Red (1 %): dilute 0.02 g of Nuclear Fast Red in 2 mL of deionized water, mix, and then filter through filter paper to eliminate large particles.
4. Absolute ethanol.
5. Ethanol (70 and 30 %): dilute 70 or 30 mL of absolute ethanol with 30 or 70 mL of deionized water, respectively.
6. Xylene.
7. Ethanol/xylene mixture (1:1, v/v): made by mixing equal volume of absolute ethanol and xylene.
8. Entellan mounting medium (Merck KGaA, Darmstadt, Germany, cat. # UN 1866).

2.3.2. Immunostaining with Anti-dextran Antibody (See Note 5)

1. PBS with 0.1 % Triton X-100: dilute 200 μL of Triton X-100 in 200 mL of PBS and mix well.
2. Normal goat serum (5 %): dilute 1 mL of normal goat serum in 19 mL of PBS and mix well. This solution can be stored in aliquots at −20 °C, if thawed it can be refrozen and thawed several times.
3. Anti-dextran antibody (1:1,000): dilute the anti-dextran mouse monoclonal antibody (Stem Cell Technologies, Vancouver, BC, USA, cat. # 01403) 1:1,000 (v:v) with PBS/0.1 % Triton

X-100. Prepare the primary antibody fresh each time, but leftover material can be stored protected from light at 4 °C and used within a week.

4. Goat anti-mouse IgG secondary antibody: dilute the goat-anti-mouse IgG (e.g., Alexa Fluor 488 labeled, available from Invitrogen Inc.) in PBS/0.1 % Triton X-100 at 1:400 (v:v) (see Note 7). The ratio of antibody to PBS depends on the antibody brand. Prepare the secondary antibody fresh each time, but the leftover can be stored protected from light at 4 °C and used within a week.
5. 4′,6-Diamidino-2-phenylindole (DAPI) stock solution (1 mg/mL): dissolve 1 mg of DAPI in 1 mL of distilled water.
6. DAPI working solution (10 μg/mL): add 0.1 mL of 1 mg/mL DAPI stock solution into 9.9 mL of PBS (see Note 8). These solutions can be kept at 4 °C for months, protected from light.
7. Vectashield mounting medium for fluorescence (Vector Laboratories Inc., Burlingame, CA, USA, cat. # H-1000) (see Note 8).
8. Nail polish for sealing the coverslips.

2.3.3. Preparing Cells for In Vitro MRI

1. Use 1.5 mL tubes, 96- or 24-well plates (see Note 9).
2. 15 % Gelatin: dilute 1.5 g of gelatin in 10 mL of distilled water. Use a water bath, at 60 °C, and the vortex to dissolve this solution. With the gelatin still hot, quickly make aliquots of 200 μL in 1.5 mL tubes, 96- or 24-well plates. Always prepare a fresh solution and only prepare it when the cells are ready to be mixed (as described in Subheading 3.3.3).

2.4. Analyzing the Labeling Efficacy by Microscopy

1. Light microscope equipped with a color camera for recording Prussian Blue images.
2. Fluorescence microscope (e.g., Axiovert 200M, Zeiss, GmbH) equipped with a camera for immunostaining experiments. For quantification, we recommend using a ×20 magnification objective.

2.5. Analyzing the Labeling Efficacy by MRI

1. For MRI an MRI scanner with appropriately sized imaging coils is required. We used a 9.4 T vertical bore system with a microimaging accessory but a horizontal research magnet or lower field clinical magnet could also be used (see Note 10).
2. Dedicated spectrometer software or other analysis programs such as Image J (from U.S. National Institutes of Health, http://rsbweb.nih.gov/ij/), and MATLAB (The Mathworks, Natick, MA) based programs are required to evaluate intensity differences between samples.

3. Methods

Carry out all procedures at room temperature unless otherwise specified.

3.1. Labeling of Adherent Cells (See Note 11)

The experimental procedures are summarized in Fig. 1. The effect of protamine on cell labeling is illustrated in Fig. 2.

1. Preparing the labeling solution: the final concentrations of SPIONs and protamine are 50 and 5 μg/mL respectively. These should always be prepared fresh as described below. Figure 2 illustrates the importance of using the recommended concentrations of SPIONs and protamine.

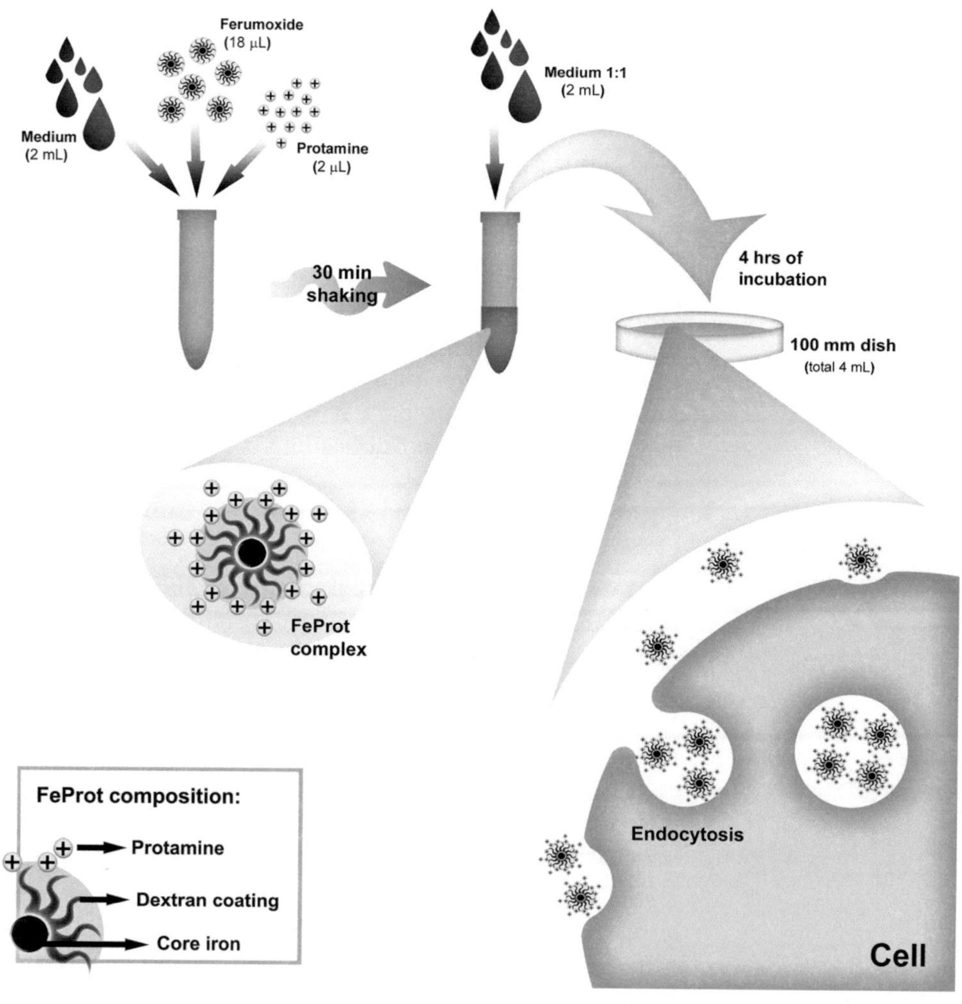

Fig. 1. Derivatization of SPION nanoparticles and labeling of cells. Schematic diagram representing the steps optimizing FeProt complex formation and incorporation by cells.

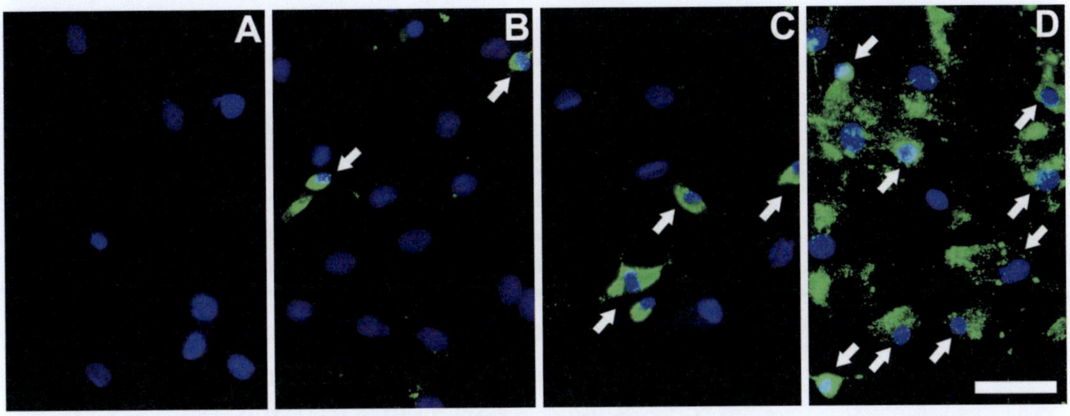

Fig. 2. Importance of facilitator agents for labeling cells with SPIONs. Representative images showing that cationic compounds, such as protamine, are necessary for the efficient uptake of SPIONs. Detection of SPIONs is with anti-dextran antibody. (**a**) Mesenchymal stem cells incubated only with protamine without ferumoxides. (**b**) Cells incubated with ferumoxides in the absence of facilitating agent. (**c**) Cells exposed to FeProt complexes using low protamine concentration (1 μg/mL), which is five times less than the recommended concentration (5 μg/mL). (**d**) Mesenchymal cells incubated with FeProt complexes with 5 μg/mL protamine. The *arrows* are showing some of the ferumoxide-labeled cells. Cell nuclei are labeled *blue* with DAPI; no cells with SPIONs are detected in (**a**), approximately 7 % cells contain SPIONs in (**b**), and 95 % in (**d**). We did not quantify the percentage of labeled cells in group (**c**). Scale bar = 50 μm.

2. For preparation of 4 mL of labeling solution (see Note 12): transfer 2 mL of FBS-supplemented culture medium (see Note 2) to a 15 mL conical tube, add 18 μL of SPIONs and 2 μL of protamine. Transfer the tube containing the labeling solution to a shaker and gently shake for 30 min at room temperature. Subsequently, add 2 mL more of culture medium to the labeling solution (1:1, v:v) and very gently mix the solution using a P1000 pipette.
3. Replace the culture medium with labeling solution and maintain the cells in the incubator in 5 % CO_2 atmosphere at 37 °C for 4 h.
4. Gently and quickly wash three times with PBS for elimination of non-incorporated SPIONs (no extended incubation with PBS is necessary). At this point the cells are ready to be fixed for Prussian Blue and/or immunostaining or to be transplanted (see Note 1) or detected in vitro by MRI.

3.2. Fixing Cells for Evaluation by Staining

1. Transfer coverslips into the wells of a 24-well plate, add 300 μL of 0.2 % gelatin to cover the coverslips and incubate for 10 min at room temperature or in the incubator at 37 °C. Aspirate the gelatin and maintain the plate for another 10 min at room temperature or in the incubator at 37 °C, at this point the plate is ready to receive the culture medium and the cells (see Note 13).

2. For SPION incorporation in 24-well plate use the same protocol as detailed above for 100 mm dish plate, but scale all components down. The final volumes to use are 300 μL of medium, 1.3 μL of SPIONs and 0.15 μL of protamine for each well.
3. To fix the cells, add 300 μL of 4 % formaldehyde to cover the coverslips in the 24-well plate and incubate for 20 min at 37 °C or at room temperature. Remove the formaldehyde and gently and quickly wash two times with PBS. At this point the cells are attached to coverslips and are ready to be stained.
4. To store the fixed cells, fill the wells with PBS and store plates at 4 °C for up to several months (see Note 14).

3.3. Determining Labeling Efficiency

Below we describe how to detect the SPIONs by Prussian Blue and/or immunostaining (Fig. 3), and how to prepare the cells for in vitro MRI (Fig. 4) (see Note 4).

3.3.1. Prussian Blue Staining

1. Wash the coverslips with fixed cells in 24-well plate twice with deionized water for 5 min and incubate with Perl's solution for 20 min.
2. Quickly wash the cells once with deionized water, incubate with 1 % Nuclear Fast Red for 5 min for nuclei counterstaining, and then quickly wash once more.
3. Dehydrate the cells using quick washes with 35 % ethanol, followed by 70 % ethanol and absolute ethanol. Immediately after that, quickly wash the cells once with the absolute ethanol:xylene (1:1) solution and twice with xylene alone (see Note 15) to remove the alcohol.
4. Mount the cover slips in slides with a drop of Entellan mounting medium, avoid having thick Entellan layer (see Note 16). Allow the mounting medium to dry (usually 20–30 min) at which time the cells are ready for imaging.

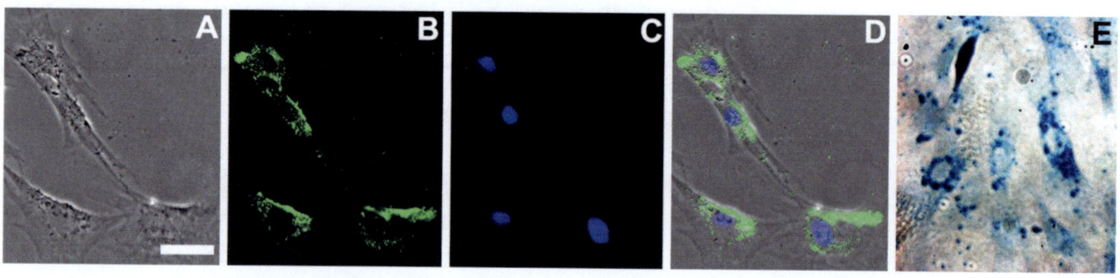

Fig. 3. Anti-dextran and Prussian Blue staining. Representative images of cells labeled with FeProt complexes for 4 h. (**a**–**d**) Mesenchymal stem cells immunostained for anti-dextran antibody. (**a**) Brightfield; (**b**) anti-dextran antibody; (**c**) nuclei counterstaining with DAPI; and (**d**) merged images. (**e**) Labeled cells stained with Prussian Blue showing the presence of iron. Note the characteristic perinuclear distribution of SPIONs revealed by both staining methods. Scale bar = 50 μm.

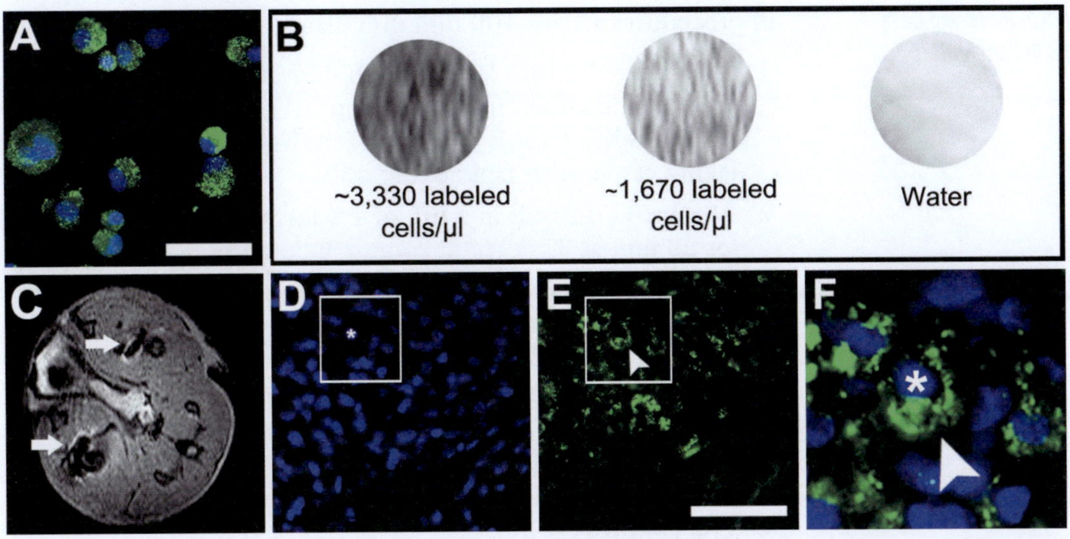

Fig. 4. Detecting SPION labeled cells using MRI in vivo and in vitro. (**a**) Appearance of FeProt labeled cells after trypsinization and immunolabeling for dextran. (**b**) In vitro MRI of cells labeled for 4 h at different concentrations of cells/μL gelatin. (**c**) Representative image of in vivo MRI (transverse plane) showing hypointense (*black*) spots corresponding to ferumoxide-labeled cells injected in the leg muscles (*white arrows*). (**d–f**) Dextran immunocytochemistry confirming the presence of ferumoxide-labeled cells in sections of leg muscle; (**d**) nuclear counterstaining with DAPI; (**e**) dextran; (**f**) merged images in higher-magnification of the area indicated by the *box* in (**d**) and (**e**) showing a transplanted cell (nucleus marked with an *asterisk*) labeled with ferumoxide (*arrowhead*). Scale bar = 50 μm.

3.3.2. Immunostaining with Anti-dextran Antibody

Here we describe an immunostaining protocol adapted for immunofluorescence staining of SPIONs.

1. Wash the coverslips with fixed cells three times with PBS/0.1 % Triton X-100 for 5 min, incubate with 5 % normal goat serum for 30 min, and then incubate with the anti-dextran antibody overnight at 4 °C in a humid chamber to avoid the antibody drying.
2. Wash the coverslips three times in PBS/0.1 % Triton X-100 for 5 min and then incubate with an anti-mouse secondary antibody for 1 h 30 min at room temperature, in a humidified chamber.
3. Wash three times with PBS without Triton for 5 min, incubate the cells for 5 min with DAPI for nuclei counterstaining (see Note 8) and wash once more with PBS without Triton.
4. Remove the excess PBS from the coverslips with an absorbent paper (do not touch the cells with the paper) and mount with a drop of Vectashield mounting medium per coverslip. Remove the excess mounting medium with an absorbent paper and seal the edges of the coverslips with nail polish (Fig. 3a–d). This technique can be used for localization of labeled cells after transplant, such as demonstrated in Fig. 4d–f.

3.3.3. Preparing Cells for In Vitro MRI

All procedures are for 100 mm dish plates.

1. Trypsinize the cells labeled with derivatized SPION nanoparticles (as described in Subheading 3.1), collect and centrifuge cells at 300×*g* for 5 min, and remove all of the supernatant taking care not to aspirate the cells.
2. Resuspend the cells in 1 mL of 4 % formaldehyde and incubate for 20 min at 37 °C or at room temperature.
3. Centrifuge the cells once more, remove all supernatant, resuspend in 300 μL of PBS and count the cells. From this point, samples with different concentrations of cells should be prepared, for example, dilute 10^6, 10^5, 10^4 and 10^3 cells per 100 μL of PBS (Fig. 4b).
4. Cool down the 200 μL aliquots of 15 % gelatin (see Note 9) from 60 °C down to 40 °C by keeping the tubes at room temperature.
5. Immediately add the 100 μL solution containing PBS and fixed cells to the gelatin aliquots and mix the solutions well. The final concentration of gelatin is 10 % and the final concentration of cells is 3,330, 333, 33 and 3.3 cells/μL.
6. At this point, the cells are ready to be imaged by MRI (see Note 17). Cells can be stored at 4 °C for future imaging if necessary.

3.4. Analyzing the Labeling Efficiency by Microscopy

1. Acquire random images of each sample (cells on coverslip) using a 20× objective (see Note 18).
2. Count the total number of cells within view by counting stained nuclei (nuclear fast red for Prussian Blue, and DAPI for immunofluorescence) (see Note 19).
3. Count the number of labeled cells. To estimate the percentage efficiency, divide the number of labeled cells by the total number of cells counted.

3.5. Analyzing the Labeling Efficiency by MRI

Contrast in MRI depends on the relaxation properties of the ^1H nuclei, mainly in water, in the cells and tissues. Longitudinal relaxation, T1, is the time it takes for the ^1H nuclei to relax back to their equilibrium value after perturbation with a radiofrequency pulse. Transverse or spin-spin relaxation, T2 and T2* (T2* includes additional effects caused by inhomogeneities in the static magnetic field), results from molecular interactions and causes a loss of detectable magnetization. T1, T2 and T2* weighted imaging methods are designed to detect differences in the relaxation properties, thus, a fluid filled cyst will appear bright on a T2-weighted image and dark on a T1-weighted image because it has long T2 and T1 relaxation times. On the other hand, a solid mass, having a short T2 and long T1, will appear bright on a T2-weighted image but dark on a T1-weighted image. In contrast, fat appears bright in T1-weighted images and dark in T2-weighted images because it has short T1 and T2 relaxation times. SPIONs, such as ferumoxides, are negative contrast agents. They cause a reduction in the T2 and T2* relaxation time of the nearby water (see Note 20) and appear dark in T2-weighted images. Thus detection of SPION labeled cells by MRI is typically accomplished using T2-weighted imaging methods.

1. Optimize magnetic field homogeneity of the cell sample and adjust MRI parameters (echo time and repetition time for optimal contrast). Although a spin-echo imaging sequence can be used, the T2* weighted gradient echo imaging sequence is recommended, particularly at higher field (~9.4 T) (see Note 21).
2. Acquire several thin (0.5 mm) contiguous image slices with optimized imaging parameters (see Note 10).
3. Quantify mean intensity of the cells to determine labeling efficiency. Use software provided by the MRI manufacturer, or free software, such as ImageJ. The image file is read into the software and the region of interest containing the labeled cells is selected. The intensities of all pixels within the region of interest are calculated and minimum, maximum and mean intensity is reported. Depending on the sample configuration several image slices may be analyzed and the results averaged.

4. Notes

1. Here we focus on cell labeling. Cell transplantations techniques and in vivo cell tracking procedures as such are beyond the scope of this paper.
2. Most authors report preparing the labeling solution using culture medium without FBS. We routinely use culture medium supplemented with FBS, and this has not affected cell labeling.
3. Protamine chlorhydrate can be used instead. For example, different countries use different types of protamine salts, for example, in USA, protamine sulfate is clinically approved by Food and Drug Administration (FDA). However, in Brazil, protamine chlorhydrate is clinically approved by The National Health Surveillance Agency (ANVISA) and it has the same clinical function as protamine sulfate to rescue the heparin effect.
4. Different types of cells can capture different amounts of SPIONs. The incorporation rate depends on the characteristics of the cells, such as the size of the cytoplasm and capacity for endocytosis. It is important to determine whether the cells can be detected by the MRI technique.

5. To detect SPION incorporation into cells, either Prussian Blue and immunostaining for anti-dextran antibody are very efficient. The Prussian Blue technique detects the iron nanoparticles, while anti-dextran antibody recognizes the dextran coating. For quantification we usually use the immunostaining technique because we consider it more convenient. However, it has been suggested that the dextran coating can undergo degradation when taken up by macrophages (33). Thus, a limitation of this technique is a potential underestimation of the number of labeled cells. If immunostaining is used for quantification, we recommend that Prussian Blue staining be done in some samples to confirm the iron presence inside the cells.
6. In this step is important to put the water into the vessel before the acid.
7. Other secondary antibody brands can be use but the ratio of antibody to PBS depends on the chosen antibody brand.
8. As an alternative for a separate DAPI solution, the Vectashield mounting medium can be replaced by Vectashield mounting medium for fluorescence containing DAPI (Vector Laboratories Inc., cat. # H-1200).
9. We have prepared the gelatin aliquots in 1.5 mL tubes because our MRI system is a vertical bore with a limited coil size that does not accommodate 24- or 96-well flat plates. However for a horizontal MRI system a plate may be more convenient.
10. For our study we used a 9.4 T magnet and imaged four 1.5 mL tubes of labeled cells at a time by mounting the tubes within a 50 mL conical tube and positioning that in a 40 mm birdcage imaging coil. We used a spin-echo sequence with a slice thickness of 0.5 mm, TR = 1 s, TE = 15 ms, and acquired four contiguous slices (Fig. 4b).
11. We recommend using the cells at 80–90 % confluence, regardless of the size of the culture plates.
12. 4 mL of labeling solution is needed for each 100 mm Petri dish plate.
13. For immunostaining or Prussian Blue staining of cells we used a 24-well plate with gelatin-coated coverslips. Other cell types require a specialized culture plate coating, for example, PLL, laminin or fibronectin. In that case, instead of gelatin, treat the coverslips with the coating required.
14. There is no need to maintain sterile conditions at this stage of experiment.
15. In this step, work in a fume hood exhaust system and perform the coverslip washes in a glass container using a glass pipette because xylene corrodes plastic.

16. Entellan is a hydrophobic medium that preserves the reaction for a longer time. However, the Entellan layer cannot be too thick because of the objective focal distance. High magnification objectives have a quite small working distance and a thick Entellan layer can prevent obtaining focused images. To obtain a thin layer, a light pressure can be applied to the coverslips and a small weighted object (smaller than the diameter of the coverslip) can be left on the coverslips until the Entellan dries.
17. We suggest that non-labeled cells be prepared for in vitro MRI as a background control.
18. For labeling efficiency and statistical analyses we recommend that at least three independent experiments be performed. In each experiment two to three samples (coverslips) should be prepared. From each sample we recommend that 6–8 images from different areas be acquired for quantification.
19. For immunofluorescence quantification we strongly recommend the acquisition of brightfield images. These images can be used to determine the membrane boundaries and to distinguish dextran-positive cells from the background.
20. An excellent review with detailed information on the practical issues involved in the MRI detection and quantification of cells labeled with SPIO nanoparticles and other contrast agents is cited in ref. (34).
21. At 9.4 T Smirnov et al. showed that the gradient echo sequence was superior to the spin echo sequence which had poor sensitivity at the higher field (35).

References

1. Guzman R et al (2007) Long-term monitoring of transplanted human neural stem cells in developmental and pathological contexts with MRI. Proc Natl Acad Sci U S A 104: 10211–10216
2. Sykova E, Jendelova P (2007) Migration, fate and in vivo imaging of adult stem cells in the CNS. Cell Death Differ 14:1336–1342
3. Hoehn M et al (2002) Monitoring of implanted stem cell migration in vivo: a highly resolved in vivo magnetic resonance imaging investigation of experimental stroke in rat. Proc Natl Acad Sci U S A 99:16267–16272
4. Dodd SJ et al (1999) Detection of single mammalian cells by high-resolution magnetic resonance imaging. Biophys J 76:103–109
5. Jendelova P et al (2004) Magnetic resonance tracking of transplanted bone marrow and embryonic stem cells labeled by iron oxide nanoparticles in rat brain and spinal cord. J Neurosci Res 76:232–243
6. Shapiro EM et al (2004) MRI detection of single particles for cellular imaging. Proc Natl Acad Sci U S A 101:10901–10906
7. Farrell E et al (2008) Effects of iron oxide incorporation for long term cell tracking on MSC differentiation in vitro and in vivo. Biochem Biophys Res Commun 369: 1076–1081
8. Modo M et al (2002) Tracking transplanted stem cell migration using bifunctional, contrast agent-enhanced, magnetic resonance imaging. Neuroimage 17:803–811
9. Lewin M et al (2000) Tat peptide-derivatized magnetic nanoparticles allow in vivo tracking and recovery of progenitor cells. Nat Biotechnol 18:410–414
10. Dodd CH et al (2001) Normal T-cell response and in vivo magnetic resonance imaging of T cells loaded with HIV transactivator-peptide-derived superparamagnetic nanoparticles. J Immunol Methods 256:89–105

11. Ahrens ET et al (2003) Receptor-mediated endocytosis of iron-oxide particles provides efficient labeling of dendritic cells for in vivo MR imaging. Magn Reson Med 49: 1006–1013
12. Frank JA et al (2003) Clinically applicable labeling of mammalian and stem cells by combining superparamagnetic iron oxides and transfection agents. Radiology 228:480–487
13. Weissleder R et al (1989) Superparamagnetic iron oxide: pharmacokinetics and toxicity. AJR Am J Roentgenol 152:167–173
14. Molday RS, MacKenzie D (1982) Immunospecific ferromagnetic iron-dextran reagents for the labeling and magnetic separation of cells. J Immunol Methods 52:353–367
15. Kim HS et al (2010) The effects of clinically used MRI contrast agents on the biological properties of human mesenchymal stem cells. NMR Biomed 23:514–522
16. Modo M et al (2004) Mapping transplanted stem cell migration after a stroke: a serial, in vivo magnetic resonance imaging study. Neuroimage 21:311–317
17. Brekke C et al (2007) The in vitro effects of a bimodal contrast agent on cellular functions and relaxometry. NMR Biomed 20:77–89
18. Greisberg JK et al (2001) Gadolinium inhibits thymidine incorporation and induces apoptosis in chondrocytes. J Orthop Res 19:797
19. Babic M et al (2008) Poly(L-lysine)-modified iron oxide nanoparticles for stem cell labeling. Bioconjug Chem 19:740–750
20. Arbab AS et al (2003) Intracytoplasmic tagging of cells with ferumoxides and transfection agent for cellular magnetic resonance imaging after cell transplantation: methods and techniques. Transplantation 76:1123–1130
21. Jasmin et al (2011) Optimized labeling of bone marrow mesenchymal cells with superparamagnetic iron oxide nanoparticles and in vivo visualization by magnetic resonance imaging. J Nanobiotechnology 9:4
22. Arbab AS et al (2004) Efficient magnetic cell labeling with protamine sulfate complexed to ferumoxides for cellular MRI. Blood 104: 1217–1223
23. Janic B et al (2009) Optimization and validation of FePro cell labeling method. PLoS One 4:e5873
24. van Buul GM et al (2009) Ferumoxides-protamine sulfate is more effective than ferucarbotran for cell labeling: implications for clinically applicable cell tracking using MRI. Contrast Media Mol Imaging 4:230–236
25. Qiu B, Yang X (2008) Molecular MRI of hematopoietic stem-progenitor cells: in vivo monitoring of gene therapy and atherosclerosis. Nat Clin Pract Cardiovasc Med 5:396–404
26. Walczak P et al (2005) Instant MR labeling of stem cells using magnetoelectroporation. Magn Reson Med 54:769–774
27. Tai JH et al (2006) Imaging islets labeled with magnetic nanoparticles at 1.5 Tesla. Diabetes 55:2931–2938
28. Qiu B et al (2010) Magnetosonoporation: instant magnetic labeling of stem cells. Magn Reson Med 63:1437–1441
29. Arbab AS et al (2005) Labeling of cells with ferumoxides-protamine sulfate complexes does not inhibit function or differentiation capacity of hematopoietic or mesenchymal stem cells. NMR Biomed 18:553–559
30. Kostura L et al (2004) Feridex labeling of mesenchymal stem cells inhibits chondrogenesis but not adipogenesis or osteogenesis. NMR Biomed 17:513–517
31. Bull BS et al (1975) Heparin therapy during extracorporeal circulation. II. The use of a dose-response curve to individualize heparin and protamine dosage. J Thorac Cardiovasc Surg 69:685–689
32. Gervin AS (1975) Complications of heparin therapy. Surg Gynecol Obstet 140:789–796
33. Bourrinet P et al (2006) Preclinical safety and pharmacokinetic profile of ferumoxtran-10, an ultrasmall superparamagnetic iron oxide magnetic resonance contrast agent. Invest Radiol 41:313–324
34. Liu W, Frank JA (2009) Detection and quantification of magnetically labeled cells by cellular MRI. Eur J Radiol 70:258–264
35. Smirnov P et al (2006) Single-cell detection by gradient echo 9.4 T MRI: a parametric study. Contrast Media Mol Imaging 1:165–174

Chapter 19

Noninvasive Quantification of Magnetic Nanoparticles by Means of Magnetorelaxometry

Maik Liebl and Dietmar Eberbeck

Abstract

Meaningful applications of magnetic nanoparticles (MNP) in biomedicine require the ability to quantify MNP in a variety of environments and biological materials (cells, tissue, and organs). Here we describe a method for the integral, noninvasive quantification of MNP. Our method is based on magnetorelaxometry (MRX), it relies on the detection of their magnetic stray fields which permeate practically unaltered virtually any biological material. MRX provides a specific signal which is not affected by dia- and paramagnetic signals from the biological matrix. We describe how to conduct the MRX experiments and use MNP-assisted gene transfection as the example. The quantification is based on the comparison of the MRX signal of a sample and an appropriate reference sample.

Key words: Magnetic nanoparticles, Magnetorelaxometry, Magnetofection, MNP quantification, Assisted gene transfer

1. Introduction

Magnetic nanoparticles (MNP) find wide application in medicine and bioanalytics, for example, as MRI contrast agents, for magnetic hyperthermia or magnetic drug delivery. MNP have promising advantages for magnetic labeling because they are stable, nontoxic, and can be manipulated by a magnetic field gradient (1). The latter property make MNP suitable for assisted gene therapy, where virus–MNP complexes are guided to the target cells by a magnetic field gradient (magnetofection) (2, 3). Hofmann et al. (3) demonstrated a 50-fold increase of provirus intake into the cells with MNP-assisted transduction compared to non-MNP-assisted transduction. For a proper understanding of the mentioned methods, the quantification of magnetic labels in the relevant target is necessary. Magnetorelaxometry (MRX), introduced by Kötitz et al. (4), is a technique for the sensitive quantification of MNP

Mikhail Soloviev (ed.), *Nanoparticles in Biology and Medicine: Methods and Protocols*, Methods in Molecular Biology, vol. 906, DOI 10.1007/978-1-61779-953-2_19, © Springer Science+Business Media, LLC 2012

down to the nanogram level (5). In MRX the magnetic moments of the MNP within the sample have a preferred orientation in the direction of an external magnetic field. After switching off this field, the sample net magnetization decays over time (relaxation) with a time constant being specific for the MNP. There is no influence by diamagnetic (e.g., biological tissues) or paramagnetic (e.g., hemoglobin) components of the sample, because the external magnetic field is nearly zero during the measurement. MRX is a fast (≤2 s for a single measurement) method which measures an integral signal to which all MNP in the sample contribute. In contrast to nuclear magnetic resonance (NMR), only the MNP contribute to the MRX signal enabling accurate quantification of MNP in vitro or in vivo, including in opaque media like blood (6, 7). Here we provide stepwise description of methods for quantifying MNP in cell samples by means of MRX.

2. Materials

Ideally all measurements should be performed inside a magnetically shielded room (MSR) in order to attenuate external magnetic fields at the sample site (see Note 1).

2.1. MRX Measurements

The MRX system is illustrated in Fig. 1 and should contain the following components (see Note 2):

1. Low-T_c SQUID (see Note 3).
2. Liquid helium-filled Dewar.

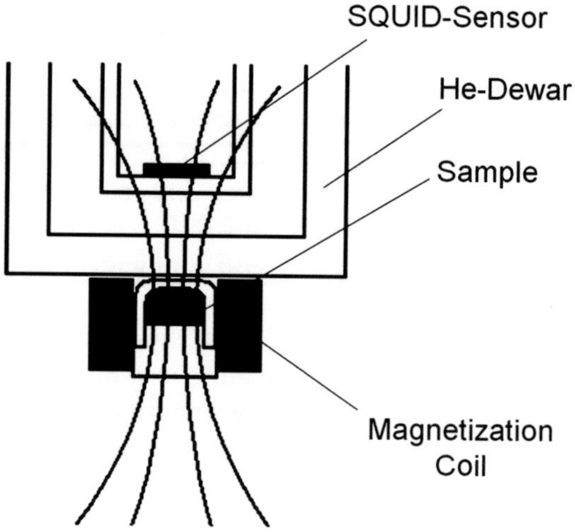

Fig. 1. MRX measurement system. Sketch of the bottom part of an MRX measurement system showing the magnetization coil, the sample, and the SQUID sensor inside the helium-filled Dewar.

3. Magnetization coil of about 10 mm bore diameter.
4. SQUID readout electronics with a bandwidth from dc to several megahertz.
5. Trigger Unit for switching the sensitive Flux-Locked-Loop (FLL) modus of the SQUIDs on, triggered by the falling slope of the current of the magnetization coil (clock frequency higher than 100 kHz) (see Note 4).
6. 150 μL polystyrene microtiter plates (Nunc).
7. Mannitol (e.g., D(-)-Mannitol) powder.
8. Freeze dryer (any commercially available freeze drying system would be suitable).
9. Endothelial cell growth medium (see Note 5).

2.2. Quantification of the MNP Amount in the Sample

The material of the sample holder should be nonmagnetic and of low conductivity in order to avoid eddy currents (see Note 6). For the quantification of the MNP a reference sample is needed. The reference sample should be prepared at the same time as the actual sample and using the same stock (batch) of MNPs. This precludes a mismatch between the signals of the particles in the sample of interest and the reference sample which might arise, for example, by aging effects of the MNP, e.g., aggregation of them. At higher MNP-concentrations, e.g., $c(\mathrm{Fe}) > 20$ mmol/L (8), the MRX-signal may depend on MNP content nonlinearly. To avoid related quantification errors, it is recommended to make a dilution series of the reference sample, spanning the targeted concentration range.

In order to ensure that the motional state of the MNP within the sample of interest as well as within the reference sample is the same, both samples should be freeze dried. Otherwise, the motional state of the MNP within the cell samples is not clearly defined, because the MNP might be either moving freely or might be partially restricted in their movements due to the viscosity of the cell matrix or might be bound and completely immobilized inside the cells. Furthermore, in order to keep the volume of the sample after freeze-drying constant, mannitol should be added to the sample prior to the freeze drying.

The samples used here to illustrate this method were virus-bound MNPs which were applied to human umbilical vein endothelial cells (HUVECs). Their preparation is described in (3) and is outside the scope of this paper. The aim of that particular experiment (3) was to evaluate the amount of nanoparticles left in the sample following the absorption of virus-MNP complexes by cells in magnetic field gradient and the subsequent washing steps. Because of the high specificity of the MRX signal of the MNPs, a wide range of MNP samples coming from different experimental setups and in a variety of buffers could be analyzed.

3. Methods

In general, the present method of MNP quantification in any sample (Fig. 2) relies on the quantitative comparison of the magnetic signals of the sample of interest with that of reference samples containing a known amount of the MNPs (typically a dilution series). Here, the magnetic signal is the magnetization decay after a polarizing field pulse, providing a specific signal. However, the MRX signal depends crucially on the binding state of the MNP, as mentioned above. In order to eliminate this influence, the MNP in both the sample of interest and in the reference sample have to be immobilized. The scheme in Fig. 2 illustrates the steps of the procedure of quantification of MNP in a sample.

3.1. MRX Measurements

By means of the MRX method the magnetization decay of an ensemble of MNP is measured. Therefore, an external static magnetic field is applied to the sample by a magnetization coil. This magnetic field polarizes the magnetic moments of the MNP within the sample along the field direction. Then the external field is switched off and the decay of the magnetization is measured by a low-T_c SQUID sensor yielding the time-dependent magnetic flux density $B(t)$ (9). The $B(t)$-curves may be parameterized by the relaxation amplitude $\Delta B = B(t_1) - B(t_{end})$, where t_1 denotes the initial measurement time and t_{end} the end of the measurement interval (Fig. 3). Under identical magnetization conditions, ΔB is directly proportional to the MNP content of the sample enabling its

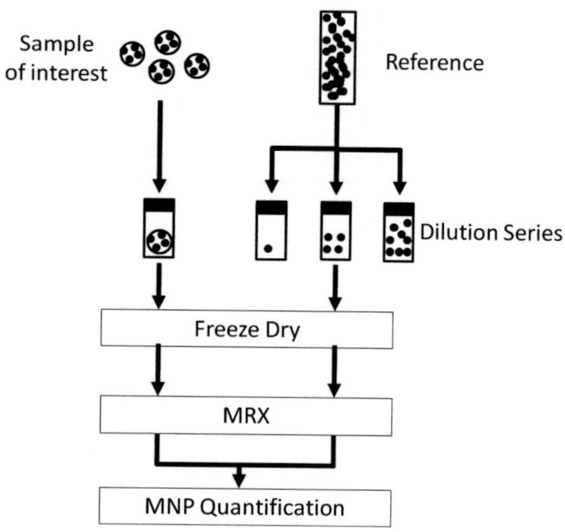

Fig. 2. Basic steps for the quantification of MNP by means of MRX. A sample of interest and a reference sample are freeze dried and MRX measurements are performed. The quantification is done by comparing the relaxation curve of the sample to those of the reference.

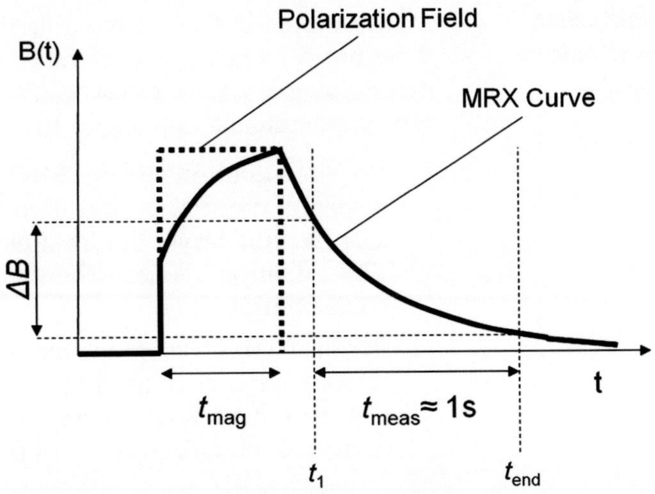

Fig. 3. MRX measurement. Schematic drawing of the MRX sequence with magnetization time t_{mag}, initial measurement time t_1, and last measurement time t_{end} determining the measurement interval t_{meas}. ΔB is the relaxation amplitude.

quantification. In order to quantify the MNP amount of a sample at least two MRX measurements have to be performed: Additionally to the sample of interest, at least one reference sample containing a known amount of MNP has to be measured. The MRX measurement data are obtained by the following steps.

1. Transfer the samples to be quantified into 150 μL polystyrene microtiter wells.
2. Add mannitol to the sample to a final concentration of 10 % by weight.
3. Freeze dry the samples.
4. Put the vials into the magnetization coil of the MRX system (Fig. 1).
5. Use the Control Unit to lock the SQUID sensors (i.e., make them insensitive to magnetic fields).
6. Apply a static magnetic field of about 2 mT magnetic flux density for e.g., $t_{mag} = 1$ s to the sample in order to align the magnetic moments of the MNP.
7. After switching off the magnetization field and a delay time of about 400 μs unlock the SQUID sensors and measure the relaxation curve $B(t)$ of the sample for about 1 s (Fig. 3, see Note 7).
8. Repeat the same procedure with an empty 150 μL polystyrene microtiter vial and measure the background signal $B_b(t)$.
9. Calculate the background-corrected relaxation curve of the sample $B_s(t) = B(t) - B_b(t)$ (see Note 8).
10. Estimate the relaxation amplitude $\Delta B_s = B_s(t_1) - B_s(t_{end})$.

3.2 Quantification of the MNP Amount in the Sample

The absolute MNP content within the sample of interest should be determined by fitting the reference curve to the sample curve and determining a scaling factor κ. The MNP amount in the sample can be quantified according to the following procedure.

1. Prepare a dilution series of MNPs with known MNP content n_r in vials of the same type as used for the cell samples for making a calibration curve. For example, a series of tenfold dilutions. The dilution series should span the targeted MNP concentration.
2. Perform MRX measurements of the sample and the reference dilution series (Subheading 3.1, see also an example in Fig. 4). Calculate the relaxation amplitude ΔB_r from the background-corrected relaxation curves of the reference samples $B_r(t)$.
3. Check that ΔB_r depends linearly on the dilution ratio enabling quantification as described below in steps 4 and 5 (see also Note 9).
4. Choose a curve $B_r(t)$ with a sufficiently high signal-to-noise ratio (SNR) for the quantification (see Note 9).
5. To get a first approximation, the scaling factor κ, that is used for the calculation of the MNP content of the samples n_s, Eq. 5, can be estimated via the relation $\kappa = \Delta B_s / \Delta B_r$.

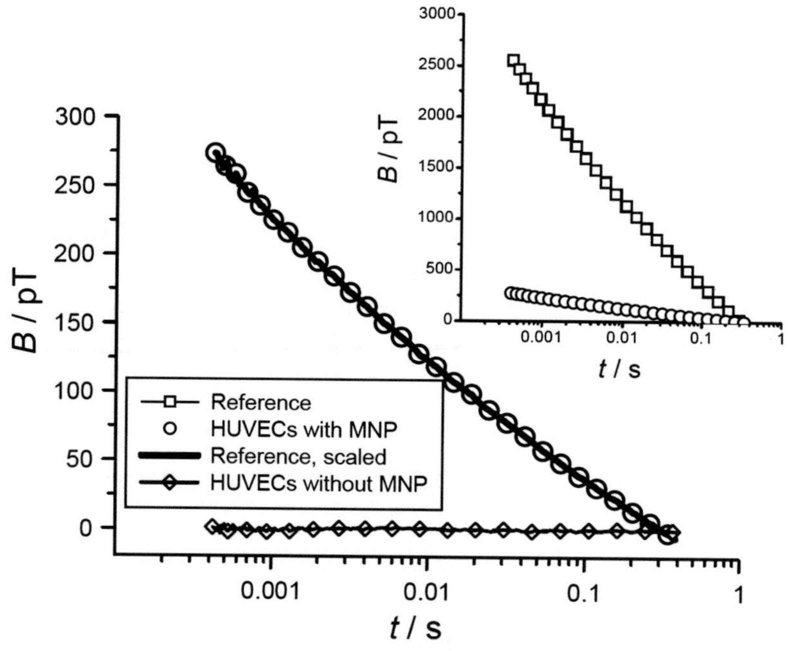

Fig. 4. MNP quantification. Decay of the sample's magnetization signal *B* in dependence of the time *t* elapsed after switching off the polarization field. For better visualization B_{off} was subtracted. The signal of cell probe (*circles*) is shown together with the reference scaled by the factor $\kappa = 0.106$ (*solid line*), where the original signal of the reference is plotted in the inset. Furthermore, it is clearly shown that cells without MNP give no signal.

6. For quantification by least square optimization, model the relaxation curve of the cell sample $B_s(t)$ by multiplying the reference signal $B_r(t)$ with the scaling factor κ and adding an arbitrary signal offset B_{off} (Fig. 4).

$$\hat{B}(t) = \kappa B_r(t) + B_{off} \tag{1}$$

In matrix notation, Eq. 1 is written as:

$$\hat{\mathbf{B}} = \mathbf{Pz} \text{ with } \mathbf{P} = [\mathbf{B}_r \mathbf{1}] \text{ and } \mathbf{z} = \begin{bmatrix} \kappa \\ B_{off} \end{bmatrix} \tag{2}$$

The sample data $\mathbf{B}_s$ is related to the model $\hat{B}$ via the vector of model deviations $\mathbf{u}$ (with the dimension of k data samples):

$$\mathbf{B}_s = \hat{\mathbf{B}} + \mathbf{u} = \mathbf{Pz} + \mathbf{u} \text{ with } \mathbf{u} = \begin{bmatrix} (B_1 - \hat{B}_1) \\ \vdots \\ (B_k - \hat{B}_k) \end{bmatrix} \tag{3}$$

The parameter vector $\mathbf{z}$ can be fitted by the least square estimator $\mathbf{z}_{OLS}$:

$$\mathbf{z}_{OLS} = \begin{bmatrix} \kappa \\ B_{off} \end{bmatrix} = \mathbf{P}^{+}\mathbf{B}_s = (\mathbf{P}^{T}\mathbf{P})^{-1}\mathbf{P}^{T}\mathbf{B}_s \tag{4}$$

with $\mathbf{P}^{+}$ being the Moore–Penrose pseudoinverse of $\mathbf{P}$ (10). The unknown MNP content inside the cell samples can then be calculated as follows:

$$n_s = \kappa n_r \tag{5}$$

7. Compare the shape between both curves in order to prove the validity of the quantification. There should be no significant differences in the shape of the relaxation curves (Fig. 4, see Note 10).

4. Notes

1. Measurements can also be made in laboratory environment, but this decreases the sensitivity due to a high level of magnetic background signal.
2. For a more detailed background of MRX system and description of its individual components see (11).
3. Alternatively, Fluxgates or atomic magnetometers can be used for MRX measurements. These sensors are less sensitive than SQUIDs, but also less expensive.

4. The normal of the pick-up-coil of the SQUID is oriented parallel to the aligned magnetic moment of the sample. This arrangement provides the highest sensitivity. Other arrangements might also be used, in particular, owing to a reduced loading of the SQUIDs by the polarization field. The distance between sensor and sample crucially determines the sensitivity. In the present case this distance amounts to 10 mm.
5. In this experiment the cells were in the endothelial cell growth medium (PromoCell), hence the same medium was used to make series of dilutions for making a dilution series of MNPs with known MNP content. For samples in different buffer, that same buffer should also be used for making MNP dilution series.
6. In order to maintain a low degree of uncertainty of the sample mass, it is favorable to prepare the samples within the wells which are finally used for the MRX measurement.
7. The SNR ratio can be improved by averaging several relaxation curves measured on the same sample. There should be a sufficient recovery time between individual measurements.
8. The data handling can be improved by logarithmic resampling of the relaxation curves combined with adaptive filtering applying filter coefficients which depend on the time t of $B(t)$ (9).
9. In the case of nonlinear dependence of ΔB_{r} on the iron content (dilution ration within the reference sample) choose a reference curve with ΔB_{r} in the vicinity of ΔB_{s} of the sample of interest.
10. The parameter $t_{1/e}$, that is the time where $B(t)$ is decayed to the value $1/e \cdot B(t_1)$, may be used as a measure of the shape of the relaxation curve. The $t_{1/e}$ values should be the same for the reference and the sample of interest. Different $t_{1/e}$ values might indicate that the cell uptake of the MNP is size selective for different particle diameters. Another reason could be the dipole–dipole interaction between MNPs in the sample of interest which arises from e.g., a dense packing of MNP within the cells. For example, we found on samples with larger MNPs (core size about 20 nm) that a deviation of 20 % between the relaxation times $t_{1/e}$ leads to a deviation in quantification (i.e., relaxation amplitudes) of 30 % (by factor 1.5).

Acknowledgements

The research was supported by the German Ministry for Education and Research (FKZ 13N11092) and Deutsche Forschungsgemeinschaft (FOR917, TR408/4-1, TR408/6-1). We thank the group of A. Pfeifer (University of Bonn) for providing virus-bound MNPs preparations.

References

1. Tartaj P, Morales MDP, Veintemillas-Verdaguer S, Gonzáles-Carreño T, Serna CJ (2003) The preparation of magnetic nanoparticles for applications in biomedicine. J Phys D Appl Phys 36:R182–R197
2. Plank C, Schillinger U, Scherer F, Bergemann C, Rémy J-S, Krötz F, Anton M, Lausier J, Rosenecker J (2003) The magnetofection method: using magnetic force to enhance gene delivery. J Biol Chem 384:737–747
3. Hofmann A, Wenzel D, Becher UM, Freitag DF, Klein AM, Eberbeck D, Schulte M, Zimmermann K, Bergemann C, Gleich B, Roell W, Weyh T, Trahms L, Nickenig G, Fleischmann BK, Pfeifer A (2009) Combined targeting of lentiviral vectors and positioning of transduced cells by magnetic nanoparticles. PNAS 106:44–49
4. Kötitz R, Trahms L, Koch H, Weitschies W (1995) Ferrofluid relaxation for biomagnetic imaging. In: Baumgartner C, Deecke L, Williamson SJ (eds) Biomagnetism: fundamental research and clinical applications. Elsevier Science, IOS Press, Amsterdam, pp 785–788
5. Wiekhorst F, Seliger C, Jurgons R, Steinhoff U, Eberbeck D, Trahms L, Alexiou C (2006) Quantification of magnetic nanoparticles by magnetorelaxometry and comparison to histology after magnetic drug targeting. J Nanosci Nanotechnol 6:3222–3225
6. Richter H, Kettering M, Wiekhorst F, Steinhoff U, Hilger I, Trahms L (2010) Magnetorelaxometry for localization and quantification of magnetic nanoparticles for thermal ablation studies. Phys Med Biol 55(3):623–633
7. Eberbeck D, Wiekhorst F, Steinhoff U, Schwarz K, Kummrow A, Kammel M, Neukammer J, Trahms L (2009) Specific binding of magnetic nanoparticle probes to platelets in whole blood detected by magnetorelaxometry. J Magn Magn Mat 321:1617–1620
8. Eberbeck D, Trahms L (2011) Experimental investigation of dipolar interaction in suspensions of magnetic nanoparticles. J Magn Magn Mat 323(10):1228–1232
9. Eberbeck D, Wiekhorst F, Steinhoff U, Trahms L (2006) Aggregation behaviour of magnetic nanoparticle suspensions investigated by magnetorelaxometry. J Phys Condens Matter 18:S2829–S2846
10. Sachs L, Hedderich J (2006) Angewandte Statistik—Methodensammlung. Springer, Berlin, Heidelberg
11. Matz H, Drung D, Hartwig S, Gro H, Kötitz R, Müller W, Vass A, Weitschies W, Trahms L (1998) A squid measurement system for immunoassays. Appl Supercond 6:577–583

Chapter 20

Synthesis, Application, and Tracking of Magnetic Carbon-Coated Nanoparticles in Plants

Alejandro Pérez-de-Luque, Zuny Cifuentes, Clara Marquina, Jesús M. de la Fuente, and M. Ricardo Ibarra

Abstract

The behavior of nanoparticles inside plants is gaining importance for its implications in research about putative applications and toxicology. Magnetic carbon-coated nanoparticles can be easily traced through plant tissues using simple and affordable histological techniques. Here we present a methodology for the synthesis of such nanoparticles. We also describe methods for growing plants using rhizotrons (in order to observe the roots), procedures for applying the nanoparticles, taking and processing the samples.

Key words: Magnetic nanoparticles, Arc-discharge furnace, Gas-phase condensation, Nanobiotechnology, Plant histology, Rizhotron, Light microscopy

1. Introduction

The development of nanodevices for applications in plant or agricultural research has become a highly topical subject (1). Specifically, nanoparticles can be used in the scope of life sciences as smart delivery systems, which can be loaded with a drug or therapeutic agent. In addition, some specific property and/or external modification of these nanoparticles allows their accumulation and/or guidance into specific areas within a living organism in which the chemical charge will be unloaded. This will allow several new applications, ranging from treatments with agrochemicals to delivery of nucleic acids for genetic transformation (2–4). There is also a clear need for detailed studies about toxicological and ecological impact of nanomaterials on the environment and the risk of their incorporation into the food chain (5). But most of the recent advances come from medical and pharmaceutical research and

Mikhail Soloviev (ed.), *Nanoparticles in Biology and Medicine: Methods and Protocols*, Methods in Molecular Biology, vol. 906, DOI 10.1007/978-1-61779-953-2_20, © Springer Science+Business Media, LLC 2012

there is clear lack of information on the behavior of nanomaterials in plants (6). The use of biocompatible suspensions of carbon-coated magnetic nanoparticles provides simple and reliable approach to study absorption, transport, and accumulation of nanomaterials in plants (7–9). The black color of the magnetic fluid allows to track the nanoparticles easily inside the plant tissues using simple histological techniques (7, 9). The magnetic core of the nanoparticles provides means for targeting the nanoparticles to specific areas by application of magnetic field gradients (7). Magnetic targeting of nanoparticles was used, for example, for delivery of nucleic acids (10) to meristematic tissues in plants by placing magnets there, for nanoharvesting of plant compounds (11) using magnetism for recovering the nanoparticles with the harvested compounds, for plant genetic transformation by inducing chromosomal changes with ferrofluids (12), and even hyperthermia may be applied against some specific plant pathogens (3).

Here we provide detailed protocols for the synthesis of the magnetic carbon-coated nanoparticles (a modification of the Kratschmer-Huffman method (13) using an arc-discharge furnace) and their application to living plants. We also describe microscopy-based methods for their detection and tracking.

2. Materials

2.1. Synthesis of Magnetic Carbon-Coated Nanoparticles

1. Iron powder, spherical, Ø < 10 μm (99.9 + % metals basis) (Alfa Aesar GmbH & Co).
2. Graphite rods, Ø 6.15 mm, 102 mm long (99.9995 % metals basis) (Alfa Aesar GmbH & Co).
3. Graphite rods, Ø 3.05 mm, 305 mm long (99.9995 % metals basis) (Alfa Aesar GmbH).
4. Power drill and drills (Ø 2.5 mm approximately).
5. He gas (global purity ≥99.9990 %).
6. Arc-discharge furnace (model 5 of Centorr Associates Inc). This model works with a 12.9 Kw power supply model 300ss (Miller Electric MFG Co). The furnace is water refrigerated. A rotary pump should be connected to purge the furnace chamber up to 10^{-2} mbar.
7. Spatulas.
8. Glass beakers.
9. Deionized water.
10. 3 M HCl.
11. Sodium dodecyl sulfate (SDS): to make 0.005 % solution dissolve 2.5 mg of SDS in 500 mL of deionized water.

12. Heating plate magnetic stirrer.
13. Oven (able to reach at least 350 °C).
14. Cylindrical (Ø 5 cm, 2 cm length) Nd-Fe-B permanent magnet (3 KOe magnetic field).
15. Sonicator (SONOREX Super RK 510H, Bandelin Electronic GmbH).

2.2. Rhizotron for Growing the Plants

1. Plastic Petri dishes (Ø 140 mm) for making rhizotrons.
2. Filter paper.
3. Expanded perlite as hydroponic medium.
4. Whatmann GF/A glass fiber filter paper disks (Ø 140 mm).
5. Sealing film (Parafilm M).
6. Plastic trays (300 × 190 × 70 mm).
7. Aluminum foil.
8. Microcentrifuge tubes (1.5 mL size).
9. Neodymium magnets (Ø 5–10 mm).
10. Calcium nitrate (0.5 M): dissolve 118 g of $Ca(NO_3)_2 \cdot 4H_2O$ in 1 L of distilled water.
11. Potassium dihydrogen phosphate (0.5 M): dissolve 68 g/L (KH_2PO_4) in 1 L of distilled water.
12. Potassium nitrate (0.5 M): dissolve 50.55 g of KNO_3 in 1 L of distilled water.
13. Magnesium sulfate (0.5 M): dissolve 123.24 g/L of $MgSO_4 \cdot 7H_2O$ in 1 L of distilled water.
14. Chelated iron (Fe-EDTA) stock solution: dissolve 9.2 g of Ethylenediaminetetraacetic acid ferric-sodium salt in 1 L of distilled water.
15. Micronutrients stock solution: 1.846 g/L, KCl, 0.773 g/L H_3BO_3, 0.169 g/L $MnSO_4 \cdot H_2O$, 0.288 g/L $ZnSO_4 \cdot 7H_2O$, 0.063 g/L $CuSO_4 \cdot 5H_2O$, 0.23 g/L Na_2MoO_4.
16. Hoagland nutrient solution: to make 1 L use 10 mL of 0.5 M calcium nitrate, 2 mL of 0.5 M potassium dihydrogen phosphate, 1 mL of 0.5 M potassium nitrate, 4 mL of 0.5 M magnesium sulfate, 2 mL of 0.025 M Fe-EDTA, and 2 mL of the micronutrients stock solution. Add water to 1 L final volume.
17. Growth chamber for controlled conditions of temperature and photoperiod.

2.3. Application of Nanoparticles on Living Plants

1. Microcentrifuge tubes (1.5 mL size).
2. Micropipette (200–1,000 μL).
3. Razor blade and scalpel.
4. Syringe (2 mL size preferred).

5. Sealing film (Parafilm M).
6. Neodymium magnets (Ø 5–10 mm).

2.4. Preparation of Samples for Light Microscopy

1. Precision tweezers.
2. Several blades and precision cutting tools (razor blades, precision blades, scalpels, etc.).
3. Phosphate buffered saline (PBS): 8 g/L NaCl, 0.2 g/L KCl, 1.14 g/L Na_2HPO_4, 0.2 g/L KH_2PO_4, pH 7.4.
4. Paraformaldehyde (PFA): to make 8 % solution add 1 g of powder PFA to 25 mL of PBS (will not dissolve and will have milky appearance). Stir the solution at 60 °C (do not overheat) until it becomes clear and transparent. If the solution remains murky, few drops of 1 M NaOH can be added (adjust later to pH 7). Cool at 4 °C and keep at 4 °C until use. For longer storage periods (i.e., more than a day), freeze and store at −20 °C.
5. 0.1 M Cacodylate buffer, pH 7.
6. 25 % Glutaraldehyde.
7. Karnovsky's fixative solution: to make 25 mL add 0.0125 g of $CaCl_2$, 6.25 mL of 0.1 M cacodylate buffer, 5 mL of 25 % glutaraldehyde and 12.25 mL of 8 % PFA solution, bring the total volume to 25 mL with distilled water.
8. Dissecting microscope.
9. Vibratome.
10. Glass slides and coverslips.
11. Optical (light) microscope.

3. Methods

3.1. Synthesis of Magnetic Carbon-Coated Nanoparticles Water Suspensions

1. Cut the 6.5 mm diameter graphite rods in small cylinders of approximately 2.5 cm length.
2. Drill a hole in the rod along its longitudinal direction, taking care of not drilling the whole length (keep a cylinder basis of about 2 mm thickness).
3. Fill the cavity with the micrometric iron powder, pressing the powder to allow a maximum load.
4. Place the rods filled with the Fe powder in the center of the stationary furnace electrode (bottom section of the arc-discharge cavity; anode).
5. Use a 3.05 mm diameter and 6 cm long graphite rod as movable furnace electrode (cathode).

6. Once the furnace is closed, purge the cavity four times (up to 10^{-2} mbar).
7. After the last purge, fill the furnace cavity with He gas up to 0.1 bar.
8. Put both furnace electrodes in contact, in a point near the lateral surface of the cylinder. Start the power supply setting a 30A electrical current in order to strike the arc between them.
9. Place the upper electrode just above the hollowed rod containing the Fe powder so that the arc sets now between the upper electrode and the rod (see Note 1). Increase the current slowly up to 50A (see Note 2), so that both graphite and Fe powder sublimate. A deposit (black soot) is formed on the inner surfaces of the furnace cavity.
10. The arc-furnace will stop by itself, once the materials completely sublimate.
11. Once the furnace is cooled down, open the furnace cavity and carefully collect with a spatula the soot deposited in the lateral and upper cavity surfaces.
12. Removal of side products (carbon nanostructures and non-coated or partially coated metallic particles) takes place by treating the soot with 50 mL of 3 M HCl. Heat in a water bath to 80 °C and stir for 90 min (see Note 3).
13. Once the mixture is cooled down, separate the soot by placing the 3 KOe cylindrical magnet under the glass beaker and pouring out the liquid.
14. Wash with 100 mL of 0.005 % SDS. Separate again the soot using the 3 KOe magnet.
15. Wash the soot again with new 100 mL of 0.005 % SDS. Repeat washing procedure five times until the color of the supernatant is pale gray or disappears completely. This procedure will remove all amorphous carbon and other carbon structures not containing magnetic nanoparticles and will enrich the content of carbon-coated magnetic nanoparticles in the powder attracted to the bottom of glass beaker by the magnet (see Note 4).
16. Wash the remaining powder at the bottom of the beaker five times with deionized water (see Note 5).
17. After the last magnetic separation place the nanopowder in an oven and heat to 350 °C and keep it in the oven for 6 h. This procedure will remove any residual carbon nanostructures, and completely evaporate the water.
18. Dissolve 100 mg of the dried powder in 100 mL of deionized water.
19. Sonicate the suspension for 15 min.

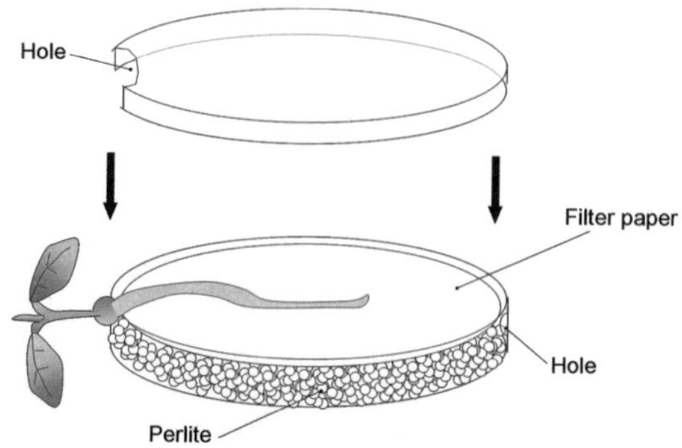

Fig. 1. Schematic drawing of the configuration of a rhizotron.

3.2. Plant Growth in the Rhizotron System

1. Germinate the seeds in Petri dishes on moistened filter paper in the dark, and keep until the seedlings reach a radicle length of 3–5 cm.
2. Prepare the rhizotrons by making two holes in opposite lateral sides of every Petri dish (see Note 6). One of the holes must be big enough for the upper part of the plant to stand out. The other hole should be small, only for allowing the nutrient solution to flux inside (see Fig. 1). Completely fill Petri dishes with perlite and place a disk of glass fiber filter paper (same diameter of the dish) on the perlite (see Note 7).
3. Place the seedling in the rizhotron with the root on the disk of paper and the aerial part standing out through the biggest hole, put the dish cover on it and seal with sealing film (see Fig. 1).
4. Wrap the rhizotrons in aluminum foil and place them vertically on a plastic tray, the hole with the plant aerial part up and uncovered to allow seedling emergence and growth.
5. Fill the tray with Hoagland nutrient solution (to reach a depth of about 2 cm) and keep the plants in a growth chamber at the required temperature and photoperiod (see Note 8).

3.3. Application of Nanoparticles to the Plants

When the plants reach the required phenological stage and the roots are well developed, proceed with the application of nanoparticles. Use 1 mg/mL nanoparticle suspension prepared earlier (see Subheading 3.1, step 19).

1. Foliar applications are made by placing a drop of the nanoparticle suspension on the leaves with a micropipette.
2. Application through the leaf petiole involves cutting the leaf and adding a few drops of the nanoparticle suspension into the pith cavity of the petiole using a syringe (see Note 9).
3. For root application, a microcentrifuge tube containing nanoparticle solution should be inserted into the rhizotron. Carefully

remove the cover of the Petri dish rhizotron. Near to the chosen root, cut the filter paper to make a hole of a size of a microcentrifuge tube. Remove the perlite from underneath the cut-out filter to allow microcentrifuge tube to fit the hole. Keeping the rhizotron as vertical as possible, put a microcentrifuge tube filled with the suspension of nanoparticles in the gap, and insert the roots into the solution. Carefully, place the Petri dish cover back and seal the rhizotron with sealing film.

4. Small magnets can be placed at different locations to concentrate and accumulate the nanoparticles. Attach the magnets to the rhizotron using sellotape.

3.4. Sampling and Observation Under Light Microscopy

1. 24–72 h after nanoparticle application, take small samples (5–10 mm length) from different parts of the plant (roots, leaves, stem) using tweezers and a razor blade.
2. Immerse the samples in fixative solution (Karnovsky) and incubate 4 h at room temperature (see Note 10). Rinse three times in 0.025 M cacodylate buffer (10–20 min each), and store samples in the same buffer at 4 °C (the samples can be stored up to several weeks).
3. Not embedded fresh cut sections of about 50–100 μm thickness can be cut by hand (see Note 11) or using a vibratome (see Note 12).
4. Place the cut sections on a glass slide, add a few drops of distilled water, cover the sample with a coverslip, and observe under the light microscope. With sections of 50–100 μm thickness, a magnification of ×200–400 can be achieved. The accumulation of carbon-coated nanoparticles in tissues can be seen as black deposits (Fig. 2). Always use negative controls (see Note 13).

4. Notes

1. Once the arc is started, separate both electrodes slowly in order to increase the length of the electrical arc. At the same time, displace the arc laterally, moving the upper electrode approaching the hollowed rod, avoiding any contact between them, until the upper electrode is on top of the rod. The movement of the arc as well as the precise control of its intensity is crucial. First of all, care should be taken in order to keep the hollowed bar in the upright position. Second, the intensity of the arc current should be enough to slowly evaporate both graphite and iron, allowing them to migrate towards the

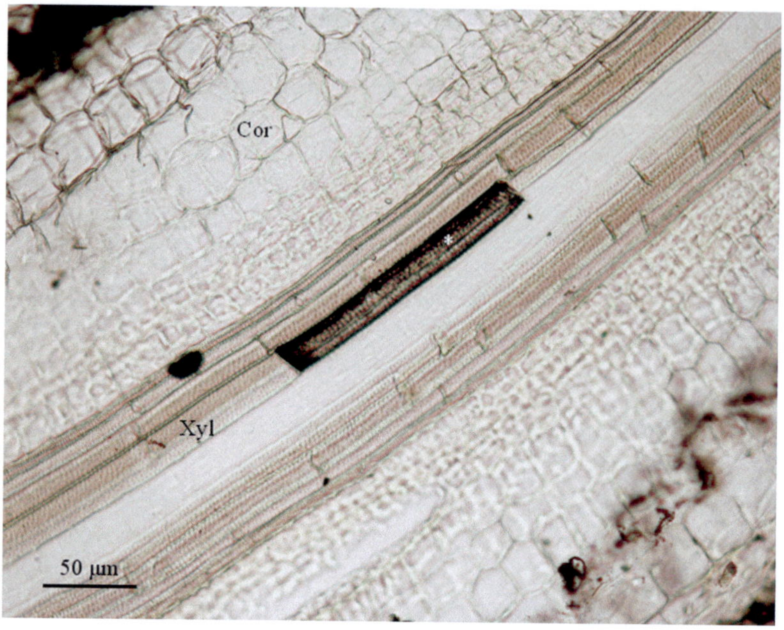

Fig. 2. Longitudinal section of a pea root. Accumulation of ferrofluid composed of carbon-coated magnetic nanoparticles, inside a xylem vessel (*asterisk*). Cortical cells (Cor); xylem vessels (Xyl).

chamber lateral and upper surfaces of the furnace, where they will condense (first the iron and subsequently the carbon, forming the coating). However, the arc intensity should be never so strong to cause quick melting of the iron powder (forming an iron button) and the evaporation of the graphite separately. For a description of the arc-furnace working principle see ref. (9).

2. The specified current value of 50A is for guidance only. The current should be sufficiently high to completely vaporize graphite and iron. But evaporation will gradually increase pressure in the chamber. Therefore, the maximum current should be limited to maintain the chamber pressure below the safety limit of the arc-furnace (0.5 bar in our furnace).
3. This chemical-etching procedure eliminates the noncoated or partially coated metallic particles from the soot. This treatment also leads to the formation of carboxylic groups on the graphitic shell, which, due to their hydrophilic nature, will contribute to the stability of the final particle suspension (9).
4. Typically it takes five washes to complete the procedure.
5. Five times is just a guiding number. Repeat the procedure until the SDS is completely removed (as confirmed by absence of any foam in the solution). Any trace of SDS will be detrimental for the biocompatibility of the final nanoparticles suspension.

6. The holes can be easily done by heating a lancet or a scalpel with a lighter. The hot tip can cut through the plastic of the Petri dishes.
7. The Petri dish must be filled completely with Perlite. Avoid making hips of perlite. The perlite surface must be leveled to get a flat surface. An excess will not allow the cover to be correctly placed, and a shortage will not support firmly the roots against the cover.
8. The controlled conditions may vary for different plant species, normally these will be 20–25 °C, and a photoperiod of 12–14 h.
9. Some plants may not be suitable for this because of the lack of petiole (e.g., wheat), or because the pith cavity does not exist.
10. Apply some vacuum to the samples to get a better penetration of the fixative into the tissues. A small vacuum chamber with a venturi pump can be used for a few minutes.
11. For manual cutting of the samples, some practise is required, but it is a suitable method when no vibratome is available or if samples are too small for such instrument. The samples are usually cut under a dissection microscope on a glass slide using precision tweezers and scalpel or razor blade. Add a few drops of fixative or distilled water. Try to produce sections as thin as possible. Always use new blades and replace them often.
12. Samples can be also embedded in different mediums (resin, paraffin) and cut using a microtome. However, such process involves several rinses during long periods of time and the nanoparticles can be eliminated by washing from the tissues (specially, xylem vessels). For that reason, such methods should be used only when high-resolution microscopy is going to be used (e.g., electronic microscopy), allowing observation of individual nanoparticles.
13. The use of negative controls is strongly recommended in order to avoid misinterpretation due to development of artifacts in tissues (e.g., shadows by refringence of air bubbles).

Acknowledgements

This work was supported by the projects granted by the Spanish Ministry of Science and Innovation (MICINN) NANOBIOMED (CONSOLIDER-INGENIO 2010 Programme), AGL2008-01467, EUI2008-00157, and EUI2008-00114. Financial support from the Autonomic Government of Aragon (DGA) is also acknowledged through ARAID foundation.

References

1. Robinson DKR, Salejova-Zadrazilova G (2010) Nanotechnologies for nutrient and biocide delivery in agricultural production. ObservatoryNANO. http://www.observatorynano.eu/project/filesystem/files/Controlled%20delivery.pdf. Accessed 18 Apr 2011
2. Torney F, Trewyn BG, Lin VSY, Wang K (2007) Mesoporous silica nanoparticles deliver DNA and chemicals into plants. Nat Nanotechnol 2:295–300
3. Pérez-de-Luque A, Rubiales D (2009) Nanotechnology for parasitic plant control. Pest Manage Sci 65:540–545
4. Pérez-de-Luque A, Cifuentes Z, Beckstead J, Ryan RO (2010) Amphotericin B nanodisks (AMB-NDs) for treatment of fungal diseases in plants. In: Ribeiro C, de Assis OBG, Mattoso LHC, Mascarenhas S (eds) International Conference on Food and Agriculture Applications of Nanotechnologies (NanoAgri-2010). Embrapa, Sao Pedro
5. FAO/WHO (2010) Expert meeting on the application of nanotechnologies in the food and agriculture sectors: potential food safety implications. Meeting report. Food and Agriculture Organization of the United Nations and World Health Organization, Rome
6. Nair R, Varghese SH, Nair BG, Maekawa T, Yoshida Y, Kumar DS (2010) Nanoparticulate material delivery to plants. Plant Sci 179: 154–163
7. González-Melendi P, Fernández-Pacheco R, Coronado MJ, Corredor E, Testillano PS, Risueño MC, Marquina C, Ibarra MR, Rubiales D, Pérez-de-Luque A (2008) Nanoparticles as smart treatment-delivery systems in plants: assessment of different techniques of microscopy for their visualization in plant tissues. Ann Bot 101:187–195
8. Corredor E, Testillano PS, Coronado MJ, González-Melendi P, Fernández-Pacheco R, Marquina C, Ibarra MR, de la Fuente JM, Rubiales D, Pérez-de-Luque A, Risueño MC (2009) Nanoparticle penetration and transport in living pumpkin plants: in situ subcellular identification. BMC Plant Biol 9:45
9. Cifuentes Z, Custardoy L, de la Fuente JM, Marquina C, Ibarra MR, Rubiales D, Pérez-de-Luque A (2010) Absorption and translocation to the aerial part of magnetic carbon-coated nanoparticles through the root of different crop plants. J Nanobiotechnol 8:26
10. Torney F, Trewyn BG, Lin VSY, Wang K (2007) Mesoporous silica nanoparticles deliver DNA and chemicals into plants. Nat Nanotechnol 2:295–300
11. Kurepa J, Paunesku T, Vogt S, Arora H, Rabatic BM, Lu J, Wanzer MB, Woloschak GE, Smalle JA (2010) Uptake and distribution of ultrasmall anatase TiO_2 alizarin red S nanoconjugates in *Arabidopsis thaliana*. Nano Lett 10:2296–2302
12. Racuciu M, Creanga D (2007) Cytogenetic changes induced by aqueous ferrofluids in agricultural plants. J Magn Magn Mater 311: 288–290
13. Kratschmer W, Lamb LD, Fostiropoulos K, Huffman DR (1990) Solid C60: a new form of carbon. Nature 347:354

Part IV

Organic, Polymeric, Biomimetic, and Biocompatible Nanoparticles and Their Applications

Chapter 21

Synthesis of Biocompatible Gelatinated Thioglycolic Acid-Capped CdTe Quantum Dots ("Jelly Dots")

Valérie A. Gérard, Yurii K. Gun'ko, Babu R. Prasad, and Yury Rochev

Abstract

Semiconductor luminescent Quantum Dots (QDs) constitute a growing area of research for biological imaging and other biomedical applications. One of the main challenges is to provide QDs with a biocompatible and easy to functionalize surface while retaining the core optical properties. Gelatine is an excellent candidate for that purpose as it is a very common natural polymer, highly biocompatible and bearing various functional groups. Here we present a simple, one-pot method for manufacturing gelatinated QDs with chosen optical properties.

Key words: Quantum dots, Cadmium telluride, Gelatine, Bioimaging

1. Introduction

Semiconductor Quantum Dots (QDs) are luminescent nanoparticles which have proved to bear great potential for biomedical applications. Their optical properties make them excellent candidates for biological imaging with a number of advantages over organic dyes such as narrow and symmetrical emission spectra, broad absorption spectra and improved resistance to photobleaching (1). They however present one major drawback which is their intrinsic toxicity. The main mechanism of cytotoxicity lies in the release of cadmium ions and production of reactive oxygen species (ROS) (2). These events may be significantly reduced by the presence of a protective layer of biocompatible material on the QD surface. Gelatine is a very common natural polymer with very good biocompatibility (3). It has been widely investigated for drug delivery applications as it can be formed into biodegradable nanoparticles (4–6). Therefore the presence of gelatine on the surface of TGA-capped

Mikhail Soloviev (ed.), *Nanoparticles in Biology and Medicine: Methods and Protocols*, Methods in Molecular Biology, vol. 906,
DOI 10.1007/978-1-61779-953-2_21, © Springer Science+Business Media, LLC 2012

CdTe QDs is believed to improve their biocompatibility and to significantly reduce their cytotoxicity (7, 8). Another advantage provided by the gelatine layer on the surface of QDs is the passivation of surface defects which leads to a higher quantum efficiency (7). It also offers a variety of functional groups available for further functionalization of QDs, owing to its amino acid residues (6, 9). They may therefore be combined to a variety of biomolecules and injected into live cells to fluorescently reveal the pathways involved in some disease, or serve as a drug carrier (10). The synthesis of such nanoparticles has been finely tuned and optimized to produce various sizes of QDs with high quantum yield (QY) and narrow emission (10). It is based on the aqueous synthesis of CdTe QDs using thiol stabilizers (11). It is thus a convenient and reproducible procedure.

2. Materials

Prepare all solutions using Millipore water and analytical grade reagents. Prepare and store all reagents at room temperature (unless indicated otherwise). Diligently follow all waste disposal regulations when disposing waste materials.

2.1. Synthesis of Quantum Dots

1. Aluminum telluride (CERAC). Store in inert atmosphere.
2. Cadmium perchlorate hydrate.
3. Thioglycolic acid (TGA).
4. Gelatine.
5. Degassed Millipore water: boil 1 L of Millipore water then cool on ice under argon.
6. Sodium hydroxide solution: 2 M solution in water.
7. Sulphuric acid solution: 2 M solution in water.
8. Schlenk line with one manifold connected to an argon source and the other to a vacuum pump.

2.2. Fractioning of Quantum Dots

1. Propan-2-ol.
2. Heidolph Laborota 4002 Rotary evaporator.
3. Hettig Universal 320R centrifuge.
4. Shimadzu UV-1601 UV–Visible Spectrophotometer.
5. Varian–Cary Eclipse Fluorescence Spectrophotometer.

2.3. Purification of Quantum Dots

1. Sephadex G25.
2. Glass chromatography column: 10 cm high, 2 cm in diameter, fitted with a rubber pump.

3. Shimadzu UV-1601 UV–Visible Spectrophotometer.
4. Varian–Cary Eclipse Fluorescence Spectrophotometer.

3. Methods

3.1. Synthesis of Quantum Dots

The whole synthesis is carried out in argon atmosphere and with constant vigorous stirring.

1. Prepare the experimental set-up as depicted in Fig. 1. Place a 100 mL three-necked glass round-bottomed flask on a magnetic stirrer and connect one neck to the Schlenk line. Place a 250 mL three-necked glass round-bottomed flask on a stirrer-heater equipped with a temperature probe and connect one neck to the exhaust flask filled with concentrated sodium hydroxide. Connect both flasks together with a glass bubbler dipping into the larger one. Close the third neck of the large flask with a stopper and the one of the small flask with a rubber septum.
2. Add 150 mL of degassed water to the 250 mL three-necked round-bottom flask equipped with a stirring bar.
3. Flush the system with argon and heat the water to 60 °C.
4. Add 0.5 g of aluminum telluride (see Note 1) to the smaller three-necked round-bottom flask (with a stirring bar) (Fig. 1).
5. Add 5.8 g of cadmium perchlorate hydrate to the warm degassed water, then add 1.4 mL of TGA (see Note 2). The solution becomes turbid.

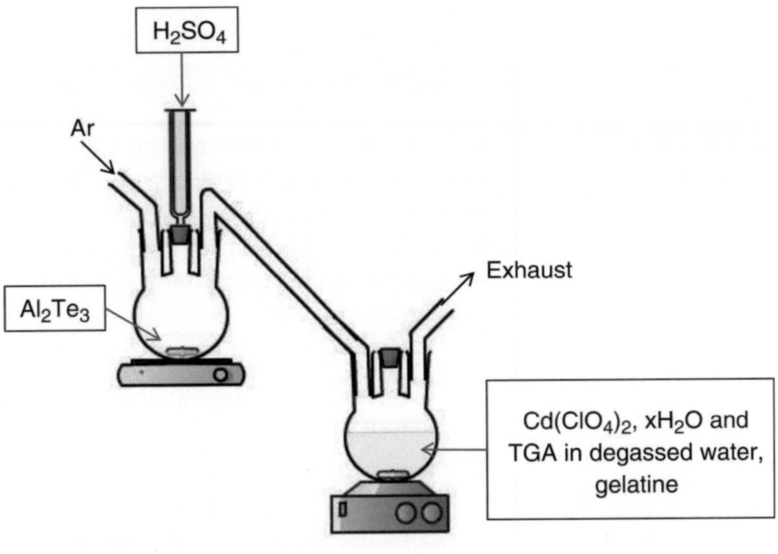

Fig. 1. Schematic of the experimental set-up for the synthesis of Jelly Dots.

6. Adjust the pH to 11 using a 2 M NaOH solution (see Note 3). Approximately 10 mL of NaOH solution will be required.
7. Add 0.3 g of gelatine and allow to dissolve.
8. Using a syringe through the rubber septum, inject 12 mL of 2 M sulphuric acid to the aluminum telluride (see Note 4).
9. Leave to stir for 5 min, until all generated hydrogen telluride has been consumed.
10. Heat the reaction mixture to 130 °C under reflux.
11. Monitor the UV–visible and Photoluminescence (PL) spectra every half an hour (see Note 5).
12. Cool down the solution of QDs when the desired maximum absorption/emission wavelength is reached.

3.2. Fractioning of Quantum Dots

The aim of this procedure is to separate the manufactured QDs into narrow emission fractions *via* size selective precipitation. The crude solution's PL emission spectrum typically exhibit a full width at half maximum (FWHM) of about 60 nm which relates to a broad size distribution of particles. After fractioning, each fraction extracted presents a FWHM of 35–40 nm indicative of a narrow size distribution (±0.1 nm).

1. Concentrate the crude solution of QDs by evaporating some of the water using a rotary evaporator (final volume 80 mL).
2. Centrifuge the resulting solution for 10 min at 1,000 × *g*. Discard that first pellet as it mostly contains unreacted materials and poorly luminescent nanoparticles.
3. Add propan-2-ol to the supernatant until it becomes cloudy (1–3 mL will be required, see Note 6), and centrifuge again for 10 min at 1,000 × *g*.
4. Resuspend the pellet in 1 mL of water.
5. Repeat steps 3 and 4 until the supernatant is pale yellow.

3.3. Purification of Quantum Dots

The use of Jelly Dots for biological application (e.g., in vitro biological imaging) requires further purification on a Sephadex column. This procedure is aimed at removing all unreacted material or free ligand left in solution. It will not alter the optical properties of the QDs. Since the fractioning step has produced a sample where all particles have a similar size, they will all progress through the gel at the same place, while small molecules will be trapped in the pores.

1. Dissolve 10 g of Sephadex G25 in 100 mL of water and leave to swell overnight.
2. Place a small piece of cotton wool in the neck of a 10 cm glass column (2 cm in diameter). Fill the column with the swollen Sephadex gel and expulse the excess water by applying pressure with a rubber pump.

3. Refill the column with water and expulse it again. Repeat this step five times to pack the gel.
4. Gently pour 1 mL of QD sample on top of the column, taking care of not disturbing the surface of the gel. The sample should enter the gel in one regular band. Add a few drops of water on top of the column without disrupting the surface. Let the QD band penetrate deeper into the gel. Once a layer of uncoloured gel at least as thick as the QD band is visible above it, fill the column with water.
5. Discard the first few drops of colored solution pouring out of the column, then collect the purified sample (approximately 3 mL). Leave the last fraction of sample in the column, to be discarded as it may contain some of the heaviest impurities.
6. Measure UV–visible absorption spectra 300–700 nm. The average diameter of QDs can be deducted from the position of the exciton peak (see Note 5) and the value of absorbance relates to the concentration through Beer's law. Record PL emission spectrum (excitation at 480 nm, emission range 490–700 nm). Calculate the QY of QDs using rhodamine 6G as a reference (See Note 7). Figure 2 displays typical spectra that are obtained for the QDs produced using our method.
7. Ideally the Jelly Dots should also be characterized by electron microscopy. Figure 3 shows typical QDs produced using our method.

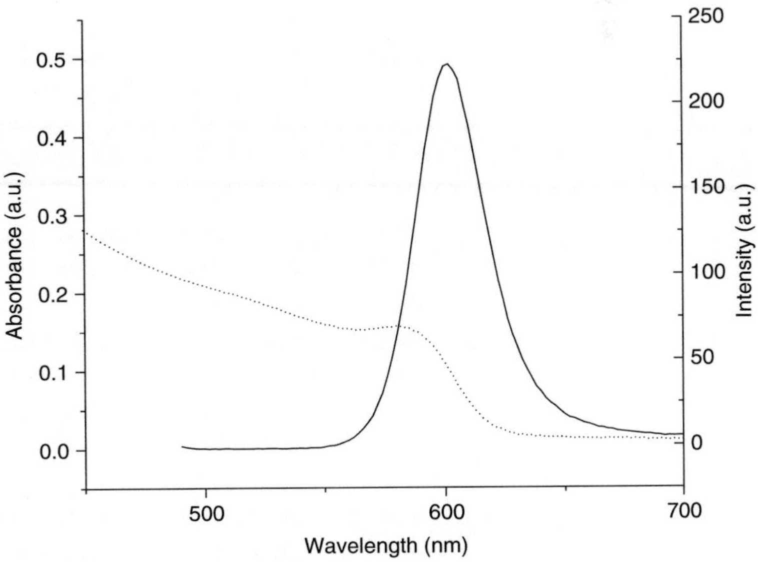

Fig. 2. Characterization of Jelly Dots. An example of UV–visible absorption (*dotted line*, left axis) and photoluminescence emission (*solid line*, right axis) spectra of Jelly Dots.

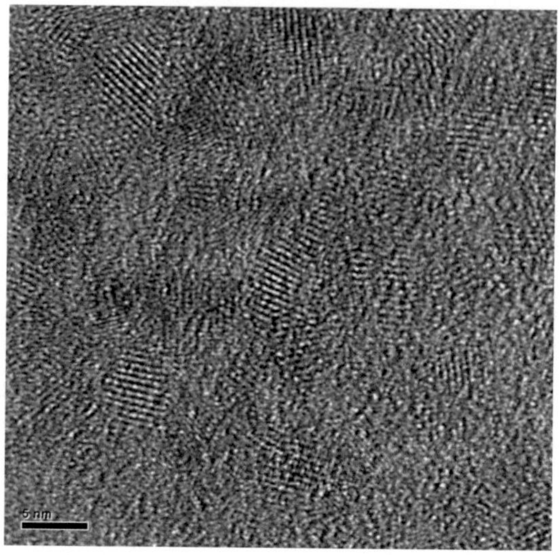

Fig. 3. High resolution transmission electron microscope (HRTEM) imaging of Jelly Dots. Scale bar 5 nm.

4. Notes

1. Aluminum telluride should be kept in inert atmosphere at all times. Use a Shlenk flask fitted with a round-bottom flask and filled with argon.
2. Since it is difficult to take out an exact mass of aluminum telluride from the Shlenk flask, an approximate amount is weighted and the mass or volumes of the other reagents are calculated based on the molar ratios of cadmium atoms to tellurium atoms to TGA molecules: 1:0.25:1.4, using the known molar masses: 436.8 g/mol for Al_2Te_3, 419.41 g/mol for $Cd(ClO_4)_2 \times H_2O$ and 92.12 g/mol for TGA with a density of 1.325 g/mL. A suitable range for aluminum telluride would be 0.4–0.8 g.
3. The cadmium–TGA complex becomes soluble at pH ~6. If the solution is not completely clear at pH 11, add more TGA dropwise until clear.
4. When injecting sulphuric acid onto the aluminum telluride, most of the hydrogen telluride gas will be generated with the first few drops. Only add a few drops initially then wait for the generated gas to flow into the second flask. The rest of sulphuric acid may be added faster if the reaction is not as vigorous.
5. QDs with different photoluminescence emission wavelengths could be manufactured. To monitor the reaction, take an aliquot after the first half hour of reflux. Use a pasteur pipette to transfer about 0.5 mL of the reaction mixture into a spectroscopy cuvette, dilute it in approximately 3 mL of water. Measure the

UV–visible absorption spectrum of the sample to obtain an indication of the particle size. Use this empirically determined formula:

$$\text{Particle size (nm)} = 0.032 * \text{maximum absorption wavelength (nm)} - 14.07$$

or a photoluminescence emission spectrum for the maximum emission wavelength. Repeat every half hour until the QDs have reached the appropriate properties.

6. The volume of propan-2-ol required to precipitate a fraction of QD depends on their sizes and the total volume. The right amount can be identified as follows: when poured into a 50 mL plastic centrifuge tube, the suspension should be too cloudy to see the graduations through.
7. Dilute QDs and rhodamine so they have the same absorbance at 480 nm and record their respective PL emission spectra. Integrate the spectra; the hereby obtained areas under the curves are referred to as $a^2_{\text{rhodamine}}$ and $a^2_{\text{sample.}}$ Use the equation: $QY = (a^2_{\text{sample}} / a^2_{\text{rhodamine}}) * 0.95$ to calculate the sample's Quantum Yield (QY).

References

1. Schumacher WC, Phipps AJ, Dutta PK (2009) Detection of *Bacillus anthracis* spores: comparison of quantum dot and organic dye labeling agents. Adv Powder Technol 20:438–446
2. Li KG, Chen JT, Bai SS, Wen X, Song SY, Yu Q, Li J, Wang YQ (2009) Intracellular oxidative stress and cadmium ions release induce cytotoxicity of unmodified cadmium sulfide quantum dots. Toxicol in Vitro 23:1007–1013
3. Draye J-P, Delaey B, Van de Voorde A, Van Den Bulcke A, De Reu B, Schacht E (1998) In vitro and in vivo biocompatibility of dextran dialdehyde cross-linked gelatin hydrogel films. Biomaterials 19:1677–1687
4. Yamamoto M, Ikada Y, Tabata Y (2001) Controlled release of growth factors based on biodegradation of gelatin hydrogel. J Biomater Sci Polym Ed 12:77–88
5. Soppimath KS, Aminabhavi TM, Kulkarni AR, Rudzinski WE (2001) Biodegradable polymeric nanoparticles as drug delivery devices. J Control Release 70:1–20
6. Wong C, Stylianopoulos T, Cui J, Martin J, Chauhan VP, Jiang W, Popovic Z, Jain RK, Bawendi MG, Fukumura D (2011) Multistage nanoparticle delivery system for deep penetration into tumor tissue. Proceedings of the United States of America PNAS 108(6): 2426–2431
7. Byrne SJ, Williams Y, Davies A, Corr SA, Rakovich A, Gun'ko YK, Rakovich YP, Donegan JF, Volkov Y (2007) "Jelly Dots": synthesis and cytotoxicity studies of CdTe quantum dot–gelatin nanocomposites. Small 3:1152–1156
8. Prasad B, Nikolskaya N, Connolly D, Smith T, Byrne S, Gerard V, Gun'ko Y, Rochev Y (2010) Long-term exposure of CdTe quantum dots on PC12 cellular activity and the determination of optimum non-toxic concentrations for biological use. J Nanobiotechnol 8:7
9. Balthasar S, Michaelis K, Dinauer N, von Briesen H, Kreuter J, Langer K (2005) Preparation and characterisation of antibody modified gelatin nanoparticles as drug carrier system for uptake in lymphocytes. Biomaterials 26(15):2723–2732
10. Byrne SJ, Corr SA, Rakovich TY, Gun'ko YK, Rakovich YP, Donegan JF, Mitchell S, Volkov Y (2006) Optimisation of the synthesis and modification of CdTe quantum dots for enhanced live cell imaging. J Mater Chem 16:2896–2902
11. Gaponik N, Talapin DV, Rogach AL, Hoppe K, Shevchenko EV, Kornowski A, Eychmuller A, Weller H (2002) Thiol-capping of CdTe nanocrystals: an alternative to organometallic synthetic routes. J Phys Chem B 106: 7177–7185

Chapter 22

Preparation and Characterization of Biomimetic Nanoparticles for Drug Delivery

Ana Maria Carmona-Ribeiro

Abstract

Biomimetic nanoparticles are hybrid nanostructures in which the uppermost layer is similar to a cell membrane. This report provides instructions for the preparation and physical characterization of three different types of biomimetic nanoparticles: (1) polystyrene sulfate nanoparticles covered with one cationic dioctadecyldimethylammonium bilayer; (2) silica nanoparticles covered with one neutral phosphatidylcholine bilayer; (3) miconazole particles covered with one anionic dihexadecylphosphate (DHP) bilayer. These examples provide the basis for the preparation and characterization of novel nanoparticles from hydrophobic or hydrophilic and organic or inorganic nanoparticle cores covered with functional outer layers. The major concepts and technical details for obtaining the optimal lipid coverage of supporting cores and for nanoparticle characterization are discussed.

Key words: Silica nanoparticles, Polymeric nanoparticles, Bilayer fragments, Drug particles, Lipids, Optimal coverage

1. Introduction

Biomimetic nanoparticles can be constructed by the use of interaction forces to drive lipid deposition from bilayers onto spherical solid cores. Nanoparticles, including polystyrene nanospheres (1–3), silica particles (4–7), or hydrophobic drug aggregates (8, 9), have been coated with lipid bilayers, yielding functional hybrid assemblies (7–13). These have found many applications in biomolecular recognition (12), drug delivery (9–13), and vaccine design (14–16).

The principal forces driving bilayer deposition onto hydrophobic or hydrophilic nanoparticles from bilayer vesicles or bilayer fragments are electrostatic attraction, van der Waals attraction, and/or the hydrophobic effect (1–6). Medium composition, pH, and the ratio

Mikhail Soloviev (ed.), *Nanoparticles in Biology and Medicine: Methods and Protocols*, Methods in Molecular Biology, vol. 906, DOI 10.1007/978-1-61779-953-2_22, © Springer Science+Business Media, LLC 2012

between the surface areas of the particles (Ap) and bilayers (Ab) also play a crucial role in achieving the optimal bilayer deposition on the particles and the resultant colloidal stability (7–13). For example, if the amount of added lipid is not sufficient to cover all particles with a bilayer, poor colloidal stability may result in aggregation of the primary particles and a broad range of particle sizes. Another important factor is the physical state of the bilayer. In general, bilayer vesicles in the rigid gel state do not disrupt upon adhesion onto solid particles, so that bilayer coverage does not take place (17). This difficulty has been circumvented by using bilayer fragments, which are prone to adsorb onto the solid particles. Charged bilayer fragments are easily obtained from vesicles by ultrasonic disruption and have been used for the production of a variety of lipid-based biomimetic particles.

The important physical properties of biomimetic particles are their size distribution, polydispersity index, and zeta-potential (18). These properties typically allow one to ascertain whether bilayer coverage indeed took place and should be determined for both bare and biomimetic nanoparticles.

If the lipid bilayers and the particles are oppositely charged, poor colloidal stability occurs at a critical lipid concentration where the sizes are maximum and the zeta-potential is zero (1, 7, 9, 13–15). Above this critical concentration, the particles exhibit zeta-potentials that are similar to that of the charged bilayer and recover colloidal stability (1, 9, 14, 15).

Aggregates of hydrophobic drugs dispersed in water have also been treated as particles and coated with lipids at high drug-to-lipid molar ratios (9, 13, 19). The formation of aggregates of hydrophobic drugs in aqueous solution can be modulated by oppositely charged bilayer fragments (Fig. 1). Under certain experimental conditions (which must be established for each particular system), it is possible to achieve coverage of the drug aggregate with a single lipid bilayer and, at the same time, colloidal stability. Thus, the most important application of this coating procedure is the production of lipid-stabilized nanoparticles of hydrophobic drug.

This chapter gives details of the preparation and characterization of three different biomimetic particles: (1) polystyrene sulfate (PSS) nanoparticles covered with a dioctadecyldimethylammonium bromide (DODAB) bilayer; (2) silica particles covered with a phosphatidylcholine (PC) bilayer; (3) miconazole (MCZ) particles covered with a sodium dihexadecylphosphate (DHP) bilayer. Examples of other functional biomimetic particles obtained from bilayer fragments and biomolecules or bilayer fragments and nanoparticles are given in Fig. 2.

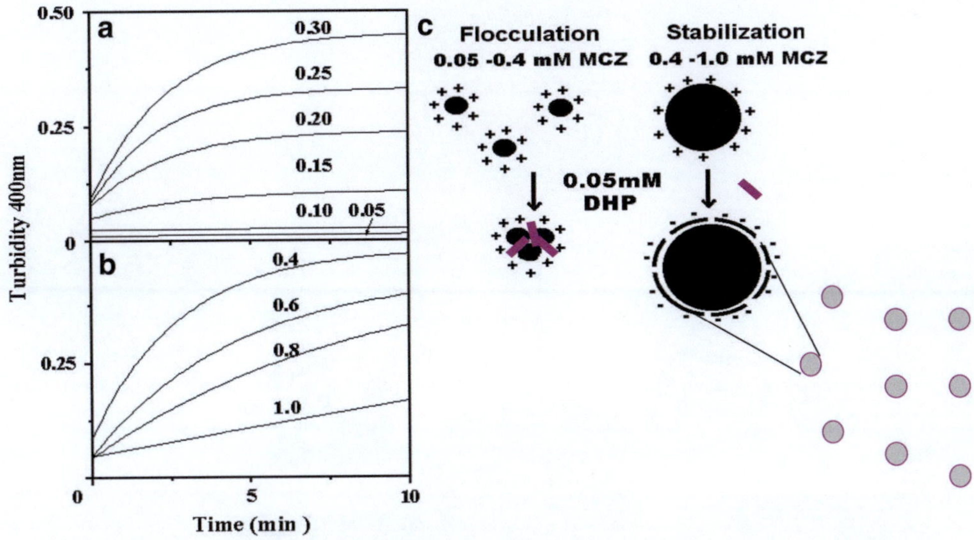

Fig. 1. The formation of aggregates of hydrophobic drugs in aqueous solution. (**a**, **b**) Turbidity as a function of time after adding water dispersion of anionic dihexadecylphosphate (DHP) bilayer fragments (at 0.05 mM DHP) to miconazole (MCZ) over a range of [MCZ]. Reproduced from (9) with permission from Academic Press. (**c**) The poor colloidal stability at low [MCZ] is due to the aggregation of small and cationic MCZ aggregates induced by the anionic DHP bilayer fragments whilst the improved colloidal stability at high [MCZ] is due to the coating of large drug particles with DHP bilayers. The DHP anionic bilayer appears in *pink*.

2. Materials

Prepare all solutions using analytical grade reagents and ultrapure water (prepared by purifying deionized water to attain a resistivity of 18 MΩ cm at 25 °C). All reagents should be prepared at room temperature unless indicated otherwise. Lipid dispersions should be freshly prepared and used within 1–2 days after dispersion. Polymeric microspheres should be stored in the fridge at 4 °C to avoid microbial growth. Silica dispersion in water solution should be prepared and used within 1 week after preparation.

2.1. Materials for Preparation and Characterization of PSS/DODAB Nanoparticles

1. Dioctadecyldimethylammonium bromide (DODAB) white powder (Sigma).
2. 1 mM NaCl solution in water.
3. 10 mM $Hg(NO_3)_2$/0.04 M HNO_3 stock solution (keep this solution protected from light in a dark bottle).
4. 0.01 N HCl solution: prepare by diluting standard solution of 0.1 N HCl tenfold (this is available form Merck). This solution will be used for titrating the previous solution and determining its analytical $Hg(NO_3)_2$ concentration.
5. 1 mM $Hg(NO_3)_2$ in 0.04 M HNO_3: make 1:10 dilution of 10 mM $Hg(NO_3)_2$ in 0.04 M HNO_3. This solution will be

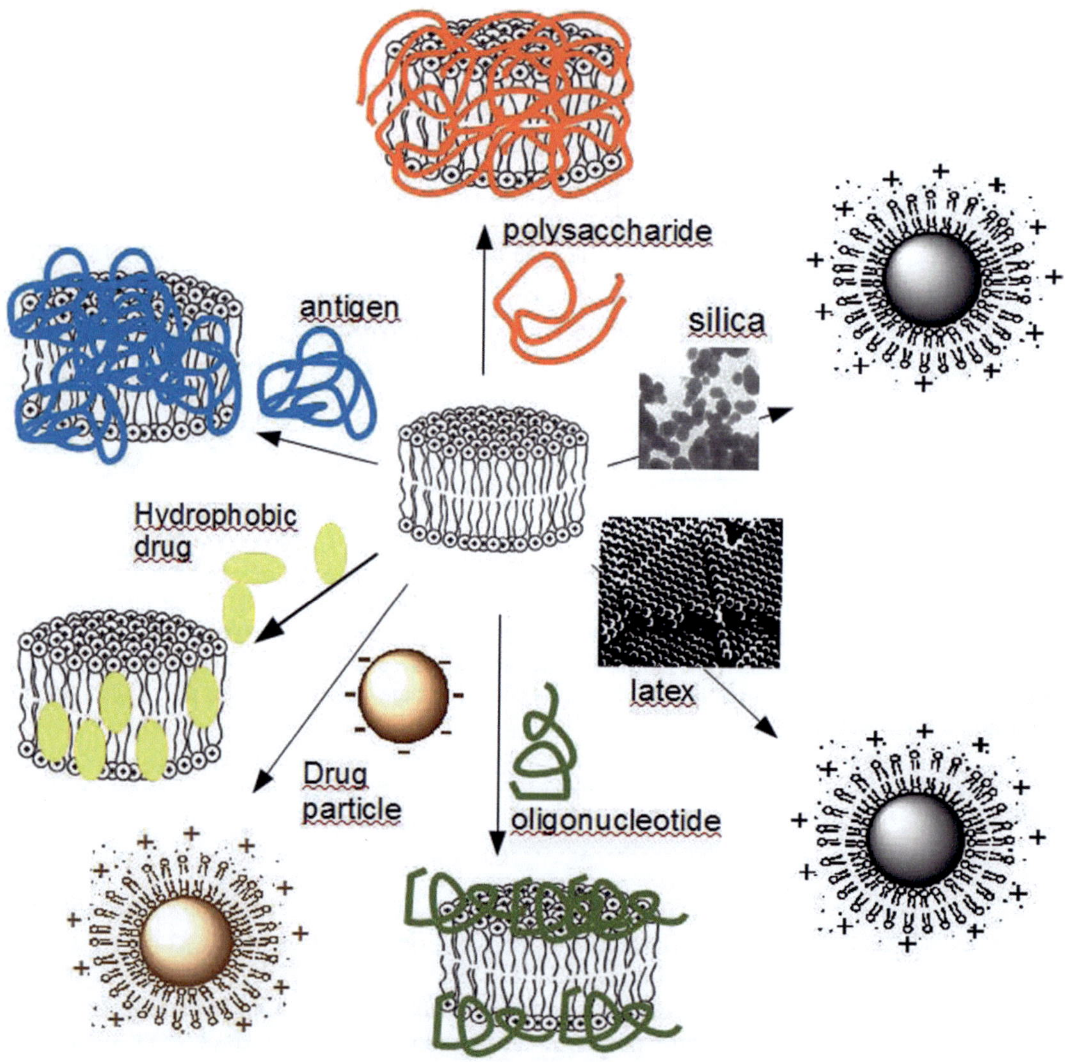

Fig. 2. The variety of functional biomimetic nanoparticles derived from the interactions between bilayer fragments and some solid nanoparticles or important biomolecules.

used to microtitrate bromide (in DODAB dispersions) or chloride (in 1 mM NaCl solution).

6. 0.01 mg/mL Diphenylcarbazone ethanolic solution: dissolve 50 μg of Diphenylcarbazone in 5 mL ethanol. This solution will be used to determine the endpoint of the halide microtitration.
7. PSS dispersion of polymeric particles (Interfacial Dynamics Corporation), see Table 1 for details (see Note 1).
8. ZetaPlus Zeta-potential Analyzer (Brookhaven Instruments Corporation, Holtsville, NY) equipped with a 677-nm laser and

Table 1
Physical properties of bilayer fragments, bare particles, and biomimetic particles (particles coated with lipid bilayer)

Dispersion (medium)	Particle concentration	Lipid (μM)	Dz (nm)	ζ-potential (mV)	Polydispersity
PSS (1 mM NaCl)	5.0×10^9 part/mL	–	319 ± 3	−47 ± 1	0.025 ± 0.008
DODAB bilayer fragments (1 mM NaCl)	–	1,000	81 ± 1	45 ± 2	0.230 ± 0.006
PSS/DODAB (1 mM NaCl)	5.0×10^9 part/mL	10	329 ± 3	44 ± 1	0.037 ± 0.015
Silica (water)	0.1 mg/mL	–	326 ± 3	−45 ± 3	0.186 ± 0.025
PC vesicles (water)	–	2,000	70 ± 1	0 ± 2	0.182 ± 0.016
Silica/PC (140 mM NaCl/10 mM Tris pH 7.4)	0.1 mg/mL	30	162 ± 2	–	–
DHP bilayer fragments (10 mM Tris pH 7.4)	–	1,000	87 ± 3	−49 ± 2	–
MCZ (10 mM Tris pH 7.4)	1.0 mM MCZ	–	510 ± 3	0 ± 2	–
MCZ/DHP (10 mM Tris pH 7.4)	1.0 mM MCZ	50	372 ± 4	−43 ± 4	–

Particles and bilayer fragments interacted for 1 h at room temperature before measurements. Data were compiled from references (1), (9), (11), (12) and (22)

dynamic light-scattering at 90 ° for particle sizing, zeta-potential analysis, and determination of polydispersities (see Note 2).

9. High speed refrigerated centrifuge capable of 20,000 × *g*, e.g., Model SCR20B (Hitachi) equipped with a RPR20-2-3079 rotor.

2.2. Materials for Preparation and Characterization of Silica Nanoparticles Coated with a PC Bilayer

1. 20 mM Egg phosphatidylcholine (PC) type XIII-E ethanolic solution: make 1 mL by mixing 150 μL of 100 mg/mL PC stock solution (Sigma) and 850 μL of Ethanol.
2. Aerosil OX-50 silica (Degussa).
3. Tris-hydroxymethylaminomethane hydrochloride (Tris).
4. Tris buffered saline: 140 mM NaCl, 10 mM Tris–HCl, pH 7.4.
5. Cellulose acetate dialysis bags (10 kDa molecular cut-off): boil for 0.5 h in 0.1 M ethylenediamine tetracetic acid (EDTA) before use. Wash thoroughly with plenty of doubly-distilled water afterwards.
6. Hamilton microsyringe (0.500 mL).
7. 70 % Perchloric acid.
8. 1.25 % Ammonium molybdate solution.
9. 5 % Ascorbic acid solution in water. Prepare fresh or store frozen at −4 °C.
10. 1 mM NaH_2PO_4 standard aqueous solution.
11. Magnetic stirrer.
12. High speed refrigerated centrifuge capable of 20,000 × *g*, e.g., Model SCR20B (Hitachi) equipped with a RPR20-2-3079 rotor.
13. Pyrex assay tubes, clean without detergent prior to use (see Note 3).
14. Oven capable of 100 °C for drying samples.
15. Assay tube heater, capable of 180 °C positioned inside a fume hood.
16. ZetaPlus Zeta-potential Analyzer (Brookhaven Instruments Corporation, Holtsville, NY) equipped with a 677-nm laser and dynamic light-scattering at 90 ° for particle sizing, zeta-potential analysis and determination of polydispersities (see Note 2).

2.3. Materials for Preparation and Characterization of MCZ Drug Particles coated with a DHP Bilayer

1. Sodium dihexadecylphosphate (DHP) white powder from Sigma.
2. 10 mg/mL miconazole nitrate (MCZ) solution in methanol (24 mM MCZ): dissolve 20 mg of MCZ powder in 2 mL of methanol.
3. 10 mM Tris–HCl, pH 7.4.

4. ZetaPlus Zeta-potential Analyzer (Brookhaven Instruments Corporation, Holtsville, NY) equipped with a 677-nm laser and dynamic light-scattering at 90 ° for particle sizing, zeta-potential analysis, and determination of polydispersities (see Note 2).
5. Sonicator with a macroprobe (~1.5 cm diameter).

3. Methods

3.1. Preparation and Characterization of PSS/DODAB Nanoparticles

1. For PSS particles with sizes around 300 nm of mean diameter, prepare a dispersion of PSS nanoparticles at 2×10^{10} particles/mL by diluting the aqueous dispersion of PSS nanoparticles in 1 mM NaCl solution. Make 20 mL of PSS nanoparticles at 2×10^{10} particles/mL (see Note 1).
2. To prepare 2 mM DODAB bilayer fragments add 20 mL of 1 mM NaCl to 25.2 mg of the DODAB powder. Disperse by sonication with a macrotip (80–90 mW, 10 min) (see Note 4).
3. Centrifuge the sonicated DODAB dispersion at 10,000 × *g* in a refrigerated centrifuge at 15 °C for 40 min to precipitate titanium residues ejected from the macrotip.
4. Determine DODAB and NaCl analytical concentrations in the dispersion by halide microtitration using the analytical solution of $Hg(NO_3)_2$ and two or three droplets of the ethanolic diphenylcarbazone solution as indicator of endpoint (see Note 5).
5. Mix PSS nanoparticles and DODAB bilayer fragments in the following order: (1) 0.75 mL of the stock PSS dispersion, (2) 0.03 mL of 1 mM DODAB bilayer fragments dispersion, and (3) 2.220 mL of 1 mM NaCl solution. Allow the components to interact for 1 h at 25 °C. The final concentrations are 5×10^9 PSS particles/mL and 0.01 mM DODAB. This latter is approximately the concentration required to allow each particle to be covered with a DODAB bilayer.
6. Determine zeta-average diameters, zeta-potentials and polydispersities for all dispersions by means of a dynamic light scattering apparatus. Results should be very similar to those in Table 1 (see Note 6).

3.2. Preparation and Characterization of Silica Coated with a PC Bilayer

3.2.1. Formation of PC Bilayer

1. Prepare a stock silica dispersion at 4 mg/mL silica in Tris buffered saline. This silica stock dispersion should be routinely dispersed by sonication with a titanium macrotip (85 W/10 min) to minimize the silica trend to aggregate in the aqueous solution.
2. Following the sonication keep the tube for 1 h to allow titanium particles ejected from the tip, to precipitate at the bottom of the Falcon tube.

3. Separate the silica dispersion from the supernatant (see Note 7).
4. To prepare small PC vesicles slowly inject 0.5 mL of the 20 mM PC ethanolic solution into 10 mL of Tris buffered saline using the 0.5 mL Hamilton syringe (at about 0.1 mL/min). Stir the mixture continuously on a magnetic stirrer. This will yield small PC vesicles at about 2 mM PC.
5. Remove ethanol from the produced PC dispersion by dialysing it in boiled cellulose acetate dialysis bags against 0.5 L of Tris buffered saline for 2 h. Change the buffer once after 1 h of dialysis.
6. Centrifuge the PC dispersion (from step 5) in a refrigerated microcentrifuge (1 h, 10,000 × *g*, 15 °C). The pellet will contain multilamellar vesicles and should be discarded.
7. Immediately following the centrifugation use the fractionated dispersion to react with silica (see Note 8) by mixing 0.25 mL of the silica stock dispersion with 0.15 mL of the PC small vesicles. Add 0.6 mL of the Tris buffered saline, pH 7.4. Final concentrations will be 1 mg/mL silica and 0.3 mM PC. Allow 1 h for bilayer deposition at 25 °C (see Note 9).
8. Incubate the biomimetic nanoparticles for 48 h at 25 °C. No sedimentation should take place over an observation period. This will confirm that PC bilayer was formed on all silica particles.
9. Dilute the biomimetic nanoparticles tenfold in Tris buffered saline, pH 7.4; dilute the original silica dispersion ten times (to 0.1 mg/mL) in the same buffer.
10. Determine the zeta-average diameter of the original and the coated silica nanoparticles. Results should be similar to the values shown in Table 1 (see Notes 6 and 10).

3.2.2. Determining PC Concentration in the Dispersion

1. Prepare ten cleaned Pyrex assay tubes (see Note 3). Add 0, 10, 30, 50, 70, 90, and 100 μL of standard 1 mM NaH_2PO_4 solution to the first seven tubes to make a suitable calibration curve. Add 10, 30, and 50 μL of the PC dispersion to the remaining three tubes.
2. Dry all assay tubes completely in an oven at 100 °C.
3. Add 0.3 mL of 70 % percloric acid per tube to digest all molecules and release the inorganic phosphorus. Cover tubes with glass beads to avoid evaporation of the acid.
4. Place all tubes in an assay tube heater at 180 °C inside a fumehood. Incubate for 1 h, then turn off the heater and allow the tubes to cool down to room temperature.
5. Add 1 mL of water and 0.4 mL of 1.25 % ammonium molybdate to each tube.
6. Stir each tube by vortexing. Add 0.4 mL of 5 % ascorbic acid to each tube. Stir again by vortexing.

7. Cover tubes with glass beads to avoid evaporation and place all tubes in a boiling water bath. After 5 min incubation allow the tubes to cool to room temperature.
8. Read absorbance of the samples at 795 nm (blank the spectrometer against the first blank sample prepared with 0 μL of standard phosphate solution, see step 1).

3.3. Preparation and Characterization of MCZ Drug Particles Coated with a DHP Bilayer

Suspension of MCZ particles in water is obtained by adding water to monomeric MCZ solubilized in a small volume of methanol (9). In the absence of any lipid and at low methanol concentrations, formation of MCZ aggregates (drug particles) in pure water is practically instantaneous and aggregate size barely increases with time over the first hour after adding water (9). Removal of methanol can be achieved by dialysis.

1. To prepare 1 mM dispersion of bare MCZ drug particles transfer 0.1 mL of 10 mg/mL MCZ methanol solution to the bottom of a 3.0 mL cuvette and then add 2.3 mL of 10 mM Tris buffer.
2. Incubate for 1 h at 25 °C.
3. Place the cuvette in the cell holder of the light scattering apparatus and determine zeta-average diameter and zeta-potential of the MCZ dispersion. The measurements should be similar to those in Table 1 (see Note 6).
4. To prepare 1 mM dispersion of DHP bilayer fragments add 20 mL of 10 mM Tris–HCl to 11 mg of DHP powder. Sonicate with the macrotip (80–90 mW, 20 min). Maintain the temperature of the dispersion above 50 °C at all times (see Note 4).
5. To prepare 1 mM dispersion of biomimetic MCZ drug particles transfer 0.1 mL of 10 mg/mL MCZ methanol solution to the bottom of a 3.0 mL cuvette and then add 2.3 mL of 0.05 mM DHP bilayer fragments dispersion in Tris buffer.
6. Incubate for 1 h at 25 °C.
7. Place the cuvette in the cell holder of the light scattering apparatus and determine zeta-average diameter and zeta-potential of the MCZ biomimetic particles. The measurements should be similar to those shown in Table 1 (see Note 6).

4. Notes

1. The PSS original particles purchased from Interfacial Dynamics Corporation have their particle density, size, surface charge density, and polydispersity specified in a specifications sheet, which is provided by the supplier. The stock dispersion containing 2×10^{10} PSS particles/mL should be prepared in 1 mM NaCl because this represents an adequate ionic strength to

assemble DODAB bilayer fragments as a single bilayer onto PSS particles by fusion of adsorbed and adjacent bilayer fragments (10).

2. Mean hydrodynamic diameters should be obtained by fitting intensity of light scattered to log-normal size distributions which do not discriminate between one, two, or more different populations, and consider all scattering particles as belonging to one single Gaussian population. On the other hand, for the size distribution data, fitting should be done using the non-negatively constrained least squares (NNLS) algorithm available from the light scattering apparatus, which is a model independent technique allowing to achieve multimodal distributions (15). ζ is usually determined from electrophoretic mobility μ in 1 mM NaCl and the Smoluchowski's equation: $\zeta = \mu\eta/\varepsilon$, where η is the medium viscosity and ε the medium dielectric constant: this is also provided by the apparatus software. The electrode of the apparatus could be used only over a low range of ionic strength (0–5 mM monovalent salt). Higher conductivities will damage the electrode.

3. Do not use detergent for cleaning the assay tubes. Many commercially available detergents contain phosphates. If they are used to clean the assay tubes the analytical determinations of inorganic phosphorus may be overestimated and wrong. Prior to Pi determination, the assay tubes should be cleaned with 0.1 mL of 70 % perchloric acid (1 h/180 °C) in the test tube heater (each tube covered with its respective glass bead) and abundantly washed several times with ultrapure Milli-Q water, dried in the oven and kept in a very clean cartoon box until use.

4. Bilayer fragments have to be always prepared above the mean gel-to-liquid crystalline phase transition temperature of the bilayer. Typical values are 45 and 47 °C for DODAB and DHP, respectively (20, 21). Thus, DODAB or DHP should be dispersed in water solution always at temperatures above these values, e.g., 50 °C at all times. Sonication will raise the temperature above 50 °C, so there is no need to additionally heat the sample. Other important parameter to note is their low stability in moderate to high ionic strength solutions. Above 10 mM concentrations of monovalent salts, DODAB bilayer fragments or DHP bilayer fragments will fuse or aggregate to yield larger assemblies ((2) and references therein).

5. The knowledge of the accurate concentrations of DODAB in dispersion should allow calculation of Ab and the ratio Ab/Ap.

6. Note that upon coverage of each particle with a bilayer, the zeta-average diameter should increase by the equivalent of two bilayer thicknesses (one bilayer thickness is 4–5 nm).

7. Titanium particles ejected by the titanium macroprobe will pellet at the bottom of the Falcon tube whilst the silica nanoparticles dispersion will remain in the supernatant. This supernatant can then be separated from the pellet by using a Pasteur pipette.
8. This will minimize formation of multilamellar vesicles that continuously takes place in dispersions of small PC vesicles.
9. The specified concentrations mean that Ab/Ap will be 1.3 as calculated from the specific surface area for silica, 47 m^2/g, and 0.7 nm^2 as the mean molecular area at the air-water interface for PC.
10. Notice that the zeta-potential cannot be determined at 150 mM of salt (the electrode will be damaged by such high conductivity).

Acknowledgments

This work was supported by FAPESP and CNPq. Professor Frank Quina is gratefully acknowledged for reviewing the manuscript.

References

1. Carmona-Ribeiro AM, Midmore BR (1992) Synthetic bilayer adsorption onto polystyrene microspheres. Langmuir 8:801–806
2. Carmona-Ribeiro AM (1992) Synthetic amphiphile vesicles. Chem Soc Rev 21: 209–214
3. Carmona-Ribeiro AM, Herrington TM (1993) Phospholipid adsorption onto polystyrene microspheres. J Colloid Interface Sci 156: 19–24
4. Rapuano R, Carmona-Ribeiro AM (1997) Physical adsorption of bilayer membranes on silica. J Colloid Interface Sci 193:104–109
5. Rapuano R, Carmona-Ribeiro AM (2000) Supported bilayers on silica. J Colloid Interface Sci 226:299–307
6. Carmona-Ribeiro AM (2001) Bilayer vesicles and liposomes as interface agents. Chem Soc Rev 30:241–247
7. Moura SP, Carmona-Ribeiro AM (2003) Cationic bilayer fragments on silica at low ionic strength: competitive adsorption and colloidal stability. Langmuir 19:6664–6667
8. Carmona-Ribeiro AM (2003) Bilayer-forming synthetic lipids: drugs or carriers? Curr Med Chem 10:2425–2446
9. Pacheco LF, Carmona-Ribeiro AM (2003) Effects of synthetic lipids on solubilization and colloid stability of hydrophobic drugs. J Colloid Interface Sci 258:146–154
10. Pereira EMA, Vieira DB, Carmona-Ribeiro AM (2004) Cationic bilayers on polymeric particles: effect of low NaCl concentration on surface coverage. J Phys Chem B 108: 11490–11495
11. Moura SP, Carmona-Ribeiro AM (2005) Biomimetic particles: optimization of phospholipid bilayer coverage on silica and colloid stabilization. Langmuir 21:10160–10164
12. Moura SP, Carmona-Ribeiro AM (2006) Biomimetic particles for isolation and reconstitution of receptor function. Cell Biochem Biophys 44:446–452
13. Lincopan N, Carmona-Ribeiro AM (2006) Lipid-covered drug particles: combined action of dioctadecyldimethylammonium bromide and amphotericin B or miconazole. J Antimicrob Chemother 58:66–75
14. Lincopan N, Espíndola NM, Vaz AJ et al (2007) Cationic supported lipid bilayers for antigen presentation. Int J Pharm 340: 216–222

15. Lincopan N, Santana MRA, Faquim-Mauro E et al. (2009) Silica-based cationic bilayers as immunoadjuvants. BMC Biotechnol 9:5. http://www.biomedcentral.com/1472-6750/9/5
16. Lincopan N, Espíndola NM, Vaz AJ et al (2009) Novel immunoadjuvants based on cationic lipid: preparation, characterization and activity in vivo. Vaccine 27:5760–5771
17. Moura SP, Carmona-Ribeiro AM (2007) Adsorption behavior of DODAB/DPPC vesicles on silica. J Colloid Interface Sci 313:519–526
18. Grabowski E, Morrison I (1983) Particle size distribution from analysis of quasi-elastic light scattering data. In: Dahneke B (ed) Measurements of suspended particles by quasi-elastic light scattering. Wiley-Interscience, New York, pp 199–236
19. Vieira DB, Carmona-Ribeiro AM (2008) Cationic nanoparticles for delivery of amphotericin B: preparation, characterization and activity in vitro. J Nanobiotechnology 6:6. http://www.jnanobiotechnology.com/content/6/1/6
20. Nascimento DB, Rapuano R, Sesso A et al (1998) Counterion effects on properties of cationic vesicles. Langmuir 14:7387–7391
21. Carmona-Ribeiro AM, Castuma CE, Sesso A et al (1991) Bilayer structure and stability in dihexadecyl phosphate dispersions. J Phys Chem US 95:5361–5366
22. Leão-Silva AC, Naves AF, Pereira EMA et al (2011) Assembly of horse radish peroxidase within supported cationic bilayers. Biotechnol Prog 27(5):1433–1441

Chapter 23

Preparation of Nanoparticles by Spray-Drying and Their Use for Efficient Pulmonary Drug Delivery

Hao-Ying Li and Fan Zhang

Abstract

The use of nanoparticulate systems for pulmonary drug delivery offers a number of advantages including the significantly improved delivery efficiency to the deep lung and the improved bioavailability. The traditional nanoparticle manufacturing process such as ball/jet milling may generate large aggregates, which could detrimentally inhibit the effective delivery of drug particles to the lower respiratory way. Here we report an alternative technique of spray-drying nanoemulsion to produce nanoparticles (<100 nm) that can be dispersed homogenously in the propellant to form an extremely stable pressurized metered dose inhaler formulation. Such nanoparticulate formulations dramatically improve the efficiency of drug delivery to the lungs, and therefore provide an ideal tool for pulmonary drug delivery.

Key words: Microemulsion, Nanoparticle, Spray-drying, pMDI, Pulmonary drug delivery

1. Introduction

Pulmonary drug delivery provides attractive route for the administration of pharmaceutically active agents to treat the local and systemic diseases (1). The size and morphology of drug particles used in the pulmonary drug delivery have dramatic effect on the site of lung delivery. Normally, the particles with the size larger than 4.7 μm will be impacted in the oral and pharynx region, particles with the size ranging from 0.43 to 2.1 μm can be inhaled into peripheral lung that is the ideal site for lung delivery (2, 3). Traditional methods for nanoparticle preparation include milling (4), precipitation (5) and the high pressure homogenization (6). These methods are not very efficient and have a poor control over the particle size and morphology, and produce particles with broad size distribution and severe aggregation (7, 8), unavoidably leading to the central and oropharyngeal deposition. Therefore, the efforts are being made to explore new approaches to further reduce

Mikhail Soloviev (ed.), *Nanoparticles in Biology and Medicine: Methods and Protocols*, Methods in Molecular Biology, vol. 906, DOI 10.1007/978-1-61779-953-2_23,

the particle size for the improvement of delivery efficiency to the lungs. One other approach which results in the increased aerosolization of drug particles relies on the use of nanoemulsions (9, 10). Micro- or nano-emulsions are thermodynamically stable dispersions, which consist of two immiscible liquids stabilized by surfactants to produce an optically transparent and isotropic system. Because the uncontinuous phase is monodispersed in the immiscible liquid (droplet size can be less than 100 nm), micro- and nano-emulsions are therefore considered as effective technology for the preparation of nanosized particles (11). Here we describe methods for the preparation of nanoparticle using spray-drying microemulsions. The spray-dried nanoparticles are subsequently suspended into propellant to prepare nanoparticulate pressurized metered dose inhaler (pMDI) formulations for efficient pulmonary delivery.

2. Materials

2.1. Preparation of Microemulsion

1. Surfactant component: dissolve 5 g of egg lecithin is 15 g of isopropanol (1:3 w/w). Filter through 220 nm PTFE filter membrane to obtain a clear solution. Store at 4 °C (see Note 1).
2. Water phase: add 15 g of salbutamol sulfate into distilled water to a final volume of 100 mL.
3. Oil phase: 100 % isooctane.
4. Vortex.

2.2. Preparation of Spray-Dried Nanoparticles

1. Nitrogen (gas).
2. Mini spray-dryer (Büchi-290 mini spray-dryer, Büchi Labortechnik) (see Note 2).
3. Nanoparticle analyzer (N4 plus, Coulter N4MD) (see Note 3).
4. Transmission electronic microscopy (TEM) (Philips EM208S, Philips, Eindhoven, The Netherlands) (see Note 4).
5. Isooctane.
6. Sonicator.
7. Sealable quartz cuvette.

2.3. Preparation of Nanoparticulate pMDI Formulations

1. Pamasol P2002 small scale crimper (Pamasol, Switzerland).
2. Pamasol P2011 propellant filler (Pamasol Willi Mäder AG, Pfäffikon, Switzerland).
3. 1,1,1,2,-tetrafluoroethane (HFA 134a) (Solvay Fluor, Hannover, Germany) (see Note 5).
4. Canister (14 mL)—a glass vial coated by polymer is used for containing the propellant and drug particles.

5. Valve (BK357 100 μL) for sealing the canister and to release a dose for patient to inhale.
6. Vortex.
7. Pentane.

2.4. In Vitro Deposition of Nanoparticle Formulation Using Aerosols

1. Non-viable cascade impactor: eight stages, operated at constant flowrates that allow the characterization of an aerosol in terms of its particle size distribution.
2. High performance liquid chromatography (HPLC) with UV detector (detection wavelength: 280 nm).
3. HPLC column: 5 μm C18, 150×4.6 mm, 80 Å.
4. Mobile phase: 1.1013 g/L heptane sulfonic acid solution in water, pH 3 (adjusted by acetic acid):methanol (60:40 v/v), use 1 mL/min flow rate.
5. Collection solution: 7 pg/mL bamethane solution in water:ethanol (40:60 v/v).

3. Methods

3.1. Preparation of Microemulsion

1. Weigh and add 4,500 mg surfactant component into 3,000 mg isooctane in a capped tube, which are subsequently vortexed for 30 s to achieve a homogeneous solution (see Note 6).
2. Gradually add 2,500 mg water phase of 15 % salbutamol sulfate solution into the oil phase, vortex for 30 s to obtain a clear, stable, isotropic suspension (see Note 7).

3.2. Preparation of Spray-Dried Nanoparticles

1. Set up and configure the spray-dryer: set inlet temperature to 150 °C, aspirator to max. power, nitrogen flow rate to 600 L/h. Adjust the feeding speed to maintain the outlet temperature at 70 °C (see Note 8).
2. Collect the spray-dried nanoparticles and lecithin from the spray-dryer. The typical yield of product is 60 % w/w (see Note 9).
3. Disperse a small amount of nanoparticulate components (approximately 0.2 mg) in 3 mL of isooctane and sonicate for 2 min. Transfer the solution into a quartz cuvette and seal with a cap to prevent the evaporation of isooctane.
4. Insert the cuvette into a nanoparticle analyzer for the particle size analysis. The expected average size of spray-dried nanoparticles should be in the range of 20–40 nm.
5. Put one drop of suspension on the TEM grid and dry at room temperature. The size and morphology of the nanoparticles are visualized using TEM (see Fig. 1 for typical example).

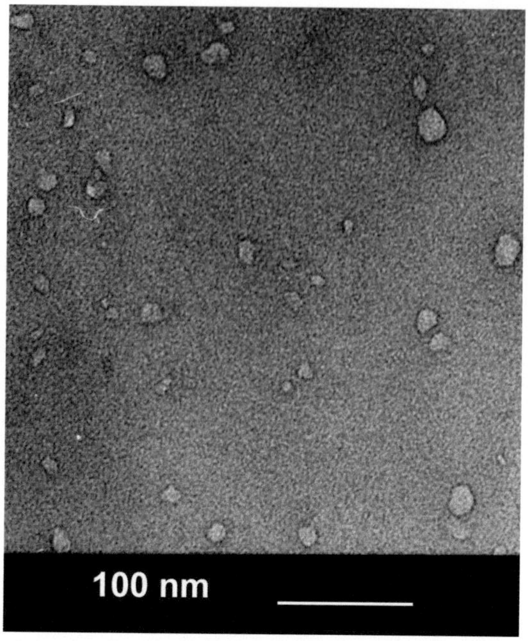

Fig. 1. TEM of spray-dried salbutamol sulfate nanoparticles. The nanoparticles are dispersed on a holey carbon film supported by a copper grid for TEM observation.

3.3. Preparation of Nanoparticlate pMDI Formulations

1. Weigh the spray-dried nanoparticles of salbutamol sulfate (22.8 mg, equivalent to salbutamol sulfate 5.4 mg), transfer to the canister and disperse in pentane (400 μL) (see Note 10).
2. Crimp a valve using a Pamasol P2002 small scale crimper (see Note 11).
3. Add 5 mL of HFA-134a through the valve using Pamasol P2011 propellant filler.
4. Vortexes the formulation for 1 min, store at room temperature.

3.4. In Vitro Deposition of Nanoparticle Formulation Using Aerosols

1. Whilst under the air flow (28.3 L/min) spray the nanoparticles by pressing the valve twice (two actuations should release ~200 μg of salbutamol sulfate) into eight-stage non-viable cascade impactor. (see Notes 12 and 13).
2. Following the deposition of nanoparticles in the impactor, collect salbutamol sulfate nanoparticles from the actuator, throat, pre-separator, impactor plates of stages 0–7 using 10 mL of the collection solution for each separate fraction.
3. Analyze salbutamol sulfate quantity by HPLC (approximate retention time ~4.5 min) (see Note 14).
4. Calculate the mass of salbutamol sulfate deposited in actuator, throat, pre-separator, and all stages. Calculate and draw the deposition pattern with different parts of non-viable cascade impactor as the *X* axis, and the mass as the *Y* axis, to estimate the drug delivery efficiency to the lungs.

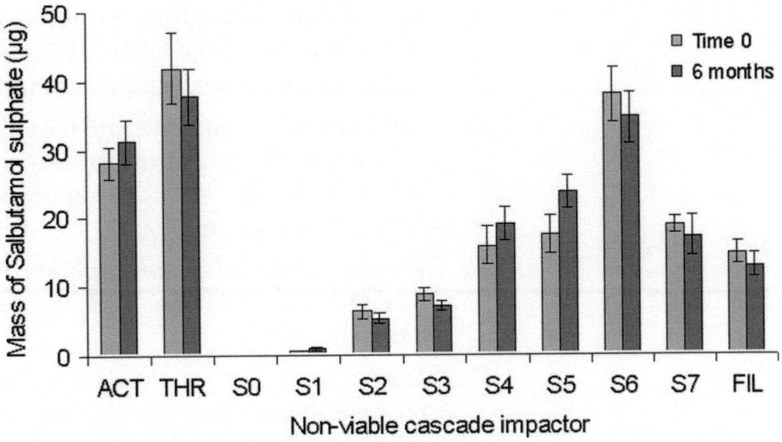

Fig. 2. The aerosolisation performance of nanoparticulate pMDI formulations (freshly prepared and 6-month stored) determined by 8-stage non-viable cascade impactor. For lung deposition, one actuation was fired from the nanoparticle formulation under the flow rate of 28.3 L/min. Then the nanoparticle formulation was shaked for 30 s and a second actuation was launched. The nanoparticles deposited on each stage were collected by collect solution and the amount of drugs was determined by HPLC for the construction of in vitro deposition pattern.

5. After storing the formulation for 6 months at room temperature, repeat the process of step 1–4 to test the aerosolisation of nanoparticle pMDI formulation. Compare the deposition data of freshly-prepared and stored samples observe any change of drug delivery efficiency to the lung and the stability of nanoparticulate pMDI formulations (Fig. 2).

4. Notes

1. The dissolution of lecithin in isopropanol may take long time. Convenient option is to leave lecithin in isopropanol overnight under constant stirring.
2. Other spray-dryers may also work well to prepare the nanoparticles from microemulsions in case of the whole system can be protected by inert gas. The brand of Büchi-290 is a mini spray-dryer that is commonly used in laboratory to prepare small scale of products, and the gas of nitrogen is usually employed as the protection medium for spray-drying organic solvents.
3. Other nanoparticle analyzer brand such as Malvern should also work provided they have the capacity for testing those particles with the size less than 100 nm.

4. Other brands of transmission electronic microscopes such as Hitachi and JEOL in Japan, TESCAN in EU, FEI in USA should work for visualizing the nanoparticles.
5. HFA 134a is an environment-friendly chemical that is newly developed as the propellant for the preparation of pMDI. The company of Solvay Fluor can supply high quality HFA 134a of pharmaceutical grade.
6. The tube should be capped tightly to avoid the lost of materials during the vortexing, and to prevent the evaporation of solvent afterwards.
7. The salbutamol sulfate solution is divided evenly as 3–5 parts. After adding each part into the oil/surfactant phase, the tube is capped and agitated for a few seconds till to see a clear solution. Then a second part is followed, and this process is repeated till all water phase is homogeneously dispersed in the microemulsion.
8. Ensure that the whole system is under the protection of nitrogen to avoid any risk of explosion. The freezing system should be running simultaneously to get rid of the evaporated organic solvent, ensuring the progress of spray-drying process.
9. The spray-dried nanoparticles are embedded in the surfactant. All spray-dried materials are sticky and adhere to the spray-dryer. An easy way to collect the spray-dried nanoparticles is to use isooctane to dissolve the lecithin and the nanoparticles are suspended in the pentane. The solvent is subsequently evaporated under vacuum to get the nanoparticles.
10. Pentane must be used to dissolve the surfactant before the addition of HFA 134a. Otherwise it is difficult to get a homogeneous suspension of nanoparticles.
11. The valve is used to seal the canister and release a dose the drug for pulmonary delivery. The crimper should be adjusted correctly to ensure the valve is capped on the canister cup tightly and in the right place. Otherwise it may cause the leakage of the propellant very possibly.
12. When 400 μL of pentane and 5 mL of HFA are added to the 22.8 mg of spray-dried product, containing 5.4 mg of salbutamol sulfate (as in step 1, Subheading 3.3), the amount of salbutamol sulfate in each actuation (singe spray) is approximately 100 μg.
13. The aerosol will follow the air flow and enter into the non-viable cascade impactor. The drug particles will be deposited in different areas of the impactor (throat, pre-separator, impactor plates of stages 0–7) according to their particle size.
14. For the HPLC test, all samples are required to be filtered through filter membrane (200 nm pore size) to avoid blocking the column. The mobile phase must be degassed.

References

1. Patton JS, Byron PR (2007) Inhaling medicines: delivering drugs to the body through the lungs. Nat Rev Drug Discov 6:67–74
2. Colthorpe P, Farr SJ, Taylor G, Smith IJ, Wyatt D (1992) The pharmacokinetics of pulmonary-delivered insulin: a comparison of intratracheal and aerosol administration to the rabbit. Pharm Res 9:764–768
3. Dolovich MA (2000) Influence of inspiratory flow rate, particle size, and airway caliber on aerosolized drug delivery to the lung. Respir Care 45:597–608
4. Mochalin VN, Sagar A, Gour S, Gogotsi Y (2009) Manufacturing nanosized fenofibrate by salt assisted milling. Pharm Res 26:1365–1370
5. Rasenack N, Müller BW (2002) Dissolution rate enhancement by in situ micronization of poorly water-soluble drugs. Pharm Res 19: 1894–1900
6. Keck CM, Müller RH (2006) Drug nanocrystals of poorly soluble drugs produced by high pressure homogenisation. Eur J Pharm Biopharm 62:3–16
7. Chow AH, Tong HH, Chattopadhyay P, Shekunov BY (2007) Particle engineering for pulmonary drug delivery. Pharm Res 24: 411–437
8. Thi THH, Danède F, Descamps M, Flament MP (2008) Comparison of physical and inhalation properties of spray-dried and micronized terbutaline sulphate. Eur J Pharm Biopharm 70:380–388
9. Jacobs C, Muller RH (2002) Production and characterization of a budesonide nanosuspension for pulmonary administration. Pharm Res 19:189–194
10. Rabinow BE (2004) Nanosuspensions in drug delivery. Nat Rev Drug Discov 3:785–796
11. Eastoe J, Hollamby MJ, Hudson L (2006) Recent advances in nanoparticle synthesis with reversed micelles. Adv Colloid Interface Sci 128–130:5–15

Chapter 24

Synthesis and Evaluation of Airway Targeted PLGA Nanoparticles for Drug Delivery in Obstructive Lung Diseases

Neeraj Vij

Abstract

Chronic airway inflammation is a hallmark of chronic obstructive airway diseases, including asthma, COPD (chronic obstructive pulmonary disease), and CF (cystic fibrosis). It is also a major challenge in delivery and therapeutic efficacy of nano-based delivery systems in these chronic airway conditions as nanoparticle (NP) need to bypass airways defense mechanisms as we recently discussed. NPs which are capable of overcoming airways defense mechanisms should allow targeted drug delivery to disease cells. Over the last decade there has been increasing interest in development of targeted NPs for cancer but relatively little effort on designing novel systems for treating chronic inflammatory and obstructive airway conditions. Here we describe methods for preparing drug loaded multifunctional nanoparticles for targeted delivery to specific cell types in airways. The formulations and methods for selective drug delivery, discussed here are currently under preclinical development in our laboratory for treating chronic airway conditions such as COPD, CF, and asthma.

Key words: PLGA nanoparticles, Therapeutics, CF, COPD, Asthma, Inflammation and lung disease

1. Introduction

We recently evaluated and discussed the efficacy of *poly-lactic-co-glycolic acid* (PLGA) based nano-systems for selective drug delivery to treat chronic obstructive lung diseases (1, 2). We have also recently described the diagnostic and therapeutic (theranostic) potential of these nanosystems in treating COPD (chronic obstructive pulmonary disease) and CF (cystic fibrosis) lung diseases (3).

A major drawback of PLGA nanoparticles (NPs) is that when formulated with the commonly used emulsifier polyvinyl alcohol (PVA), they have a high negative charge on their surface (3). As a result, such a system, when administered in experimental animals, is

Mikhail Soloviev (ed.), *Nanoparticles in Biology and Medicine: Methods and Protocols*, Methods in Molecular Biology, vol. 906, DOI 10.1007/978-1-61779-953-2_24, © Springer Science+Business Media, LLC 2012

rapidly cleared by airway defense system (1, 2, 4). The best way to overcome this challenge is coating of the drug delivery system with the outer layer of *poly-ethylene-glycol* (PEG) that endow these NPs with airway defense evading properties. PEGylation also increases the retention time of the NPs, thereby enhancing their propensity of accumulation in target diseased cells (1). NP mediated drug delivery presents with the added advantage of targeting the drug to specific diseased cells in the body, for example by conjugating it with an antibody that will target the system to the specific airway cells which over express the complementary antigen (our ongoing studies, as we recently discussed) (3).

In this chapter we describe methods and formulations of NPs as well as in vitro models to evaluate and standardize the targeted drug delivery to diseased cells. The nanosystem discussed here can be used to overcome airway defense mechanisms (i.e., macrophage, neutrophils, and mucus etc.) in chronic obstructive lung diseases and in variety of other pulmonary conditions.

2. Materials

2.1. Polymer Nanoparticle Synthesis and Conjugation

All chemical regents used should be cell culture or pharmacological grade.

1. PLGA with L:G ratio 50:50 and MW: 5,000–15,000.
2. PEGylated phospholipids: Distereolylphos-phatidylethanolamine-$mPEG^{2000}$ (DSPE-$mPEG^{2000}$) (Avanti Polar Lipids).
3. Antibody: immunoglobulin A (IgA) (1 μg/mL) in PBS.
4. 7.5 mM 2-mercaptoethylamine (2-MEA), store in a desiccator at 4 °C.
5. 25 % HEPES in 0.9 % NaCl (pH 7.2).
6. Acetone.
7. Amicon Centrifugal Filter Device (Ultracel YM-100, Regenerated Cellulose 100 kDa MWCO, Millipore, YM-100).
8. Sonicator—32 W.
9. Refrigerated Rotary Evaporator (Buchi R220 Rotavap).
10. Antibiotics: ibuprofen or other suitable antibiotics.
11. CF corrector drug: VX-809/770 etc.
12. Nile RED (NR) dye (Invitrogen).
13. Rhodamine 6G (R6G) dye (Invitrogen).

2.2. Polymer Nanoparticle Characterization

1. 1 mM HEPES buffer for zeta potential measurements.
2. Formvar-coated copper grids for transmission electron microscopy (TEM) (Electron Microscopy Sciences Inc).
3. ZETAPLUS particle electrophoresis system or NanoDLS (Brookhaven Instruments Inc).
4. Polystyrene standards (Brookhaven Instruments Inc) for calibrating particle electrophoresis system.
5. PHILIPS EM 420 TEM.
6. 2 % (w/v) uranyl acetate solution in ethanol.
7. FACsCalibur flow cytometer (BD Biosciences).
8. CellQuest software (BD Biosciences).

2.3. Nanoparticle Mediated Drug Delivery In Vitro

1. Secondary antibody to IgA (0.4 μg/mL in PBS) for flow cytometry.
2. The cell permeabilization and fixation Cytofix/Cytperm Kit (BD Biosciences) for intracellular staining and flow cytometry.
3. Strain controlled rheometer (ARES 100, Rheometrics Scientific or TA Instruments).
4. Laser scanning confocal microscope (Zeiss LSM 510 or Nikon A1RMP).
5. Confocal microscope coupled to multiple particle tracking (MPT, SIT Camera VE-1000 Dage-MTI) device.
6. Metamorph software (Universal Imaging Corp) for time-lapsed 3-D imaging for quantifying NP mucus transport rates.
7. Epithelial cells: CF bronchial epithelial cells (CFBE41o-) and human bronchial epithelial cells (HBEo-).
8. Cell culture media, DMEM/F12 (Invirogen).
9. 10 % Fetal bovine serum (FBS, Invitrogen).
10. 1 % Penicillin, streptomycin, and amphotericin (PSA, Invitrogen).
11. Cell culture lysis reagent, MPER (Pierce).
12. NFκB reporter plasmid and renilla luciferase internal control.
13. Lipofecatmine 2000 (Invitrogen).
14. Dual Luciferase Assay kit (Promega).
15. Standard phosphate buffer saline (PBS): dissolve 8 g NaCl, 0.2 g KCl, 1.44 g Na_2HPO_4, and 0.24 g KH_2PO_4 in 800 mL of distilled H_2O, pH 7.4.
16. Spectrophotometer, e.g., VersMax (Molecular Devices) and SoftMaxPro Software (Molecular Devices) to quantify uptake of dye/drug loaded NPs by epithelial cells.

3. Methods

3.1. Polymer Nanoparticles Synthesis and Conjugation

1. Dissolve 30 mg of PLGA in 0.5 mL of acetone (see Note 1) while stirring.
2. Dissolve 0.5 mg of antibiotic or CF corrector drug in 0.5 mL of acetone (see Note 1) in a glass vial.
3. Dissolve 0.5 mg of dye (NR or R6G) in 0.5 mL of acetone (see Note 1) in a glass vial.
4. Add 0.5 mL of PLGA solution (step 1) drop wise to the DRUG (step 2) or dye (step 3) solution and mix well.
5. Dissolve 25 mg of DSPE-mPEG^{2000} in 0.5 mL of water (see Note 1) in a glass vial with gentle stirring.
6. Add 1 mL of PLGA$^{DRUG/DYE}$ solution (step 4) drop wise to the DSPE-mPEG^{2000} solution (step 5) with vigorous stirring. A two-layered solution should form.
7. Sonicate the two-layered solution in an ice bath for 1–2 min at ~20 W until emulsification. Gradually increase the sonicating power/pulses from 0 to 30 W (see Notes 2 and 3).
8. Leave the emulsified solution overnight in cold room (4 °C) with gentle stirring to evaporate the organic solvent(s).
9. Centrifuge the resulting block copolymer (~5,000 × *g*, 20 min), wash in deionized water and dry for 30 min using refrigerated rotary evaporator to completely remove organic solvents.
10. Store PEGylated NPs (PNPs) at 4 °C. For the longer term storage freeze-dry the NPs and store at –20 °C.
11. Concentrate the PNPs (step 9) by filtration, using Amicon Centrifugal Filter Device (Ultracel YM-100; see Notes 4 and 5) at ~5,000 × *g* in 4 °C for 20 min. Equilibrate the Microcon columns with HEPES buffer prior to filtration.
12. Reduce 50 μL of 1 μg/mL IgA antibody by adding β-MEA (7.5 mM final concentration) for 30 min at 37 °C in HEPES buffer (see Note 6).
13. To conjugate the antibody, incubate the reduced antibody (step 12) with the PNP containing DSPE-mPEG (NP:ab = 5:1,v/v). Incubate at RT for 3–5 h (Fig. 1, see Note 6). Store PEGylated immune-conjugated NPs (PINPs) at 4 °C. For the longer term storage freeze-dry the NPs and store at –20 °C.

3.2. Polymer Nanoparticles Characterization

3.2.1. Zeta Potential Measurement and Dynamic Laser Scattering

1. Reconstitute the powdered NPs in 250 μL of 1 mM HEPES buffer, pH 7.4.
2. The ZETAPLUS Instrument (Brookhaven Instruments Inc) is used to calculate the dynamic laser scattering (DLS) as per instruction manual (see Notes 7 and 8). DLS data indicates the hydrodynamic size distribution and colloidal stability of NPs (Fig. 2a).

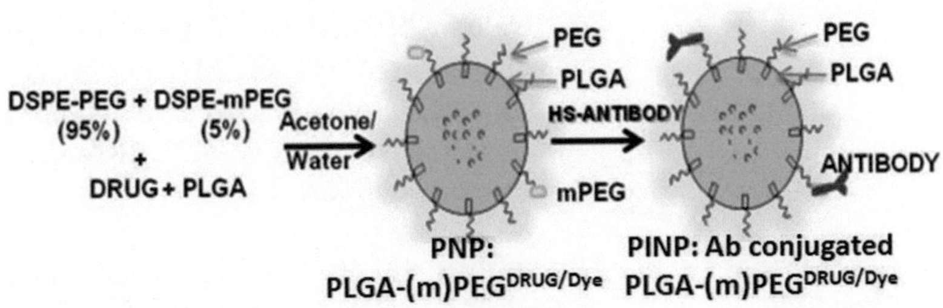

Fig. 1. Schematic of targeted nanoparticle synthesis.

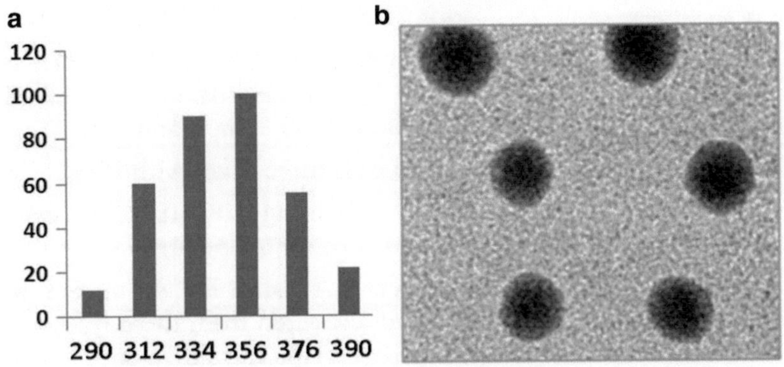

Fig. 2. Characterization of targeted NPs. Dynamic laser scattering (DLS) is employed to measure the hydrodynamic size distribution and colloidal stability of the targeted NPs while transmission electron microscopy (TEM) is used to characterize the size and shape of the NPs. (**a**) The DLS results verifies the hydrodynamic size of NPs. The hydrodynamic size of NPs varies by less than 15 %, suggesting that their colloidal stability is not affected under physiological pH. (**b**) TEM shows that targeted NPs are mono-dispersed and spherical. Moreover, it determines the NP size.

3.2.2. Transmission Electron Microscopy

TEM is used to determine the particle size, distribution, and surface morphology.

1. Dry the aqueous dispersion of the NPs on formvar-coated copper grids.
2. Stain the nanoparticles with 2 % uranyl acetate solution (negative staining) (see Note 9).
3. Visualize the nanoparticles using a PHILIPS EM 420 TEM.

3.2.3. Flow Cytometry, Protein Assay, and Fluorescence Microscopy

Flow cytometry is used to analyze the dispersion of marker dye (NR/R6G) loaded targeted NPs as compared to non-targeted NPs using BD FACsCalibur and CellQuest software.

1. Analyze the data using the FSC vs. SSC (Forward SCatter vs. Side SCatter) plots of PNPs (negative control obtained in Subheading 3.1, step 10) and antibody immunoconjugated PNPs obtained in Subheading 3.1, step 13) by FACS Caliber flow cytometer to determine the shift in FSC/SSC for PINPs (over PNPs) due to change in particle size (see Note 10).

2. Evaluate NP delivery (0.1 and 1 μg/mL) to bronchial epithelial cells cultured in DMEM-F12 media for 24–48 h, by analyzing dye (NR/R6G) loaded PINP (FL1) using flow cytometer as compared to the dye (NR/R6G) loaded PNP as a negative control (see Note 11).

3.3. Nanoparticles Mediated Drug Delivery In Vitro

Determine the in vitro specificity and cellular uptake of the dye-loaded PINPs using spectrofluorimetry.

3.3.1. In Vitro Efficacy of NP Mediated Targeted Cell Delivery

1. Seed bronchial epithelial cells at 25,000-cells/mL/well in DMEM-F12 media in a 24-well plate and allow them to attach for 24 h at 37 °C and 5% CO_2.
2. Determine efficacy of NP uptake by incubating these cells (step 1) with suspension of dye loaded PNP/PINP in growth medium for 15 min, 1 and 4 h.
3. Wash cells three times with PBS.
4. Lyse cells using 100 μL of 1× cell culture lysis reagent for 5 min at room temperature.
5. Analyze cell lysates for NP uptake by quantifying the amount of dye recovered from these lysates using the spectrophotometer. Record the changes in optical density at excitation/emission wavelengths of 525/555 to determine the uptake of dye loaded NPs. The amount of dye recovered from the lysates (the changes in optical density) directly correlate with the efficiency of PINP uptake by the target cells. The non-antibody conjugated PNPs should be used as a control to determine if antibody conjugation improves efficacy of PINP uptake by target cells.

3.3.2. Mucus Transport and In Vivo Kinetics

1. To measure mechanical properties of COPD and CF mucus use a strain-controlled rheometer. Measure elastic and loss moduli, $G'(\omega)$ and $G''(\omega)$, by applying 1 % amplitude oscillatory deformations of frequencies between 0.01 and 100 rad/s and quantify the in- and out-of-phase components of the resulting stress (see Note 12).
2. Load the layer (1–5 μm thick) of undiluted mucus on a 24 well plate followed by ~100 μL of dye loaded PINPs.
3. Use Laser scanning confocal microscopy and MPT Camera for time-lapsed 3-D imaging to determine NP transport rates through the mucus using the Metamorph software. The rates of PINP penetration through the mucus layer from various patients (at least $n = 15$) can be used to determine if these NPs can effectively penetrate mucus for drug delivery in chronic obstructive lung diseases.

3.3.3. NFκB Reporter Assays

1. Seed and transfect bronchial epithelial cells with reporter plasmids (0.1 μg of NFκB and 0.01 μg Renilla luciferase internal control).

2. Six hours post-transfection, treat the cells with $PINP^{DRUG}$ (10 and 1 μL drug, from Subheading 3.1, step 13) and incubate plates for 24 h. To determine the efficacy of PINP mediated drug delivery in controlling NFκB mediated inflammation; step 2 is also done in the presence of 10 ng/mL IL1-β/TNFα.
3. Use standard Dual Luciferase Assay Promega kit and follow manufacturer's instructions to calculate the NFκB reporter activity normalized to renilla luciferase internal control for step 2. The data will determine the efficacy of PINP mediated drug delivery in controlling NFκB mediated inflammatory response.

4. Notes

1. The acetone is used as a solvent based on the solubility of selected DRUG/Dye. Accordingly, other solvents can be used based on the solubility of selected drug/dye. Single/Double emulsion technique can be used depending on the hydrophilic/phobic nature of the drug.
2. Stirring rod should be removed prior to sonication.
3. Keep your tubes on ice while sonicating.
4. Microcon filter device column only holds 500 μL volume, use multiple spins of same sample for higher volumes. Make sure the column does not run dry.
5. Use low speed centrifugation (800 × *g*, 30 s) or microcon filter to remove unbound target antibody.
6. IgA antibody is used to selectively bind to the polymeric immunoglobin receptors (IgR) on the surface of epithelial cells in the lungs. Other target antibodies can be used based on the alveolar cell type.
7. The instrument should be calibrated using polystyrene standards (e.g., from Brookhaven Instruments Inc) prior to the measurements.
8. Serial dilution (10^{-1}–10^{-4}) may be required.
9. Uranyl acetate stains PEG (negative staining) and enhances its electron density. This staining results in a dim ring surrounding the PLGA core (Fig. 2b).
10. We developed this approach to verify antibody conjugation to PINPs, in addition to classical ELISA's or protein quantification.
11. This is an important step to verify both NPs epithelial cell delivery and efficacy.

12. The mechanical properties of mucus are variable among patients and this will help correlate the efficacy of NP transport rates through mucus with varied consistency.

Acknowledgements & Declaration of Interest

NV is currently supported by FAMRI YCSA and NIH (U54CA141868) grants. The funders had no role in decision to publish or preparation of the manuscript.

References

1. Vij N, Min T, Marasigan R et al (2010) Development of PEGylated PLGA nanoparticle for controlled and sustained drug delivery in cystic fibrosis. J Nanobiotechnology 8:22
2. Roy I, Vij N (2010) Nano-delivery in airway diseases: challenges and therapeutic applications. Nanomedicine 6(2):237–44
3. Vij N (2011) Nano-based theranostics for chronic obstructive lung diseases: challenges and therapeutic potential. Expert Opin Drug Deliv 8:1105–1109
4. Gref RY, Minamitake MT, Peracchia V et al (1994) Biodegradable long-circulating polymeric nanospheres. Science 263:1600–1603

Chapter 25

Preparation of siRNA-Encapsulated PLGA Nanoparticles for Sustained Release of siRNA and Evaluation of Encapsulation Efficiency

Panayotis Pantazis, Konstantinos Dimas, James H. Wyche, Shrikant Anant, Courtney W. Houchen, Jayanth Panyam, and Rama P. Ramanujam

Abstract

Nanoparticles (NPs) formulated using poly (D,L-lactide-*co*-glycolide) (PLGA), a biodegradable, biocompatible, and clinically approved polymer, have been widely used for targeted drug delivery. Here we provide methods for preparing PLGA NPs that encapsulate small interfering RNA (siRNA). The siRNA NPs are formulated using a double-emulsion solvent evaporation technique with the addition of a small amount of the cationic polymer, polyethyleneimine, which significantly increases siRNA encapsulation.

Key words: Poly(lactic-*co*-glycolic) acid, Cancer therapy, Cancer treatment, Drug delivery, siRNA, RNAi, Nanoparticles, Cancer therapy

1. Introduction

RNA interference (RNAi) is a cellular pathway that decreases the expression of specific genes. RNAi are small interfering RNA molecules (siRNA), also known as short interfering RNAs and silencing RNAs. siRNAs are 20–25 nucleotide-long double-stranded RNAs. Since their discovery in 2001 by Baulcombe's group (1), siRNAs have gained widespread interest in the biomedical research and drug development fields because of their ability to interfere with or "knock down" a particular gene of interest. However, a major limitation for the use siRNA, both in vitro and in vivo, is the inability of naked and nuclease-sensitive siRNA to passively diffuse through cellular membranes due to the strong anionic charge of the phosphate backbone and consequent electrostatic repulsion

Mikhail Soloviev (ed.), *Nanoparticles in Biology and Medicine: Methods and Protocols*, Methods in Molecular Biology, vol. 906, DOI 10.1007/978-1-61779-953-2_25, © Springer Science+Business Media, LLC 2012

from the anionic cell membrane surface. Therefore, the primary success of siRNA applications depends on the preparation of suitable vectors to deliver therapeutic genes to targets of interest.

In recent years, many techniques reliant on therapeutic lipids and polymers have been developed for specific gene delivery applications, but their general use is hindered by several drawbacks. These include the difficulty associated with preparing the needed copolymers, the limited density of functional groups, poor encapsulation efficiency, and targeting effects that decrease with time due to desorption or degradation of the adsorbed group as the particle or scaffold erodes (2, 3).

Poly (lactide-*co*-glycolide) or PLGA is one of the most studied synthetic polymer because of its biocompatibility, biodegradability, and approval by the United States FDA for human use. PLGA NPs have been widely used as a vehicle for pharmaceutical delivery of nucleotides, hormones, or drugs to target tissues. PLGA polymers in nanoscale have also been used for the fabrication of bioscaffolds for tissue engineering. Our studies have shown that addition of a small amount of the cationic polymer, polyethyleneimine (PEI), during NP formulation significantly increased siRNA encapsulation in NPs, an improved siRNA release profile, and improved effectiveness of gene silencing in cultured cancer cell and animal models of the human tumor/mouse xenograft system (unpublished data). We use polyvinyl alcohol (PVA) to stabilize the emulsion during nanoparticle formation, because PVA forms particles of relatively small sizes and uniform size distribution.

Patil and Panyam (3) reported a variation of the formulation protocol and used the same to formulate dual action NPs with both a siRNA targeted to P-glycoprotein and an anti-cancer drug (4). Unlike the above, the protocol described in this chapter utilizes pre-complexing of siRNA with PEI, followed by encapsulation of the complex in PLGA nanoparticles. A small amount of PEI is added to the PLGA polymer phase (30:0.1 PLGA-PEI weight ratio) to improve the encapsulation and release of siRNA.

In our opinion, the likelihood of successful therapy of cancer genes should be enhanced by the availability of drugs that could be delivered systemically to have specific antitumor targeting capability along with the ability to induce death in primary and metastatic tumor cells. In addition, we believe that the selective and robust effect of siRNA on gene expression makes it a valuable research tool, both in cell culture and in living organisms. However, administering siRNAs to living animals, and to humans, poses many challenges. In laboratory conditions, siRNAs show different effectiveness in different cell types in a manner yet poorly understood: some cells respond well to siRNAs and show a robust knockdown, whereas others show no such knockdown. Nevertheless, siRNA-encapsulated in PLGA NPs as described below can be efficiently used to systematically knockdown cancer signaling markers in tumor cells.

2. Materials

All RNA and especially single-stranded RNA oligonucleotides are susceptible to degradation by RNase during handling. Wear gloves and change them frequently. Use RNase-free pipette tips, plasticware, water, buffers, and treated glassware.

2.1. Preparation of siRNA NPs

1. siRNA: Custom-made RNA oligonucleotides. These need to be designed by the user depending on the mRNA target. The oligonucleotides used in our work are listed in Table 1 as examples only. In general, siRNAs are shipped by vendors as freeze-dried single-stranded or pre-annealed double-stranded RNAs. Centrifuge the tube(s) briefly upon receipt to ensure that the siRNA is collected at the bottom of the well. Store at −20 °C in a frost-free freezer. If necessary, siRNAs as dry pellets (unopened) can be stored at 4 °C for several weeks.
2. 5× Annealing buffer: 50 mM Tris, pH 8.0, 100 mM NaCl, prepared in RNase-free water. This is needed only if single-stranded oligonucleotides are provided. Buffer can be stored at −20 °C and freeze-thawed many times. Dilute with RNase-free water to 1× concentration before use.
3. Control siRNA: We used a random (scrambled) double-stranded RNA, see Table 1.
4. Chloroform ($CHCl_3$).
5. PEI: Average molecular weight 750,000, make 5 % solution (w/v) in Chloroform.
6. PLGA: 50:50 (by weight) monomer ratio with molecular weight of 106 kDa and viscosity of 1.05 dL/g.
7. PVA: 2 % in RNase-free water.
8. Sterile RNase-free water: Sterile. Do not use DEPC-treated water.
9. Ficoll/trehalose solution: 5 % Ficoll 400, 5 % trehalose in RNase-free water.
10. Analytical balance.

Table 1
siRNAs used for making PLGA nanoparticles

siRNA	Sequence
RNA-binding motif protein 3 (RBM3)	GGGTATGGATATGGATATG
B-catenin	CGGGAUGUUCACAACCGAATT
Scrambled RNA	UUCUCCGAACGUGUCACGUT

11. Fume hood.
12. Magnetic stir plate.
13. Probe sonicator: using continuous mode and 100 % amplitude.
14. Repeat pipettor.
15. Refrigerated fixed-angle centrifuge capable of 20,000 × *g*.
16. Freeze-drying equipment.

2.2. Characterization of siRNA NPs

1. Release buffer (TE buffer): 10 mM Tris–HCl, pH 7.4, 1 mM EDTA, prepared in RNase-free water.
2. ZetaPlus particle size and zeta potential analyzer equipped with a 35 mW solid state laser (658 nm) (Brookhaven Instruments).
3. Scanning electron microscope (SEM) Hitachi S-900 Cold Field Emission Gun SEM (Hitachi) or a similar instrument to characterize the surface morphology of nanoparticles.
4. 300-Mesh copper grid for the above.
5. Ultra sensitive fluorescent nucleic acid stain Quant-iT picogreen dsDNA reagent (Molecular Probes).
6. FLx-800 fluorescent microplate reader (BIO-TEK Instruments) or equivalent Spectrophotometer for measuring fluorescence of Quant-iT picogreen reagent.
7. Orbital shaker at 37 °C.

3. Methods

3.1. Preparation of siRNA-Encapsulated PLGA NPs

We recommend making a separate PREPARATION OF PLGA NPs with scrambled siRNA for use as a negative control for in vitro and in vivo experiments. In our studies, we use siRNA with a random (scrambled) sequence. Steps 3–7 should be performed in a fume hood. Synthesis of siRNA encapsulated poly PLGA nanoparticles using a double-emulsion solvent evaporation ($w_1/o/w_2$) technique is summarized in Fig. 1.

1. Resuspend the RNAs in RNase-free water. Anneal single stranded oligonucleotides if not supplied as double-stranded oligonucleotides (see Note 1). Briefly centrifuge the tube to bring down all droplets from the wall and lid of the tube (see Note 2).
2. Add 200 μL of siRNA to 400 μL of 5 % PEI. Incubate at room temperature for 30 min.
3. Dispense 1 mL of $CHCl_3$ into a glass vial containing a magnetic stir bar. Add 30 mg of PLGA. Stir to dissolve (see Note 3).

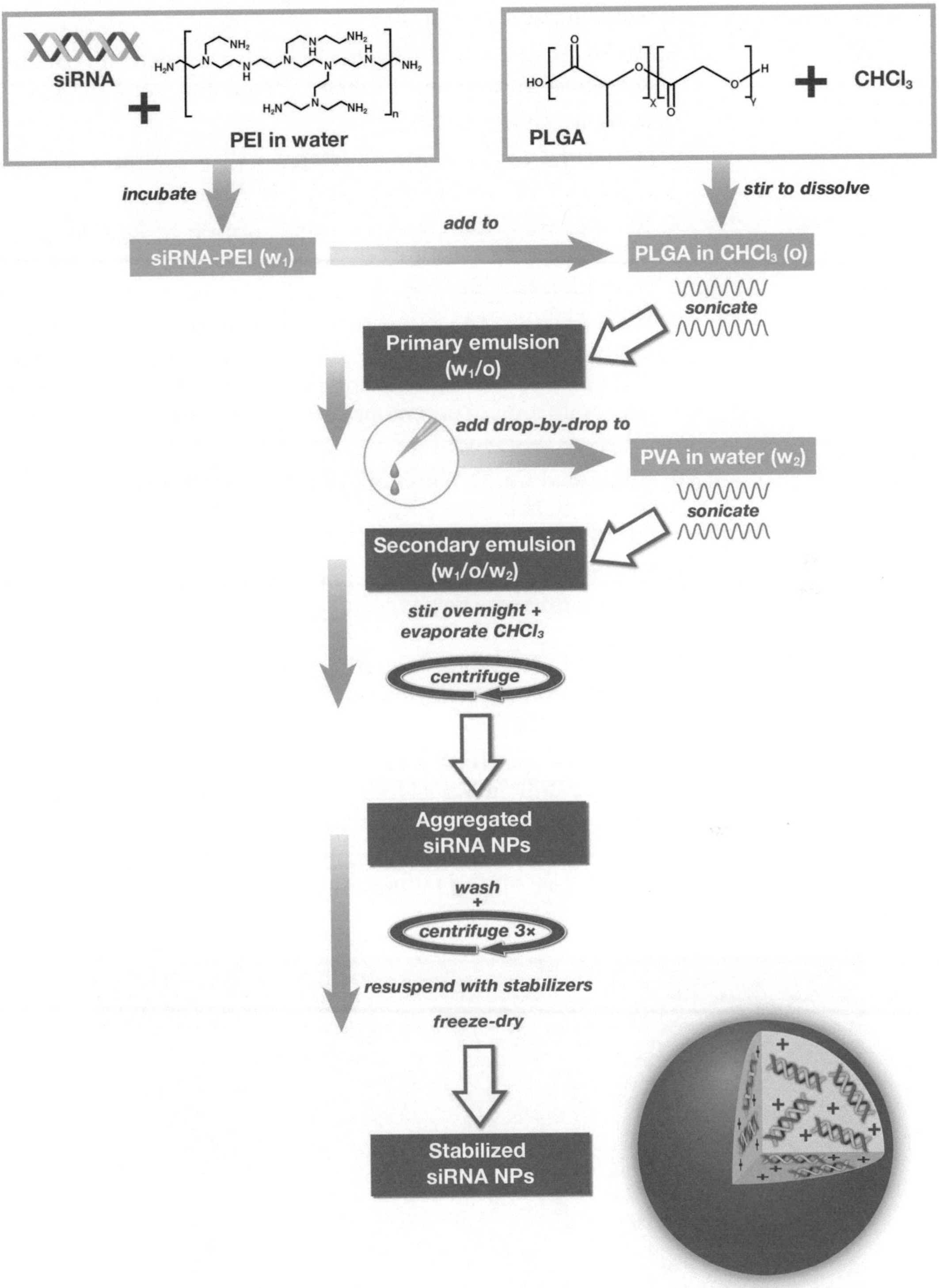

Fig. 1. Schematic representation of synthesis of siRNA encapsulated poly (D,L-lactide-*co*-glycolide) (PLGA) nanoparticles using a double-emulsion solvent evaporation ($w_1/o/w_2$) technique.

4. Sonicate on ice for 30 s.
5. Add 6 mL of 2 % PVA to a glass vial with a magnetic stir bar. Add the emulsion from step 4 to the PVA solution (see Note 4).
6. Sonicate on ice for 1 min.
7. Allow the organic solvent ($CHCl_3$) to evaporate overnight (18 h) under continuous stirring (see Note 5).
8. Recover the PLGA NPs by centrifugation at 20,000 × *g* for 20 min at 4 °C. Remove the supernatant and save it for later evaluation (see Note 6).
9. Wash the pellet three times with 10 mL of RNase-free water to remove any residual PVA and free (i.e., non-encapsulated) siRNA. Save the supernatant from each wash separately for later evaluation of un-encapsulated RNA remaining in the wash solutions (see Note 7).
10. Resuspend the NPs in 5 mL of 5 % Ficoll/5 % trehalose solution to stabilize the siRNA inside the NPs. Sonicate the NPs for 30 s over an ice bath. (see Note 8).
11. Freeze-dry the PLGA NPs. The lyophilized NPs are stable up to 6 months at −20 °C.

3.2. Characterization of the siRNA NPs

1. Use diffraction light scattering (DLS) to determine the NP size, the polydispersity index (PI) and zeta potential (see Notes 9 and 10) (Fig. 2).
2. Use a UV spectrophotometer to obtain absorbance measurements at 260 nm for all four supernatants from the NP formulation performed in Subheading 3.1.
3. Determine the siRNA release kinetics using Quant-iT picogreen reagent (see Note 11). Prepare several tubes containing 1 mg

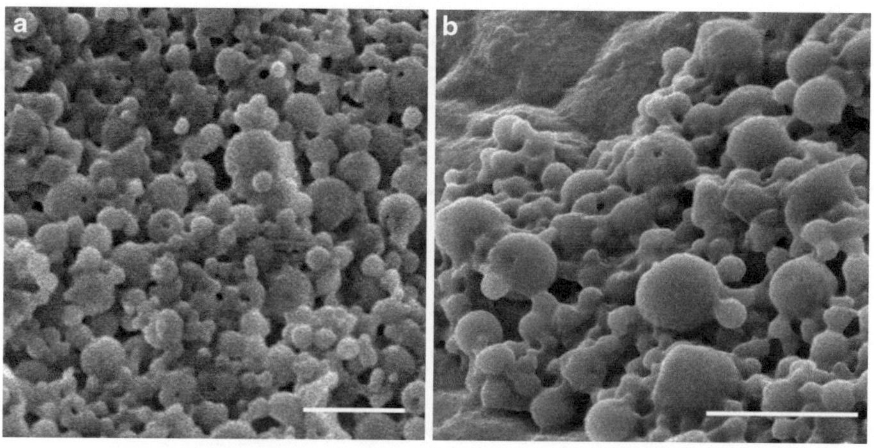

Fig. 2. SEM image of dual agent nanoparticles without surface functionalization (**a**) and with biotin functionalization (**b**). Nanoparticles were platinum-coated for visualization. Bar is 670 nm. Reproduced from (4) with permission from Elsevier.

of siRNA nanoparticles in 1 mL of release buffer for each tube. Incubate all the tubes on an orbital shaker at 100 rpm and 37 °C. At pre-defined time points (e.g., "zero incubation time," 1, 3, 5, 7, 14, 21 and 30 days), take out sets of three tubes and centrifuge at 20,000 × *g* for 20 min at 4 °C. To determine the amount of oligonucleotide or siRNA in the supernatant measure the fluorescence generated by the binding of picogreen to double-stranded siRNA in triplicate for each time point. The siRNA encapsulation in nanoparticles can be determined by subtracting the amount of siRNA recovered in the wash solutions from the initial amount of siRNA added. We typically obtain >75 % encapsulation efficiencies.

4. Notes

1. Gene silencing is triggered by using siRNA molecules, that are about 20–25 base pairs (bp) long. Upon introduction into cells, siRNAs assemble into ribonuclease-containing complexes known as RNA-induced silencing complexes (RISCs). The siRNA strands guide the RISCs to complementary RNA molecules, where they cleave and destroy the target RNA. Therefore, double-stranded RNAs are used for this procedure. To anneal single stranded oligonucleotides, dissolve them in 1× annealing buffer to a final concentration of 50 μM each. Gently pipette up and down three times to mix well, aliquot into sterile tubes, and store at −20 °C if not used immediately. For annealed double-stranded RNAs, resuspend to a final concentration of 20 μM (for each of the two oligonucleotides in the tube (5)).
2. Aliquot the annealed siRNA into RNase-free tubes and store at −20 °C. Do not freeze-thaw more than five times.
3. For water-in-oil (W/O) emulsion preparations, utilizing glass vials with magnetic stirring provides consistent results than harsh vortexing.
4. PVA acts as a surfactant.
5. Alternatively, the sample can be stirred overnight with the cap on, after which the solvent can be evaporated under vacuum.
6. Use a pipette to remove the supernatant. Do not invert the tube, as this may dislodge the aggregated NPs. These should be visible, whitish in color, at the bottom of the tube.
7. Use a pipette to resuspend the NPs, then centrifuge to sediment them after each wash. To quantify the remaining RNA in the wash solutions use Picogreen dsDNA quantitation reagent (Invitrogen), which is suitable for determining RNA ranging from 100 ng to 5 μg of siRNA/mL.

8. We use a final 5 % Ficoll/5 % trehalose concentration to achieve a T_g (glass transition temperature; (6)) of greater than 60 °C. This increases the shelf-life of the freeze-dried NPs and provides good solubility when the NPs are resuspended in water.
9. We calculate mean hydrodynamic diameters for size distribution by weight, assuming a lognormal distribution. We use Smoluchowski's equation to calculate Zeta potential values from measured velocities. In each case, distilled water may be used as solvent for the nanoparticles and the results should be expressed as means SEM of five runs. To study the surface morphology of the NPs, use SEM. Nanoparticle suspension can be in water (2 mg/mL), and a single drop should be air-dried on a 300-mesh copper grid for SEM. Figure 2 shows a typical SEM image of dual agent nanoparticles without surface functionalization (A) and with biotin functionalization (B).
10. In our experiments, DLS measurements yielded the average particle size (hydrodynamic diameter) of about ~100–250 nm. With these NPs, initially, a burst release was seen for both siRNA (15–20 % in the first 24 h); and a more sustained and continuous release was observed over the next 10–20 days.
11. siRNA can be quantified using UV absorbance methods or Quant-iT picogreen reagent (PicoGreen) which is ~10,000-fold more sensitive than UV absorbance.

    For hydrophilic drugs encapsulated inside polymeric nanoparticles, an initial burst followed by a relatively constant release over a number of days (e.g., 5–28 days) is very characteristic. This burst release could be explained by the fact that the hydrophilic drug has readily escaped or diffused into the aqueous medium under a concentration gradient. The relatively constant release that follows is primarily due to the hydrolysis of the ester bonds between the individual monomer, which causes the degradation of the nanoparticles and hence the sustained release and bioavailability of encapsulated siRNA (7–9).

Acknowledgement

This was partly funded by NIH grant 3R44AT004118-03S1 (Rama P. Ramanujam) and NIH grant 7R21CA116641 (Jayanth Panyam). We thank Fadee G. Mondalek and Sivapriya Ponnurangam for technical assistance.

References

1. Hamilton A, Baulcombe D (1999) A species of small antisense RNA in posttranscriptional gene silencing in plants. Science 286: 950–952
2. Takahashi Y, Nishikawa M, Takakura Y (2009) Nonviral vector-mediated RNA interference: its gene silencing characteristics and important factors to achieve RNAi-based gene therapy. Adv Drug Deliv Rev 61:760–766
3. Patil Y, Panyam J (2009) Polymeric nanoparticles for siRNA delivery and gene silencing. Int J Pharm 367:195–203
4. Patil Y et al (2010) The use of nanoparticle-mediated targeted gene silencing and drug delivery to overcome tumor drug resistance. Biomaterials 31:358–365
5. Elbashir SM et al (2001) Duplexes of 21-nucleotide RNAs mediate RNA interference in cultured mammalian cells. Nature 411: 494–498
6. Treml S et al. (1997) Biological reagent spheres. US Patent 5593824
7. Panyam J, Labhasetwar V (2003) Biodegradable nanoparticles for drug and gene delivery to cells and tissue. Adv Drug Deliv Rev 55: 329–347
8. Mondalek F et al (2010) Inhibition of angiogenesis- and inflammation-inducing factors in human colon cancer cells in vitro and in ovo by free and nanoparticle-encapsulated redox dye, DCPIP. J Nanobiotechnology 8:17–25
9. Sureban S et al (2011) Nanoparticle-based delivery of siDCAMKL-1 increases *microRNA-144* and inhibits colorectal cancer tumor growth via a Notch-1 dependent mechanism. J Nanobiotechnology 9:40

Chapter 26

Synthesis of Poly (2-Hydroxyethyl Methacrylate) (PHEMA) Based Nanoparticles for Biomedical and Pharmaceutical Applications

Rajesh Saini, Jaya Bajpai, and Anil K. Bajpai

Abstract

The performance of polymeric nanomaterials relies greatly upon their properties which are intimately related to the methods of fabrication of the materials. Among various synthetic polymers, the polymers of 2-hydroxyetyhyl methacrylate (PHEMA) maintain a prime position in biomedical field due to their useful physicochemical properties and suitability for controlled drug delivery applications. Here we focus on three methods of preparation of PHEMA nanoparticles, by suspension polymerization, emulsion polymerization and dispersion polymerization without the use of any surfactants.

Key words: Poly (2-hydroxyethyl methacrylate), Nanoparticles, Synthesis, Polymerization, Drug delivery

1. Introduction

Polymeric nanoparticles constitute a versatile class of drug delivery systems, which can potentially overcome physiological barriers, and guide the drug to specific cells or intracellular compartments by passive or ligands-mediated targeting approaches because of their submicron size which makes extravasations possible and occlusion of terminal blood vessel unlikely (1, 2). It also allows controlling the release pattern and sustaining drug levels for a longer time by appropriately selecting the polymeric carriers. Although numerous biodegradable polymeric nanoparticles of natural polymers such as proteins or polysaccharides are largely used as a drug delivery carrier in controlled drug delivery technology, however, synthetic polymers have received significantly more attention in

Mikhail Soloviev (ed.), *Nanoparticles in Biology and Medicine: Methods and Protocols*, Methods in Molecular Biology, vol. 906, DOI 10.1007/978-1-61779-953-2_26, © Springer Science+Business Media, LLC 2012

this area because with polymeric nanoparticles both controlled drug release and disease-specific localization can be achieved (3).

Hydrophilic polymers based on poly (2-hydroxyethyl methacrylate) (PHEMA) is particularly attractive for biomedical engineering applications because of their high water content, non-toxicity, and favorable tissue compatibility, and its physical properties can be easily manipulated through formulation chemistry and it has been extensively used in medical applications such as contact lenses (4, 5), keratoprotheses, and as orbital implants (6). The presence of a hydroxyl and carboxyl groups on each repeat unit in PHEMA makes this polymer compatible with water and the hydrophobic α-methyl groups and backbone imparts hydrolytic stability to the polymer and support the mechanical strength of the polymer matrix (7). PHEMA nanoparticles may be prepared using a variety of methods such as atom transfer radical polymerization (ATRP), reversible addition—fragmentation chain transfer (RAFT) polymerization, photochemical polymerization etc. One of the simplest yet versatile methods, however, is the suspension polymerization which does not require any surfactants. In this chapter we describe simple laboratory methods for preparing PHEMA nanoparticles using suspension polymerization, emulsion polymerization and dispersion polymerization.

2. Materials

All chemicals should be of analytical reagent grade. Use double distilled water throughout the experiments.

2.1. Purification of Monomer (HEMA)

1. 2-Hydroxyethyl methacrylate (HEMA), (Sigma Aldrich Co). Minimum purity 98 %. The monomer should be purified prior to synthesis. Store at 4 °C (see Note 1).
2. Cyclohexane: Carbon tetrachloride extraction mixture: mix 100 mL of cyclohexane and 100 mL of carbon tetrachloride.
3. Anhydrous sodium carbonate, used without further purification.
4. Anhydrous sodium sulfate, used without further purification.
5. Deionised water.
6. Magnetic stirrer plate.
7. Whatman filter paper No. 42, pore size 2.5 μm, diameter 12.5 cm.
8. 500 mL separating funnel (Borosil).
9. Vacuum pump.
10. 500 mL beaker.
11. NaCl.

2.2. Suspension Polymerization of HEMA

1. Polyvinyl alcohol (PVA, hot processed, MW 14000), used without further purification. To make 0.5 % (w/v) solution of PVA dissolve 1 g of PVA in 200 mL of deionized hot water. Incubate at 80 °C for at least 4 h to completely dissolve the polymer. PVA is used as stabilizing agent (see Note 2).
2. Ethyleneglycol dimethacrylate (EGDMA), used without further purification. EGDMA is used as a cross-linker to polymerize HEMA.
3. Benzoyl peroxide (BPO), used without further purification. BPO is used to initiate the polymerization reaction.
4. Toluene, used without further purification.
5. Acetone, used without further purification.
6. Round bottom flask, 500 mL (Borosil).
7. Magnetic stirrer with hot plate.
8. Centrifuge capable of spinning 15 mL tubes at 2,683 × *g*.
9. Vacuum desicator.

2.3. Emulsion Polymerization of HEMA

1. PVA (hot processed, MW 14000), used without further purification. To make 1 % (w/v) solution of PVA dissolve 0.5 g of PVA in 50 mL of hot water. Heat the solution to 80 °C with constant stirring on a hot plate and maintain at 80 °C for at least 4 h to completely dissolve the polymer. PVA is used as stabilizing agent (see Note 2).
2. EGDMA, used without further purification. EGDMA is used as a cross-linker to polymerize HEMA.
3. Potassium persulfate, used without further purification. Potassium persulfate is used to initiate the polymerization reaction.
4. Sonicator UP4005 (operating freq. 19–25 kHz, output power –100 W) or equivalent.
5. Thermostated shaker capable of 70 °C and controlled nitrogen atmosphere else?
6. Methanol 99.8 %.
7. Centrifuge capable of spinning 15 mL tubes at 2,683 × *g*.
8. Vacuum desicator.

2.4. Dispersion Polymerization of HEMA

1. Toluene, used without further purification.
2. Prepare 0.5 % (w/v) solution of polybutadiene by adding 0.5 g of polybutadiene to 100 mL of toluene.
3. Azo-bis-isobutyronitrile (AIBN) should be purified by recrystallisation from methanol-chloroform (1/1 v/v) at 40 °C. AIBN is used as polymerization initiator.
4. EGDMA, used without further purification. EGDMA is used as a cross-linker to polymerize HEMA.

5. Ethylic ether, used without further purification.
6. Butanol, used without further purification.
7. Magnetic stirrer with hot plate.

3. Methods

3.1. Purification of Monomer (HEMA)

HEMA monomer may degrade during transportation and storage at ambient temperatures to yield increased levels of methacrylic acid (MAA) and the naturally occurring cross-linker EGDMA, see Fig. 1. In commercially available monomer preparations, an inhibitor such as hydroquinone is often present at 300 ppm. These are added by the manufactures to avoid polymerization during transportation, storage etc. Therefore monomers need to be purified to achieve reproducible results. The described procedure aims to remove the impurities, mainly the MAA and EGDMA form HEMA monomer (see Note 3).

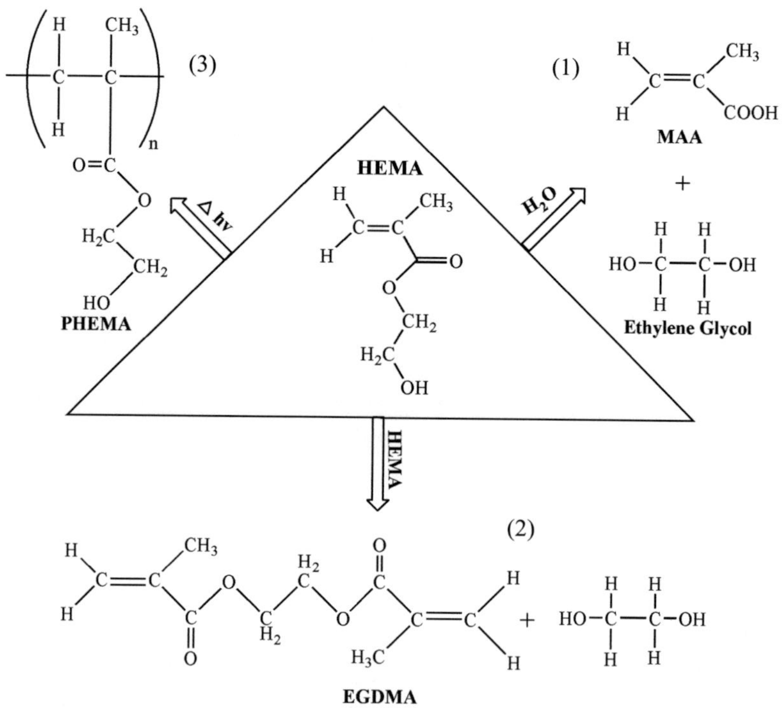

Fig. 1. Reactions of monomer 2-hydroxyethyl methacrylate (HEMA). HEMA monomer readily undergoes three common reactions: (1) HEMA may hydrolyze at the ester linkage to form MAA and ethylene glycol; (2) two molecules of HEMA may transesterify to form the cross linker and ethylene glycol; (3) monomer may polymerize at the double bond resulting in oligomer and polymer.

1. Add 15 g anhydrous sodium carbonate to 100 g of HEMA in a beaker. Stir it for 3 h at 24 °C.
2. Vacuum filter the suspension through Whatman filter papers. Filtering a 100 mL sample will normally yield about 89 mL of filtrate (see Note 4).
3. Dissolve one volume of the monomer HEMA in three volumes of water.
4. Use a separating funnel to perform four consecutive extractions of the HEMA solution as follows: transfer ~300 mL of HEMA—water mixture to a separating funnel and add 50 mL of cyclohexane and CCl_4 mixture (1:1 v/v), shake vigorously for 10 min and allow the mixture to settle for 30 min. Discard the organic layer and repeat the extraction procedure three more times. Each extraction should take approximately 40 min and there should be a gap of 30 min between each two successive extractions (see Note 5).
5. Following the last extraction place the aqueous layer containing HEMA under vacuum to remove the remaining organic solvent; incubate under the vacuum overnight.
6. Transfer the purified HEMA monomer (~250 mL) into a 500 mL beaker and 100 g of NaCl (mix with a glass rod) for 5 min to salt out the monomer.
7. Collect the upper layer of HEMA and distil it under vacuum at 55 °C. Distillation may continue for 5–6 h. Discard the fractions collected in the first and last 30 min.
8. Add 10 g of Na_2SO_4 for per 100 mL of HEMA monomer to remove any remaining water. Store the monomer at 4 °C.

3.2. Suspension Polymerization of HEMA

1. Prepare the organic phase by adding 0.6 mL of purified and distilled monomer (HEMA), 0.1 mL of crosslinker (EGDMA) and 0.8 g of BPO (initiator) to 100 mL of toluene and stir for 10 min.
2. Dissolve 1 g of PVA in 200 mL of distilled water to make 0.5 % (w/v) solution of PVA which is used as a stabilizing agent here.
3. Mix the aqueous solution of PVA (200 mL, step 2) and the toluene oil phase (100 mL from step 1) together in a flat bottom flask (500 mL capacity) using magnetic stirrer or shaking it vigorously to prepare a stable suspension (see Note 6).
4. Flush the suspension with nitrogen gas for 1 h and seal the flask (see Note 7).
5. Place the reaction mixture on a magnetic stirrer hot plate and heat it at 80 °C for 2 h with continuous stirring.
6. Raise the temperature to 90 °C and incubate the mixture for 1 h. The polymerization should be completed by the end of this incubation step.

7. Allow the reaction mixture to cool to room temperature. The solution should appear dense white which indicates the presence of PHEMA nanoparticles.
8. Separate the nanoparticles by centrifugation at 2,236 × *g* for 45 min.
9. To wash the nanoparticles transfer the pellet to a 250 mL conical flask, add 100 mL toluene, shake for 5 min, and separate the nanoparticles by centrifugation at 2,236 × *g* for 45 min. Repeat the process at least three times to remove unreacted monomer and residual organic materials.
10. Wash the nanoparticles pellet two times with acetone using the procedure described in previous step (see Note 8).
11. Wash the nanoparticles pellet with double distilled water three times using the procedure described in previous step.
12. Dry the washed nanoparticles at room temperature for 72 h in a vacuum dessicator.
13. Store the nanoparticles in air tight polyethylene bags at room temperature until use (see Note 9).

3.3. Emulsion Polymerization of HEMA

1. Prepare the organic phase by adding 0.6 mL of purified and distilled monomer (HEMA) and 0.01 mL EGDMA to the 50 mL of 1 % PVA solution.
2. Sonicate the HEMA/EGDMA mixture for 30 min at maximum power (100 W (max)) to obtain a stable emulsion (see Note 10).
3. Add 40 mg of potassium persulfate (initiator) to the sonicated emulsion and bubble nitrogen gas through the mixture for 5 min to remove the dissolved oxygen (see Note 11).
4. Transfer the emulsion into a thermostated shaker at 70 °C and shake at 300 RPM under nitrogen atmosphere for 24 h. As the polymerization proceeds, the reaction mixture turns white.
5. Separate the nanoparticles by centrifugation at 2,236 × *g* for 45 min.
6. Transfer the washed nanoparticles into a 250 mL conical flask, and 100 mL methanol is added into it. Shake the whole mixture vigorously for 5 min. Separate particles by centrifugation at 2,236 × *g* for 45 min. Repeat the process at least three times in order to remove any trace of the unreacted monomer or other organic residues.
7. Wash the nanoparticles pellet with double distilled water using the procedure described in previous step for at least two times.
8. Dry the washed nanoparticles at room temperature for 72 h in a vacuum dessicator.
9. Store the nanoparticles in air tight polyethylene bags at room temperature until use (see Note 9).

3.4. Dispersion Polymerization of HEMA

1. Add 122 mL of butanol to the 100 mL of 0.5 % (w/v) solution of polybutadiene in toluene at 40 °C under continuous stirring on a magnetic stirrer.
2. Add 10 mL of HEMA, 0.2 g of AIBN and 0.4 mL of EGDMA dropwise to above solution under continuous stirring.
3. Heat the mixture to 65 °C with continuous stirring under nitrogen atmosphere and incubate for 6.5–7 h to complete the polymerization.
4. For washing, nanoparticles are transferred in to a 250 mL conical flask, and 100 mL toluene is added into it. After that, the whole mixture is shaken for 5 min and the nanoparticles are separated by centrifugation for 45 min at 2,236 × *g*. Repeat the process at least three times to remove unreacted monomer and residual organic materials.
5. Dry the washed PHEMA nanoparticles at 37–38 °C for 24 h in a vacuum dessicator.
6. Store the nanoparticles in air tight polyethylene bags at room temperature until use (see Note 9).

4. Notes

1. HEMA is known to be a poorly stable monomer and, therefore, it is essential to purify it prior to synthesis. Always keep the monomer bottle in a refrigerator at or below 4 °C.
2. PVA is commercially available as hot and cold water soluble polymer. We recommend using hot soluble polymer.
3. This is an important step. HEMA should be free from the MAA and ethyleneglycol dimethacrylate which are ionic monomer and crosslinker, respectively. The polymerization of impure or partially pure monomer may greatly affect the extent of swelling of polymer and alter its properties. We recommend that the purity of the monomer is checked by HPLC (minimum purity 98 %).
4. If monomer HEMA is of old stock then the recovered yield of the monomer may be less.
5. In every extraction the EGDMA goes into the organic layer which is discarded after each extraction.
6. Stable suspension should not show any settlement of particles for at least 3 h.
7. Bubbling of nitrogen gas through the suspension is essential for the removal of oxygen, which may inhibit the polymerization rate and lower the yield.

8. One must ensure that all unreacted monomer, crosslinker, and the toluene are removed; as these may interfere with characterization of the prepared nanoparticles.
9. For dispersion of particles in water, Brij 35, Tween 60, Tween 80, and sodium dodecyl sulfate may be used as emulsifier.
10. Insufficient sonication may result in a nonhomogeneous nanoparticles size.
11. Instead of adding solid potassium persulfate a freshly prepared solution can be used. Persulfate solution can be stored at 4 °C for 3–4 days.

References

1. Vashir JK, Reddy MK, Labhasetwar VV (2005) Nanosystems in drug targeting: opportunities and challenges. Curr Nanosci 1:47–64
2. Barrat G (2003) Colloidal drug carriers: achievements and perspectives. Cell Mol Life Sci 60:21–37
3. Singh R, Lillard JW Jr (2009) Nanoparticles-based targeted drug delivery. Exp Mol Pathol 86:215–223
4. Yakup Arıca M, Bayramoglu G, Arıca B, Yalcın E, Ito K, Yagci Y (2005) Novel hydrogel membrane based on copoly(hydroxyethyl methacrylate/p-vinylbenzylpoly(ethyleneoxide)) for biomedical applications: properties and drug release characteristics. Macromol Biosci 5:983–992
5. Brigger I, Dubernet C, Couvreur P (2002) Nanoparticles in cancer therapy and diagnosis. Adv Drug Deliv Rev 54(5):631–651
6. Flynn L, Dalton PD, Shoichet MS (2003) Fiber templating of poly (2-hydroxyethyl methacrylate) for neural tissue engineering. Biomaterials 24:4265–4272
7. Chou KF, Han CC, Lee S (2000) Water transport in crosslinked 2-hydroxyethyl methacrylate. Polym Eng Sci 40(4):1004–1014

Chapter 27

Preparation of Lipid:Peptide:DNA (LPD) Nanoparticles and Their Use for Gene Transfection

Fan Zhang and Hao-Ying Li

Abstract

Therapeutic gene delivery systems offer the potential for the treatment of a range of inherited and acquired inherited diseases. In contrast with viral gene vectors, the nonviral gene vectors provide a safer alternative and additional advantage such as the improved delivery efficiency, low cost, and often unlimited capacity to package DNA. Here we describe preparation of nonviral gene delivery technique based on lipid:peptide:DNA (LPD) complexes. The size of LPD particles is in the nanometre range. The use of these nanoparticulate LPDs results in high efficiency transfections and a high level of gene expression in vitro. LPDs provide a convenient and efficient tool for gene therapy for the gene delivery.

Key words: Liposome, Nonviral gene vector, Gene transfection, Flow cytometry

1. Introduction

Genetic disorders are responsible for numerous human diseases such as cancer, severe combined immunodeficiency, syndromecystic fibrosis, and cystic fibrosis (1). The direct approach of curing the genetic disorders is to deliver healthy genes into patients' target cells to replace the abnormal ones that are responsible for the disease development (2). Two major delivery systems have been developed to date, such as viral and nonviral gene vectors (3, 4). A number of viral gene vectors have been developed for the gene transfection, including adenoviruses, adeno-associated viruses, retroviruses, lentiviruses, and herpes simplex viruses. Despite their high transfection efficiency, the viral gene vectors have many disadvantages including the technical difficulties of manipulation, high cost, and particularly their immunogenicity, pathogenicity, cytotoxicity, and risk of oncogenesis (5). In contrast, nonviral gene vectors are synthetic, non-biological systems, having no limitations on the amount of genetic information (DNA load capacity) and offer

Mikhail Soloviev (ed.), *Nanoparticles in Biology and Medicine: Methods and Protocols*, Methods in Molecular Biology, vol. 906,
DOI 10.1007/978-1-61779-953-2_27,

other advantages such as the lack of immunogenicity, improved safety profile, and are considered as an attractive alternative to the viral gene vector (6–8).

Cationic liposomes represent the most widely used nonviral gene delivery systems (9). The reasons for that are possibly based on two aspects: (1) the electrostatic interactions are easily generated between the cationic liposomes and the negatively charged DNA that is subsequently protected against the nuclease degradation; (2) the double layer structures of liposomes are identical to the cell membranes, which potentially assists the cellular uptake of gene vectors and consequently increase the gene delivery efficiency (10–13). The *N*-(1-(2, 3-dioleoyloxy)propyl)-*N*,*N*,*N*-trimethylammonium chloride (DOTAP) is an example of a widely used cationic lipid, widely used as the carrier for the delivery of genes (14, 15). The cationic peptides such as protamine are further utilized to aid the condensation of plasmid in order to improve the stability of plasmid DNA (DNA) in the cytoplasm and subsequently improve the transfection efficiency. Additionally, surface charge and particle size of nonviral gene vectors are also critical parameters for efficient gene delivery (16, 17). The positive surface charge and smaller particle size (<200 nm) of nonviral gene vectors play a key role in gene vector binding to the cell membrane and subsequently in cellular uptake (18).

In this chapter, we describe the preparation of nanosized particles of DOTAP:protamine:DNA (LPD) complexes and their application in delivering the reporter gene pEGFP-N1 (4.7 kb) to the mammalian cells. The particle size of such nonviral gene vectors is determined by photon correlation spectroscopy (PCS) and observed by transmission electron microscopy (TEM). Zeta potentials are determined by microelectrophoresis with a laser electrophoretic mobility instrument.

2. Materials

2.1. Preparation of 1,2-Dioleoyl-3-Trimethylammonium-Propane Cationic Liposomes

1. DOTAP (Avanti Polar Lipids).
2. Chloroform.
3. Round-bottomed glass flask (50 mL).
4. Rotary evaporator.
5. Sterilized distilled water.
6. Nitrogen (gas).
7. Polycarbonate membranes (100 nm pores, 10 mm diameter).
8. Quartz cuvettes (3 mL).
9. An extrusion apparatus (The Extruder™, Lipex Biomembranes Inc, Vancouver, Canada).
10. Water bath (40 °C).

2.2. Preparation of LPD Nanoparticles

1. Tris-EDTA buffer (TE): 10 mM Tris–HCl, 1 mM EDTA, pH 8.0.
2. Plasmid DNA pEGFP-N1: 1 mg/mL in TE buffer.
3. Protamine sulfate: 1 mg/mL in sterilized distilled water as stock solution.
4. Nanoparticle analyzer (Coulter N4 Plus, Coulter Electronics, Luton, UK).
5. Clear-sided quartz cuvettes (3 mL).
6. Malvern Nanosizer ZS or equivalent instrument.
7. Carboxy-modified polystyrene latex (+50 ± 5 mV) for the calibration of the Zetasizer.
8. Philips EM 208 transmission electron microscope (TEM).

2.3. Cell Culture

1. Human lung carcinoma cells (A549) (European Collection of Animal Cell Cultures).
2. Dulbecco's minimum essential medium (DMEM) (Gibco BRL, Life Technologies Ltd.).
3. Fetal bovine serum (FBS, Gibco BRL, Life Technologies Ltd).
4. 1 % Penicillin/streptomycin (5,000 IU/mL).
5. T-25 (25 cm^2) cell culture flasks.
6. 37 °C humidified CO_2 incubator.

2.4. In Vitro Gene Transfection Using LPD Nanoparticles

1. 24-well tissue culture plates.
2. Culture medium: DMEM, 10 % FBS, 1 % penicillin/streptomycin.
3. Phosphate buffered saline (PBS) buffer: dissolve 8 g of NaCl, 0.2 g of KCl, 1.44 g of Na_2HPO_4, and 0.24 g of KH_2PO_4 in 1,000 mL of distilled water, pH 7.4. Autoclave. Pre-warm to 37 °C before use.
4. Trypsin-EDTA (1×, containing 0.5 g/L trypsin, 0.2 g/L EDTA, Gibco BRL, Life Technologies Ltd).
5. Microscope counting chamber (hemocytometer) for counting cells.
6. 37 °C humidified CO_2 incubator.

2.5. Flow Cytometry Analysis Components and Determining Transfection Efficiency

1. PBS buffer: dissolve 8 g of NaCl, 0.2 g of KCl, 1.44 g of Na_2HPO_4, and 0.24 g of KH_2PO_4 in 1,000 mL of distilled water, pH 7.4. Autoclave. Pre-warm to 37 °C before use.
2. Trypsin-EDTA (1×, Gibco BRL, Life Technologies Ltd).
3. Flow cytometry test tubes (Becton Dickinson).
4. Sheath fluid: PBS buffer containing 1 mM EDTA, 25 mM HEPES, 1 % heat-inactivated FBS.

5. WinMDI software package.
6. Flow cytometry instrument, e.g., FACSCan (Becton Dickinson Diagnostic Systems).
7. Inverted fluorescence microscope (Olympus IX 70, Japan).

3. Methods

3.1. Preparation of 1,2-Dioleoyl-3-Trimethylammonium-Propane Cationic Liposomes

1. Dissolve 10 mg of DOTAP in 5 mL of chloroform in a clean, dry, and previously weighed round-bottomed flask (see Note 1).
2. Place the flask in a rotary evaporator at high vacuum and rotate at 60 rpm, 40 °C (higher than the phase transition temperature Tc of DOTAP) for 90 min.
3. After evaporation, introduce nitrogen into the flask to remove any residual chloroform (see Note 1).
4. Warm sterile distilled water up to 40 °C. Prepare 1 mg/mL DOTAP suspension by adding 10 mL of the pre-warmed sterilized distilled water into the flask and then mixing gently for few seconds.
5. Incubate the flask in a water-bath (40 °C) for 30 min to allow multilamellar liposomes to form.
6. Lipid extrusion by enforcing multilamellar liposomes pass through the Extruder™ (100 nm pore size polycarbonate membranes) under a pressure of 2–3 bars (see Note 2).
7. Store the extruded DOTAP liposomes in a glass bottle filled with nitrogen at 4 °C for the preparation of LPD complex (see Note 3).

3.2. Preparation of LPD Nanoparticles

1. Add 20 μL of plasmid DNA (1 mg/mL) into 880 μL of sterile distilled water in an eppendorf (1.5 mL), mix gently but thoroughly. Do not vortex the solution.
2. Add 40 μL of protamine sulfate (1 mg/mL) into the DNA solution. Tightly close and invert the tubes for 10 min to mix the solution. DNA will be condensed by protamine during this procedure.
3. Add 60 μL of the 1 mg/mL DOTAP liposomes (from Subheading 3.1) into the condensed DNA solution. Gently pipette up and down two to three times and incubate the suspension at room temperature for 10 min. Do not vortex the LPD suspension (see Note 4). The prepared LPD complexes are ready for use in gene transfection.
4. Transfer 0.1 mL of LPD suspensions into a clear-sided cuvette containing 2.9 mL of sterilized distilled water, and mix thoroughly. Insert the cuvette into nanoparticle analyzer, set the

light scattering angle to 90° and determinin the particle size (see Note 5).

5. Calibrate Zetasizer instrument using an electrophoretic standard (carboxy-modified polystyrene latex +50 ± 5 mV).
6. Dilute LPD suspensions (0.1 mL) with 2.9 mL of fresh distilled degassed water, and then introduce the diluted LPD suspension into Zetasizer to analyze the particle surface charge (see Note 6).

3.3. Cell Culture

1. Culture human lung epithelial carcinoma cells (A549) using T-25 (25 cm^2) flasks at 37 °C with 95 % air/5 % CO_2 in humidified atmosphere. Use DMEM fortified with 10 % FBS and 1 % penicillin/streptomycin (5,000 IU/mL) (see Note 7).
2. Change the medium every other day and allow the cells to grow for 5 days or untill full confluency before proceeding with gene transfection.

3.4. In Vitro Gene Transfection Using LPD Nanoparticles

1. To harvest the cells, remove the culture medium from the flask and wash twice with pre-warmed PBS.
2. Trypsinize the cells with 1 mL of trypsin-EDTA. Gently rotate the flask for 60 s, aspirate excess trypsin solution and incubate at 37 °C for 10 min (see Note 8).
3. Stop the trypsinisation by adding 10 mL of full medium (DMEM, 10 % FBS, 1 % penicillin/streptomycin) into the flask. Gently pipette the medium up and down gently several times to resuspend all the cells.
4. Count the cells.
5. For the gene transfection, seed the cells into 24-well plates at a density of 40,000 cells per cm^2. Add 1 mL of the culture medium to each well.
6. Incubate the cells at 37 °C for 72 h in 5 % CO_2 in humidified incubator until they reach 80 % confluence.
7. Aspirate the wells and wash the cells with pre-warmed PBS twice (see Note 9).
8. Refill the wells with 950 μL of DMEM (no FBS or antibiotics) and then transfer 50 μL of LPD complex solution into each well (made as described in Subheading 3.2). Keep at least one well of untreated cells as a negative control.
9. Incubate the transfected cells at 37 °C in 5 % CO_2 in humidified incubator for 6 h.
10. Aspirate the wells, and wash the cells with pre-warmed PBS twice.
11. Add 1 mL of full medium (DMEM containing 10 % FBS, 1 % penicillin/streptomycin antibiotics) into each well.
12. Culture the cells at 37 °C in a 5 % CO_2 in humidified incubator for another 42 h to allow the gene expression.

3.5. Flow Cytometry Analysis Components and Determining Transfection Efficiency

1. Wash the cells twice with pre-warmed PBS after the wells are aspirated.
2. Detach the cells by adding 0.3 mL of trypsin into each well and gently shaking for no longer than 30 s at room temperature (see Note 10).
3. Remove excess trypsin solution and incubate the plates at 37 °C for 10 min.
4. Stop the trypsinisation by adding 0.3 mL of the culture medium into each well. Gently pipette the medium up and down gently several times to resuspend all the cells (see Note 11).
5. Transfer the cells from each well into individual flow cytometry test tubes. Use untreated cells as negative control. Keep cells on ice at all times prior to the analysis.
6. On FACscan equipment, set the excitation wavelength to 488 nm and the band pass filter (510/40), these settings will be suitable for detecting green fluorescence emitted from GFP.
7. Collect 10,000 events (cells) per sample for statistic analysis (Fig. 1). Always run negative controls first.
8. Use winMDI software to analyze the data collected from the flow cytometry. Estimate the efficiency of transfection with LPD complexes.
9. Use inverted fluorescence microscope to visualize Green Fluorescence Protein confirm the gene transfection and expression.

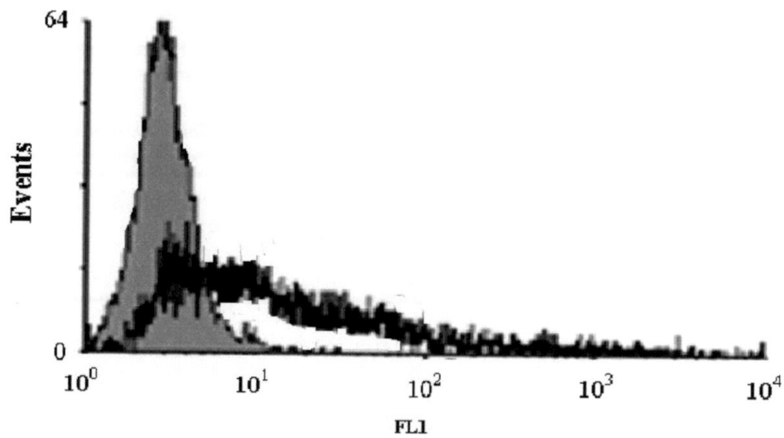

Fig. 1. Green fluorescent protein (GFP) reporter gene expression and GFP-associated fluorescence (FL1-H) quantified by flow cytometry. The percentage of A549 cells displaying GFP-associated fluorescence (FL1-H) is quantified by flow cytometry. A fluorescence region is established containing 1 % of the population of untreated control cells (*gray* filled histogram), and the percentage of cells moving into this marked region for treated cells (*black* unfilled histogram) can be calculated.

4. Notes

1. Prior to adding DOTAP into the flask, use the hair dryer to blow the round-bottomed flask thoroughly to ensure it is completely dried. After the evaporation of chloroform, introduce the nitrogen to DOTAP that is attached to the flask wall for enough long time to completely get rid of chloroform, which can be verified by putting the flask containing the DOTAP on the balance and no mass loss can be observed.
2. The gas pressure should be increased slowly form 0 to 2–3 bars to avoid accidental loss of the liposome suspension.
3. The storage time for extruded DOTAP at 4 °C should not normally exceed 2 weeks.
4. Do not vortex the LPD suspension. Only gentle pipetting is allowed. Vortexing can cause severe aggregation of LPD complexes to form precipitates, which will reduce the cellular uptake and therefore the transfection efficiency.
5. The larger particles (>200 nm) are difficult for cell to uptake, whilst smaller (~100 nm) particles are most suitable for gene transfection.
6. The surface charge for LPD complex should be positive. The zeta potential is anticipated to be within a range +10–40 mV. LPD with negative zeta potential will not be suitable for cell uptake and gene transfection.
7. Approximately 8–10 mL of culture medium should be used for T-25 flasks. As a general rule, use 0.1–0.5 mL of medium per cm^2 area of the flask used.
8. One minute of trypsin treatment is usually taken for cell detachment for elongated time of treatment might kill the cells. During the incubation time, remove the flask from the incubator and observe the cells under the microscope from time to time. When cells appeared round, distinct and free flowing, remove the flask from the incubator, and promptly add full culture medium to the cells to stop the trypsinization reaction.
9. Aspirate the wells completely after the last wash. The trace of serum in the PBS wash may inhibit the gene transfection.
10. Absolutely control the trypsinization time. Long time treatment by trypsin can cause cells detached from the well surface and subsequent loss when remove the excessive trypsin.
11. Ensure that all the cells are dispersed in medium. Any aggregates of cells may block the flow cytometer nozzle during the flow cytometry test.

References

1. Emery DE (2004) Gene therapy for genetic diseases: on the horizon. Clin Appl Immunol Rev 4:411–422
2. Hashida M, Nishikawa M, Yamashita F, Takakura Y (2001) Cell-specific delivery of genes with glycosylated carriers. Adv Drug Deliv Rev 52:187–196
3. Schwiebert EM, Boyd C, Scholte BJ (2004) Gene delivery systems-gene therapy vectors for cystic fibrosis. J Cystic Fibrosis 3:203–212
4. Patel DH, Misra A (2011) Gene delivery using viral vectors. Challenges Deliv Ther Gen Proteom 5:207–270
5. Lee RJ, Huang L (1997) Lipidic vector systems for gene transfer. Crit Rev Ther Drug Carrier Syst 14:173–206
6. Park TG, Jeong JH, Kim SW (2006) Current status of polymeric gene delivery systems. Adv Drug Deliv Rev 58:467–486
7. Pouton CW, Seymour LW (2001) Key issues in non-viral gene delivery. Adv Drug Deliv Rev 46:187–203
8. Liu F, Huang L (2002) Development of non-viral vectors for systemic gene delivery. J Control Release 78:259–266
9. Woodle MC, Scaria P (2001) Cationic liposomes and nucleic acids. Curr Opin Colloid Interface Sci 6:78–84
10. Zabner J (1997) Cationic lipids used in gene transfer. Adv Drug Deliv Rev 27:17–28
11. Pedroso de Lima MC, Simões S, Pires P, Faneca H, Düzgüne N (2001) Cationic lipid-DNA complexes in gene delivery: from biophysics to biological applications. Adv Drug Deliv Rev 47:277–294
12. Zhdanov RI, Podobed OV, Vlassov VV (2002) Cationic lipid-DNA complexes-lipoplexes-for gene transfer and therapy. Bioelectrochemistry 58:53–64
13. Morille M, Passirani C, Vonarbourg A, Clayreul A, Benoit JP (2008) Progress in developing cationic vectors for non-viral systemic gene therapy against cancer. Biomaterials 29:3477–3496
14. Moret I, Peris E, Guillem VM, Benet M, Revert F, Dasí F, Crespo A, Aliño SF (2001) Stability of PEI–DNA and DOTAP–DNA complexes: effect of alkaline pH, heparin and serum. J Control Release 76:169–181
15. Remaut K, Sanders NN, Fayazpour F, Demeester J, DeSmedt SC (2006) Influence of plasmid DNA topology on the transfection properties of DOTAP/DOPE lipoplexes. J Control Release 115:335–343
16. Ahearn A, Malone R (1999) Models of cationic liposome mediated transfection. Gene Ther Mol Biol 4:159–170
17. Zauner W, Farrow NA, Haines AM (2001) *In vitro* uptake of polystyrene microspheres: effect of particle size, cell line and cell density. J Control Release 71:39–51
18. Tsai JT, Furstoss KJ, Michnick T, Sloane DL, Paul RW (2002) Quantitative physical characterization of lipid-polycation-DNA lipoplexes. Biotechnol Appl Biochem 36:13–20

Chapter 28

Crotamine: A Novel Cell-Penetrating Polypeptide Nanocarrier with Potential Anti-Cancer and Biotechnological Applications

Mirian A.F. Hayashi, Eduardo B. Oliveira, Irina Kerkis, and Richard L. Karpel

Abstract

Crotamine is a basic, 42-residue polypeptide derived from snake venom that has been shown to possess cell-penetrating properties. Crotamine forms nanoparticles with a variety of DNA and RNA molecules, and crotamine-plasmid DNA nanoparticles are selectively delivered into actively proliferating cells in culture or in mice. As such, these nanoparticles could form the basis for a nucleic acid drug-delivery system. Here we describe the preparation, purification, and biochemical and biophysical analysis of venom-derived, recombinant, chemically synthesized, and fluorescent-labeled crotamine; the formation and characterization of crotamine-DNA and -RNA nanoparticles; and the delivery of these nanoparticles into cells and animals.

Key words: Cell-penetrating peptides, Amphiphatic peptides, Protein-DNA nanoparticles

1. Introduction

Progress in anticancer drug delivery has been hindered by cell membrane impermeability and the lack of a selective marker for tumor cells. A number of therapeutic gene- and siRNA-delivery vehicles have been developed but their utility is often restricted by entrapment following endocytosis, the need for tight association with the cargo, elusive cell entry mechanisms, and/or inadequate pharmacokinetic behavior, as well as nonspecific penetration in most organs and rapid clearance from the circulation as a result of binding to anionic plasma (1, 2). One type of carrier, cell-penetrating peptides (CPPs), can transport drugs and even nanoparticles into mammalian cells without requiring specific receptors, both in vitro

Mikhail Soloviev (ed.), *Nanoparticles in Biology and Medicine: Methods and Protocols*, Methods in Molecular Biology, vol. 906, DOI 10.1007/978-1-61779-953-2_28,

a

```
  1  CGGCACGAGCCAGAACCAGTCTCAGCATGAAGATCCTTTATCTGCTGTTCGCATTTCTTT
                                M   K   I   L   Y   L   L   F   A   F   L     11
 61  TCCTTGCATTCCTGTCTGAACCAGGGAATGCCTATAAACAGTGTCATAAGAAAGGAGGAC
      F   L   A   F   L   S   E   P   G   N   A   Y   K   Q   C   H   K   K   G   G     31
121  ACTGCTTTCCCAAGGAGAAAATATGTATTCCTCCATCTTCTGACTTTGGGAAGATGGACT
      H   C   F   P   K   E   K   I   C   I   P   P   S   S   D   F   G   K   M   D     51
181  GTCGATGGAGATGGAAATGCTGTAAAAAGGGAAGTGGAAAATAATGCCATCTCCATCTAG
      C   R   W   R   W   K   C   C   K   K   G   S   G   K   *                         65
241  GACCATGGATATCTTCAAGATATGGCCAAGGACCTGAGAGTGCCGCCTGCTATCGCTTTA
301  TCTTTCTTTATCTAAATAAAATTGCTACCTATCAACGCTAAAAAAAAAA
```

b

Fig. 1. Crotamine sequences. (**a**) Nucleotide and deduced amino acid sequences of crotamine precursor full-length cDNA clone AF044674 [http://www.ncbi.nlm.nih.gov/nuccore/AF044674]. The cDNA (349 bp), is numbered at the *left side* of the sequences, and the encoded crotamine precursor sequence is presented with the corresponding amino acid sequence, with one-letter symbols and numbered, at the right side. The 22 amino acid signal peptide is *underlined*, and the 42 amino acid sequence constituting the mature crotamine with the terminal lysine (K), which is post-translationally removed, is in *bold letters*. (**b**) Sequence of the mature crotamine indicating the three disulfide bridges.

and in vivo, providing advantages for diagnostic and therapeutic applications (3–9). Thus, CPPs may have important roles in the development of novel treatments directed against a growing variety of targets and disease areas, including cardiology, pain, and stroke (10).

Crotamine is a 42-residue, highly basic CPP derived from the venom of the South American rattlesnake *Crotalus durissus terrificus* (11–14), see Fig. 1.

It has a unique specificity for actively proliferating (AP) cells (11, 12), and is internalized essentially by endocytosis even when carrying a cargo (14), with subsequent concentration-dependent leakage to the cell cytoplasm (14, 15). Crotamine selectively delivers plasmid DNA into AP mammalian cells (2, 4). The ability to transport cargoes without need of covalent bond formation (14) and the preferential cytotoxic effect against tumoral cells (15) suggests a potential application of crotamine for cancer therapy and/or diagnosis. Recent experiments have shown that remote subcutaneous tumors in nude mice are efficiently targeted by crotamine (Nascimento et al., in preparation). In the context of nanocarrier technology, the potential of crotamine to assist in the specific delivery of a variety of macromolecules has brought much attention to this natural nanocarrier (13).

When interacted with DNA, crotamine forms high molecular weight nanoparticles. Dynamic light scattering (DLS) measurements indicate particle diameters of 200–500 nm, which increase with increasing DNA length and time of incubation, and decrease with increasing ionic strength (Karpel et al., in preparation). The formation of high molecular weight nanoparticles is likely a consequence of the very basic character of the protein (net charge at neutral pH is 8+), so that the protein could simultaneously bind more than one DNA molecule. The size of the nanoparticles could be crucial to the DNA delivery properties of the protein (16). In the following sections, we detail the preparation and use of crotamine nanoparticles.

2. Materials

All solutions should be prepared using ultrapure water (18 MΩ cm at 25 °C) and analytical grade reagents. Animals are housed in a temperature (21 ± 2 °C) and light-controlled (12/12 h light/dark cycle) room and are allowed to have free access to water and food *ad libitum*. All animals should be handled under ethical conditions according to international rules of animal care, stated by the International Animal Welfare Recommendations, and in accordance with the Guidelines for the Use of Animals in Biochemical Research (17).

2.1. Purification of Native Crotamine

1. Dry *C. d. terrificus* venom.
2. 0.25 M ammonium formate buffer pH 3.5.
3. 1 M Tris base solution.
4. Loading buffer: 0.04 M Tris buffer, containing 0.1 M NaCl pH 8.8.
5. Elution buffer: 0.04 M Tris buffer, containing 0.65 M NaCl pH 8.8.
6. 4 mL of CM-Sepharose Fast Flow resin (GE Healthcare) packed into a 1.2 × 6.0 cm disposable plastic column.
7. Refrigerated centrifuge for 50 mL disposable plastic tubes (Sorvall™ Legend RT Centrifuge Series, Thermo Fisher Scientific Inc.) (see Note 1).
8. Benzoylated dialysis tubing (molecular weight cutoff 3,000).
9. UV spectrophotometer, capable of 280 nm, and 1 cm quartz cuvette.
10. Freeze dryer (Labconco, 4.5 L model).

2.2. Expression and Purification of Recombinant Crotamine

1. *C. d. terrificus* venom gland total cDNA.
2. Expression plasmid vector TAGZyme pQE-1.
3. Heat shock competent bacteria *E. coli* M15[pREP4].
4. Ni-NTA magnetic agarose beads (Qiagen).

5. Crotamine sequence-specific PCR primers: sense (5′-AGAGGA GTATACAAACAGTG-3′) and antisense (5′-ATGAGAAA GCTTTTATCCACTTC-3′). These are based on the cDNA encoding full-length crotamine [GenBank: AF044674] and contain the restriction sites for *BstZ*17 and *Hind* III, respectively. Restriction enzymes *BstZ*17, *Pvu* II and *Hind* III.
6. Taq DNA polymerase and PCR buffer (Invitrogen), 50 mM $MgCl_2$, and 10 mM dNTPs for PCR.
7. Eppendorf Mastercycler® gradient thermal cycler or similar instrument.
8. Plasmid expression vector TAGZyme pQE-1(Qiagen).
9. GFX PCR DNA and Gel Band Purification kits (GE Healthcare) (see Note 2).
10. Versatile high-capacity refrigerated centrifuge CR7 (Hitachi Koki Co. Ltd., Tokyo, Japan) (see Note 1).
11. Liquid LB medium.
12. 100 mg/mL ampicilin.
13. 25 mg/mL of kanamycin.
14. 1 M isopropyl-β-D-thiogalactopyranoside (IPTG) in sterile Milli-Q water.
15. Resuspension buffer: 20 mM Tris–HCl pH 8.5, 100 mM KCl, 20 mM imidazole, 10 % (v/v) glycerol.
16. Elution buffer: 20 mM Tris–HCl pH 8.5, 100 mM KCl and 100 mM imidazole.
17. Centricon filter system YM-3 and YM-10 (Millipore).
18. Sonicator (Vibra-Cell™ Ultrasonic Processor, SONICS). Typical use involves 6–8 cycles of 30 s each, with an amplitude of 60–90 %, all procedures should be carried in ice bath.
19. Ni-NTA agarose beads Sepharose® CL-6B (Qiagen). One mL of resin should be sufficient for the binding of approximately 5–10 mg of 6×His-tagged protein.
20. Resuspension buffer for Ni-NTA agarose beads: 20 mM Tris–HCl pH 8.5, 100 mM KCl, 20 mM imidazole and 10 % (v/v) glycerol.
21. Elution buffer for Ni-NTA agarose beads: 20 mM Tris–HCl pH 8.5, 100 mM KCl and 100 mM imidazole.

2.3. Crotamine-Nucleic Acid Nanoparticle Formation and Characterization

1. Genetic material such as DNA, RNA or siRNA (for instance, a circular or linear plasmid vector pEGFP-N1, Clontech).
2. Pure crotamine (98 % purity or higher).
3. PBS buffer for crotamine-DNA complexes: 50 mM potassium phosphate containing 150 mM NaCl pH 7.2.

4. Agarose gel electrophoresis system and UV transluminator (UVP) (see Note 3).
5. Bench vortex.
6. DLS instrument (Zetasizer Nano ZS) or equivalent.
7. DLS solvent: 20 mM HEPES pH 7.7, 10 mM NaCl and 0.1 mM EDTA. Filter through 0.22 μm Millex-GV filters (Millipore Corp.).

2.4. Delivery of Crotamine-Nucleic Acid Nanoparticles into Cells and Animals

2.4.1. In Vitro Cell Transfection

1. Tissue culture plastic plates 6 × 35 mm well (Corning) suitable for fluorescence microscopy or confocal microscopy.
2. Murine melanoma cells B16-F10: culture in Dulbecco's modified Eagle's medium (DMEM) supplemented with 10 % Heat Inactivated Fetal Bovine Serum (FBS) from Gibco®.
3. Inverted fluorescence microscope Axio observer A1 (Carl Zeiss) or a confocal microscope, for example, a LSM 510 (Carl Zeiss) or similar microscopes.

2.4.2. In Vivo Delivery

1. Animals: C57BL6 or nude mice. Maintain the animals at (21 ± 2°C) and in a light-controlled environment (12/12 h light/dark cycle).
2. Tissue-Tek embedding matrix (Sakura).
3. Cryo-microtome (Model CM 1100, Leica).
4. Inverted fluorescence microscope for example, an Axio observer A1 (Carl Zeiss) or a confocal microscope, for example, a LSM 510 (Carl Zeiss).

3. Methods

Crotamine peptide may be purified from the venom of *C. d. terrificus*, or obtained by recombinant means or by chemical synthesis. The use of recombinant or chemical synthetic methods offers the possibility of obtaining crotamine variants with altered and, possibly, improved functionalities. Solid phase chemical synthesis of crotamine has been achieved, and after re-folding under thiol-disulfide exchange conditions, a circular dichroism (CD) spectrum very similar to the native crotamine could be obtained for synthetic crotamine (see Fig. 2), suggesting an overall similar secondary structure for both. Successful folding of synthetic polypeptides similar to crotamine has been described under thiol-disulfide exchange conditions (18, 19). At least under the employed conditions, the recombinant crotamine expressed in bacteria showed a profile more characteristic of a disordered peptide chain, suggesting re-folding may also be necessary, see Fig. 2.

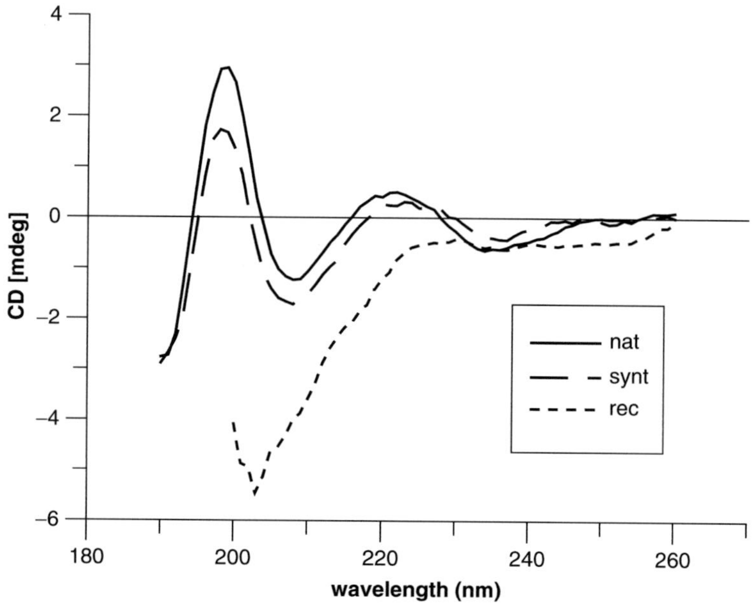

Fig. 2. Circular dichroism (CD) spectra profile of the native, synthetic, and recombinant crotamine. CD spectra of native (nat), full-length synthetic (synt) and recombinant (rec) crotamine.

3.1. Purification of Native Crotamine

Natural crotamine is obtained chromatographically from pooled dry *C. d. terrificus* venom (see Note 4).

1. Prepare the venom solution by dissolving 100 mg of dry *C. d. terrificus* venom in 2 mL of 0.25 M ammonium formate buffer pH 3.5, at room temperature.
2. Centrifuge at 500 × *g* for 15 min, at 4°C, to remove particulate matter.
3. Add 8 mL of ice-cold water dropwise to the magnetically stirred venom solution to precipitate the contaminant proteins.
4. Incubate for 2 h at 4 °C, then centrifuge the solution at 500 × *g* for 10 min. Discard the precipitate containing the bulk of contaminant venom proteins.
5. Add 1 M Tris base solution dropwise to raise the pH of the crotamine-rich supernatant to 8.8. Conduct this procedure at room temperature.
6. Equilibrate the 1.2 × 6.0 cm column packed with 4 mL of CM-Sepharose Fast Flow resin using 20 mL of loading buffer.
7. Add the crotamine-rich solution to the column, at flow rate of 1 mL/min, followed by 25 mL of loading buffer, to wash out unbound proteins.
8. Elute crotamine by adding 35 mL of elution buffer. Collect 1 mL fractions throughout the experiment. Crotamine is recovered in a narrow protein peak in a highly purified form (see Fig. 3, lane 3).

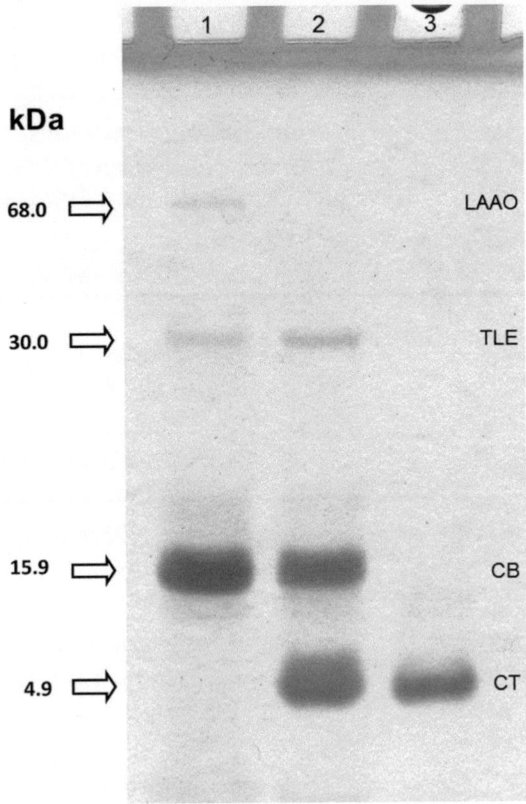

Fig. 3. SDS-polyacrylamide gel (SDS-PAGE) analyses of purified crotamine and two individual *Crotalus durissus terrificus* venoms. Twenty microgram of venom extracted from specimens that do not secrete (lane 1) or secrete (lane 2) crotamine, and 10 μg of purified crotamine (lane 3) were separated on 16 % SDS-PAGE gel under reducing conditions. Protein bands were stained with Coomassie Blue. The most prominent protein bands of the venoms are indicated: L-amino acid oxidase (LAAO), thrombin-like enzyme (TLE), crotoxin, basic chain (CB) and crotamine (CT).

9. Estimate protein concentration by measuring absorbance at 280 nm (OD_{280nm} = 0.62 corresponds to crotamine concentration of 0.25 mg/mL using 1 cm quartz cuvette) (see Note 5).
10. Dialyse crotamine solution at 4 °C against 100 volumes distilled water for 15 h, with two changes, using benzoylated dialysis tubing.
11. Freeze the salt-free crotamine solution in dry ice/ethanol bath and lyophilize it.

Lyophilized pure crotamine can be stored at 4 or −20 °C.

3.2. Expression and Purification of Recombinant Crotamine

1. Amplify by PCR the coding region of crotamine cDNA using crotamine sequence-specific primers.
2. Assemble PCR reaction in 50 μL reaction volume with 1× PCR buffer in the presence of 3 μL of $MgCl_2$ (50 mM), 1 μL of

dNTPs (10 mM), 1 μL of crotamine specific reverse and forward primers (7.5 pmol/μL) and 2.5 U of Taq DNA polymerase.

3. Amplification conditions: one pre-denaturation cycle of 94 °C for 3 min; 35 amplification cycles of 50 °C for 45 s and 72 °C for 1.5 min, followed by a final extension cycle of 7 min at 72 °C.
4. Digest the PCR product with the restriction enzymes *BstZ*17 and *Hind* III.
5. Digest the plasmid expression vector TAGZyme pQE-1 (see Note 6) with the restriction enzymes *Pvu* II and *Hind* III.
6. Purify both the double-digested crotamine cDNA insert and the linearized expression plasmid vector using GFX PCR DNA and Gel Band Purification kit.
7. Subclone the cDNA encoding the full-length crotamine in frame into the linearised expression plasmid vector TAGZyme pQE-1.
8. Transform *E. coli* competent cells with the ligation mixture and plate them on LB plates containing 100 μg/mL ampicilin and 25 μg/mL kanamycin (see Note 7).
9. Following the transformation of competent *E. coli* cells with the above ligation mixture and purification of plasmidial DNA from isolated bacteria colonies, select the recombinant clones by double-digestion with *Eco* RI and *Hind* III.
10. Confirm the in frame insertion of the crotamine cDNA and the entire cDNA nucleotide sequence by DNA sequencing.
11. For production of the recombinant crotamine transform competent bacteria *E. coli* M15[pREP4] with the expression clone or use positive clones from step 9 (above) and grow bacteria in liquid LB medium under antibiotic selection with 100 μg/mL of ampicilin and 25 μg/mL of kanamycin to OD_{600nm} ~ 0.6.
12. Induce the expression of recombinant crotamine by addition of IPTG to a final concentration of 0.4 mM.
13. Incubate in a shaker for 3 h at 30 °C, 200 rpm.
14. Harvest the bacteria by centrifugation at 4,000 × *g* for 20 min.
15. Discard the supernatant. Store the bacteria pellet at −20 °C until use.
16. For purification of the recombinant crotamine, resuspend the stored bacteria in resuspension buffer, and lyse the cells by two cycles of freezing and thawing followed by sonication.
17. Remove bacterial debris by centrifugation at 15,000 × *g* for 20–30 min.

18. Purify the recombinant crotamine by affinity chromatography using Ni-NTA agarose beads (see Note 8). Load the supernatant of the culture medium on a column packed with 1 mL Ni-NTA agarose resin. Wash the column with 25 mL of resuspension buffer.
19. Elute crotamine using 1 mL of elution buffer (see Note 9).
20. Remove the imidazole and other salts using centricon YM-3 filter system.
21. The purity and quality of the obtained crotamine sample should be determined by SDS-PAGE or MALDI-TOF MS analysis. Crotamine with 98 % purity or higher is suitable for in vitro or in vivo gene delivery.

3.3. Crotamine-Nucleic Acid Nanoparticle Formation and Characterization

1. To produce crotamine-DNA complexes observable by electrophoresis, incubate 2–4 μg of DNA, RNA, or siRNA (see Note 10) with 10 μg of crotamine in 200 μL of PBS buffer for 5 min at room temperature with occasional vortexing.
2. Confirm the complex formation by electrophoresis in an agarose gel and stain with ethidium bromide (EtBr) (see Note 11). Use 20 μL aliquot of the nanocomplex. Figure 4 shows results of a typical experiment in which different amounts of Cy5-crotamine were used to form nanocomplexes with 1 μg of 4.7-kb circular plasmid pEGFP-N1.
3. To determine nanoparticle size by DLS, use the Zetasizer and follow the instrument protocol. Transfer 1 mL of DLS solvent to a 3 mL disposable polystyrene cell. Add a small (<20 μL) aliquot of a concentrated DNA solution to a final residue concentration of nucleic acid 5–10 μM.
4. Determine size distribution as per the Zetasizer protocol. Prior to the addition of crotamine, there will be no evidence of nanoparticle formation.
5. Add 5 μL aliquots of a concentrated (80 μM) stock crotamine solution, incubate for 5 min after each addition (see Note 12), and perform size measurements. Under these conditions, unique, reproducible nanoparticle size distributions begin to appear when the [crotamine]:[DNA residue] exceeds 0.1. A typical result, with a 63-residue single-stranded DNA oligonucleotide, and a [crotamine]:[DNA residue] = 0.18, is shown in Fig. 5 (see Note 12).

3.4. Delivery of Crotamine-Nucleic Acid Nanoparticles into Cells and Animals

3.4.1. In Vitro Cell Transfection

1. Plate the cells (for instance, 4×10^5 cells for B16-F10) in 2 mL of medium into 6×35 mm well plate, using DMEM containing 10 % heat inactivated FBS. Grow the cells for ~24 h until about 70 % confluence.
2. Wash the cells with PBS once and add 1 mL of fresh medium.

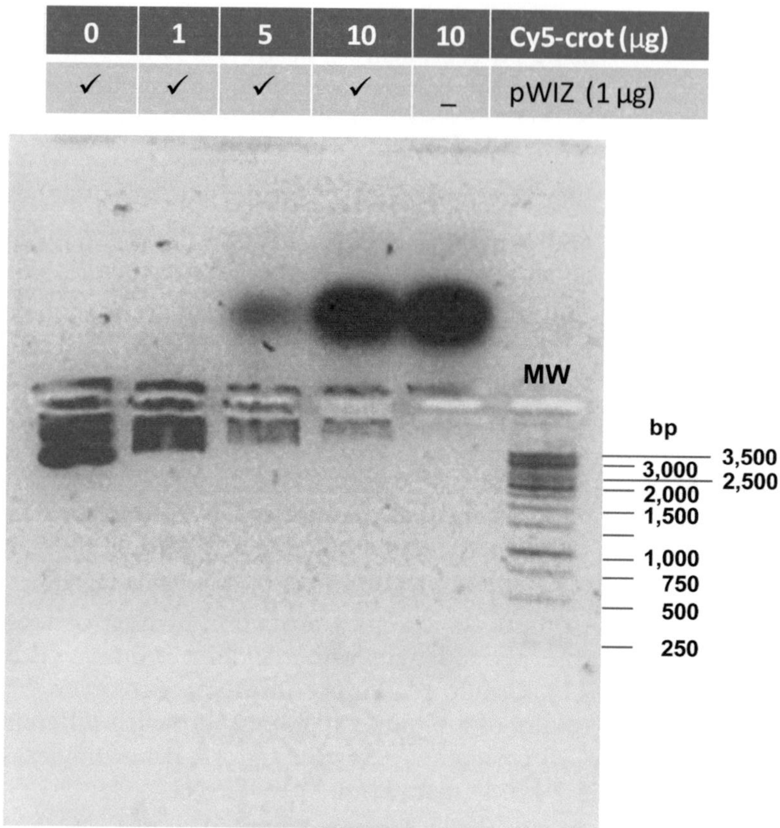

Fig. 4. Agarose gel shift assay illustrating nanoparticle formation. The amount of circular DNA pEGFP-N1 (~4.7 kb) used was fixed (1 μg) and increasing amounts of fluorescent labeled-crotamine (Cy5-crot, 0, 1, 5, 10 μg) were added as indicated in the *upper panel* of the figure. The ethidium bromide stained DNA (*lower side* of gel) was detected on UV transillumination under 302 nm midrange wavelength, while the fluorescence of Cy5-linked to crotamine was visualized at 650 nm excitation and 670 nm emission (*upper side* of gel). MW indicates the 250 bp DNA ladder.

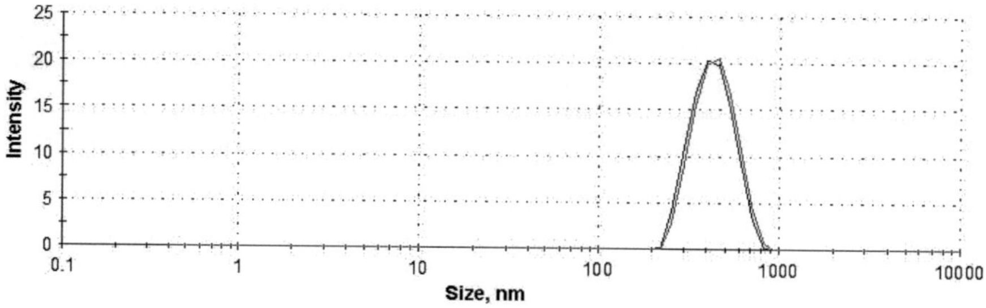

Fig. 5. Size distribution of a crotamine—single stranded DNA (63-mer) nanoparticle observed by dynamic light scattering. DNA residue concentration = 11 μM, [crotamine] = 2.0 μM. Average size of the nanoparticle formed (after two consecutive determinations by DLS) = 425 nm.

3. Add 200 μL of the peptide-DNA complexes (made as described in Subheading 3.3 using 2–4 μg of DNA, for example, pEGFP-N1 and 10 μg of crotamine) to the cultured cells dropwise.
4. Add the same amount of plasmid DNA (2–4 μg of DNA) to negative control wells.
5. Incubate these cells with the peptide-DNA complex (or DNA alone) for 6 and 24 h to allow internalization of the complex into the cell. The efficiency of the internalization may vary with the DNA-crotamine complex size and also with the cell type.
6. Following the incubation wash the cells with PBS and replace 2 mL of fresh DMEM medium containing 10 % heat inactivated FBS per well, and incubate for another 24 h.
7. Observe the cells under inverted microscope to verify the morphology, and the expression of fluorescent reporter gene as green fluorescent protein (GFP) expressed from the pEGFP-N1 vector can be observed under confocal microscope (as shown in Fig. 6) (see Notes 10 and 13).
8. Use an inverted fluorescence microscope or a confocal microscope to acquire the images under the following configurations: objective Plan-Neofluar 40×/1.3 il; Pinhole: 2.55 Airy Units; Frame Size: 512 × 512 to 1,024 × 1,024; Scan Speed: 11. The confocal optical image thickness is of about 1 μm.

3.4.2. In Vivo Delivery

1. For the in vivo transfection assays, inject 0.2 mL of complex solution (prepared as described in step 1 in Subheading 3.3) into mice intraperitoneally (i.p.). Inject control group with 0.2 mL nucleic acid solution alone. Use two groups of four animals each for experimental and control.
2. Sacrifice mice 24 h after the injection of the crotamine-DNA nanocomplexes (see Note 14).
3. The isolation and examination of living cells from bone marrow (BM) and peritoneal liquid allows direct observation of the expression of the green fluorescent reporter gene under the confocal microscope (see Note 10).
4. For organ tissues dissect and prepare 6–10 μm cryostat sections using a cryo-microtome and Tissue-Tek embedding matrix. Mount the microscope slides with antifade reagent and observe under confocal laser scanning microscope to observe the expression of the green fluorescent reporter gene (see Note 14).

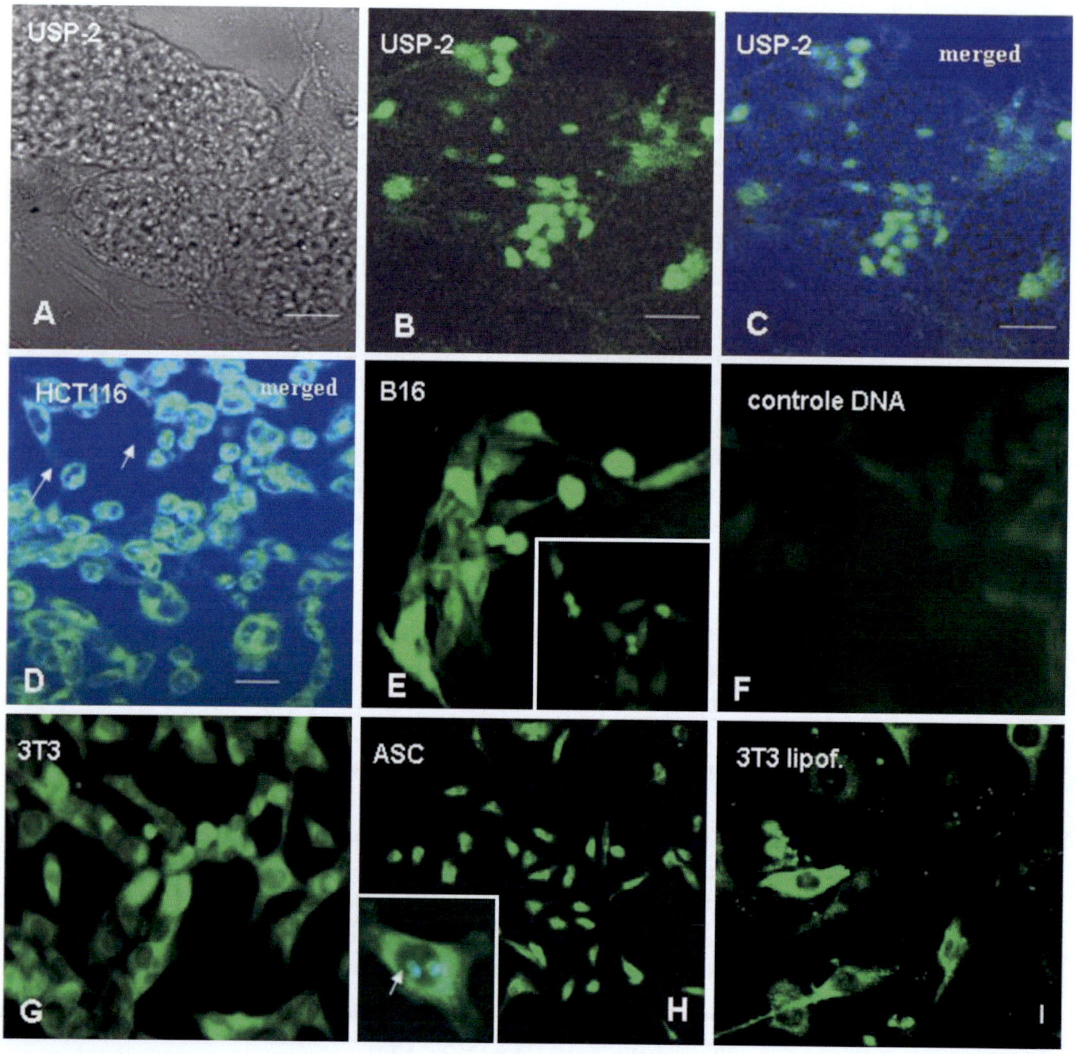

Fig. 6. In vitro GFP transfection using crotamine-DNA nanocomplexes. Different cell types and incubation times are shown. All images are obtained using epifluorescent microscopy, unless specified otherwise. (**a**–**b**) In vivo transfection (4 h) of mouse stem cells (mES), showing GFP expression in approximately 20 % of the mES cells; (**a**) differential interference contrast (DIC); (**b**) fluorescent confocal microscopy (Fcm); (**c**) merged DIC and Fcm. (**d**) Human colorectal carcinoma HCT116 cells (24 h). (**e**) Mouse melanoma B16-F10 cells (24 h). (**f**) Negative control with B16-F10 cells (24 h) treated only with plasmidial DNA. (**g**) Mouse embryo fibroblast 3T3 cells (48 h), showing the expression of the reporter protein in all cells. (**h**) Adult human ES cell (48 h) showing a homogeneous expression of the green protein. Inset showing a cell with expression in both cytoplasm and nucleus (*arrow*). (**i**) Mouse embryo fibroblast 3T3 cells (6 h) transfected with the same amount of DNA using Lipofectamine™ transfection reagents. Some cells are strongly fluorescent while others have fluorescence comparable to those in the negative control (**f**). (**a**–**d**) Bars = 100 μm. Magnification is 600× in (**e**–**g**, **i**) or 250× in (**h**).

4. Notes

1. Any alternative equipment that allows centrifugation 500 ×*g* at 4°C can be used.
2. Any alternative kit or method that allows purifying the DNA ($OD_{260/280}$ ratio around 1.8–2.0) can be used instead.
3. Visualization of DNA bands in gels stained with ethidium bromide(EtBr) can be easily performed on Benchtop UV Transilluminators under 365 nm longwave for extended periods of time avoiding photonicking damage, but for increasing the fluorescence for photodocumentation (for instance, using a digital camera or appropriate workstation), the 302 nm midrange is recommended.
4. Individual *C. d. terrificus* snakes secrete variable amounts of crotamine into the venom.
5. Purity of 98 % or higher can be attained with yield up to 15 % on a dry weight basis.
6. The TAGZyme system utilizes recombinant exoproteases, which efficiently and precisely remove N-terminal His-tags from proteins. The small size and low immunogenicity of the His-tag means that its removal is not usually required. However, there are some applications, such as structure-determination studies by X-ray or NMR, or the production of therapeutics or nanovectors, where a protein product free from vector-derived amino acids may be preferred. For more information see: http://www.qiagen.com/products/protein/purification/tagzymesystem/default.aspx.
7. For amplification of the obtained expression vector in TAGzyme, any *E. coli* strain routinely used in the laboratory can be employed for plasmidial DNA amplification. However, for expression of the recombinant protein specific strains harboring the repressor (pREP4) plasmids are suggested, for instance, bacteria *E. coli* M15[pREP4] and SG13009[pREP4]. Other strains lacking the repressor (pREP4) plasmids may be less efficient and expression will be regulated less tightly.
8. The pQE vectors, such as the TAGZyme pQE-1 used here, contain 6×His sequence 5′ upstream to the cloning region and allow fast and efficient production of heterologously expressed 6×His-tagged proteins. One-step purification of the His-tagged protein can be achieved by affinity chromatography on nickel-nitrilotriacetic acid (Ni-NTA) column.
9. Alternatively, release the recombinant crotamine by proteolytic digestion with DAPase, Qcyclase, and pGAPase (Qiagen). The method is based on the use of dipeptidyl aminopeptidase I (DAPase Enzyme) alone or in combination with glutamine

cyclotransferase (Qcyclase Enzyme) and pyroglutamyl aminopeptidase (pGAPase Enzyme). These recombinant enzymes contain a C-terminal His tag and can therefore be bound to Ni-NTA matrices. This allows their removal from the reaction solution by Immobilized-Metal Affinity Chromatography (IMAC). This feature has been utilized in the design of a simple process consisting of aminopeptidase cleavage followed by subtractive IMAC. The use of exopeptidases in the TAGZyme System eliminates the risk of endoproteolytic cleavage within the body of the protein, and the rapid rate of reaction means that sensitive proteins can be processed even at 4 °C in a matter of hours. These proteolytic enzymes can be removed from the samples by filtration using centricon YM-10 filter system.

10. We have found that saturation of nucleic acids can be achieved with one crotamine molecule per five nucleotide residues, so the amount of crotamine and genetic material might be adjusted depending of the size of the gene [for instance, the amounts suggested in Subheading 3.3 is for a 4.7-kb circular or linear plasmid vector pEGFP-N1 (Clontech) containing a reporter gene].
11. The analysis in EtBr stained agarose gel electrophoresis to confirm the crotamine-DNA complex formation is recommended to ensure formation of the complexes.
12. Nanoparticle size was found to increase with increasing [crotamine]:[DNA] and DNA length, as well as with time of incubation. For this reason, all samples were incubated a standard 5 min prior to size determination. Nanoparticles have been observed to form with both double- and single-stranded DNA. Their diameters were dependent on ionic strength.
13. It is also possible to plate the cells on glass slides and after the transfection, fix them and observe the fixed cells instead of living cells. The fixed cells have the advantage of allowing the observations at higher magnification, and this is the only way to observe the intracellular localization of cargo and the transport vector crotamine, if it is fluorescently labeled.
14. For visualization of biodistribution of fluorescently labeled crotamine (nanocarrier alone), animals can be sacrificed as early as 15 min after i.p. injection, and it might be detectable even 5 h after injection (14, 20), as suggested by non-invasive real time optical imaging (Nascimento et al., in preparation).

Acknowledgements

M.A.F.H. is grateful for the continuous grant support from FAPESP (Fundação de Amparo à Pesquisa do Estado de São Paulo) and CNPq (Conselho Nacional de Desenvolvimento Científico e Tecnológico), Brazil. R.L.K. thanks the University of Maryland Baltimore County Designated Research Initiative Fund (DRIF) for support, and Ms. Elizabeth Plum for expertly providing the DLS data.

References

1. Fonseca SB, Pereira MP, Kelley SO (2009) Recent advances in the use of cell-penetrating peptides for medical and biological applications. Adv Drug Deliv Rev 61:953–964
2. Andaloussi SEL, Lehto T, Mäger I, Rosenthal-Aizman K, Oprea II, Simonson OE, Sork H, Ezzat K, Copolovici DM, Kurrikoff K, Viola JR, Zaghloul EM, Sillard R, Johansson HJ, Hassane FS, Guterstam P, Suhorutšenko J, Moreno PMD, Oskolkov N, Hälldin J, Tedebark U, Metspalu A, Lebleu B, Lehtiö J, Smith CIE, Langel Ü (2011) Design of a peptide-based vector, PepFect6, for efficient delivery of siRNA in cell culture and systemically *in vitro*. Nucleic Acids Res 39:3972–3984
3. Howl J, Nicholl ID, Jones S (2007) The many futures for cell-penetrating peptides: how soon is now? Biochem Soc Trans 35:767–769
4. Deshayes S, Morris MC, Divita G, Heitz F (2005) Cell-penetrating peptides: tools for intracellular delivery of therapeutics. Cell Mol Life Sci 62:1839–1849
5. Gupta B, Levchenko TS, Torchilin VP (2005) Intracellular delivery of large molecules and small particles by cell-penetrating proteins and peptides. Adv Drug Deliv Rev 57:637–651
6. Lee JY, Choi YS, Suh JS, Kwon YM, Yang VC, Lee SJ, Chung CP, Park YJ (2011) Cell penetrating chitosan/doxorubicin/TAT conjugates for efficient cancer therapy. Int J Cancer 128:2470–2480
7. Liu BR, Li JF, Lu SW, Leel HJ, Huang YW, Shannon KB, Aronstam RS (2010) Cellular internalization of quantum dots noncovalently conjugated with arginine-rich cell-penetrating peptides. J Nanosci Nanotechnol 10:6534–6543
8. Sawant R, Torchilin V (2011) Intracellular delivery of nanoparticles with CPPs. Methods Mol Biol 683:431–451
9. Choi YS, Lee JY, Suh JS, Lee SJ, Yang VC, Chung CP, Park YJ (2011) Cell penetrating peptides for tumor targeting. Curr Pharm Biotechnol 12(8):1166–1182
10. Johnson RM, Harrison SD, Maclean D (2011) Therapeutic applications of cell—penetrating peptides. Methods Mol Biol 683:535–551
11. Kerkis A, Kerkis I, Radis-Baptista G, Oliveira EB, Vianna-Morgante AM, Pereira LV, Yamane T (2004) Crotamine is a novel cell-penetrating protein from the venom of rattlesnake *Crotalus durissus terrificus*. Faseb J 18:1407–1409. Full text at http://www.fasebj.org/cgi/reprint/03-1459fjev1
12. Kerkis A, Hayashi MA, Yamane T, Kerkis I (2006) Properties of cell penetrating peptides (CPPs). IUBMB Life 58:7–13
13. Kerkis I, Silva FS, Pereira A, Kerkis A, Radis-Baptista G (2010) Biological versatility of crotamine—a cationic peptide from the venom of a South American rattlesnake. Expert Opin Investig Drugs 19:1515–1525
14. Nascimento FD, Hayashi MA, Kerkis A, Oliveira V, Oliveira EB, Radis-Baptista G, Nader HB, Yamane T, dos Santos Tersariol IL, Kerkis I (2007) Crotamine mediates gene delivery into cells through the binding to heparan sulfate proteoglycans. J Biol Chem 282:21349–21360
15. Hayashi MA, Nascimento FD, Kerkis A, Oliveira V, Oliveira EB, Pereira A, Radis-Baptista G, Nader HB, Yamane T, Kerkis I, Tersariol IL (2008) Cytotoxic effects of crotamine are mediated through lysosomal membrane permeabilization. Toxicon 52:508–517
16. Choi HS, Kim HH, Yang JM, Shin S (2006) An insight into the gene delivery mechanism of the arginine peptide system: role of the peptide/DNA complex size. Biochim Biophys Acta 1760:1604–1612

17. Giles AR (1987) Guidelines for the use of animals in biomedical research. Throm Haemost 58: 1078–84
18. Wu Z, Hoover DM, Yang D, Boulègue C, Santamaria F, Oppenheim JJ, Lubkowski J, Lu W (2003) Engineering disulfide bridges to dissect antimicrobial and chemotactic activities of human beta-defensin 3. Proc Natl Acad Sci USA 100:8880–8885
19. Wu Z, Powell R, Lu W (2003) Productive folding of human neutrophil β-defensins in vitro without the pro-peptide. J Am Chem Soc 125:2402–2403
20. Pereira A, Kerkis A, Hayashi MA, Pereira AS, Silva FS, Oliveira EB, Prieto da Silva AR, Yamane T, Rádis-Baptista G, Kerkis I (2011) Crotamine toxicity and efficacy in mouse models of melanoma. Expert Opin Investig Drugs 20(9):1189–1200

Chapter 29

Elastin-Based Nanoparticles for Delivery of Bone Morphogenetic Proteins

Raul Machado, Paulo C. Bessa, Rui L. Reis, José Carlos Rodriguez-Cabello, and Margarida Casal

Abstract

Biocompatibility, stability, and efficiency remain among the major goals in the development of new drug delivery systems. Herein we describe a facile method for encapsulation of Bone Morphogenetic Protein-2 (BMP-2) at high concentrations and in mild conditions by exploiting the thermo-responsiveness of an elastin-like polymer. Chemically synthesized poly(VPAVG) demonstrates excellent biocompatibility, stability at room temperature, and allows the use of mild conditions and harmless solvents, such as water, for its production. Here, we describe how to use a recombinantly produced poly(VPAVG) polymer to entrap BMP-2 with a high encapsulation efficiency and the methods for in vitro bioactivity assays.

Key words: Elastin-like polymer, Poly(VPAVG), Nanoparticles, Self-assembly, Encapsulation, BMP-2

1. Introduction

Discovered in 1965, bone morphogenetic proteins (BMPs) are a group of cytokines belonging to the transforming growth factor-β (TGF-β) superfamily playing an important role in bone and cartilage formation. During the process of bone formation, BMPs trigger the differentiation of stem cells into bone forming cells (1). In recent years, BMP-2 has demonstrated a successful regenerative potential for applications in spinal disk regeneration, healing of bone fractures and periodontal repair (2). Several different materials have been proposed as delivery carriers for BMPs (3) to increase lifetime, stability, and bioactivity. In this sense, nano- and microparticles offer one of the most promising approaches for bone tissue engineering as they potentially overcome fundamental issues such as insufficient mechanical strength of the scaffolds and the lack of stability or bioactivity of the growth factor at the defect site (4).

Mikhail Soloviev (ed.), *Nanoparticles in Biology and Medicine: Methods and Protocols*, Methods in Molecular Biology, vol. 906, DOI 10.1007/978-1-61779-953-2_29, © Springer Science+Business Media, LLC 2012

Fig. 1. Representative micrograph of self-assembled VPAVG nanoparticles.

Additionally, the use of microparticles allows targeting the delivery into a specific injury site, with no dispersion into surrounding tissues while using low amounts of material.

Elastin-like polymers (ELPs) are a new class of synthetic biopolymers inspired by the sequence of mammalian elastin; the composition of which is based on repetitions of the pentapeptide VPGVG. The most striking feature of ELPs is their inverse temperature transition property. Below a certain critical temperature and in the presence of water, the polymer is soluble with polymer chains extended in a disordered state but above this critical temperature the polymer chains hydrophobically fold and segregate from solution through a self-assembling process. The polymer poly(VPAVG) is an ELP where the central glycine is substituted by an L-alanine. A recombinantly produced poly(VPAVG), consisting of 200 repetitions of the main pentamer VPAVG, was shown to self-assemble into spherical nanoparticles (Fig. 1) at temperatures above 33 °C, re-solubilizing only when the solution was cooled down to about 10 °C (5). As a result of this hysteresis behavior, the particles, once formed, are stable at room temperature and can be used for drug delivery purposes. Additionally, chemically synthesized poly(VPAVG) demonstrated excellent biocompatibility with no inflammatory response or toxic effect being observed when injected subcutaneously (6). The major advantages of this system include biocompatibility, stability at room temperature, and the use of mild conditions and harmless solvents, such as water, for its production.

Here, we describe how to use a recombinantly produced poly(VPAVG) polymer to entrap BMP-2 with a high encapsulation efficiency as well as all the required tests for in vitro bioactivity assays. The poly(VPAVG) polymer is produced by recursive directional ligation (5, 7) using conventional genetic engineering techniques and biologically produced in *Escherichia coli*. By exploiting the thermo-responsiveness of this polymer, purification is easily achieved by simple hot and cold cycles (5). From our experience, up to 94 % of the active BMP-2 protein could be successfully encapsulated in nano-sized particles (237.5 ± 3.0 nm) and released over a period of 14 days (8).

This self-assembling behavior of poly(VPAVG) can be further explored as a method for encapsulation of various hydrophobic drugs or proteins. During the self-assembling process, a high amount of aqueous solvent remains entrapped inside the newly formed particles and thus it is a reasonable idea that any solute present in the aqueous media will become encapsulated as the particles are formed. In general, the amount of solute to be encapsulated will depend on the concentration of poly(VPAVG) and some examples of this are described in literature (8, 9). Significant amounts of the drug dexamethasone phosphate (DMP) were encapsulated in particles of a chemically synthesized poly(VPAVG) (9) and encapsulation efficiencies of about 94 and 99 % were observed with BMP-2 and BMP-14, respectively, in recombinantly produced poly(VPAVG) (8). The poly(VPAVG) system can therefore be used as a method for encapsulation and controlled release of drugs for use in regenerative medicine.

2. Materials

Prepare all solutions with ultrapure water in sterile conditions and use analytical grade chemicals. Prepare and store all reagents according to manufacturer's instructions. Diligently follow all waste disposal regulations when disposing of waste materials.

2.1. Production of Elastin-Like Nanoparticles

1. ELP $(VPAVG)_{200}$ (see Note 1).
2. Phosphate Buffered Saline (PBS): dissolve 8 g of NaCl, 0.2 g of KCl, 1.44 g of Na_2HPO_4, and 0.24 g of KH_2PO_4 in 800 mL of ultrapure water. Adjust to pH 7.4. Bring the volume up to 1 L with ultrapure water. Sterilize by autoclaving. Cool down on ice before use.
3. Sterile 1.5 and 2 mL microcentrifuge tubes.
4. Make stock solutions of 0.1 % VPAVG (w/v) by reconstituting 1.5 mg of pure lyophilised VPAVG polymer in 1.5 mL of ice-cold sterile PBS in a 2 mL microcentrifuge tube. Mix with mild agitation or by inverting 8–12 times; incubate the solution on ice for 1 h with mild agitation or by inverting regularly until complete dissolution is achieved (see Notes 2 and 3). Leave the solution on ice or at 4 °C.

5. Prepare 250 μL aliquots of the VPAVG stock solution in sterile 1.5 mL microcentrifuge tubes and store at −20 °C until use. Whenever a working aliquot is needed, thaw the solution on ice.
6. Mini Spin® plus benchtop centrifuge (Eppendorf) for centrifugations at ≥10,000 × *g* and at room temperature.
7. Refrigerated centrifuge *5804 R* (*Eppendorf*) for poly(VPAVG) centrifugations at different temperatures (4 and 40 °C).

2.2. Encapsulation of Bone Morphogenetic Protein-2

1. ELP $(VPAVG)_{200}$ (see Note 1).
2. PBS: dissolve 8 g of NaCl, 0.2 g of KCl, 1.44 g of Na_2HPO_4, and 0.24 g of KH_2PO_4 in 800 mL of ultrapure water. Adjust to pH 7.4. Bring the volume up to 1 L with ultrapure water. Sterilize by autoclaving. Cool down on ice before use.
3. Recombinant Human Bone Morphogenetic Protein-2 (BMP-2).
4. Sterile 1.5 mL microcentrifuge tubes.
5. Mini Spin® plus benchtop centrifuge (Eppendorf) for centrifugations at ≥10,000 × *g* and at room temperature.

2.3. Determining Encapsulation Efficiency

1. Recombinant Human BMP-2.
2. BMP-2 sandwich-type ELISA kit (Eubio) or other suitable ELISA kit (e.g., from Biosensis).
3. Microplate reader.

2.4. Bioactivity Assays

2.4.1. MTT Cell Viability Assay

1. *Mus musculus* C2C12 cells (mouse myoblast cell line).
2. Dulbecco's Modified Eagle's medium (DMEM) without antibiotics.
3. Fetal calf serum (FCS).
4. MTT (3-(4,5-Dimethylthiazol-2-yl)-2,5-diphenyltetrazolium bromide) cell viability assay.
5. Humidified CO_2 incubator.

2.4.2. Alkaline Phosphatase Activity

1. *Mus musculus* C2C12 cells (mouse myoblast cell line).
2. DMEM without antibiotics.
3. Fetal calf serum (FCS).
4. PBS: dissolve 8 g of NaCl, 0.2 g of KCl, 1.44 g of Na_2HPO_4, and 0.24 g of KH_2PO_4 in 800 mL of ultrapure water. Adjust to pH 7.4. Bring the volume up to 1 L with ultrapure water. Sterilize by autoclaving. Cool down on ice before use.
5. Substrate buffer: 1 M diethanolamine HCl (pH 9.8), containing 2 mg/mL of *p*-nitrophenyl phosphate (*p*NPP).
6. Stop solution: 2 M NaOH, 0.2 mM EDTA in deionised water.

2.4.3. Calcium Mineralization Assay (Alizarin Red Assay)

1. *Mus musculus* C2C12 cells (mouse myoblast cell line).
2. DMEM without antibiotics.
3. Fetal calf serum (FCS).
4. PBS: dissolve 8 g of NaCl, 0.2 g of KCl, 1.44 g of Na_2HPO_4, and 0.24 g of KH_2PO_4 in 800 mL of ultrapure water. Adjust to pH 7.4. Bring the volume up to 1 L with ultrapure water. Sterilize by autoclaving. Cool down on ice before use.
5. Osteogenic Media—DMEM supplemented with 100 μM ascorbic acid and 10 mM β-glycerophosphate (see Note 4).
6. Formaldehyde.
7. Alizarin Red 1 % solution: dissolve 1 g of Alizarin Red S in 100 mL of distilled water and mix well. Adjust pH to 4.1–4.3 using 0.5 % ammonium hydroxide and filter before use. Solution is stored at 2–8 °C for a maximum advised period of 1 month. The pH is critical so check the pH if the solution is more than 2 weeks old.
8. Orbital shaker.

3. Methods

3.1. Production of Elastin-Like Nanoparticles

All procedures should be carried out in sterile conditions.

1. Thaw a working aliquot containing 250 μL of 0.1 % VPAVG on ice.
2. Warm sterile PBS to 37 °C.
3. Transfer VPAVG aliquot from ice to 37 °C and incubate for 30–45 min to allow the self-assembly to occur. Hereafter **always** keep the solution at room temperature (see Notes 5 and 6).
4. Collect the particles by centrifugation at ≥10,000 × *g* for 10 min at room temperature (see Note 7).
5. Carefully remove the supernatant; avoid touching or removing the white pellet.
6. Wash the pellet containing the particles with 250 μL of pre-warmed PBS (37 °C).
7. Collect the particles by centrifugation ≥10,000 × *g* for 10 min at room temperature and carefully discard the supernatant.
8. Ressuspend the self-assembled particles in 250 μL of pre-warmed PBS and store at room temperature (about 25 °C). Use as soon as possible.

3.2. Encapsulation of Bone Morphogenetic Protein-2

All procedures should be carried out under sterile conditions.

1. Thaw a working solution as prepared in Subheading 3.1 and gently add 5 μg of BMP-2 to achieve a concentration of 20 μg/mL (see Note 8).
2. Keep the mixture on ice. Mix the components with mild agitation or by inverting the tube until complete dissolution of BMP-2. Avoid harsh agitation and do not vortex to avoid possible denaturation of BMP-2 and foam formation.
3. Warm up the solution to 37 °C and incubate for 30–45 min to allow the self-assembling process to occur. From this point onwards, always keep the solution at room temperature (see Note 5).
4. Warm sterile PBS to 37 °C.
5. Collect the particles by centrifugation at ≥10,000 × *g* for 10 min at room temperature (see Note 7).
6. Carefully remove the supernatant, avoid touching or removing the white pellet. Transfer the supernatant into a fresh tube for determination of encapsulation efficiency (see Subheading 3.3).
7. Wash the pellet containing the particles with 250 μL of pre-warmed PBS (37 °C). Collect the particles by centrifugation at ≥10,000 × *g* for 10 min at room temperature (see Note 7). Carefully remove and discard the supernatant.
8. Resuspend the loaded particles in 250 μL of pre-warmed PBS and use them as soon as possible (see Note 9).

3.3. Determining Encapsulation Efficiency

Encapsulation efficiency is measured by comparing the total amount of BMP-2 with the free unloaded BMP-2, after particle production. Determination of this value allows quantification of the amount of BMP that is encapsulated and will be used for in vitro experiments.

Encapsulation efficiency (E.E.) is expressed as a percentage and calculated according to the following mathematical expression:

$$\text{E.E.\%} = 100 * \frac{\text{total BMP}(\mu\text{g}) - \text{free BMP}(\mu\text{g})}{\text{total BMP}(\mu\text{g})} \tag{1}$$

The most important factor in determining the encapsulation efficiency relies on the precise quantification of BMP-2. This quantification is determined by comparison with a calibration curve obtained with standard BMP-2 preparations of known concentrations. For this purpose the use of commercial kits is advised and an enzyme-linked immunosorbent assay (ELISA) can be employed to assure an accurate quantification. Making the measurements in triplicate is also advisable.

1. Make a calibration curve with known concentrations of BMP-2 by preparing a range of standards expected to include the

probable concentration of BMP-2. Concentrations should lie within the working range of the measuring instrument. Quantify with a BMP-2 sandwich-type ELISA development kit according to the manufacturer's instructions.

2. Quantify the amount of free BMP-2 remaining in the supernatant obtained in step 6 (Subheading 3.2) using the ELISA kit by comparison signal measured with the calibration curve.
3. Calculate the encapsulation efficiency using the mathematical expression Eq. 1 above (see Note 10).

3.4. Bioactivity Assays

All procedures must be carried out in sterile conditions and performed in triplicate.

3.4.1. MTT Cell Viability Assay

1. Seed C2C12 cells at a concentration of 5×10^4 cells/mL in a 24-well tissue culture plate (1 mL per well), in DMEM supplemented with 1 % (v/v) FCS (see Note 11) and no antibiotics, grow cells at 37 °C and 5 % CO_2 in a humidified environment.
2. Transfer 10 and 100 μL of unloaded elastin-like nanoparticles (produced as described in Subheading 3.1) to cells in two separate wells in the cell culture plate.
3. Incubate the cells at 37 °C and 5 % CO_2 in humidified atmosphere for 5 days or until 90–100 % confluency is achieved.
4. Perform the MTT assay according to the manufacturer's instructions. Use standard culture medium without particles as a positive control for cell viability. Express cell viability for treated cell cultures as a percentage viability relative to the control untreated cells (cells maintained in normal culture media).

3.4.2. Alkaline Phosphatase Activity

Alkaline phosphatase (ALP) hydrolyses phosphate esters and is involved in the initial processes of bone extracellular matrix mineralization. It is widely used as an indicator of osteoblastic differentiation. Bioactivity of BMP-2 released from elastin nanoparticles is assessed by determining the ability to induce ALP enzymatic activity over basal levels in C2C12 cells.

All procedures are carried out in sterile conditions and performed in triplicate.

1. Seed C2C12 cells at a concentration of 5×10^4 cells/mL in a 24-well tissue culture plate (1 mL per well), in DMEM supplemented with 1 % (v/v) FCS (see Note 11) and no antibiotics, grow cells at 37 °C and 5 % CO_2 in a humidified environment.
2. Transfer 133 μL of BMP-2 loaded elastin-like nanoparticles (produced as described in Subheading 3.2) to cells in a well in the cell culture plate. For positive control, add the equivalent amount of free BMP-2 to culture media in another well (see Note 12). Add the same volume of unloaded elastin-like nanoparticles (produced as described in Subheading 3.1) to another well in the cell culture plate for use as negative control.

3. Incubate the cells at 37 °C and 5 % CO_2 in humidified atmosphere for 5 days or until 90–100 % confluency is achieved.
4. ALP activity is quantified by the specific conversion of *p*NPP into *p*-nitrophenol (*p*NP) (see Note 13). For the study of ALP activity vs. time, carefully collect supernatants from cell cultures without disturbing cells at various time points (in triplicate) and freeze at −20 °C until quantification.
5. Set up the enzymatic reactions in a 96-well plate, by mixing 50 μL of the collected supernatants with 150 μL of substrate buffer.
6. Incubate solutions at 37 °C for 1 h.
7. Stop the reaction by adding 200 μL of stop solution.
8. Measure absorbance of *p*NP at 405 nm, and at 620 nm for reference (see Note 14).
9. Make a standard curve with *p*NP values ranging from 0 to 200 nmol/mL. Express results in nmol of *p*NP produced/mL/mg protein.

3.4.3. Calcium Mineralization Assay (Alizarin Red Assay)

1. Carry steps 1–3 (Subheading 3.4.2) to prepare cells and to incubate them with unloaded (133 μL) and loaded nanoparticles (133 μL) and controls (equivalent amount of BMP-2), (see Note 15).
2. Very carefully aspirate the medium without disturbing cells.
3. Wash cells at least three times with PBS, to ensure that no particles are retained in the cell monolayer. Aspirate PBS very carefully after each wash.
4. Add fresh osteogenic medium (DMEM supplemented with 100 μM ascorbic acid and 10 mM β-glycerophosphate) to the cells in wells.
5. Add the second dose of fresh particles (same amounts of loaded and unloaded) as well as the same dose of fresh BMP-2, to the same culture wells as in step 1 above.
6. Incubate the cells for another 5 days at 37 °C and 5 % CO_2 in a humidified environment.
7. Repeat steps 2–5 once again and incubate the cells for another 4 days at 37 °C and 5 % CO_2 in a humidified environment.
8. After the total of 14 days of cell culture (steps 1–7 with three nanoparticles loadings in total) carefully aspirate the medium. Wash cells three times with sterile PBS, aspirating after each wash. Be careful not to aspirate cells.
9. Fix cells in 4 % (v/v) formaldehyde for 30 min.
10. Remove the fixative and wash the cells twice with distilled water.
11. Cover cells with 1 % Alizarin Red solution, followed by gentle agitation in an orbital shaker for 10 min.

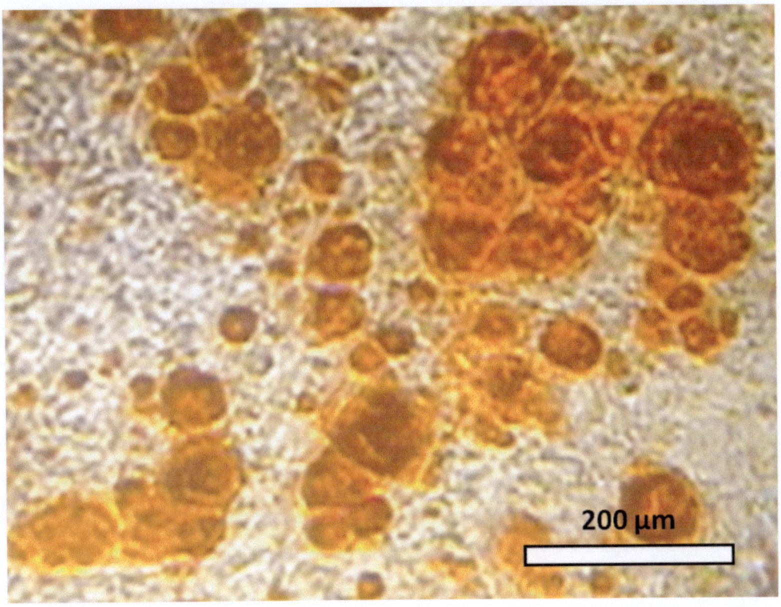

Fig. 2. Alizarin Red staining showing sites of calcium-phosphate deposits. Reproduced from (8) with permission from Elsevier.

12. Remove the Alizarin Red solution by carefully aspirating the solution.
13. Gently wash cells four times with distilled water, aspirating after each wash. Proceed to visual inspection or image acquisition. Osteocytes containing calcium deposits will be stained orange-red by the Alizarin Red Solution (Fig. 2) (see Note 16).

4. Notes

1. The ELP poly(VPAVG) is prepared by recursive directional ligation using conventional genetic engineering techniques. For a more detailed description see (5, 7). Before gene synthesis, optimize the codon usage for the expression system used in order to maximize production. Briefly, a synthetic DNA coding for 10 repetitions of VPAVG and flanked by inverted restriction enzyme recognition sites, is subjected to concatenation until production of a concatemer with 200 repeats of VPAVG. The construct with 200 repetitions is subcloned into the expression vector pET25 (NOVAGEN) and used to transform *Escherichia coli* strain BL21 (DE3). Purification is achieved by exploring the thermo-responsive property of the polymer and based on simple hot and cold cycles. The purified polymer fraction is lyophilised and stored at –80 °C until use.

2. The transition temperature of the VPAVG polymer in PBS is around 30 °C. When handling low volumes of solution, especially when using microcentrifuge tubes, the solution may easily reach this temperature just by holding it in the hands. For this reason, avoid holding the solution for long periods.
3. The use of mild agitation is advisable as harsh mixing leads to the formation of foam in the solution. The VPAVG polymer is highly soluble in water and complete dissolution is achieved within a few minutes, producing an optically clear solution. Maintaining the polymer solution on ice for an additional 30–45 min is advisable to allow the complete disorganization of the polymeric chains at the molecular level.
4. Ascorbic acid degrades quickly, so enough media for 1 week should be prepared at any time.
5. The self-assembling process can be visually followed by an increase in the opacity as the solution turns white. Due to the hysteresis behavior of the VPAVG polymer, the particles are stable at room temperature and only solubilize when cooled to about 10 °C.
6. After this step, the polymer is already self-assembled into nanoparticles. The following steps are designed to maintain the same experimental conditions as used in Subheading 3.2 and use as control for the in vitro experiments.
7. If the supernatant contains floating particles after centrifugation, centrifuge the supernatant again before proceeding to the next step.
8. In this protocol we describe the encapsulation of BMP-2. However, other compounds can be used for encapsulation. Add the new compound using a range of ELP concentrations and establish a correlation between the concentration of the VPAVG polymer and the encapsulation efficiency (see Subheading 3.3).
9. BMP-2 is released by diffusion and so it is advisable to use the loaded particles as soon as possible.
10. Determination of encapsulation efficiency allows for estimation of the amount of BMP-2 that is to be used in the bioactivity assays. In previously reported experiments (8), about 94 % of total BMP-2 (20 μg/mL of solution) was encapsulated. Therefore, one can assume that 18.8 μg of BMP-2 are encapsulated in 1 mL of solution.
11. Since C2C12 cells become confluent rapidly (ultimately leading to cell detachment), cells should be seeded at low density and a low serum concentration (1 %) should be used to slow down cell proliferation.
12. Typically the encapsulation efficiency is about 94 % and thus the added volume would contain approximately 2.5 μg of BMP-2. This amount produces best results in our experience,

however, the determination of optimum concentration is recommended. Optimum loading may be determined by using a range of BMP-2 amounts (e.g., 0.1, 0.25, 0.5 and 2.5 μg of BMP-2). Take into account the encapsulation efficiency determined in Subheading 3.3 to calculate the required volumes of the loaded nanoparticles.

13. ALP activity is quantified by the specific conversion of *p*NPP into *p*NP and determined according to standard procedures as in Salgado et al. (10).
14. Colored samples may interfere with O.D. 405 nm readings, so use a sample background reference at 620 nm. For calculations, correct background by subtracting the value derived from the reference at 620 nm.
15. From our experience, the best results are obtained using 2.5 μg of BMP-2 and hence this quantity was used in the calcium mineralization assay. We recommend to always determine the optimum concentration, as described in Subheading 3.4.2, and to use controls, as described above.
16. Mineralization can be quantified by the method proposed by Gregory et al. (11).

References

1. Bessa PC, Casal M, Reis RL (2008) Bone morphogenetic proteins in tissue engineering: the road from laboratory to the clinic, part I (basic concepts). J Tissue Eng Regen Med 2:1–13
2. McKay WF, Peckham SM, Badura JM (2007) A comprehensive clinical review of recombinant human bone morphogenetic protein-2 (INFUSE® Bone Graft). Int Orthop 31:729–734
3. Bessa PC, Casal M, Reis RL (2008) Bone morphogenetic proteins in tissue engineering: the road from laboratory to the clinic, part II (BMP delivery). J Tissue Eng Regen Med 2:81–96
4. Kim K, Fisher JP (2007) Nanoparticle technology in bone tissue engineering. J Drug Target 15:241–252
5. Machado R, Ribeiro AJ, Padrão J, Silva D, Nobre A, Teixeira JA, Arias FJ, Cunha AM, Rodríguez-Cabello JC, Casal M (2009) Exploiting the sequence of naturally occurring elastin: construction, production and characterization of a recombinant thermoplastic protein-based polymer. J Nano Res 6:133–145
6. Rincón AC, Molina-Martinez IT, de las Heras B, Alonso M, Baílez C, Rodíguez-Cabello JC, Herrero-Vanrell R (2006) Biocompatibility of elastin-like polymer poly(VPAVG) microparticles: *in vitro* and *in vivo* studies. J Biomed Mater Res A 78:343–351
7. Meyer DE, Chilkoti A (2002) Genetically encoded synthesis of protein-based polymers with precisely specified molecular weight and sequence by recursive directional ligation: examples from the elastin-like polypeptide system. Biomacromolecules 3:357–367
8. Bessa PC, Machado R, Nürnberger S, Dopler D, Banerjee A, Cunha AM, Rodriguez-Cabello JC, Redl H, van Griensven M, Reis RL, Casal M (2010) Thermoresponsive self-assembled elastin-based nanoparticles for delivery of BMPs. J Control Release 142: 312–318
9. Herrero-Vanrell R, Rincón AC, Alonso M, Reboto V, Molina-Martinez IT, Rodríguez-Cabello JC (2005) Self-assembled particles of an elastin-like polymer as vehicles for controlled drug release. J Control Release 102: 113–122
10. Salgado AJ, Gomes ME, Chou A, Coutinho OP, Reis RL, Hutmacher DW (2002) Preliminary study on the adhesion and proliferation of human osteoblasts on starch-based scaffolds. Mater Sci Eng 20:27–33
11. Gregory CA, Gunn WG, Peister A, Prockop DJ (2004) An alizarin red-based assay of mineralization by adherent cells in culture: comparison with cetylpyridinium chloride extraction. Anal Biochem 329:77–84

Chapter 30

Synthesis, Surface Modification, Characterization, and Biomedical In Vitro Applications of Organically Modified Silica (ORMOSIL) Nanoparticles

Diksha and Indrajit Roy

Abstract

Organically modified silica (ORMOSIL) nanoparticles, owing to the presence of lipophilic organic groups, can host active lipophilic molecules in their interior, as well as form electrostatic complex with therapeutic agents such as genes on their surface. Owing to this fact, they have numerous potential applications in diagnostic imaging, as well as drug, and gene delivery. Here we describe the synthesis, characterization, and biomedical applications of ORMOSIL nanoparticles encapsulated with either an anticancer drug (for cancer chemotherapy), or a photosensitizer drug (for photodynamic therapy), or complexed with a plasmid DNA (for gene therapy).

Key words: ORMOSIL nanoparticles, Fluorescence, Confocal imaging, Cancer chemotherapy, Photodynamic therapy, Gene therapy

1. Introduction

The rapidly emerging research field of nanotechnology, and the processes used to generate, manipulate, and deploy nanomaterials, provides exciting new possibilities for advanced development of new analytical tools and instrumentation for bioanalytical and biotechnological applications (1–3). Among the various types of nanomaterials used, silica-based nanoparticles are among most promising owing to the ease of synthesis, ultrafine size (diameter below 50 nm), monodispersity, biological inertness, non-antigenicity, and ability to resist microbial attack (4). When amorphous silica nanomaterials are conjugated to functional molecules or another nanomaterial, their applications extend broadly to bioimaging, biosensing, drug delivery, cancer therapy, and gene delivery.

Mikhail Soloviev (ed.), *Nanoparticles in Biology and Medicine: Methods and Protocols*, Methods in Molecular Biology, vol. 906,
DOI 10.1007/978-1-61779-953-2_30,

Owing to the rich and versatile chemistry of organically modified silica (ORMOSIL) nanoparticles, they provide the ideal nanoplatform for the fabrication of multimodal imaging probes, combining a wide variety of contrast agents that are suitable for large number of different imaging applications such as florescence (optical) imaging, magnetic resonance imaging (MRI), positron emission tomography (PET), optical coherence tomography (OCT), to name just a few (1, 5). These fluorescent nanoparticles are based on the incorporation of organic dye molecules or luminescent semiconductor nanoparticles (quantum dots) within the optically transparent silica-based nanoparticles, which can be easily synthesized in the microemulsion media, to which other imaging probes can be physically or chemically amalgamated (5, 6).

ORMOSIL nanoparticles have hydropobic core, tunable mesoporosity, and optical transparency that may have rich surface chemistry and are suitable for multimodal imaging. ORMOSIL nanoparticles are gaining prominence as nanoprobes for a number of biomedical applications as drug-carriers for drug delivery, photodynamic therapy, and gene therapy (7–10). New beneficial features of these nanoparticles are fast emerging in the literature, which includes their ability to release drug in a controlled manner, systemic non-toxicity and clearance from the body of small animals (11–13). In conjunction with co-incorporated diagnostic agents, these nanoparticles provide the ideal platform for the fabrication of theranostic nanoprobes (14).

Here we describe the synthesis of ORMOSIL nanoparticles, with or without termination with amino-groups on their surface, and the optional drug encapsulation (using doxorubicin as example of conventional chemotherapy and 2-[1-hexyloxyethyl]-2-devinyl pyropheophorbide (HPPH) for light activated photodynamic therapy). We also describe how plasmid DNA can be attached to the amino-terminated nanoparticles for use in gene delivery/therapy. We describe a range of characterization techniques suitable for determining nanoparticles' size, surface charge, optical properties, and time-dependent release-pattern of encapsulated drugs (in case of drug-doped nanoparticles), ability to generate singlet oxygen (in case of photosensitizer-drug doped nanoparticles only), and their ability to form complex with plasmid DNA. Finally, we elaborate some of their in vitro biomedical applications, which primarily rely on their ability to infiltrate and distribute within cancer cells in culture. We investigate their ability to induce expression of foreign genes (transfection) inside target cells. In the end, we discuss cytotoxicity properties of regular (for doxorubicin-doped) and light-activated (for HPPH-doped) nanoparticles.

2. Materials

2.1. Microemulsion Mediated Synthesis, Surface Modification, and Purification of ORMOSIL Nanoparticles

1. Double distilled water should be used throughout.
2. Any regular magnetic stirrer and a range of magnetic stir bars.
3. Surfactant: dioctylsulfosuccinate (Aerosol OT, or AOT; 98 %).
4. Co-surfactant: 1-butanol (99.8 %).
5. *N*,*N*-Dimethylformamide (DMF).
6. Glass vials (25 mL capacity).
7. ORMOSIL precursor 1: triethoxyvinylsilane (TEVS), use neat.
8. ORMOSIL precursor 2: 3-aminopropyltriethoxysilane (APTES), use neat.
9. Drug: make 10 mg/mL solution of Doxorubicin hydrochloride in DMF.
10. Photosensitizer: HPPH, this e.g., could be made available on request from Roswell Park Cancer Institute, New York. Make 15 mM solution of HPPH in DMF.
11. 18 M Ammonium hydroxide solution in water.
12. Cellulose membrane for dialysis (12–14 kDa cut-off), from Spectrum Laboratories, Inc.
13. Sterile membrane filters (0.2 μM), Nalgene.
14. Large (1 L) beaker for dialysis.

2.2. Surface Modification of ORMOSIL Nanoparticles with Plasmid DNA for Gene Delivery/Therapy

1. Plasmid DNA pEGFP (Clontech).
2. Sterile plastic centrifuge tubes (500 μL) for mixing.

2.3. Characterization of ORMOSIL Nanoparticles

2.3.1. Generic Approaches

1. PHILIPS EM 420 TEM instrument or any other standard TEM, such as JEOL-JEM-2020.
2. Grids for transmission electron microscopy (TEM): formvar-coated copper grids (obtained from Electron Microscopy Sciences Inc).
3. ZETA-PLUS Dynamic Light Scattering (DLS) Instrument (Brookhaven Instruments) or a similar instrument for measuring particle size and zeta potential.
4. Standard plastic/glass/quartz cuvettes.
5. Shimadzu UV-3600 or any other standard UV-Visible Spectrophotometer for measuring absorption spectra of the drug-doped nanoparticles.

6. Fluorolog-3.11 (JobinYvon) or any other standard Spectrofluorimeter for measuring the emission spectra of the drug-doped nanoparticles.

2.3.2. Release Kinetics of the Drug from the Nanoparticles

1. Surfactant Tween-80 (98 % purity): make 1 %w/v aqueous solution by dissolving 1 g of Tween-80 in 100 mL water.
2. Cellulose membrane for dialysis (12–14 kDa cut-off), Spectrum Laboratories, Inc.
3. Fluorolog-3.11 (JobinYvon, Longjumeau, France) or equivalent standard Spectrofluorimeter.
4. Standard glass/quartz cuvettes.
5. Any regular magnetic stirrer.
6. Large (1 L) beaker.

2.3.3. Generation of Singlet Oxygen for HPPH-Encapsulated ORMOSIL Nanoparticles

1. Disodium salt of 9, 10-anthracenedipropionic acid (ADPA): make 5.5 mM solution in water.
2. Photosensitizer: 15 μM HPPH.
3. 650 nm laser source Verdi (Coherent), or any other equivalent solid-state diode-pumped laser. To be operated at a power of 50 mW.
4. Standard glass/quartz cuvette.
5. Shimadzu UV-3600 or any other standard Spectrophotometer.

2.3.4. Agarose Gel Electrophoresis to Study the Complex Formation of Plasmid DNA with Nanoparticles

1. 5 mg/mL DNase 1.
2. Electrophoresis buffer: 10× Tris-Acetate-EDTA, or TAE, pH 7.2. Dilute to 1× before use.
3. Ethidium bromide (EtBr): Prepare stock solution (10 mg/mL) in water.
4. Use agarose (Molecular Biology grade, 99 % pure) and 1× TAE buffer to make 1 % agarose gel.
5. Any standard DNA gel-electrophoresis chamber.
6. Any standard UV DNA gel imaging system, e.g., LM-20E Ultraviolet BenchtopTransilluminator (UVP), in conjunction with an Olympus Digital Camedia C-4000 with a UV filter and lens.
7. Image analysis software, e.g., DOC-IT system software.

2.4. Biomedical Applications of ORMOSIL Nanoparticles

In addition to specific reagents and materials listed in each of the following subsections, these standard cell culture reagents are required for all experimental steps where cell culturing is required.

1. Phosphate buffered saline (PBS), 1×, pH 7.2. Prepare this buffer solution by dissolving 8 g NaCl, 0.2 g KCl, 1.44 g Na_2HPO_4 and 0.24 g K_2HPO_4 in 800 mL of distilled water. Adjust the

pH to 7.2 by adding 0.1 M HCl solution. Then bring up the total volume of the solution to 1 L by adding water.

2. DMEM cell culture medium (Invitrogen).
3. Heat inactivated fetal bovine serum (FBS).
4. Penicillin-streptomycin: 5,000 units of penicillin (base) and 5,000 μg of streptomycin (base)/mL water in 0.85 % saline (Invitrogen).
5. Trypsin-EDTA: This can be prepared by dissolving the following components in 1 L water: 400 mg Potassium Chloride (KCl), 60 mg Potassium Phosphate monobasic (KH_2PO_4), 350 mg Sodium Bicarbonate ($NaHCO_3$), 8,000 mg Sodium Chloride (NaCl), 90 mg Sodium Phosphate dibasic (Na_2HPO_4-$7H_2O$), 1,000 mg D-Glucose (Dextrose), 10 mg Phenol Red, 380 mg Disodium EDTA (Na_2-EDTA) and 2,500 mg Trypsin.
6. Cell culture incubator (37 °C, 5 % CO_2).

2.4.1. Confocal Microscopic Analysis of Uptake of Nanoparticles in Treated Cells in Vitro

1. Sterile cell culture dishes (35 mm).
2. Sterile 15 and 50 mL falcon tubes.
3. Cell growth medium: DMEM medium, supplemented with 10 % FBS and 1 % penicillin-streptomycin.
4. Cell line MiaPaCa-2: maintain in Cell growth medium.
5. Nikon Eclipse TE2000 confocal microscope equipped with Nuance GNIR imaging system (Cambridge Research & Instrumentation Inc), or any other standard confocal microscope.

2.4.2. Quantitative Estimation of Cellular Uptake of the Nanoparticles Using Spectrofuorimetry

1. Sterile cell culture dishes in 24-well format.
2. 1× cell culture lysis reagent: Prepare this by dissolving 1 % (mass/volume) of surfactant Triton-X 100 in 1× PBS, pH 7.2.
3. Fluorolog-3.11 (JobinYvon) or other standard Spectrofluorimeter.
4. Standard glass/quartz cuvettes.
5. Sterile cell scraper.
6. 1× cell culture lysis reagent (Promega).
7. Sterile 1.5 mL microcentrifuge tubes.

2.4.3. In Vitro Transfection of DNA-Coated ORMOSIL Nanoparticles

1. 60–70 % confluent MiaPaCa-2 cells: plate one day before transfection in a 60-mm culture plate.
2. Plasmid DNA pEGFP (Clontech).
3. OPTI-MEM medium (Invitrogen).
4. Sterile microcentrifuge tubes.

2.4.4. Cell Viability (MTS) Assay to Probe Chemotherapeutic Efficacy of Doxorubicin-Doped Nanoparticles

1. MTS assay (CellTiter 96® AQueous One Solution Cell Proliferation Assay) reagent (PROMEGA).
2. 96 well tissue culture plates.

2.4.5. Cell Viability (MTS) Assay to Probe Photo-dynamic Therapeutic (PDT) Efficacy of HPPH-Doped Nanoparticles

1. 650 nm laser source Verdi (Coherent), or any other equivalent solid-state diode-pumped laser. To be operated at a power of 1.5 mW/cm^2.
2. MTS assay (CellTiter 96® AQueous One Solution Cell Proliferation Assay) reagent, available from PROMEGA.

3. Methods (See Note 1)

3.1. Microemulsion Mediated Synthesis, Surface Modification, and Purification

ORMOSIL nanoparticles, encapsulating active agents, can be synthesized by the alkaline condensation-polymerization of their organosilane precursors within the confined dimensions of both oil-in-water (normal) and water-in-oil (reverse) microemulsion. However, owing to the widespread literature reporting nanosynthesis in the normal microemulsion, we chose to limit our methods to the description of this system only (see Note 2). Our methods describe the synthesis of bare or drug-loaded nanoparticles in the nonpolar core of AOT/1-butanol/water microemulsion. The synthesis procedures are summarized in Fig. 1.

1. To prepare AOT/1-butanol/water microemulsion dissolve 0.44 g of AOT and 800 μL (0.56 g) of 1-butanol in 20 mL of doubly distilled water by vigorous magnetic stirring for about 15–20 min (see Note 3). The emulsion should be optically clear.
2. *To make drug-doped ORMOSIL nanoparticles add* 30 μL of 15 mM HPPH in DMF or 10 mg/mL Doxorubicin in DMF to the optically clear microemulsion. Stir vigorously for about 10 min. Go to the step 4 below.

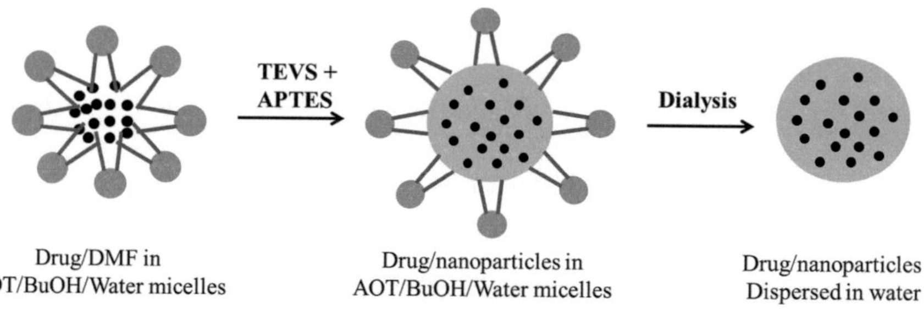

Fig. 1. Scheme depicting the synthesis and purification of drug-doped ORMOSIL nanoparticles in a normal microemulsion medium.

3. *To make void ORMOSIL nanoparticles* add 30 μL of DMF (without HPPH or Doxorubicin) to the optically clear microemulsion from step 2 above. Stir vigorously for about 10 min.
4. Then, add 200 μL of neat TEVS (ORMOSIL precursor) to the micellar system, and stir the resulting solution for about 1 h.
5. Add 30 μL of 18 M ammonium hydroxide solution to the system and continue stirring for another 30 min.
6. Add 30 μL of neat APTES and continue stirring for 20–24 h. This will yield amino-terminated ORMOSIL nanoparticles (with or without drug, see steps 2 and 3 above). The entire process should to be carried out at room temperature (see Note 4).
7. To purify ORMOSIL nanoparticles form the excess of surfactants (AOT and 1-butanol), dialyze the solution against water using dialysis membrane for 48 h. Use about 1 L of water per ~10 mL of the nanoparticles, and keep changing the water regularly at 4–6 h intervals.
8. Sterile filter the dialyzed solution through a 0.2 μm cut-off membrane filter and store it at 4 °C (see Note 5).

3.2. Surface Modification of ORMOSIL Nanoparticles with Plasmid DNA for Gene Delivery/Therapy

For the purpose of gene delivery/therapy, the amino-terminated (positively charged) ORMOSIL nanoparticles may be electrostatically condensed with negatively charged plasmid DNA by simple incubation.

1. Gently mix the aqueous dispersion of the amino-terminated ORMOSIL nanoparticles (250 μL) with 4 μL of plasmid DNA encoding EGFP (pEGFP) (0.5 μg/μL) at room temperature.
2. Incubate overnight at 4 °C for the formation of the DNA–nanoparticle complex (nanoplex) (see Note 6).

3.3. Characterisation of ORMOSIL Nanoparticles

In this section we list a few of the standard characterization techniques applicable to ORMOSIL nanoparticles as well as many other types of nanoparticles. A user should follow standard operating procedures and follow the available instrument manuals.

3.3.1. Generic Approaches

1. Transmission electron microscopy provides useful and direct information on particle size. Typically 2–3 drops of aqueous dispersion of 0.1–0.5 mg/mL NPs should be added to a formvar-coated copper grid. Allow the sample to dry for about 1 h in air, and observe the grids using a standard Transmission Electron Microscope instrument and follow standard operating procedures.
2. DLS is suitable for measuring particle size and zeta potential. Typically 3 mL of aqueous dispersion of 0.1–0.5 mg/mL NPs is required. Transfer the sample into a standard glass/quartz cuvette (capacity about 3 mL), and follow the instrument manual to obtain nanoparticle hydrodynamic size and

zeta-potential using a ZETA-PLUS DLS Instrument (see Notes 7 and 8).

3. Use conventional UV-Visible absorption spectroscopy to measure the absorption spectra of the drug-doped nanoparticles. This measurement should be carried out to ensure that the characteristic absorption spectrum of the drug is not altered following nanoencapsulation.

   Typically 0.5–1 mL of aqueous dispersion of ORMOSIL nanoparticles at 0.1–0.5 mg/mL NPs is needed. The whole visible range of 400–700 nm needs to be scanned, since the absorption peaks for HPPH or Doxorubicin lie in this range.

4. Use conventional fluorescence spectroscopy to measure the emission spectra of the drug-doped nanoparticles. This measurement should be carried out to ensure that the characteristic fluorescence spectrum of the drug is not altered following nanoencapsulation.

   Typically 0.5–1 mL of aqueous dispersion of 0.1–0.5 mg/mL NPs is required. Set excitation wavelength to 590 nm for Doxorubicin and 632 nm for HPPH. Absorption spectra measured earlier may indicate the appropriate wavelength for other compounds.

3.3.2. Release Kinetics of the Drug from the Nanoparticles

The time dependent release (release kinetics) of Doxorubicin or HPPH from the ORMOSIL nanoparticles is determined by dialyzing them against an aqueous solution containing Tween-80 micelles for about 2 weeks (see Note 9).

1. Dialyze the solution of nanoparticles (40–50 μM "drug" and 0.1–0.5 mg/mL nanoparticles concentrations) against water using dialysis membrane for 2 weeks. Use 100 mL of 1 % Tween-80/water for 5 mL of the sample. Because of the low cut-off pore size of the dialysis membrane (12–14 kD) only free Doxorubicin or HPPH molecules can come out of the membrane into the solution outside.
2. To estimate the amount of released drug, measure the Doxorubicin or HPPH concentration in the bulk solution as a function of dialysis time (see Subheading 3.3.1, step 4). A plot of the cumulative percentage of drug released vs. time (in days) provides the release kinetics data, assuming the original amount of drug within the ORMOSIL nanoparticles (prior to dialysis) as 100 %.

3.3.3. Generation of Singlet Oxygen for HPPH-Encapsulated ORMOSIL Nanoparticles

Singlet oxygen can be detected chemically, using the disodium salt of ADPA as a singlet oxygen sensor. ADPA (a water-soluble anthracene derivative) is bleached by singlet oxygen to its corresponding endoperoxide. The reaction is monitored spectrophotometrically by recording the decrease in optical density at 400 nm (which is the adsorption max of ADPA).

1. Transfer 3 mL of 15 μM HPPH entrapped in nanoparticles (nanoparticle concentration of ~0.2 mg/mL water)—to a standard quartz cuvette.
2. Add 150 μL of 5.5 mM ADPA in D_2O to the cuvette. For the control experiment, use the same amount of ADPA mixed with void nanoparticles dispersed in water.
3. Irradiate the cuvettes using a 650 nm laser source (solid-state diode-pumped laser) for 30 s.
4. Record their optical densities at 400 nm every 10 min for a total time period of 1 h.
5. Plot the measured optical density at 400 nm (*Y*-axis) as a function of time in minutes (*X*-axis) (see Note 10).

3.3.4. Agarose Gel Electrophoresis to Study the Complex Formation of Plasmid DNA with Nanoparticles

1. Transfer two 50 μL aliquots from each of the samples prepared in Subheading 3.2 into sterile microcentrifuge tubes.
2. Mix one of the two samples with 1 μL of 5 mg/mL DNase 1; do not add anything to the second tube in each set. Incubate at 37 °C for 30 min.
3. Load 10 μL of each of the nanoparticles samples into individual wells. Run the gel at 100 V for 1.5 h. Subsequently stain with ethidium bromide (at a final concentration of 10 μg/mL) and record the image using the gel imaging system (see Note 11).

3.4. Biomedical Applications of ORMOSIL Nanoparticles

Here we are focusing on three applications of ORMOSIL nanoparticles: conventional drug delivery (using Doxorubicin-encapsulated nanoparticles), photodynamic therapy (using HPPH-encapsulated nanoparticles), and gene delivery (using void nanoparticles conjugated with plasmid DNA on the surface). The applications are investigated in vitro only in cancer cells in culture. As these nanoparticles are fluorescently doped (both Doxorubicin and HPPH are fluorescent), the efficiency of uptake of these nanoparticles in the cells, which is the first step for any in vitro application, are recorded and quantified using confocal microscopy of live cells, as well as by spectrofluorimetric analysis of cell lysates. In addition, the gene delivery (transfection) efficiency of the nanoparticles complexed with the gene encoding the fluorescent protein pEGFP is recorded and quantified using confocal microscopy of treated cells. Following these uptake and transfection studies, the functional (therapeutic) effects of these nanoparticles on the treated cells are studied by cell viability (MTS) assay. Below we describe the conditions for chemotherapy, photodynamic therapy and gene therapy individually, even though they can be combined to provide synergistic therapeutic effect.

3.4.1. Confocal Microscopic Analysis of Uptake of Nanoparticles in Treated Cells in Vitro

1. For studying the cellular uptake and distribution of the targeted nanoparticles, plate the cells in two 35 mm tissue culture dishes, each containing 2 mL of Cell growth medium. Grow cells to approximately 50 % confluency.
2. Aspirate the growth medium and replace with 2 mL of fresh Cell growth medium per each culture dish.
3. Treat the cells with drug-doped nanoparticles by adding 100 μL of 0.1–0.5 mg/mL nanoparticles to the culture dish. Mix the contents in the culture dish by gentle swirling.
4. Treat a control set of cells using the same procedure as above but use 100 μL of sterile PBS to the "control" culture dish instead of the nanoparticles. Mix the contents in the culture dish by gentle swirling.
5. Incubate the treated cells for variable times (15 min, 1 and 4 h) in cell culture incubator (37 °C, 5 % CO_2).
6. Rinse the plates with sterile PBS, replace with fresh medium and immediately image the plates on a Nikon Eclipse TE2000 confocal microscope equipped with Nuance GNIR imaging system. This microscope is capable of multispectral (wavelength-resolved) imaging in the range of 500–950 nm (see Note 12).

3.4.2. Quantitative Estimation of Cellular Uptake of the Nanoparticles Using Spectrofluorimetry

1. Seed the cells (50,000 cells/mL/well) in a 24-well plate and allow them to attach and grow for 24 h.
2. Remove the growth medium, add 1 mL of complete growth medium containing 50 μL of 0.1–0.5 mg/mL of the aqueous dispersion of drug-doped nanoparticles and incubate for 15 min, 1 and 4 h.
3. Wash the cells with PBS thrice.
4. Lyse cells using 1× cell culture lysis reagent (100 μL of lysis buffer per well) for 30 min at room temperature.
5. Analyze the cell lysates for nanoparticle uptake by quantifying the fluorescence recovered from the lysates. For this, scrape the cells (in the wells containing the lysis buffer) with a sterile scraper, add further 1 mL of PBS/well and mix well with the lysed cells in lysis buffer and transfer the contents in 1.5 mL microcentrifuge tubes (containing approximately 1.1 mL of lysed cells in buffer per tube).
6. Separate the cell debris from the lysate (containing the fluorescent nanoparticles) by centrifugation (3,000 rpm or 1,000 × *g*, 5 min) (see Note 13).
7. Examine the amount of the fluorescent material in the supernatant in a spectrofluorimeter, as described before (Subheading 3.3.1). The amount of fluorescence recovered from the lysates directly correlates with the efficiency of nanoparticle uptake by the cells. Run the experiment in triplicate, and calculate the mean and SD.

3.4.3. In Vitro Transfection of DNA-Coated ORMOSIL Nanoparticles

To confirm the prospect of using the ORMOSIL nanoparticles as nonviral vectors for gene therapy, preform transfection in cells, using the plasmid DNA pEGFPcomplexed with the amino-functionalized ORMOSIL nanoparticles.

1. Add 20 μL of 0.3 μg/μL pEGFP stock solution to 0.25 mL of OPTI-MEM medium in a microcentrifuge tube at room temperature.
2. Add 50 μL of 0.1–0.5 mg/mL nanoparticles (Subheading 3.1) to 0.25 mL of OPTI-MEM medium in another microcentrifuge tube at room temperature.
3. Mix the contents of the two tubes together gently and incubate at room temperature for 30 min.
4. Add all of the DNA–ORMOSIL suspension to a 60-mm culture plate of 60–70 % confluent cells containing 5 mL of the cell growth medium and incubate for 48 h (37 °C, 5 % CO_2).
5. Rinse the transfected cells with PBS, add 5 mL per plate fresh PBS, and image the cells in a confocal microscope to check for the expression of the encoded green fluorescent protein (see Note 14).

3.4.4. Cell Viability (MTS) Assay to Probe Chemotherapeutic Efficacy of Doxorubicin-Doped Nanoparticles

1. Seed 2,000–5,000 cells per well in 96 well tissue culture plate in appropriate growth medium (100 μL per well) and allow to attach for 24 h.
2. Aspirate the media and replace with fresh growth medium containing varying dosages of Doxorubicin nanoparticles (1, 2 and 5 μM of final concentration of the drug). This can be achieved by mixing 11 μL of Doxorubicin/nanoparticles in water (at Doxorubicin conc. of 50 μM and nanoparticle concentration of 2 mg/10 mL) with 100 μL of medium (for 5 μM of final concentration of Doxorubicin), 4.2 μL of Doxorubicin/nanoparticles in water with 100 μL of medium (for 2 μM of final concentration of Doxorubicin), and 2.2 μL of Doxorubicin/nanoparticles in water with 100 μL of medium (for 1 μM of final concentration of Doxorubicin). Use equivalent amounts of Doxorubicin dissolved in DMSO for comparison. Also, use void nanoparticles to evaluate non-specific toxicity, if any. All samples (Doxorubicin/nanoparticles) and controls (drug alone and void nanoparticles) should be run in triplicates.
3. Incubate the cells with the drug/nanoparticles for 4 h (37 °C, 5 % CO_2).
4. Aspirating the existing media and replacing with fresh growth media. Return to the incubator (37 °C, 5 % CO_2). Grow cells for 3 days.

5. Analyze the cells for viability by a modified MTS assay:
   - Aspirate the media and replace with fresh serum-free media (200 μL/well).
   - Add 40 μL of the MTS assay reagent to each well, mix, and return to the incubator. Dim the lights when performing this experimental step.
   - Incubate the cells in the incubator for 1 h.
   - Transfer 100 μL of the liquid mixture from each well to the individual wells in a 96 well plate, and read the plate using any spectrophotometric plate reader at wavelength 490 nm. Run the experiment in triplicates to access the statistical significance of the data, see Note 15.

3.4.5. Cell Viability (MTS) Assay to Probe Photodynamic Therapeutic (PDT) Efficacy of HPPH-Doped Nanoparticles

1. Seed 5,000 cells per well in 96 well tissue culture plates in appropriate growth medium (100 μL per well) and allow to attach for 24 h.
2. Aspirate the media and replace with fresh growth medium containing varying dosages of HPPH containing nanoparticles (1, 2 and 5 μM of HPPH in nanoparticles) This can be achieved by mixing 12.5 μL of HPPH/nanoparticles in water (at HPPH conc. of 45 μM and nanoparticle concentration of 2 mg/10 mL) with 100 μL of medium (for 5 μM of final concentration of HPPH), 4.8 μL of HPPH/nanoparticles in water with 100 μL of medium (for 2 μM of final concentration of HPPH), and 2.4 μL of Dox/nanoparticles in water with 100 μL of medium (for 1 μM of final concentration of HPPH). Use equivalent amounts of HPPH dissolved in DMSO for comparison. Also, employ void nanoparticles to evaluate non-specific toxicity, if any. All samples (HPPH/nanoparticles) and controls (drug alone and void nanoparticles) are run in triplicates.
3. Incubate the cells with the drug/nanoparticles for 4 h (37 °C, 5 % CO_2).
4. Aspirating the existing media and replacing with fresh growth media. Return to the incubator (37 °C, 5 % CO_2) for 4 h (see Note 16).
5. Irradiate the plates with a 650 nm laser light source (at a power level of 1.4 mW/cm^2) for 10 min each.
6. Return cells to the incubator. Grow cells for 2 days (37 °C, 5 % CO_2).
7. Analyze the cells for viability by a modified MTS assay protocol, as described previously (Subheading 3.4.4).

4. Notes

1. In this chapter, the various therapeutic effects of the drugs in cells are described separately. However, the same nanoparticle can be made to incorporate more than one therapeutic modality (e.g., combine encapsulated HPPH with complexed plasmid DNA). To study the combined therapeutic effect on cultured cells, one would need to modify the protocol accordingly to suit both the therapeutic procedures.
2. Most drug molecules (e.g., doxorubicin, paclitaxel, etc.) or fluorescent probes (e.g., fluorescein, quantum dots, etc.) are lipophilic, or poorly water soluble. Therefore, a possible reason for the abundance of literature regarding the use of ORMOSIL nanosynthesis in normal microemulsion is that this medium, and not reverse microemulsion medium, is used for the encapsulation of lipophilic agents.
3. The surfactant AOT will not dissolve in water without the presence of the co-surfactant *n*-butanol. Add AOT in water first, and let the solution vigorously stir for about 10 min, when the AOT is more-or-less uniformly dispersed (not dissolved) in water. At this point, add *n*-butanol, followed by further vigorous stirring for about 5 min or so, till the solution becomes optically clear.
4. The addition of APTES results in the formation of amino-terminated ORMOSIL nanoparticles. If non-amino terminated nanoparticles are desired to be prepared, then addition of APTES is not necessary, and instead the solution should be allowed to stir for 20–24 h following addition of aqueous ammonia.
5. The ORMOSIL solution can be kept at room temperature, or at 4° for long-term storage. This solution must not be frozen.
6. The presence of encapsulated drug/s in the core of the nanoparticles does not interfere with the formation of the DNA–nanoparticle complex at the surface.
7. While TEM gives the actual size of the nanoparticles, DLS gives the hydrodynamic size, which combines the size of the nanoparticles with that of the surrounding hydration layer. Therefore, the size obtained from TEM is always less than that obtained from DLS for all nanoparticles.
8. Avoid any air bubbles in the sample while performing DLS measurements to avoid light scattering by the bubbles.
9. As both Doxorubicin and HPPH are lipophilic, and are nestled within the hydrophobic core of the ORMOSIL nanoparticles, dialyzing them against pure water will not provide the driving force for their release from the nanoparticles. Therefore, the

dialysis is carried out against an aqueous solution of Tween-80 (1 % w/volume), whose hydrophobic core attracts the released Doxorubicin or HPPH.

10. The optical density at 400 nm (characteristic absorption maxima of ADPA) should decrease as a function of time in the case of the HPPH/nanoparticle, owing to the HPPH-induced bleaching of ADPA. However, in case of ADPA mixed with void nanoparticles (no HPPH encapsulated), the optical density at 400 nm should remain constant with time.

11. Here, the electrophoretic mobility of the investigational (DNA/RNA-nanoparticle complex) sample is compared to that of DNA/RNA-only sample. Retarded electrophoretic mobility of the DNA/RNA present in the investigational sample, as compared to that of the DNA/RNA only sample, indicates complex formation in the former. On the other hand, same electrophoretic mobility of the DNA/RNA present in the investigational sample, as compared to that of the DNA/RNA only sample, indicates no complex formation in the former.

12. With confocal microscopy, we get a semi-quantitative estimation of the uptake, as well as visualize the intracellular localization, of all fluorescent species (fluorescent nanoparticles in this case) in cells in culture. This uptake should increase with exposure time till a saturation point is attained. In this overall protocol, this step serves as a preliminary step of verifying nanoparticle-uptake of cells, prior to a quantitative estimation as described in the next section.

13. The Centrifugation should not be carried at a speed higher than 3,000 rpm or 1,000 × *g*, as higher speed may lead to the settling down of some quantum dots, along with the cell debris, from the solution. This will introduce errors in the measurement.

14. The expression of the encoded protein (green fluorescent protein) indicates the success of transfection. Free DNA cannot transfect cells as they have poor permeation through cellular membranes. However, DNA can enter cells when complexed with nanoparticles, such as ORMOSIL. Once they enter the cells, the DNA frees itself from the nanoparticle, enters the nucleus of the cell, and then expresses the encoded protein.

15. The reagent reacts with only live cells and produces water soluble purple crystals that absorb at 490 nm. Therefore, the value of absorbance would directly correlate with the number of viable cells present in each well.

16. Throughout this experiment, care should be taken that the experiment takes place in minimal light conditions. Exposure of cells to stray light should be limited as much as possible, as this might trigger photodynamic therapy action prior to treatment with laser light.

Acknowledgements

The authors would like to acknowledge the Start-up R&D grant and seed-money provided by Delhi University, India.

References

1. Prasad PN (2004) Nanophotonics. Wiley, New York
2. Whitesides GM (2003) The 'right' size in nanobiotechnology. Nat Biotechnol 21:1161–1165
3. Alexis F, Pridgen EM, Langer R, Farokhzad OC (2010) Nanoparticle technologies for cancer therapy. Handb Exp Pharmacol 197:55–86
4. Jain TK, Roy I, De TK, Maitra AN (1998) Nanometer silica particles encapsulating active compounds: a novel ceramic drug carrier. J Am Chem Soc 120:11092–11095
5. Yong KT, Roy I, Swihart MT, Prasad PN (2009) Multifunctional nanoparticles as biocompatible targeted probes for human cancer diagnosis and therapy. J Mater Chem 19:4655–4672
6. Das S, Jain TK, Maitra A (2002) Inorganic-organic hybrid nanoparticles from n-octyltriethoxysilane. Colloid Interface Sci 252(1): 82–88
7. Liu Y, Lou C, Yang H, Shi M, Miyoshi H (2011) Silica nanoparticles as promising drug/gene delivery carriers and fluorescent nanoprobes: recent advances. Curr Cancer Drug Targets 11(2):156–163
8. Rosenholm JM, Sahlgren C, Lindén M (2010) Towards multifunctional, targeted drug delivery systems using mesoporous silica nanoparticles-opportunities & challenges. Nanoscale 2(10):1870–1883
9. Roy I, Ohulchansky T, Pudavar HE, Bergey EJ, Morgan J, Oseroff AR, Dougherty TJ, Prasad PN (2003) Ceramic-based nanoparticles entrapping water-insoluble photosensitizing anticancer drugs: a novel drug-carrier system for photodynamic therapy (PDT). J Am Chem Soc 125:7860–7865
10. Roy I, Ohulchansky TY, Bharali DJ, Pudavar HE, Mistretta RA, Kaur N, Prasad PN (2005) Optical tracking of organically modified silica nanoparticles as DNA carriers: a nonviral, nanomedicine approach for gene delivery. Proc Natl Acad Sci USA 102(2):279–284
11. Vivero-Escoto JL, Slowing II, Trewyn BG, Lin VS (2010) Mesoporous silica nanoparticles for intracellular controlled drug delivery. Small 6(18):1952–1967
12. Roy I, Stachowiak MK, Bergey EJ (2008) Non viral gene transfection nanoparticles: functions and applications in CNS. Nanomedicine 4(2):89–97
13. Kumar R, Roy I, Ohulchanskyy TY, Vathy LA, Bergey EJ, Sajjad M, Prasad PN (2010) In vivo biodistribution and clearance studies using multimodal ORMOSIL nanoparticles. ACS Nano 4(2):699–708
14. Xie J, Lee S, Chen X (2010) Nanoparticle-based theranostic agents. Adv Drug Deliv Rev 62(11):1064–1079

Chapter 31

Extraction of Organic Nanoparticles from Plants

Scott C. Lenaghan, Qin Zhu, Lijin Xia, and Mingjun Zhang

Abstract

Despite the wealth of studies related to the isolation and synthesis of metallic nanoparticles, few studies have been conducted to extract nanoparticles from plants. Nanoparticles isolated from plants have been demonstrated to have unique properties for biomedical and industrial applications. In this chapter, we report a method to extract nanoparticles from plants. The proposed method consists of multiple purification steps, including pre-filtration, dialysis, and size exclusion chromatography.

Key words: Organic nanoparticles, Plant culture, Dialysis, Size exclusion chromatography, Nanotechnology

1. Introduction

Nanomaterials, nanoparticles in particular, have emerged as important materials for drug delivery, medical adhesives, molecular sensing, and imaging. The nanoscale size feature of the nanoparticles often leads to unique optical and material properties, as well as, the ability to traverse cell membranes. One of the primary goals in the production of nanoparticles is to eliminate the toxicity associated with chemically synthesized nanoparticles. One emerging strategy is to manufacture nanoparticles using live plants or their extracted parts as factories (1–5). In addition to "green" metallic nanoparticles, natural polymeric nanoparticles have been used heavily in biomedical applications due to their biocompatibility, low toxicity, and reduced environmental footprint (6, 7).

In an attempt to find new organic nanoparticles, we turned our efforts to the extraction of nanomaterials directly from plants. To achieve this goal, our group has investigated multiple polymers secreted by natural systems in order to inspire the design of new organic polymer nanostructures. In 2008, our group first discovered

Mikhail Soloviev (ed.), *Nanoparticles in Biology and Medicine: Methods and Protocols*, Methods in Molecular Biology, vol. 906, DOI 10.1007/978-1-61779-953-2_31, © Springer Science+Business Media, LLC 2012

that organic nanoparticles were secreted with the adhesive of both Boston and English ivy (8). It was believed that these nanoparticles contributed to the high strength of the ivy adhesive, allowing it to attach to surfaces through both the physical and chemical cross-linking of nanostructures (9–11). After further examination of the nanoparticles, we discovered that the nanoparticles had unique optical properties, making them promising candidates for UV fillers in sunscreens (12). Continued study of other plants has yielded the discovery of various nanomaterials, including nanoparticles, which may be used for a wide-range of biomedical applications (13–16).

In the following sections, we report a method for extraction of nanoparticles from plants using three basic procedures: collection and preparation of plant tissue, homogenization, and separation of nanoparticles by size. The isolated nanoparticles are analyzed for both size and morphology using atomic force microscopy (AFM), dynamic light scattering (DLS), and Zeta Potential analysis.

2. Materials

All solutions were prepared using Ultrapure or double-distilled water and reagents were electron microscopy grade when available. Prepared solutions were stored at 4 °C unless otherwise indicated. Follow proper waste disposal procedures, and store all toxic waste in approved containers.

2.1. Plant Sterilization Components

1. Roots soaking solution: 1.25 % sodium hypochlorite, 0.05 % Tween 20.
2. Murashige and Skoog (MS) medium (Sigma).
3. Magenta GA7 (Magenta Corporation, Chicago, IL) container. Drill four holes in the corners of the GA7 boxes using a 1/8″ boring drill bit, this will allow the shoots to stand upright in the boxes.
4. A plant growth chamber equipped with humidity, light, and temperature control is necessary to ensure uniform growth conditions. As an alternative, plants could be grown in a highly controlled greenhouse.
5. 70 % Ethanol for sterilization of scalpels and forceps.
6. 50 mL BD Falcon conical centrifuge tubes or a Tupperware container for larger batches.
7. Rotary shaker and shake for 20 min.
8. 100 mm Petri dishes.

2.2. Homogenization Components

1. Fisher Scientific Model 505 Sonic Dismembrator with ½″ probe. Other probes can be attached, depending on the volume of solution.

2. Eppendorf 5702 R centrifuge with a swinging bucket rotor or similar centrifuge capable of 1,000 × *g*.
3. BD 5–50 mL syringe with Luer-Lok tip (Becton Dickinson).
4. Whatman Anotop 10 Plus 0.2 μm inorganic syringe filter.
5. Whatman Puradisc 25 NYL syringe filter with a 1.0 μm nylon membrane and polypropylene housing for filtration of nanoparticles.
6. 300 kDa Cellulose Ester dialysis tubing (Spectrum laboratories, Rancho Dominguez, CA) and two dialysis tubing clamps (Bio design, Cat. No. NC9951362) for dialysis. Tubing should be maintained in a hydrated state, and cut to 6 cm in length.

2.3. Size Separation Components

1. Make sucrose solutions ranging from 10 to 90 % in Ultrapure water. Make a 10 % concentration by mixing 5 g of sucrose with 50 mL of dH_2O in a beaker and mix it with constant stirring on a stir plate. Once the sucrose has dissolved, transfer the solution to a 50 mL conical tube and store it at 4 °C. This method should be repeated for all concentrations, i.e., 10 g sucrose in 50 mL of dH_2O for a concentration of 20 % and so on (see Note 1).
2. 15 mL Falcon conical tube.
3. Refrigerated centrifuge, capable of spinning 15 mL tubes at 5,800 × *g*.
4. Benchtop microcentrifuge capable of 14,000 rpm.
5. Phenomenex BiosSep SEC-S 4000 Size exclusion chromatography (SEC) column.
6. Agilent HPLC system with a UV detector, 200 μL sample loop, and auto sampler or a similar system.

2.4. Analysis of Nanoparticle Size and Morphology

1. ZetaSizer Nano ZS (Malvern Instruments Ltd.). This will allow for DLS and zeta potential analysis of the nanoparticles in solution.
2. Disposable semi-micro cuvettes (Fisher Scientific) for DLS and 4.5 mL disposable polystyrene cuvette (Fisher Scientific) for zeta potential analysis.
3. Prepare fresh Piranha solution by mixing concentrated sulfuric acid (H_2SO_4) and hydrogen peroxide (H_2O_2) at a 3:1 ratio (see Note 2).
4. Dip Cell (Malvern).
5. Agilent AFM 6000 ILM/AFM system (Agilent Technologies) for AFM.
6. Budget Sensors® Tap150AL-G (Innovative Solutions Bulgaria) with a resonant frequency of 150 kHz and a force constant of 5 N/m for imaging nanoparticles.

3. Methods

3.1. Collection and Sterilization of Plants

1. Collect plants for analysis either from the environment or grow from seedlings in a greenhouse. Greenhouse grown plants are desirable because they can be better controlled to prevent disease, specifically epiphytes, making further sterilization easier (see Note 3).
2. After harvesting plants, cut the shoots into 6 cm segments using pruning shears and remove the leaves.
3. Soak the shoots in a root soaking solution: 1.25 % sodium hypochlorite, 0.05 % Tween 20 to sterilize the shoots. Tween 20 acts as a detergent to help wash the shoots and remove large debris. Place the cut shoots into 50 mL BD Falcon conical centrifuge tubes (Becton Dickinson) and agitate for 20 min using a rotary shaker. For larger batches, place the shoots into a Tupperware container and shake for 20 min (see Note 4).
4. After sterilization place the shoots upright into Magenta GA7 containers containing Murashige and Skoog (MS) medium.
5. Expose shoots to a 16:8 h day/night cycle with 82 μmol/m^2/s irradiance in a plant growth chamber for at least 10 days.
6. Harvest the nanoparticle containing tissue of interest, leaves, roots, petioles, etc. after a significant amount of tissue emerges, by removing the shoots from the Magenta boxes with flame sterilized forceps and placing them in sterile 15 × 100 mm petri dishes (see Note 5).
7. After placing the shoot in the petri dish, use a sterile scalpel to remove the specific tissue of interest and place into a 50 mL Falcon tube for homogenization (see Note 6).

3.2. Homogenization of Harvested Plant Tissue

1. After collecting all of the tissue for homogenization, fill a Falcon tube with Ultrapure water and shake vigorously by hand for 1 min. Pour off the water and repeat the process three times (see Note 7). After the final wash, resuspend the tissue in Ultrapure water to a level just above the tissue.
2. Homogenize the tissue using the sonic dismembrator. Using the ½″ probe, 10–250 mL of solution can be sonicated. Sonicate the samples contained in the 50 mL conical tubes at maximum amplitude of 75 % for 15 min. To make sure that the sample is not heated during sonication, the sample should be contained in an ice bucket with crushed ice for the duration of the process.
3. After sonication, centrifuge the 50 mL conical tube at 1,000 ×g in a swinging bucket rotor.

4. Transfer the supernatant to a sterile 50 mL conical tube for further processing. Leave large debris and tissue fragments at the bottom of the tube.
5. Pour the homogenized sample into the syringe and pass through the Whatman Puradisc 25 NYL syringe filter. This removes debris larger than 1,000 nm.
6. Collect the filtrate in either a 15 or 50 mL conical tube (see Note 8).
7. Repeat steps 5 and 6 using the Whatman Anotop 10 Plus 0.2 μm syringe filter to filter out debris larger than 200 nm. At this point only nanoparticles less than 200 nm are expected to remain in the filtrate (see Note 9).
8. Once the filtration has been completed, dialyze the samples through 300 kD dialysis tubing to remove salts and small biomolecules from the samples. Clamp the tubing at one end using a dialysis tube clamp. With one end clamped, fill the dialysis tubing with the sample until the tubing is taut and then clamp the open end using another tube clamp.
9. Place the sealed dialysis tubing into a 1,000 mL beaker with Ultrapure water and suspend from the top of the beaker with fishing line. Once the dialysis tube has been properly suspended, the beaker can be placed onto a stir plate and stirred with gentle mixing overnight.

3.3. Size-Dependent Separation of Nanoparticles

Nanoparticles extracted from plants often vary in size, it is therefore necessary to further fractionate them in order to obtain uniform nanoparticles. Both the size and shape of nanoparticles can be highly variable. Here we describe two methods for separating nanoparticles of different sizes.

3.3.1. Sucrose Density Gradient Centrifugation

1. Make the step gradients of sucrose (10–90 %) by first adding 1 mL of the highest concentration of sucrose, 90 %, to the bottom of a 15 mL Falcon conical tube using a 1 mL serological pipette.
2. Add 1 mL of the second highest concentration, 80 %, to the top of the first layer by *gently* pipetting the solution along the wall of the tube. At the completion of this you should see two distinct layers. The process will then be continued for all subsequent layering (see Note 10).
3. Once the gradient has been formed, the content of the dialysis tubing (containing the nanoparticles) should be layered on top of the gradient (see Note 11).
4. After layering the nanoparticle solution on top of the gradient, transfer the tube to a centrifuge to pull the different fractions into the gradient for separation. Centrifuge the gradients at 5,800 × *g* for 30 min at 4 °C to effectively separate different nanoparticles (see Note 12).

5. Upon completion of centrifugation remove each layer of the gradient, 1 mL at a time, and place them in 1.5 mL Eppendorf microfuge tubes. Typically the layers containing metallic nanoparticles have a slight color, which gives a hint that there are many nanoparticles in that fraction. With organic nanoparticles, however, there may be no color in each layer, despite the presence of nanoparticles.
6. Centrifuge each collected layer at max speed for at least 30 min to pellet the nanoparticles.
7. Remove and discard the supernatant.
8. Resupend the nanoparticle pellet in Ultrapure water and repeat steps 6–8 three times to wash the sucrose from the nanoparticles.
9. After the final wash, resuspend the nanoparticles in 100 μL Ultrapure water prior to analysis by DLS and AFM.

3.3.2. Size Exclusion Chromatography

SEC can accurately and repeatedly be used to separate the target particles from other components in the extract. SEC relies on the use of a gel filtration column of known pore size to separate particles of different sizes. This technique requires limited sample preparation, and can be easily automated for high throughput. Many columns can be used; however, in this procedure we will describe the use of the Phenomenex BiosSep SEC-S 4000 column for separation.

1. Connect the inflow and outflow of the column to an Agilent HPLC system with a UV detector, 200 μL sample loop, and auto sampler.
2. Allow the UV detector to warm up for at least 1 h before use.
3. Wash the storage solution, typically methanol or sodium azide, from the column by connecting the inflow tubing to a 1 L bottle containing Ultrapure water. To remove all storage solution, the column should be run overnight at a flow rate of 0.5 mL/min.
4. Use the autosampler to load 200 μL of the dialyzed solution into the column (see Note 13).
5. Elute the sample from the column with running buffer (Ultrapure water).
6. Monitor the elution of nanoparticles using the UV detector at wavelengths of 230, 280, and 340 nm. Peaks in the detector indicate that the fraction contains structures that contain polyphenols, peptides, proteins, or sugars respectively. This does not directly indicate the presence of nanoparticles, but could provide information for subsequent chemical analysis should nanoparticles be present in these fractions.

7. Collect fractions every 1 min, by setting of the autosampler software, and analyze the fractions by DLS and AFM for the presence of nanostructures (see Note 14).

3.4. Analysis of Nanoparticle Size and Morphology

3.4.1. Dynamic Light Scattering

This technique can be used to quickly determine the size of the nanoparticles in solution, but cannot provide any data on the morphology of the nanoparticles. This method is used prior to AFM imaging, since AFM imaging is more laborious and technically difficult.

1. Add the fractions to be analyzed, either from the SEC or sucrose gradient separation, to disposable semi-micro cuvettes. If necessary, dilute the samples to a final volume of 1.0 mL with Ultrapure water to ensure that the optical cell is filled (see Note 15).
2. Once the sample has been loaded into the semi-micro cuvette, place the cuvette into a Malvern ZetaSizer Nano DLS detector, with the optical path facing the front of the machine.
3. Using the ZetaSizer Nano software, a new SOP should be created indicating the cuvette type, low volume disposable cuvette, temperature, 25 °C, the material type, protein, and the equilibration period, 30 s (see Note 16).
4. After saving the SOP, select the Measurement tab and run the SOP that was created. The ZetaSizer Nano will automatically optimize the detection range and the number of measurements to provide statistically significant data.
5. Upon completion of the run, the software will display a report indicating the size range of the nanoparticles, and the reliability of the data. If the report indicates a "good value," then the sample can be set aside for AFM. If the report indicates that the data is questionable then the sample must be rerun.
6. If no nanoparticles are present in the fraction, then the correlation curve will be discontinuous and the sample can be discarded.

3.4.2. Zeta Potential Analysis

Zeta potential analysis provides an accurate measure of the nanoparticles tendency to aggregate and precipitate from solution, and thus can be coupled with DLS to get a more complete picture of the nanoparticles in solution.

1. Transfer fractions that were shown to contain nanoparticles by DLS to a standard polystyrene cuvette for Zeta Potential analysis.
2. Once the sample has been loaded into the cuvette, insert the Dip Cell into the cuvette with the triangles on the Dip Cell facing the triangle on the cuvette. Place the entire Dip Cell and cuvette into the sample holder on the ZetaSizer Nano.

3. Using the ZetaSizer Nano software, a new SOP should be created indicating the cuvette type, low volume disposable cuvette, temperature, 25 °C, the material type, protein, and the equilibration period, 30 s. This procedure is the same as for sizing, except that the measurement type selected from the menu will be for the Zeta Potential analysis (see Note 17).
4. After saving the SOP, select the Measurement tab and run the SOP that was created. The ZetaSizer Nano will automatically optimize the detection range and the number of measurements to provide statistically significant data.
5. Upon completion of the test, the software will display a report indicating the Zeta Potential of the sample. A Zeta Potential greater than 30 mv or less than −30 mv indicates a stable suspension. Values outside this range indicate that the nanoparticles are aggregating, or falling out of solution.

3.4.3. Atomic Force Microscopy

AFM is a common approach for determining the nanoscale morphology of nanomaterials, and can image nanostructures up to 100 μm and below 1 nm.

1. After determining which samples from the separation techniques contain nanoparticles, use an AFM to determine the particle morphology.
2. Clean glass coverslips that will be used as the substrate for imaging by dipping in freshly prepared Piranha solution for 15 min to remove any organic debris on the glass surface.
3. Place the coverslips into a sterile petri dish until needed.
4. Pipet 50 μL of the nanoparticles solution onto the center of cleaned coverslips.
5. Air-dry samples overnight at ambient temperature in a Biological Safety Cabinet or a hood.
6. Image air-dried samples using AFM operating in air in AC mode with specified cantilevers.
7. Scan at less than 1 ln/s with a resolution of 1,024 × 1,024 pixels to achieve publication quality scans.
8. The morphology and size of the scanned nanoparticles can be determined with the PicoView® software package.

4. Notes

1. At concentrations higher than 80 % it may be necessary to gently heat the solution to get it in to solution faster.
2. Piranha solution is *extremely* corrosive and requires full personal protective equipment including a full face shield, heavy

duty rubber gloves, and an acid apron. The solution should only be mixed in an appropriate fume hood, and the hydrogen peroxide should always be slowly added to the acid. The reaction generates substantial heat, and the solution should be allowed to cool before transferring into a secondary container.

3. Contamination control: The isolation of nanoparticles from the natural environment requires special consideration due to the existence of nanoparticles in the environment from manufacturing processes. Environmental samples often carry dust and other tightly attached particles, and these samples have natural mixtures of particles with a wide size range. All of these will complicate the later purification steps. To minimize contamination, plants selected for isolation should be grown aseptically in a sterile environment, or in a well controlled greenhouse. This will minimize the potential for contamination from circulating air.
4. Some plants have epiphytes that lead to contamination of the MS media. If this occurrs, Aliette® WDG can be added to the sterilization dip to inhibit fungal growth.
5. Harvesting varies between plants, however, roots tend to emerge within the first 48 h, while leaves are much slower to emerge.
6. For adequate homogenization, it was found that mincing the tissue with the sterile scalpel provided more homogenous samples.
7. To ensure that no tissue is lost during pouring, the opening of the tube should be covered with a Millipore 224 Circular Ashless filter paper (Millipore). This paper retains particles greater than 16 μm.
8. The choice of filter depends on the size of the particles that are desired. The use of smaller filters will result in less larger sized nanoparticles, however, it is often difficult to filter these materials through a syringe filter. Filtering through a decreasing size series of filters will make the filtration easier. It is also necessary to carefully select the membrane for the filter to make sure that the nanoparticles don't bind to the membrane and clog the filter.
9. If the goal is to isolate nanoparticles less than 100 nm, then the process can be repeated using a Whatman Anotop 10 0.1 μm inorganic syringe.
10. It is crucial that the pipetting be well controlled, if the solutions mix, then the gradient will not form. If you are not familiar with the technique, add a dye, such as bromothymol blue to every other layer to make sure you have the technique perfected before attempting separation.
11. Since the nanoparticles solution is often colored, it is easy to make sure that the gradient has been formed.

12. To separate smaller nanoparticles it may be necessary to spin at higher speeds. If ultracentrifugation is required, make sure that the Falcon tubes are replaced with Corning centrifuge tubes that can withstand the higher G-forces.

13. Depending on the sample loop chosen, and the volume of nanoparticle solution after dialysis, the loading volume may vary.

14. After the initial screening of the fractions, the time that the fractions can be collected can be adjusted based on the results from the nanoparticle analysis. Oftentimes, a more narrow collection time can be used over a shorter duration to collect finer distributions of nanoparticles.

15. Make sure that no bubbles are present in the cuvette, the presence of bubbles will drastically affect the DLS measurement.

16. Depending on the type of nanoparticles that are expected to be isolated, all of these factors can be adjusted, however, these are the settings typically used. The selection of protein as the material type does not factor into the size measurement, and would only be used in the case of attempting to measure molecular weight.

17. It is important to determine the pH of the solution prior to analysis of the Zeta Potential. Without this data the Zeta measurements will not be valid.

References

1. Bar H, Bhui DK, Sahoo GP et al (2009) Green synthesis of silver nanoparticles using seed extract of *Jatropha curcas*. Colloids Surface A 348:212–216
2. Chandran SP, Chaudhary M, Pasricha R et al (2006) Synthesis of gold nanotriangles and silver nanoparticles using *Aloevera* plant extract. Biotechnol Prog 22:577–583
3. Kumar V, Yadav SK (2009) Plant-mediated synthesis of silver and gold nanoparticles and their applications. J Chem Technol Biotechnol 84:151–157
4. Harris A, Bali R (2008) On the formation and extent of uptake of silver nanoparticles by live plants. J Nanopart Res 10:691–695
5. Gardea-Torresdey JL, Parsons JG, Gomez E, Peralta-Videa J et al (2002) Formation and growth of Au nanoparticles inside live *Alfalfa* plants. Nano Lett 2:397–401
6. Wen ZS, Xu YL, Zou XT et al (2011) Chitosan nanoparticles act as an adjuvant to promote both Th1 and Th2 immune responses induced by ovalbumin in mice. Mar Drugs 9: 1038–1055
7. Ofokansi K, Winter G, Fricker G et al (2010) Matrix-loaded biodegradable gelatin nanoparticles as new approach to improve drug loading and delivery. Eur J Pharm Biopharm 76:1–9
8. Zhang M, Liu M, Prest H et al (2008) Nanoparticles secreted from ivy rootlets for surface climbing. Nano Lett 8:1277–1280
9. Zhang M, Liu M, Bewick S et al (2009) Nanoparticles to increase adhesive properties of biologically secreted materials for surface affixing. J Biomed Nanotechnol 5:294–299
10. Wu Y, Zhao X, Zhang M (2010) Adhesion mechanics of ivy nanoparticles. J Colloid Interf Sci 344:533–540
11. Xia L, Lenaghan S, Zhang M et al (2011) Characterization of English ivy (*Hedera helix*) adhesion force and imaging using atomic force microscopy. J Nanopart Res 13: 1029–10379
12. Xia L, Lenaghan S, Zhang M et al (2010) Naturally occurring nanoparticles from English ivy: an alternative to metal-based nanoparticles for UV protection. J Nanobiotechnology 8:12

13. Arzt E, Gorb S, Spolenak R (2003) From micro to nano contacts in biological attachment devices. Proc Natl Acad Sci USA 100:10603–10606
14. Lenaghan S, Xia L, Zhang M (2009) Identification of nanofibers in the Chinese herbal medicine: Yunnan Baiyao. J Biomed Nanotechnol 5:472–476
15. Xia L, Lenaghan S, Wills AB et al (2011) Evaluation of the nanofibrillar structure of *Dioscorea* opposite extract for cell attachment. Colloids Surface B 88:425–431
16. Zhang M, Lenaghan S, Xia L et al (2010) Nanofibers and nanoparticles from the insect-capturing adhesive of the Sundew (*Drosera*) for cell attachment. J Nanobiotechnology 8:1–10

Part V

Nano-Toxicity and Nano-Safety

Chapter 32

Assessment of Cytotoxicity of Carbon Nanoparticles Using 3-(4,5-Dimethylthiazol-2-yl)-5-(3-Carboxymethoxyphenyl)-2-(4-Sulfophenyl)-2H-Tetrazolium (MTS) Cell Viability Assay

Amanda M. Schrand, Jonathan B. Lin, and Saber M. Hussain

Abstract

Ever since the discovery of carbon nanotubes, there has been an increasing interest in technologies that rely upon these incredibly small particles for their unique properties. However, assessment of their biological consequences has been riddled with assay limitations. Here, we describe application of a 3-(4,5-dimethylthiazol-2-yl)-5-(3-carboxymethoxyphenyl)-2-(4-sulfophenyl)-2H-tetrazolium cell viability assay to study cytotoxicity of various carbon-based nanomaterials on cells and discuss some pitfalls of this method.

Key words: Carbon, Nanotube, Diamond, Nanomaterial, Cell viability, MTS assay

1. Introduction

Ever since the discovery of carbon nanotubes (CNTs) (1), there has been an increasing interest in technologies that rely on these incredibly small particles for their unique properties, e.g., enhanced mechanical, chemical, and electrical properties. Relatively recently, there has been a great shift in focus on how carbon nanomaterials may be used in biomedicine based on their diverse surface chemistry and structure (2). The toxic potential of CNTs depends upon a number of factors including surface chemistry, purity, synthesis method, dosing concentration, exposure time, biochemical assay, and dimensions (3–5). As the use of CNTs in living systems becomes more feasible, it becomes increasingly important to assess their potential for doing harm within a living system. The most common route of exposure seems to be through the inhalation of NPs; therefore, it is very common to examine potential cytotoxicity through lung epithelial cells, human keratinocytes, or lung macrophages that provide the nonspecific immune response

Mikhail Soloviev (ed.), *Nanoparticles in Biology and Medicine: Methods and Protocols*, Methods in Molecular Biology, vol. 906, DOI 10.1007/978-1-61779-953-2_32,

in these fragile areas (6–9). Other carbon nanomaterials such as nanodiamonds are finding immense biomedical applications (10–12).

Cytotoxicity can be evaluated in many ways, both in vitro and in vivo. One of the most common cell viability assays is the 3-(4,5-dimethylthiazol-2-yl)-5-(3-carboxymethoxyphenyl)-2-(4-sulfophenyl)-2H-tetrazolium (MTS) assay. This assay consists of a cell-permeable, soluble tetrazolium salt, and electron coupling reagent (phenazine methosulfate, PMS) that is added to the cellular environment after the appropriate dosing periods. During incubation with MTS, the dehydrogenase enzymes present within the mitochondria of healthy cells reduce the MTS into a dark purple, water-soluble formazan compound, which can be quantified with a microplate reader at 490 nm. Absorbance values measured at zero incubation time are subtracted from the final readings to account for background absorbance due to the NPs or the solution. The changes are compared to controls (cells not exposed to NPs), and percent reduction is calculated. Further, the inherent reactivity of the carbon nanomaterials with the MTS reagent under acellular conditions is also assessed as a control measure to ensure the validity of the results.

2. Materials

Wear appropriate personal protective equipment, i.e., gloves, eye goggles, and lab coat when handling nanomaterials and chemicals. Perform all procedures in a sterile environment such as a laminar flow biohood with aseptic techniques. Prepare and store all chemicals in the appropriate environment, e.g., freezer, refrigerator. Diligently follow all waste disposal regulations for nanoparticles and cell cultures.

2.1. Cell Culture and Plating Cells

1. Neuro-2a (N2A) cell line or any appropriate cell line (see Note 1) grown in T-75 flasks in a CO_2 incubator set at 37 °C.
2. Dulbecco's Modified Eagle's Medium (DMEM) nutrient mixture F-12 Ham's supplemented with 10 % Fetal Bovine Serum (FBS) and 1 % Penicillin-Streptomycin, or other appropriate media depending on the cell line.
3. 96-Well tissue culture plates.
4. Phosphate buffered saline (PBS): 2.97 mM Na_2HPO_4, 1.06 mM KH_2PO_4, and 155.17 mM NaCl, pH 7.4.
5. 1 % Trypsin in 1× PBS.
6. Hemacytometer.

2.2. Preparing Nanoparticle Dosing Solutions (See Note 2)

1. Carbon nanoparticles dry powder, e.g., from Cheap Tubes Inc (see Note 3).
2. Nanoparticles stock solution: weigh 3.3 mg of carbon nanoparticles and transfer to a 15 mL polystyrene conical tube. Add 3.3 mL of sterile water and vortex to disperse. At this step it may be important to characterize the particles in solution (see Note 4).
3. DMEM nutrient mixture F-12 Ham's.

2.3. MTS Assay for Cell Viability

1. CellTiter 96® AQ_{ueous} One Solution Cell Proliferation Assay (MTS Assay, Promega), contains a tetrazolium compound MTS, inner salt, and an electron coupling reagent phenazine ethosulfate (PES). This is normally shipped at room temperature but should be aliquoted into 2–4 mL vials and stored at −20 °C. Avoid excessive freeze-thaw cycles.
2. BD Tek Microplate Reader or other similar instrument that can measure absorbance at 490 nm.
3. Softmax 3.1 software (Softmax Pro).

2.4. Data Analysis

1. Statistical analysis software in Microsoft Excel or other programs such as PhSTAT, IBM SPSS 17.0 (see Note 5).

3. Methods

In short, the MTS assay is performed by adding a small amount of the CellTiter 96® AQ_{ueous} One Solution Reagent directly to culture wells, incubating for 1–4 h and then recording absorbance at 490 nm with a 96-well plate reader. The quantity of formazan product "as measured by the amount of 490 nm absorbance" is directly proportional to the number of living cells in culture. A background reading is taken at time "zero" immediately upon the addition of nanoparticles to account for absorbance changes.

3.1. Cell Culture and Plating Cells

1. Remove flask containing grown N2A cells from the incubator and place under the laminar flow hood.
2. Remove media with glass pipette connected to a vacuum pump.
3. Add 1–2 mL 1 % trypsin to the flask and tap with hand to release the cells, incubate for approximately 5 min.
4. Add 5–10 mL of media to neutralize the trypsin, mixing the cells up and down with the pipette to break up any clumps.
5. Transfer a small amount (30 μL) of the cells in media to the hemacytometer and place a coverslip on top to count them.

Estimate the cells/mL by counting the total number of cells in one of the 4×4 grids and multiplying that number by 10^4.

6. Dilute cells using growth media (with serum) for plating. For the next day analysis (24 h of cell growth) make ~500,000 cells/mL suspension, for a 3 day growth make ~250,000 cells/mL suspension to ensure that the cells reach ~80 % confluence by the time of analysis.
7. Add the diluted cells to the plates. Add 0.2 mL/well for a 96 well plate (14.4 mL total volume needed), use 1 mL/well for 6-well plates (6 mL in total), or 0.5 mL/well for 24-well plates (12 mL in total). It is useful to use boats and a micropipetter for transferring aliquots of cells to the plate wells. The different volumes assist in plating optimal cell numbers and liquid volumes to coat the bottoms of the wells. Under these conditions, cells should be seeded at a sufficient density to achieve 80 % confluency by the time of treatment (see Notes 6 and 7).
8. Label new flask and plates with information such as cell type, passage number, and date/time. Transfer the cells to a CO_2 incubator and grow at 37°C until 80 % confluent.

3.2. Nanoparticle Dosing Solutions

1. Prepare nanoparticle dosing solutions. Use serial dilutions to make 0, 5, 10, 25, 50, 100 μg/mL NPs in DMEM/F12 medium. Do not use serum for N2A cells. For control wells (0 μg/mL), use cell culture media without serum (see Note 8).
2. For dosing, allow the nanoparticle solution to equilibrate to room temperature if stored in the refrigerator and mix the solutions well by vortexing and pipetting.
3. Remove the 96-well plates with 80 % confluent cells from the incubator and place under the laminar flow hood.
4. Remove media with glass pipette connected to the vacuum pump. Add 200 μL of each of the dose solutions to the 96-well plate. Allow sufficient number of replicates, see Fig. 1.
5. Set up positive and negative controls for cells viability, for the negative control use dosing solution without NPs (zero concentration). Use known toxic materials such as $AgNO_3$ or CdO for positive control. If the positive control is reactive, e.g., H_2O_2, make sure that is prepared fresh.
6. Note the time and the date dosing solutions were added to the cells, label the plate and return the plate to the incubator.

3.3. MTS Assay for Cell Viability

1. Transfer all plates to the laminar fume hood and add 20 μL of MTS stock per well (for 96-well plates). Do not aspirate the media, add MTS directly onto the side of the well into the media in each well.
2. Immediately take a reading at 490 nm at the 0 h time point (see Note 9).

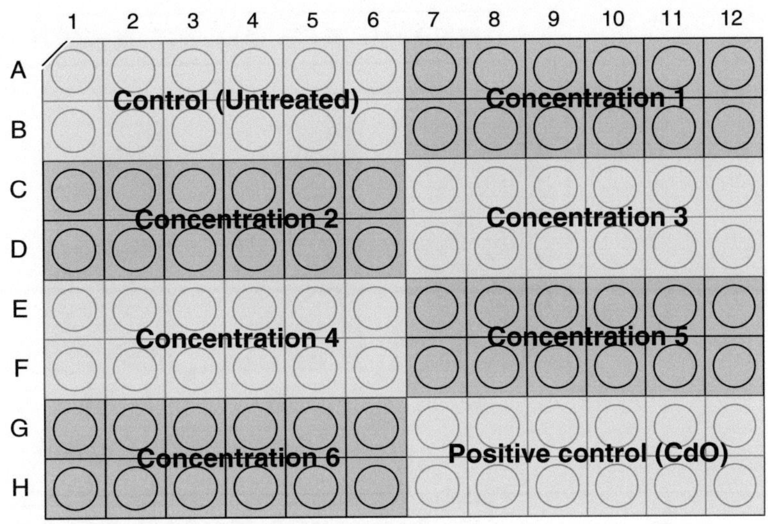

Fig. 1. Representative set-up for 96-well plate.

3. Transfer the plates back to the CO_2 incubator and incubate at 37 °C for 1–4 h (some wells should turn purple).
4. Take another reading at 490 nm.
5. Appropriately dispose of waste materials.

3.4. Data Analysis

1. Take raw microplate readings for each cell condition and subtract the background readings taken at "zero" incubation time. Do not average the readings at this stage. Treat readings from individual wells separately.
2. Determine percent viability for all treated cells using this simple formula:

$$\% \text{ viability} = 100 \times (\text{OD}_t / \text{OD}_c),$$

where

OD_t = Optical Density of treated samples (with the background subtracted).

OD_c = Optical Density of control (with the background subtracted.

1. Average these values across all wells for each treatment condition.
2. Conduct statistical tests (*t* test, one-way ANOVA with post-hoc test) to determine the significance of differences using any of many statistical analysis software programs (e.g., IBM SPSS 17.0, PhSTAT, see Note 5). Figure 2 shows an example of the data obtained for carbon-based NPs.

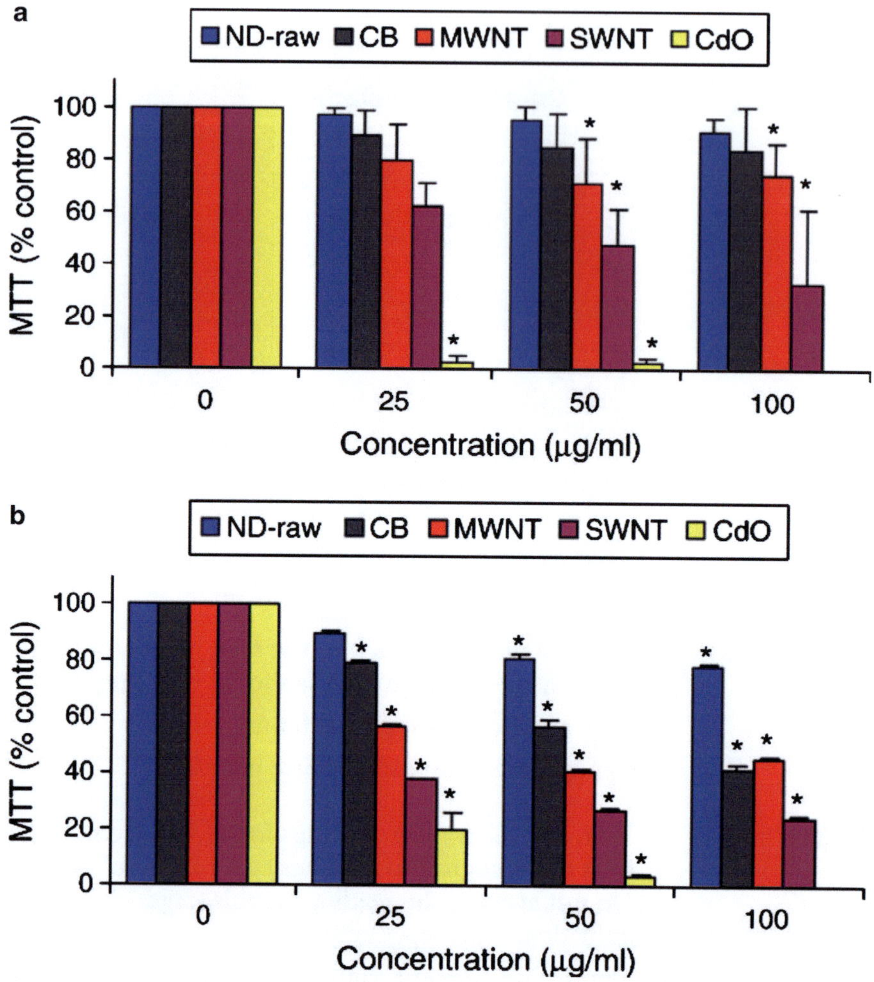

Fig. 2. Cytotoxicity evaluation through an assay similar to MTS (MTT) after 24 h of incubation with a variety of carbon-based NPs in (**a**) neuroblastoma cells and (**b**) macrophages. Reproduced from (13) with permission from Elsevier.

4. Notes

1. Other cell lines may be more appropriate depending on the application and the model system desired.
2. Nanoparticles are typically weighed and suspended in deionized water, at ~1 mg/mL. When weighing dry nanoparticles use a scale in a hood and close the doors to prevent wind/air from affecting your measurements. Tare the scale with weighing paper. If NPs are received as a solution, dilute to ~1 mg/mL with water or the recommended solvent.
3. Dry nanoparticles should be examined in the scanning electron microscope (SEM) or transmission electron microscope (TEM) for size and morphology estimation prior to making any

dilutions. Other characterization for impurities, crystallinity, etc. may be desirable prior to cellular dosing.

4. Procedures to enhance nanomaterial dispersion may change surface or other physicochemical properties. Keep good notes of any processing to the materials prior to dosing, e.g., sonication, surfactants added. Other relevant solution properties to assess include dynamic size in solution and pH.
5. Other statistical analysis software can also be used, as long as it can perform all of the necessary functions.
6. Too many or two few cells may skew the results of the MTS assay. It is suggested to seed cells at high enough density to achieve 80 % confluency by time of dosing.
7. During cell plating, be sure to re-agitate the cell solution before each addition to assure that the cells are dispersed and have not settled to the bottom of the boat.
8. Doses for soluble control materials are typically lower such as $AgNO_3$: 0, 0.5, 1, 2.5, 5, 10 μg/mL or Mn acetate: 0, 1, 2.5, 5, 10, 25 μg/mL.
9. We used the BD Tek Microplate Reader and Softmax 3.1 and followed the instructions provided with the instrument. Another plate reader and software can be used, always follow the manufacturer's instructions.

Acknowledgement

A.M.S. received funding from the National Research Council (NRC) and J.B.L. received funding from the S.T.E.P. program at AFRL/RHPB.

References

1. Ijima S (1991) Helical microtubules of graphitic carbon. Nature 354:56–58
2. Dai L (ed) (2006) Carbon nanotechnology: recent developments in chemistry, physics, materials science and device applications. Elsevier, Amsterdam
3. Bottini M, Bruckner S, Nika K, Bottini N, Bellucci S, Magrini A, Bergamaschi A, Mustelin T (2006) Multi-walled carbon nanotubes induce T lymphocyte apoptosis. Toxicol Lett 160:121–126
4. Magrez A, Kasas S, Salicio V, Pasquier N, Seo JW, Celio M, Catsicas S, Schwaller B, Forro L (2006) Cellular toxicity of carbon-based nanomaterials. Nano Lett 6(6):1121–1125
5. Tian F, Cui D, Schwarz H, Estrada GG, Kobayashi H (2006) Cytotoxicity of single-wall carbon nanotubes on human fibroblasts. Toxicol In Vitro 20:1202–1212
6. Fiorito S, Serafino A, Andreola F, Bernier P (2006) Effects of fullerenes and single-wall carbon nanotubes on murine and human macrophages. Carbon 44:1100–1105
7. Sato K, Imai Y, Higashi N, Kumamoto Y, Mukaida N, Irimura T (2005) Redistributions of macrophages expressing the macrophage galactose-type C-type lectin (MGL) during antigen-induced chronic granulation tissue formation. Int Immunol 17:559–568

8. Jia G, Wang H, Yan L, Wang X, Pei R, Yan T, Zhao Y, Guo X (2005) Cytotoxicity of carbon nanomaterials: single-wall nanotube, multi-wall nanotube, and fullerene. Environ Sci Technol 39:1378–1383
9. Murr LE, Garza KM, Soto KF, Carrasco A, Powell TG, Ramirez DA, Guerrero PA, Lopez DA, Venzor J III (2005) Cytotoxicity assessment of some carbon nanotubes and related carbon nanoparticle aggregates and the implications for anthropogenic carbon nanotube aggregates in the environment. Int J Environ Res Public Health 2:31–42
10. Schrand AM, Citan SA, Shenderova OA (2009) Nanodiamond particles: properties and perspectives for bioapplications. Crit Rev Solid State Mater Sci 34:18–74
11. Schrand AM, Huang H, Carlson C, Schlager JJ, Ōsawa E, Hussain SM, Dai L (2007) Are diamond nanoparticles cytotoxic? J Phys Chem B 111:2–7
12. Schrand AM, Lin JB, Hens CS (2011) Temporal and mechanistic tracking of cellular uptake dynamics with novel surface fluorophore-bound nanodiamonds. Nanoscale 3:435–445
13. Schrand AM, Dai L, Schlager JJ, Hussain SM, Ōsawa E (2007) Differential biocompatibility of carbon nanotubes and nanodiamonds. Diamond Relat Mater 16:2118–2123

Chapter 33

Luciferase Reporter System for Studying the Effect of Nanoparticles on Gene Expression

Min Ding, Linda Bowman, and Vincent Castranova

Abstract

Nanotechnology exploits the fact that nanoparticles exhibit unique physicochemical properties, which are distinct from larger particles of the same composition. It follows that nanoparticles may also express distinct bioactivity and unique interactions with biological systems. Therefore, it is essential to assess the potential health risks of exposure to nanoparticles to allow development and implementation of prevention measures. One of the biggest challenges facing the field of nanotoxicology is the huge variety of different nanoparticle types possessing a variety of properties. Genetic Luciferase Reporter System or Reporter gene assay has become an invaluable tool in studies of gene expression. This is achieved by linking the firefly luciferase gene to a promoter sequence. Luciferase assays are quick, highly sensitive, have wide dynamic range, and are cheap to perform. Because of their simplicity and versatility, and because of the absence of endogenous luciferase activity in most cell types, this test can be applied for testing a large variety of nanomaterials for their pathogenic or carcinogenetic effects on a wide range of mammalian cells. This system is an ideal early-stage toxicology tool for screening nanomaterials. Here we describe the Genetic Luciferase Reporter System as the method for detecting alteration of gene expression in response to external stimuli (e.g., nanoparticles).

Key words: Nanoparticles, Firefly luciferase, Reporter genes, Gene expression, Promoter

1. Introduction

Nanotechnology includes manipulating matter on a near-atomic scale to produce new structures, materials, and devices. It is essential to conduct hazard and exposure assessment to support risk assessment and allow the development of recommendations for exposure limits and the implementation of prevention strategies (1). A major challenge in hazard assessment in nanotechnology is the large and rapidly growing number of possible nanoparticles to be tested for biological activity. For example, carbon nanotubes (CNT) can vary in width, length, be single-walled, double-walled, or multi-walled; CNT can be functionalized with a variety of chemical

Mikhail Soloviev (ed.), *Nanoparticles in Biology and Medicine: Methods and Protocols*, Methods in Molecular Biology, vol. 906, DOI 10.1007/978-1-61779-953-2_33, © Springer Science+Business Media, LLC 2012

groups. Thus, for CNT alone, there can be hundreds of different test materials with different physicochemical properties, possibly exhibiting substantially different bioactivity.

Reporter gene assay provide an easy to perform, fast, and affordable way to test nanomaterials for a wide variety of biologic effects in mammalian cells. Direct quantitation of changes in gene expression may rely on the measurement of specific mRNAs, using techniques such as northern blot hybridization, RT-PCR, or a variety of array-based techniques. These procedures are not always practical for the analysis of many different gene constructs. An alternate approach to gauge changes in transcription is to link the presumed *cis*-acting sequences from the gene of interest to the coding sequence of an unrelated reporter gene (2–6) (Fig. 1).

The most versatile and common reporter gene is the luciferase of the North American firefly *Photinus pyralis*, (see Fig. 1). The protein requires no posttranslational modification for enzyme activity; it is not toxic at high concentration (in vivo) and can be used in prokaryotic and eukaryotic cells (7–9). A reporter protein (i.e., luciferase, see Fig. 2). is easily detectable and quantifiable.

The assay procedure is simple and includes transfection of cells with luciferase expression plasmids, followed by cell lysis using Triton X-100 to release the expressed reporter protein luciferase. Both ATP and the substrate luciferin are added to the lysate in a luminometer. The enzyme catalyzes a rapid, ATP-dependent oxidation of the substrate, which then emits light (Fig. 3). Total light output is proportional to the amount of luciferase present over a wide range of enzyme concentrations (10, 11). It is important to recognize that these assays use reporter constructs to measure protein level or activity and not RNA level. The two levels are frequently, but not always, correlated with one another.

The most commonly used luciferase reporter systems in our lab are pre-oncogene/promoter-luciferase reporters, such as AP-1-luciferase, NF-κB luciferase, p53 luciferase, COX-2 luciferase, NFAT luciferase (11–14). Alterations of these promoter activation or gene expression are linked to neoplastic transformation, tumor

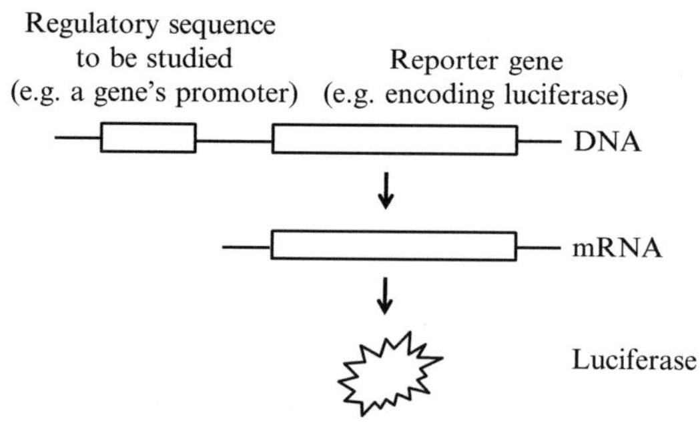

Fig. 1. The use of luciferase reporter gene system to study a regulatory sequence.

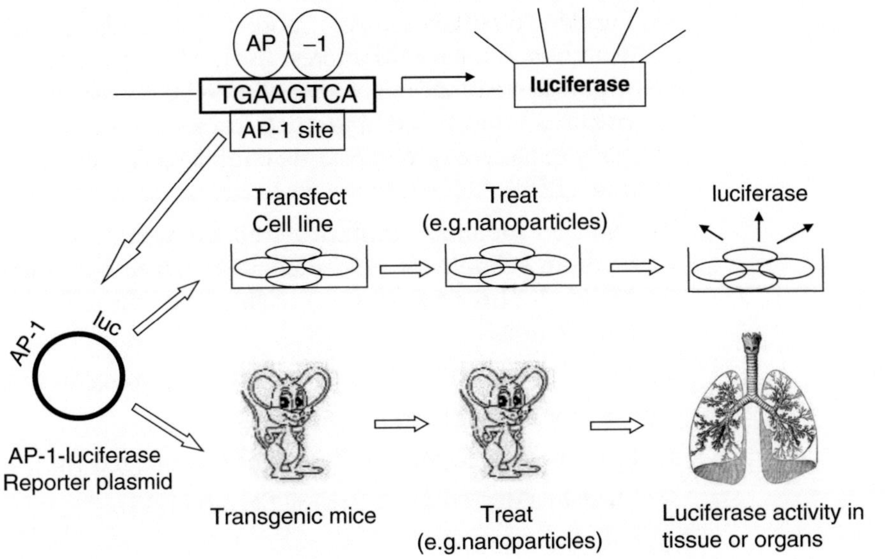

Fig. 2. Overview of luciferase reporter system. A diagram of how a reporter system is created in cell lines or animals to study a gene alteration/expression.

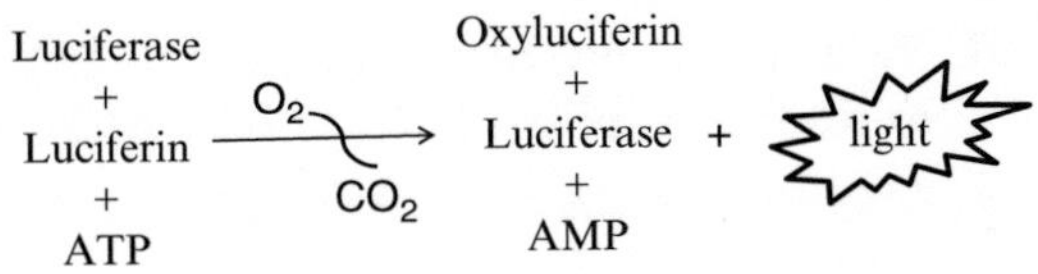

Fig. 3. The firefly luciferase catalyzes bioluminescent oxidation of the luciferin in the presence of ATP, magnesium, and oxygen.

progression, metastasis, inflammation, proliferation, and apoptosis. The assays described here would be suitable for studying the effects of nanomaterials on AP-1, NF-κB, p53 activity, or COX-2 expression and should help to assess the potential health risks of exposure to nanoparticles.

2. Materials

2.1. Development of Stable Transfected Cell Lines

All reagents and equipment coming into contact with live cells must be sterile, and proper sterile technique should be followed accordingly. Selection of stably transfected colonies requires that cells be grown essentially in isolation. A cell line that does not grow as an isolated cell will not be able to be stably transfected.

1. Parental adherent cell lines (e.g., JB6).
2. Minimum Essential Medium Eagle (EMEM) supplemented with FBS (5 %), Penicillin-Streptomycin (1 %), and L-glutamine (2 mM) suitable for the cell line use (JB6 in this case).
3. 37 °C, 5 % CO_2,80 % humidity incubator for cell culturing.

4. Purified plasmid DNAs: reporter plasmids containing the promoter sequences of interest (e.g., AP-1, NF-kB, p53) and the antibiotic-resistance gene that allows the transfected cells to survive in a culture containing the corresponding antibiotic. The ready to use vector plasmids are commercially available, e.g., plasmid 11783, "NFAT/AP-1 3x luciferase from Addgene Inc".
5. Dilution medium: antibiotic-free and serum-free cell culture medium suitable for lipid-mediated transfection such as Opti-MEM I (Invitrogen Inc.) yields the highest efficiency of transfection.
6. Transfection Medium as above but containing serum.
7. Cell counter or hemocytometer.
8. Cationic Lipofectamine™ reagent (Invitrogen Inc).
9. Heat-inactivated Fetal Bovine Serum (FBS).
10. Cloning cylinder.
11. Sterile vacuum grease.
12. Trypsin-EDTA: 0.5 % Trypsin, 5.3 mM EDTA. Dilute to 0.01 % in 1× PBS for JB6 cells.
13. Phorbol-12-mysristate-13-acetate (TPA): 20 μg/mL solution in dimethyl sulfoxide (DMSO). Aliquot and store at −80 °C.
14. Luciferase Assay System (Promega), containing Luciferase assay substrate, Luciferase assay buffer, and Cell culture lysis 5× reagent (30 mL/vial). Add Luciferase Assay Buffer (10 mL) to the vial containing the lyophilized Luciferase Assay Substrate. Reconstituted luciferase assay reagent should be stored in aliquots at −20 °C for up to 1 month or at −70 °C for up to 1 year. Add 4 volumes of water to 1 volume of 5× Luciferase Cell Culture Lysis Reagent to make 1× lysis buffer (see Note 1).
15. Complete medium for JB6 cells: EMEM supplemented with FBS (5 %), Penicillin-Streptomycin (1 %), and L-glutamine (2 mM) for the cell line of JB6.
16. Antibiotics for selection of transfected cells: 300 mg/mL G418 for use with JB6 cells.

2.2. Nanoparticles-Induced Alterations of Gene Expression

2.2.1. Preparation of Nanomaterials

1. Nanoparticles to be investigated, e.g., TiO_2 nanoparticles. To make stock solutions, dissolve nanoparticles in sterile PBS at 10 mg/mL, and sonicate the solution. Working solutions are prepared by diluting the stock solution of nanoparticle with cell growth medium for cell treatment. The final concentration of the particle should be 0–400 μg/cm^2. Use 1× sterile PBS to dilute the stock solution of nanoparticles for animal.
2. Phosphate Buffered Saline (PBS): 137 mM NaCl, 2.7 mM KCl, 4.3 mM Na_2HPO_4, 1.47 mM KH_2PO_4, pH of 7.4.

3. 15 mL Sterile conical centrifuge tubes.
4. Sonicator with microprobe set at 5 output (approximately 150 W output) and constant duty cycle.

2.2.2. Firefly Luciferase Reporter Gene Assay

1. Eagle's Minimum Essential Medium (EMEM) supplemented with FBS (5 %).
2. 1 % Penicillin-Streptomycin solution (10,000 units/mL penicillin G sodium and 10,000 μg/mL streptomycin sulfate) in 0.85 % saline.
3. 200 mM L-glutamine solution.
4. Phosphate Buffered Saline (PBS): 137 mM NaCl, 2.7 mM KCl, 4.3 mM Na_2HPO_4, 1.47 mM KH_2PO_4, pH of 7.4.
5. Trypsin-EDTA: 0.5 % Trypsin, 5.3 mM EDTA. Dilute to 0.01 % in 1× PBS for JB6 cells.
6. Cell counter or hemocytometer.
7. Triton/glysine luciferase cell culture lysis buffer (1×): 25 mM Tris-phosphate (pH 7.8), 2 mM DTT 2 mM 1, 2-diaminocyclohexane-*N,N,N′,N′*-tetraacetic acid, 10 % glycerol, 1 % Triton® X-100. This buffer can be prepared from commercially available 5× concentrated stocks. Equilibrate the lysis buffer at room temperature before use.
8. Single-Tube Luminometers with auto injectors and printer.
9. Luminometer cuvettes 12×75 mm.
10. Luciferase Assay System (Promega), containing Luciferase assay substrate, Luciferase assay buffer, and Cell culture lysis 5× reagent (30 mL/vial). Add Luciferase Assay Buffer (10 mL) to the vial containing the lyophilized Luciferase Assay Substrate. Reconstituted luciferase assay reagent should be stored in aliquots at −20 °C for up to 1 month or at −70 °C for up to 1 year. Add 4 volumes of water to 1 volume of 5× Luciferase Cell Culture Lysis Reagent to make 1× lysis buffer (see Note 1).

3. Methods

3.1. Development of Stable Transfected Cell Lines

Since each parental cell line has a different sensitivity to G418, hygromycin, or puromycin, one should determine the optimal concentration of drug for selection. We typically use 300 mg/mL of G418 for selecting the transfected clones using JB6 cells.

3.1.1. Titrating G418, Hygromycin, or Puromycin

1. Culture JB6 cells in a 75 cm² flask with EMEM containing FBS (5 %), glutamine (2 mM), and penicillin/streptomycin (1 %) and incubate at 37 °C, 80 % humidity, 5 % CO_2 until 80 % confluent.
2. Remove medium, rinse 1× with 10 mL 1× PBS and aspirate.

3. Add 5 mL trypsin/EDTA (0.01 %).
4. Incubate at 37 °C for 2–5 min until cells become detached.
5. Separate cells by pipetting up and down several times.
6. Add 5 mL fresh medium.
7. Count cells using cell counter or hemocytometer and adjust volume to 1×10^5 cells/mL.
8. Seed cells into a 24-well plate (0.5 mL, 5×10^4 cells per well) and incubate overnight.
9. On day 2, Pre-warm the complete media: EMEM + FBS (5 %) + Penicillin-Streptomycin (1 %) + L-glutamine (2 mM).
10. Make serial dilutions of G418 in complete media yielding eight concentrations of G418: 0, 50, 100, 200, 300, 350, 400, 500 mg/mL. Run triplicate wells for each concentration (see Note 2).
11. Remove cells from incubator.
12. Remove media from the cells.
13. Replace media with antibiotic containing media and return the plate to incubator.
14. The minimum concentration of G418 resulting in complete cell death after 3 days of titration should be used for selection (see Note 3).

3.1.2. Cationic Lipid-Mediated Transfection of JB6 Cells and Selection of Stable Transfected Clones (See Note 4)

1. Grow JB6 cells in 60-mm culture dishes using complete medium until ~80 % confluence by the day of transfection (see Note 5). Make at least four dishes for each plasmid to allow for sufficient number of transfected colonies to grow.
2. On the day of transfection, add 10 μg plasmid DNA into 250 μL dilution medium in a polystyrene tube, mix well. This amount is sufficient for use with one 60 mm culture dish (see Note 6).
3. In another tube mix 50 μL of cationic lipid reagent with 250 μL of dilution medium (see Note 6).
4. Combine diluted DNA and diluted cationic lipid reagent, mix, and incubate for 15 min at room temperature.
5. While DNA-lipid complexes are forming, replace the cell medium in the culture dishes with 3.5 mL of fresh transfection medium per 60 mm dish.
6. Slowly and gently add DNA-lipid complexes to the culture dish(es) containing cells. Mix complexes into the medium gently, holding the plate at an angle.
7. Incubate cells for at least 5 h at 37 °C in 5 % CO_2 (up to overnight).
8. Following the incubation add 6 mL fresh medium and serum (to reach final concentration of 5 % serum) per dish (each dish

already contains ~4 mL transfection mixture with plasmid DNA). Incubate cells at 37 °C in 5 % CO_2.

9. One day after the start of transfection passage cells onto fresh culture medium containing 300 mg/mL G418.
10. Two days post-transfection, change medium and add G418 to 300 mg/mL final concentration to select for expression of the transfected antibiotic-resistance gene (see Note 2).
11. Observe cell growth every 2–3 days and change medium containing the selected antibiotic once every week. Only stably transfected cells will survive and grow to form colonies in the presence of the selective antibiotics. Isolated colonies should begin to appear under a microscope after about 2–4 weeks (see Note 7).
12. Find large, healthy, and well separated colonies containing ~ 500–1,000 cells. Select the colonies to be picked. Circle the colonies with a laboratory marker to determine where to place the cloning cylinders.
13. Coat one end of cylinder with sterile vacuum grease by touching the cylinder to grease that has been autoclaved in a glass Petri dish. Gently place colony cylinder around the colony to be picked (see Fig. 4).
14. Using a sterile Pasteur pipet, rinse the colony with trypsin/EDTA by filling and emptying the cloning cylinder.
15. Add three drops of trypsin/EDTA to the cloning cylinder. Wait for 1–2 min. Fill the cloning cylinder with medium and repeatedly run the contents of the cylinder in and out through a Pasteur pipet in order to remove the trypsinized cells from the dish and disperse them.
16. Plate the collected cells in a 6-well plate (one colony per well) and add fresh medium with the selective antibiotic (300 mg/mL G418).
17. Maintain the cells in culture and split the cells every 4–5 days, until they reach 80 % confluency (see Note 8).
18. Split clones that reach ~80 % confluence into triplicate 6-well plates. Use one or two plates for checking protein expression/induction and save one plate for expansion of selected positive clone.
19. Check the clones for the highest luciferase protein expression by treating with inducing reagent (a positive control). We recommend using 20 ng/mL TPA to treat JB6 cells for inducing expression of AP-1, NF-κB, or COX-1 genes. Following 24 h incubation with the inducing reagent pick six colonies to check for protein expression using luciferase activity test.
20. Expand the best clones to create a stable transfected cell line. The cell line can be stored in liquid N_2 (see Note 9).

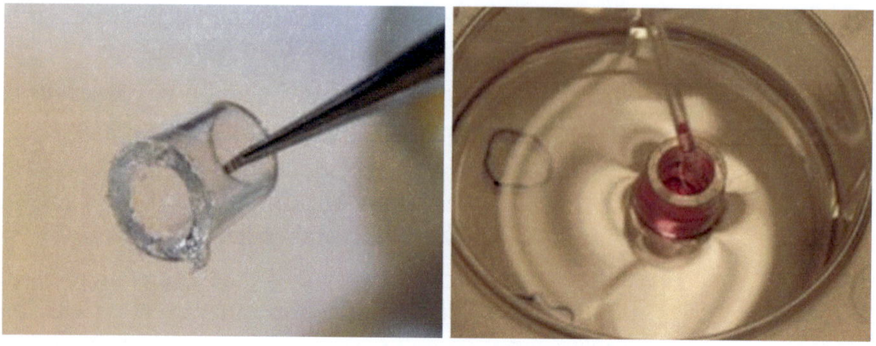

Fig. 4. Isolation of transfected colonies with a cloning cylinder.

3.2. Nanoparticles-Induced Alterations of Gene Expression

3.2.1. Preparation of Nanoparticles

1. Add 10 mg of nanoparticles (see Note 10) to 1 mL sterile PBS in 15 mL sterile conical centrifuge tube. Keep on ice.
2. While on ice, sonicate for 30 s, then pause for 15 s.
3. Repeat the sonication cycle five more times (see Note 11).
4. Dilute the nanoparticles with medium or PBS to desired concentration for cells or animal exposure (see Notes 10 and 11).

3.2.2. Firefly Luciferase Reporter Gene Assay

1. Culture the JB6 cell line stably transfected with a gene of interest in luciferase reporter plasmid (e.g., AP-1, p53, or NF-kB etc.) at 37 °C in a humidified atmosphere of 5 % CO_2 using complete medium, until 90–100 % confluency.
2. Remove culture medium. Rinse once with 10 mL of 1× PBS and aspirate. Add 5 mL trypsin/EDTA (0.01 %). Incubate at 37 °C for 2–5 min until cells become detached. Separate cells by pipetting up and down several times. Add 5 mL EMEM media containing FBS (5 %), glutamine (2 mM) and penicillin/streptomycin (1 %) to neutralize trypsin.
3. Count cells using cell counter or hemocytometer.
4. Resuspend 5×10^4 viable cells per 1 mL of Eagle's MEM Media (see step 2). Add 1 mL of cell suspension per well to a 24-well plate.
5. Incubate plates at 37 °C in a humidified atmosphere of 5 % CO_2 overnight.
6. Prepare nanoparticle solutions in fresh growth media. To find optimal dose test a range of nanoparticle concentrations (see Note 10).
7. Remove growth medium from the cells. Add 1 mL of nanoparticles resuspended in fresh growth medium per well. Incubate the cells for 12–24 h.
8. Remove growth medium from cells, rinse cells with PBS being careful not to dislodge attached cells. Remove as much PBS as possible.

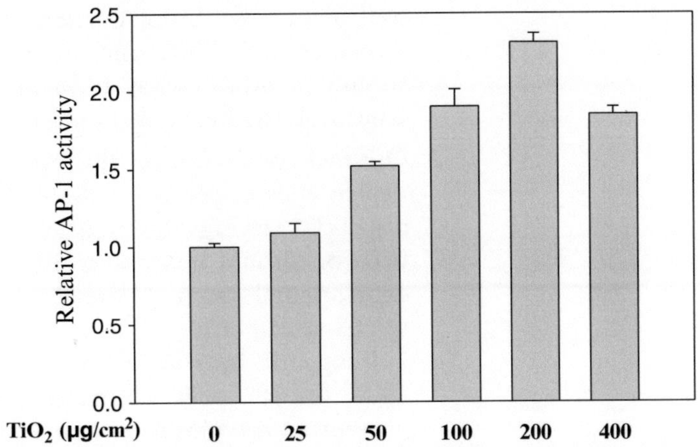

Fig. 5. Dose-dependent induction of AP-1 activation by TiO_2 nanoparticle. JB6 P^+ cells, stably transfected with AP-1 luciferase reporter plasmid, were exposed to various concentrations of TiO_2 for 24 h and the luciferase activity was measured. Results are means ± SEM of four assay wells.

9. Add 150 μL of (1×) luciferase cell culture lysis buffer per well to cover the cells.
10. Rock culture dishes several times to ensure that all cells are covered with the lysis buffer. Leave at room temperature for 30 min (see Note 12).
11. Mix well by pipetting up and down several times.
12. Program the luminometer to inject 60 μL luciferase assay substrate and perform a measurement cycle containing a 2-s measurement delay followed by a 10-s measurement read for luciferase activity.
13. Connect the reconstituted luciferase assay reagent vial (10 mL) to the injector of luminometer. Prime the luminometer injector at least three times with Luciferase Assay Reagent or as recommended in the owner's manual.
14. Dispense 40 μL of samples into luminometer cuvette, one cuvette per sample.
15. Place the tube in the luminometer and initiate reading as recommended in the owner's manual (see Note 13).
16. Express the readings as relative values (activity in treated samples compared to untreated controls) (see Fig. 5).

4. Notes

1. Do not store Luciferase Assay Reagent with dry ice. Thaw Luciferase Assay Reagent at temperatures below 25 °C and mix

well before use. Store Luciferase Assay Substrate in the dark. Reporter Lysis Buffer may be stored at room temperature and should be stored away from direct sunlight. Luciferase Cell Culture Lysis Reagent should be stored at −20 °C.

2. Optimal concentration for resistant clones selection in mammalian cells depends on the cell line used as well as on the plasmid carrying the resistance gene. Therefore antibiotic titration should be done to find the best condition for every experimental system. Optimal concentrations, as recommended by Clontech are 400 mg/mL G418 for HeLa cells and 200 mg/mL hygromycin for CHO cells. We use 300 mg/mL G418 for JB6 cells. In mammalian cells the optimal level of puromycin is typically around 1 mg/mL.
3. The chosen selective conditions (the minimum antibiotics level) may still allow mammalian cells to divide once or twice but will eventually kill them.
4. Different transfection techniques might be suitable for different cell lines. We recommend that DEAE-dextrans, calcium phosphate, or liposome-mediated transfection are used with adherent cell lines and that electroporation or liposome-mediated transfection is used for nonadherent cell lines.
5. Avoid antibiotics at the time of plating and during transfection.
6. Prepare a master mix if multiple identical transfection are performed.
7. Densely plated cells will require more frequent media changes. Un-transfected cells will not survive beyond ~3 days following the addition of antibiotic. Unstably transfected cells will survive for a while but will eventually die if they lost the transfected plasmid. Only stable transfected cells will survive and proliferate to form a clone and a cell line. In our experience, only a few cells are left in each dish after 2–3 weeks of incubation with antibiotic. One to two months later, each stably transfected cell becomes a colony. Reduce the frequency of medium changes to once per 2 weeks to promote the growth of colonies.
8. The clones need to be grown in order to obtain enough cells for further tests and/or storage. It may take 2–4 weeks for a clone (~50–100 cells) to reach 80 % confluence in a 6-well plate.
9. Development of a stable transfected mammalian cell line takes 2–3 months. Stable cell line development services are also available by GenScript Inc.
10. The amount of nanoparticles used will depend on the particular nanomaterial. The nanoparticle doses should be relevant to anticipated workplace exposures or to the amounts used in cellular or animal test systems. We used 0.5–100 μg/cm^2 for JB6 cells.

11. Nanoparticles need to be suspended in physiological solutions or medium use in the study. Since nanoparticles in physiological media tend to form coarse agglomerates, sonication is necessary to help nanoparticles disperse in solution. Many commercially available nanoparticles, such as Titanium dioxide (TiO_2) nanoparticle, Single-walled carbon nanotubes (SWCNT), and Tungsten carbide-cobalt nanoparticles (WC-Co) are water insoluble. For the tests with adherent cells, we recommend using these at 0–400 μg/cm^2 concentration.
12. At this point the plates can be transferred to 4 °C and stored for 2–3 days. The supernatant/cell lysate also can be stored at −70 °C for weeks.
13. If the samples were stored at 4°C or −70 °C, then Luciferase assay reagent and samples should be brought to ambient temperature prior to performing a luciferase assay. Generally, luciferase activity is stable for several hours at room temperature in 1× Luciferase Cell Culture Lysis Reagent. Assaying cold samples (0–4 °C) using standard assay volumes may result in a 5–10 % decrease in enzyme activity.

Acknowledgements

The plasmids used in the development of this method were gifts from our collaborators Dr. Nancy Colburn and Dr. Chuanshu Huang.

Disclaimer

The findings and conclusions in this report are those of the author(s) and do not necessarily represent the views of the National Institute for occupational Safety and Health.

References

1. Geraci CL, Castranova V (2010) Challenges in assessing nanomaterial toxicology: a personal perspective. Wiley Interdiscip Rev Nanomed Nanobiotechnol 2:569–577
2. Alam J, Cook JL (1990) Reporter genes: application to the study of mammalian gene transcription. Anal Biochem 188:245–254
3. Bronstein I, Fortin J, Stanley PE, Stewart GS, Kricka LJ (1994) Chemiluminescent and bioluminescent reporter gene assays. Anal Biochem 219:169–181
4. Chalfie M, Tu Y, Euskirchen G, Ward WW, Prasher DC (1994) Green fluorescent protein as a marker for gene expression. Science 263:802–805
5. Cullen BR, Malim MH (1992) Secreted placental alkaline phosphatase as a eukaryotic reporter gene. Methods Enzymol 216:362–368
6. De Wet JR, Wood KV, DeLuca M, Helinski DR, Subramani S (1987) Firefly luciferase gene: structure and expression in mammalian cells. Mol Cell Biol 7:725–737

7. Nguyen VT, Morange M, Bensaude O (1988) Firefly luciferase luminescence assays using scintillation counters for quantitation in transfected mammalian cells. Anal Biochem 171:404–408
8. Nordeen SK (1988) Luciferase reporter gene vectors for analysis of promoters and enhancers. Biotechniques 6:454–457
9. Thompson JF, Hayes LS, Lloyd DB (1993) Modulation of firefly luciferase stability and impact on studies of gene regulation. Gene 103:171–177
10. Yang J, Thomason DB (1993) An easily synthesized, photolyzable luciferase substrate for in vivo luciferase activity measurement. Biotechniques 15:848–850
11. Ding M, Kisin ER, Zhao J, Bowman L, Lu Y, Jiang B, Leonard S, Vallyathan V, Castranova V, Murray AR, Fadeel B, Shvedova AA (2009) Size-dependent effects of tungsten carbide-cobalt particles on oxygen radical production and activation of cell signaling pathways in murine epidermal cells. Toxicol Appl Pharmacol 241:260–268
12. Zhang XD, Zhao JS, Bowman B, Shi XL, Castranova V, Ding M (2010) Tungsten carbide-cobalt particles activate Nrf2 and its downstream target genes in JB6 Cells possibly by ROS generation. J Environ Pathol Toxicol Oncol 29:31–40
13. Ding M, Shi XL, Dong ZG, Chen F, Lu YJ, Castranova V, Vallyathan V (1999) Fresh fractured crystalline silica induces activator protein-1 activation through Erks and p38 mitogen-activated protein kinase. J Biol Chem 274:30611–30616
14. Murray AR, Kisin E, Leonard SS, Young SH, Kommineni C, Kagan VE, Castranova V, Shvedova AA (2009) Oxidative stress and inflammatory response in dermal toxicity of single-walled carbon nanotubes. Toxicology 257:161–171

Chapter 34

Single Cell Gel Electrophoresis Assay (Comet Assay) for Evaluating Nanoparticles-Induced DNA Damage in Cells

Linda Bowman, Vincent Castranova, and Min Ding

Abstract

Comet assay provides a rapid and reliable way to screen genotoxic effects of a wide variety of nanoparticles. Single cell gel electrophoresis assay (SCGE) or comet assay gained in popularity as a standard technique and is widely used for testing novel particles or chemicals for genotoxicity, monitoring environmental contamination with genotoxins, human biomonitoring and molecular epidemiology, and fundamental research in DNA damage and repair. Here, we describe the method of alkaline comet assay to detect nanoparticle-induced DNA damage.

Key words: Nanoparticles, DNA damage, Genotoxicity, Reactive oxygen species, Comet assay

1. Introduction

Nanotechnology is the manipulation of matter on a near-atomic scale to produce nanoparticles with unique properties, allowing new commercial applications. Since nanoparticles exhibit unique physicochemical properties, evaluation of the genotoxic effects of nanoparticles is essential. Genotoxic activities may be produced by direct interaction of particles with the genetic material or by secondary damage from particle-induced reactive oxygen species (ROS). Both pathways may relate to surface properties, the presence of transition metals, intracellular iron mobilization, or lipid peroxidation processes. Other aspects relevant to primary genotoxicity are particle size, shape, particle uptake, and the presence of mutagens carried with the particles (1). Secondary genotoxicity is characterized by excessive and persistent formation of ROS by inflammatory cells. In vitro genotoxicity tests, such as the comet assay, are sensitive tools for the detection of the mutagenic potential of a test material (2). Nanomaterials that stimulate generation of ROS are suggested to test its genotoxicity.

Mikhail Soloviev (ed.), *Nanoparticles in Biology and Medicine: Methods and Protocols*, Methods in Molecular Biology, vol. 906, DOI 10.1007/978-1-61779-953-2_34,

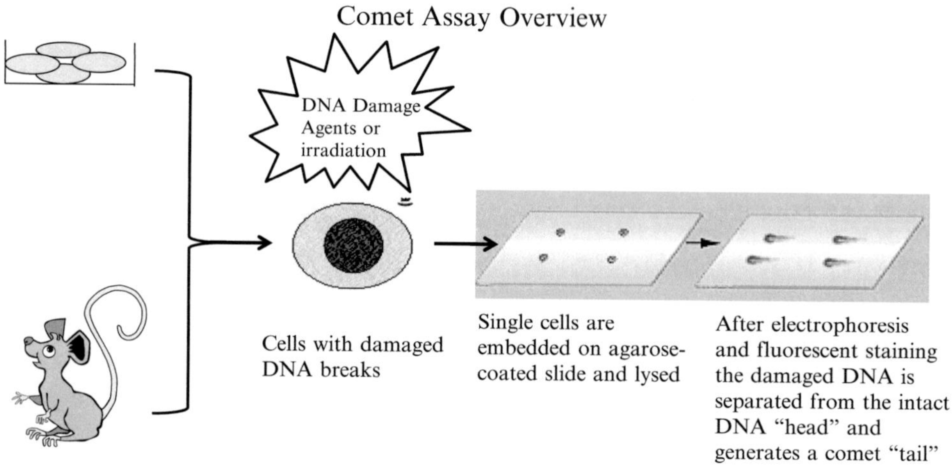

Fig. 1. An overview of single cell gel electrophoresis assay (comet assay).

Single cell gel electrophoresis assay (SCGE) or comet assay was first described by Singh et al. in 1988 (3). The principle of the comet assay is that the damaged DNA fragments will migrate out of the cell when an electric current is applied, whereas the undamaged DNA will remain in the cell nucleus. The results resemble a comet with a head (intact DNA) and a tail (damaged DNA). The size and shape of the tail and the distribution of DNA within the comet correlate with the extent of DNA damage (4–6). Treated cells are embedded in an agarose gel on a microscope slide. The cells are lysed to remove cellular protein. Cells are then treated with alkali to denature the DNA. The damaged DNA is allowed to migrate away from the nucleus by undergoing electrophoresis. The samples are stained with a DNA-specific fluorescent dye. The gel is then analyzed for the amount of fluorescence in the head and tail and the tail length (6–11) as shown in (Fig. 1).

The image analysis measures the overall intensity of the fluorescence for the whole nucleoid and the fluorescence of the migrated DNA and compares the two signals. The stronger the signal from the migrated DNA, the more damage there is present (12). The brighter and longer the tail, the higher the level of damage (Fig. 2).

Comet assay is an extremely sensitive DNA damage assay. This sensitivity needs to be handled carefully as it is also vulnerable to physical changes, which can affect the reproducibility of the results. Essentially, anything that can cause DNA damage or denaturation except the factor(s) being researched is to be avoided (4). Due to its simple and inexpensive setup, it can be used in conditions where more complex assays are not available.

Comet assay kits, reagents, control cells, and electrophoresis systems are available from Trevigen, Inc. Trevigen has developed a complete comet assay system which includes comet assay kits,

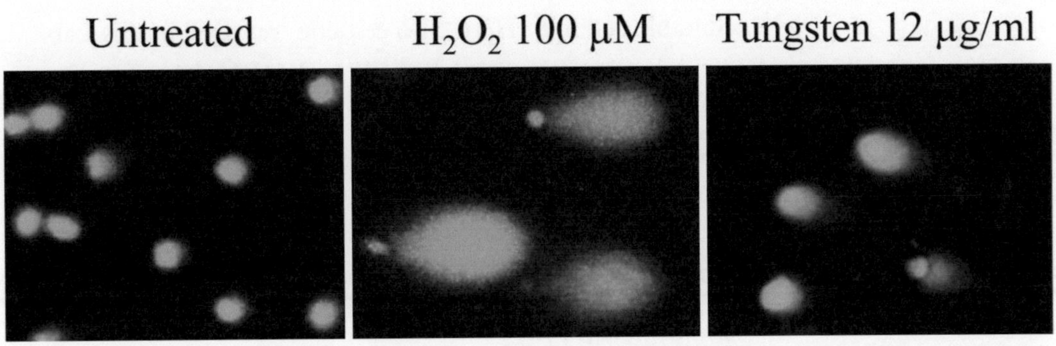

Fig. 2. Photo micrographs showing the DNA damage of JB6 cells. *Left*: untreated cells *center*: damage by H_2O_2 (positive control); *Right*: damage by nanoparticles of tungsten carbide-cobalt (WC-Co).

comet slides, comet assay control cells, and a specialized electrophoresis unit. These kits and reagents were successfully used in our lab for screening the genotoxicity of occupational related dusts, chemicals, and nanoparticles.

2. Materials

1. Comet assay kit (Trevigen) containing Lysis Solution (2×500 mL, store at room temperature), comet LM agarose (LMA) (15 mL, store at room temperature), Trevigen Comet Slide™ (25 slides, store at room temperature), 200 mM EDTA (pH 10) (12.5 mL, store at room temperature), and SYBR® Green I nucleic acid gel stain (store at −20 °C).
2. DMSO.
3. Phosphate Buffered Saline (PBS 1×, 137 mM NaCl, 2.7 mM KCl, 4.3 mM Na_2HPO_4, 1.47 mM KH_2PO_4, adjust to a final pH of 7.4), calcium and magnesium free.
4. 500 mM EDTA solution, pH 8.
5. Alkaline DNA denaturing solution (200 mM NaOH, 1 mM EDTA, pH >13). To prepare 50 mL of alkaline unwinding solution, combine: 0.4 g of NaOH pellets, 100 μL of 500 mM EDTA, 49.75 mL distilled H_2O. Stir until fully dissolved. The solution will warm during preparation. Allow to cool to room temperature before use (see Note 1).
6. Alkaline electrophoresis solution (NaOH 200 mM, EDTA 1 mM, pH >13). To prepare 1.5 L, dissolve 12 g of NaOH pellets in distilled H_2O, add 3 mL of 500 mM EDTA (pH = 8), and make up to 1.5 L with H_2O (see Note 1).
7. TE buffer: 10 mM Tris–HCl, 1 mM EDTA, pH = 7.5.
8. SYBER Green 1 staining solution. Add 1 μL of SYBER Green I to 10 mL of TE Buffer, pH = 7.5

9. Suitable cell line. Here, JB6 cell line was used in this Lab.
10. Eagle's minimal essential medium (EMEM).
11. 200 mM L-glutamine.
12. Penicillin–Streptomycin (10,000 penicillin/mL, 10 mg streptomycin/mL).
13. Heat-inactivated fetal bovine serum (FBS).
14. Complete EMEM medium for JB6 cells: EMEM medium containing FBS (5 %), L-glutamine (2 mM), and penicillin–streptomycin (1 %).
15. Trypsin–EDTA: 0.5 % Trypsin, 5.3 mM. Dilute to 0.01 % in 1× PBS for JB6 cells.
16. Tungsten carbide-cobalt nanoparticles (Inframat Advanced Materials): prepare 10 mg/mL stock solution in sterile PBS.
17. Boiling water bath (or a large beaker) and 37 °C water bath.
18. Horizontal DNA electrophoresis apparatus.
19. Epifluorescence microscope equipped with fluorescein filter or light transmission microscope when using a silver staining kit (Olympus AX-70).
20. 15 mL sterile conical centrifuge tubes.
21. 1 L graduated cylinder.
22. 4 °C refrigerator/cold room.
23. Hemacytometer.
24. Heating block.
25. Sonicator with microprobe set at output voltage of 150 W and constant duty cycle (Branson Sonifier 450).

3. Methods

3.1. Preparation of Nanoparticles (See Note 2)

1. Add 10 mg of nanoparticles to 1 mL sterile PBS in 15 mL sterile conical centrifuge tube to prepare a stock solution.
2. Keeping sample on ice, sonicate for 30 s, then rest 15 s. Repeat cycle to sonicate for a total of 3 min.
3. Make serial dilutions of the nanoparticles with complete EMEM medium to cover a range of concentration (typically 0–100 μg/cm²) for cell treatment (see Notes 3–5).

3.2. Cell Treatment and Preparation

1. Seed JB6 cells at 5×10^4 cells/well in 24-well (growth area, 1.9 cm²/well) tissue culture plate in complete medium (EMEM medium with 5 % FBS, 2 mM L-glutamine, 1 % penicillin–streptomycin).

2. Incubate the cells overnight in a 37 °C, 5 % CO_2, 80 % humidity incubator.
3. Prepare nanoparticles and make a series dilutions of nanoparticles in complete EMEM medium (see Subheading 3.1).
4. Remove cells from incubator.
5. Remove media from the cells by aspiration.
6. Replace medium with desired concentration of nanoparticle suspended in complete EMEM medium, return the plate to incubator, and incubate cells for 0.5–24 h (see Note 6).
7. After incubation, remove the medium and wash 1× with PBS, remove PBS.
8. Add 0.5 mL 0.01 % Trypsin–EDTA and incubate at 37 °C for 5–10 min to detach the cells. Pipette up and down several times to separate cells, and then add 0.5 mL complete EMEM medium containing FBS (5 %), L-glutamine (2 mM), penicillin–streptomycin (1 %).
9. Transfer cells to 1.5 mL Eppendorf microcentrifuge tube, obtain a cell count by using hemacytometer, and then pellet cells by centrifuging at 1,000 rpm for 3 min.
10. Remove supernatant. Wash once in ice cold 1× PBS (Ca^{++} and Mg^{++} free). Resuspend cells at 1×10^5 cells/mL in ice cold 1× PBS (see Note 7).

3.3. Alkaline Comet Assay

All steps are performed at room temperature unless otherwise specified. Work under dimmed or yellow light to prevent damage from UV.

1. Prepare lysis solution without DMSO and chill on ice for at least 20 min before use (see Note 8).
2. Melt LMA agarose in a beaker of boiling water for 5 min with cap loosened, then place bottle in a 37 °C water bath for at least 20 min to cool (see Note 9).
3. Add 500 μL melted agarose to 1.5 mL microcentrifuge tube and keep at 37 °C.
4. Add 50 μL cell (~500 cells) suspensions to the Eppendorf microcentrifuge tube containing melted agarose and keep at 37 °C. Mix well with a pipette (see Note 10).
5. Immediately pipette 50 μL cell suspensions with agarose on Comet Slide that is kept at 37 °C by placing on top of heating block or incubator. Cover entire sample area of slide by spreading with pipette tip (see Note 11).
6. Place slide flat in box to keep in the dark at 4 °C (e.g., place in refrigerator or on ice) for 30 min. A 0.5 mm clear ring appears at edge of Comet Slide area.

7. Immerse slide in prechilled lysis solution and leave at 4 °C in the dark for 30–60 min.
8. Drain excess buffer from slides and immerse in freshly prepared alkaline DNA denaturing solution (see Note 12).
9. Leave Comet Slide in alkaline DNA denaturing solution for 40 min at room temperature, in the dark.
10. For the comet assay ES tank, add 950 mL prechilled alkaline DNA denaturing solution, place slides in electrophoresis slide tray, and cover with slide tray overlay. Set power supply to 21 V and apply voltage for 30 min.
11. Gently drain excess electrophoresis solution, rinse by dipping several times in distilled H_2O.
12. Then place in 70 % ethanol for 5 min.
13. Air dry samples at room temperature in the dark for 10–15 min. Drying brings all the cells in a single plane to facilitate observation. Samples may be stored at room temperature with desiccant prior to scoring at this stage.
14. Place 100 μL of diluted SYBR® Green I onto each circle of dried agarose and place in refrigerator for 5 min. Gently tap slide to remove excess SYBR solution. Allow slides to dry completely at room temperature in the dark.
15. View slide under fluorescence microscope equipped with Fluorescein filters in the dark (see Note 13 and 14).

4. Notes

1. Use of freshly made solution is recommended. Prechill the alkaline electrophoresis solution at 4 °C. The alkaline DNA denaturing solution should be kept at room temperature. Wear gloves when preparing and handling the alkaline solution.
2. Most commercially available nanoparticles, such as titanium dioxide (TiO_2) nanoparticle, single-walled carbon nanotubes (SWCNT), and tungsten carbide-cobalt nanoparticles (WC-Co), are water-insoluble. For in vitro testing, nanoparticles should be suspended in culture medium. Since nanoparticles tend to form coarse agglomerates in physiological media, nanoparticles have to be sonicated to improve dispersion.
3. We recommend using 0–100 $\mu g/cm^2$ in adhesion cell system. To do so, the volume of particle-containing medium added to each well should be adjusted according to the growth area of each well. For example, a final nanoparticle concentration of 5 $\mu g/cm^2$ may be achieved by adding 1 mL of medium with the nanoparticle concentration of 10 μg/mL or 2 mL of

medium containing 5 μg/mL nanoparticles. These values should also be adjusted for different plates, i.e., for a 24-well plate, the growth area of each well is 2 cm^2 if purchased from Falcon Inc., but it is 1.9 cm^2 if obtained from Costar Inc.

4. The nanoparticle doses chosen for cellular or animal testing systems need to be relevant to anticipated workplace exposures. We used 0–100 μg/cm^2 for JB6 or lung fibroblast V79 cells.
5. Solutions of nanoparticles need to be freshly prepared on the day of treatment.
6. For best results, start with the range of doses (0–100 μg/cm^2) and incubation times (0.5–24 h). A sample of untreated cells should always be processed as a negative control for assay. To generate positive samples for comet tails, we used H_2O_2 (100 μm) to treat JB6 cells for 20 min.
7. The media in the cell culture needs to be washed away since it can reduce the adhesion of the agarose on the Comet Slide. Cell samples should be prepared immediately before starting the assay and should be handled under dimmed or yellow light to prevent DNA damage from ultraviolet light. All buffers should be chilled to 4 °C or on ice to inhibit endogenous damage occurring during sample preparation and to inhibit repair in the unfixed cells. PBS should be calcium and magnesium free to inhibit endonuclease activities. The appropriate controls should also be included (see below).
8. The addition of DMSO is optional and is required only for samples containing heme, such as blood cells or tissue samples. The buffer formulation is proprietary.
9. The temperature of the agarose is critical or the cells may undergo heat shock. Heat blocks are not recommended for regulating the temperature of the agarose. However, once the agarose and cells have been spread on the slides, it works very well to maintain 37 °C.
10. For best results, use 500–1,000 cells per Comet Slide sample area. Using 50 μL of a cell suspension at 1×10^5 cells/mL combined with 500 μL of LM agarose will provide the correct agarose concentration and cell density for optimal results when spreading 50 μL per well. Control cells and treated cells should be handled in an identical manner. All steps are performed at room temperature unless otherwise specified.
11. When working with many samples, it may be convenient to place aliquots of the molten agarose into prewarmed microcentrifuge tubes and place the tubes at 37 °C. Add cells to one tube, mix by gently pipetting once or twice, then transfer 50 μL aliquots onto each sample area as required. Then proceed with the next sample of cells. If sample is not spreading evenly on the slide, warm the slide at 37 °C before application.

12. Wear gloves when working with alkaline solution.
13. Qualitative analysis: The comet tail can be scored according to DNA content (intensity). The control (untreated cells) should be used to determine the characteristics of data for a healthy cell. Scoring can then be made according to nominal, medium, or high intensity tail DNA content. At least 50 cells should be scored per sample.
14. Quantitative analysis: There are several image analysis systems that are available for quantitation of comet assay data from Trevigen Inc. The more sophisticated systems include the microscope, camera, and computer analysis package. These systems can be set up to measure the length of DNA migration, image length, nuclear size and calculate DNA damage parameters. At least 50 randomly selected cells should be analyzed per sample.

Disclaimer

The findings and conclusions in this report are those of the author(s) and do not necessarily represent the views of the National Institute for Occupational Safety and Health.

References

1. Geraci CL, Castranova V (2010) Challenges in assessing nanomaterial toxicology: a personal perspective. Wiley Interdiscip Rev Nanomed Nanobiotechnol 2:569–577
2. Kisin ER, Murray AR, Keane MJ, Shi XC, Schwegler-Berry D, Gorelik O, Arepalli S, Castranova V, Wallace WE, Kagan VE, Shvedova AA (2007) Single-walled carbon nanotubes: geno- and cytotoxic effects in lung fibroblast V79 cells. J Toxicol Environ Health A 70:2071–2079
3. Singh NP, McCoy MT, Tice RR, Schneider EL (1988) A simple technique for quantitation of low levels of DNA damage in individual cells. Exp Cell Res 175:184–191
4. Fairbairn DW, Olive PL, O'Neill KL (1995) The comet assay: a comprehensive review. Mutation Res 339:37–59
5. Lemay M, Wood KA (1999) Detection of DNA damage and identification of UV-induced photoproducts using the Comet Assay® kit. Biotechniques 27(4):846–851
6. Malyapa RS, Bi C, Ahern EW, Roti JL (1998) Detection of DNA damage by the alkali comet assay after exposure to low dose gamma radiation. Radiat Res 149:396–400
7. Henderson L, Wolfreys A, Fedyk J, Bourner C, Windeback S (1998) The ability for the comet assay to discriminate between genotoxins and cytotoxins. Mutagenesis 13:89–94
8. Visvardis EE, Tassiou AM, Piperakis SM (1997) Study of DNA damage induction and repair capacity of fresh cryopreserved lymphocytes exposed to H_2O_2 and γ-irradiation with the alkaline comet assay. Mutat Res 383:71–80
9. Collins AR, Ma AG, Duthie SJ (1995) The kinetics of repair of oxidative DNA damage (strand breaks and oxidized pyrimidine dimers) in human cells. Mutation Res 336:69–77
10. Singh NP, Stephens RE (1997) Microgel electrophoresis: sensitivity, mechanisms, and DNA electrostretching. Mutation Res 383:167–175
11. Östling O, Johanson KJ (1984) Microelectrophoretic study of radiation-induced DNA damage in individual cells. Biochem Biophys Res Commun 123:291–298
12. Morris EJ, Dreixler JC, Cheng K-Y, Wilson PM, Gin RM, Geller HM (1999) Optimization of single-cell gel electrophoresis (SCGE) for quantitative analysis of neuronal DNA damage. Biotechniques 26:282–289

Chapter 35

Assessment of Toxicity of Nanoparticles Using Insects as Biological Models

Yan Zhou, Aracely Rocha, Carlos J. Sanchez, and Hong Liang

Abstract

Nanomaterials have become increasingly important in medicine, manufacturing, and consumer products. The fundamental understanding in effects of nanoparticles (NPs) on and their interactions with biomolecules and organismal systems have yet to be achieved. In this chapter, we firstly provide a brief review of the interactions between nanoparticles and biological systems. We will then provide an example by describing a novel method to assess the effects of NPs on biological systems, using insects as a model. Nanoparticles were injected into the central nervous system of the discoid cockroach (*Blaberus discoidalis*). It was found that insects became hyperactive compared to negative control (water injections). Our method could provide a generic method of assessing nanoparticles toxicity.

Key words: Nanomaterials, Toxicity, Insects, Nervous system, Magnetic nanoparticles, Cockroach walking distance

1. Introduction

Nanomaterials are known to have a wide spectrum of applications in biomedical, chemical, electronical, material, optical, and physical sciences. The properties of nanoparticles (NPs) are unique from their bulk due to the quantum effects (1–3). In bio-applications, NPs have been used to detect genetic disorders, cell labeling, gene therapy, and drug delivery (4–6). Here are a few examples of those commonly used in biomedical research and applications.

The core-shell nanoparticles are generally spherical cores encapsulated by a shell of another material. The shell is generally a few nanometers in thickness and has been found to absorb a certain wavelengths in order to increase energy. For example, a core of silica coated by gold shell is used to kill surrounding cells by generating heat after radiation (7).

Mikhail Soloviev (ed.), *Nanoparticles in Biology and Medicine: Methods and Protocols*, Methods in Molecular Biology, vol. 906, DOI 10.1007/978-1-61779-953-2_35, © Springer Science+Business Media, LLC 2012

Dendrimers are highly branched structures with molecular "hooks" attaching cell-identification tags, dyes, and enzymes. Applications include catalysts, chemical sensors, drug and gene delivery agents/cells, imaging agents, micelle mimics, and reaction vessels (8).

Fullerenes-Bucky balls and carbon nanotubes (CNT) are carbon-based nanoparticles having well-defined lattice structures. These structures are sized from nanometers to as much as millimeters. Carbon tubes can be used as carriers for drugs, vaccines, and other molecules, in biomedical applications (9, 10).

Liposomes are lipid-based nanoparticles being used in cosmetic and drug delivery (11). These NPs are able to break cells. One issue remaining in such applications is their tentativeness to agglomerate in aqueous environment (12).

Nanoparticles and nanorods have at least one dimension less than 100 nm. Most materials are made of semiconductors and used as contrast and imaging agents. Nanorods are often found made of gold, silicon, or inorganic phosphate, among other materials.

Quantum dots (QDs) are mostly semiconductor nanocrystals. Depending on their size, they emit light at various colors upon excitation. The quantum effects confine conduction-band electrons, valence-band holes, and/or excitations in spatial directions. QDs have been reported to be applied to cell labeling and imaging for cancer (13).

Superparamagnetic nanoparticles are NPs attracted to magnetic field without retaining residual magnetism when the field is removed. Iron oxide nanoparticles (Fe_3O_4) have been reported to be applied for drug and gene delivery. Such type of NPs have also been used as the contrast agents for MRI (14).

The interactions between organisms and nanoparticles are crucial for understanding the toxicity of the latter. We have recently demonstrated a novel method to study the effects of nanoparticles on insects' behavior. In this approach, nanoparticles are injected into the central nervous system of the discoid cockroach (*Blaberus discoidalis*). This is followed by digital observation over a period of time in order to quantitatively evaluate insects' locomotion and statistical analysis is conducted to find cockroach activity levels. Cockroaches are hyperactive if they are stimulated by nanoparticles injected into their nervous system; this is also an indication of nanotoxicity on insect nervous system. Food consumption results often correlated with activity levels and hyperactive group consumes more food. However, survival rate reflects organism level toxicity and may be different from nervous system toxicity reflected by activity levels.

2. Materials

1. Select adult discoid cockroaches *B. discoidalis* from stock cultures. Randomly divide them into groups of 10 individuals (5 male, 5 female), and place each group in a plastic container (416 mm L × 327 mm W × 165 mm H).
2. Maintain containers at 30 °C and 40% relative humidity, and set a 12/12 h light/dark cycle. Provide them 10 g of dog food (Beneful Dry Dog Food) once a week and water twice a week, as shown in Fig. 1.
3. Carbon dioxide gas for use as anesthetic before injection.
4. Microapplicator (ISCO, Inc.) to control NP injection.
5. Nanoparticles for toxicity testing. Cautions in NPs solution selection and preparation are necessary (see Notes 1 and 2). Here, we provide examples of nanoparticles suitable for this assessment, and they will also be used in following examples. These include positively charged gold nanoparticles (+AuNPs) which have spherical shape with a diameter of 50 nm and are well dispersed in the cetyl-trimethylammonium bromide (CTAB) solution at a nanoparticle concentration 0.1 mg/mL; the positive charge is provided by CTA+ surrounding the

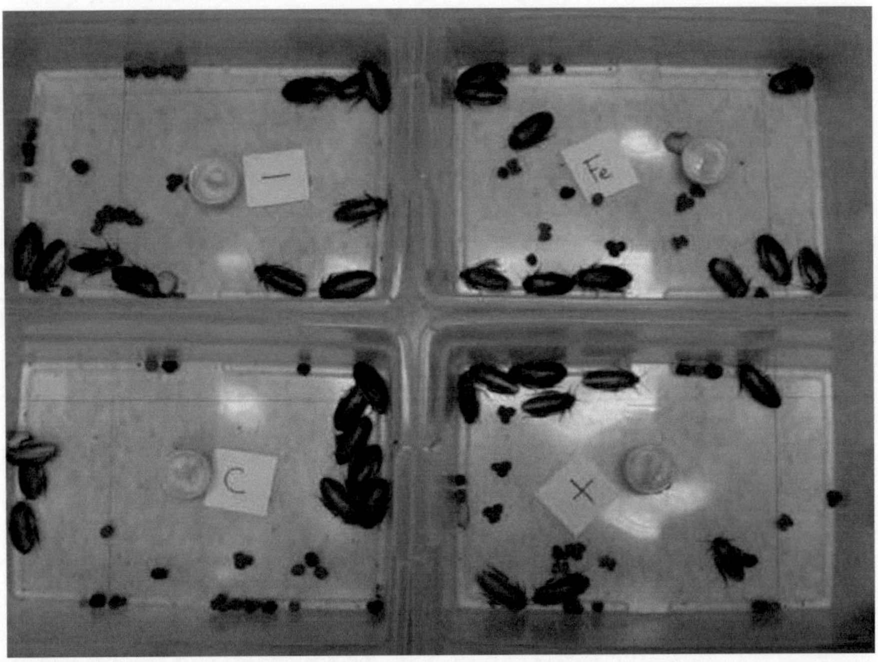

Fig. 1. Experimental setup. *Blaberus discoidalis* were divided into four groups and kept in separate boxes. The notation in each group indicates example NPs that are injected: "–" means negative charged gold NPs, "+" stands for positively charged gold NPs, "C" stands for control which is water-injected, and "Fe" is iron oxide NPs.

nanoparticle. Negatively charged gold nanoparticles (–AuNPs) have average diameter of 60 nm at a nanoparticle concentration 0.02 mg/mL; negative charge is generated by citrate ions.

6. Iron oxide nanoparticles (FeONPs), average diameter of 13 nm at a nanoparticle concentration 0.6 mg.
7. Suitable transmission electron microscopy for NP size/shape confirmation.
8. 29G 1 mL syringe (BD).
9. Dissecting microscope.
10. Video camera (Sony DSC-T70) for recording roaches' walking distance, or any digital camera has video recording feature can be used here.
11. Software ImageTool 3.0 (UTHSC) for distance tracking.
12. Statistical software (SPSS) to determine the statistical significance of the distance measurements between groups.
13. 2% (*v/v*) glutaraldehyde.

3. Methods

1. Apply carbon dioxide as anesthetic before injection, place carbon dioxide cylinder hose into plastic container, release carbon dioxide until no roach stop moving.
2. Inject 7 μL of NPs solution into each roach using a 1 mL syringe with 29G needle driven by ISCO microapplicator at a constant injection speed.
3. Insert the needle into the head capsule through the dorsal region of the "neck" membrane between the head and pronotum. The position has to be accurate to make sure injected solution reaches cockroach's nervous system. A dissecting microscope should be used during injection in order to find the correct location.
4. Record Roaches' walking distance using a video camera. Take a 2 min video clip for each group at the same time on selected days (once every 2 or, e.g., 5 days for 2 months) (see Note 3).
5. Analyze Roaches' walking distance. Use any suitable software, or manually transfer Roaches' tracks into computer (see Note 4). To manually transfer tracks, play video clip on a computer, cover computer screen with a piece of A4 paper, and draw the track on the paper by following individual walking roaches in the video clip in play. Replay video clip in order to get all the tracks. Mark the box length on the paper to get a reference. Scan the paper and Import scanned image into software ImageTool. Count pixels of the track made by each roach.

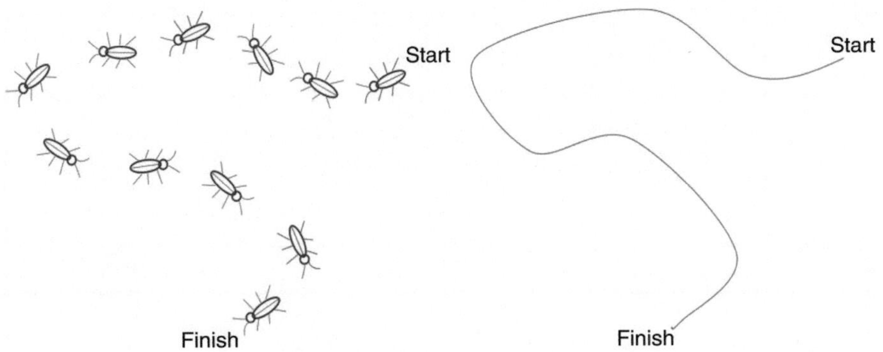

Fig. 2. Tracking individual cockroaches. *Left*: walking cockroach in the video clip. *Right*: drawn track.

Table 1
Statistic pattern of experimental groups

Treatment	Mean	Standard error	95% Confidence interval	
			Lower bound	Upper bound
Control	1,190.498	105.560	966.720	1,414.276
FeONPs	1,778.411	79.796	1609.251	1,947.571
-AuNPs	1,254.300	79.796	1085.140	1,423.460
+AuNPs	1,283.039	149.285	966.569	1,599.508

The actual walking distance can then be calculated using the following equation. Add up individual distance measurements in the same group at the same day. An example of analysis is shown in Fig. 2.

$$\text{Walking distance (mm)} = \text{Distance in video (pixel)} \times \frac{\text{Box length (mm)}}{\text{Box length in video (pixel)}}$$

6. Use statistical software (SPSS) to determine the statistical significance of the distance measurements between groups. Perform pairwise comparison to evaluate the statistical difference between groups. Usually, a p value less than 0.05 is considered to be statistically significant. An example of calculations is shown in Table 1 and Fig. 3. The control group (where water was applied as control injection agent) has the lowest average walking distance. The more active group shows longer walking distance. Due to the roaches' cold-blooded nature which makes them highly environmentally sensitive, as shown in Fig. 3, big variations in roaches' activities are expected. Use each group's average walking distance on every monitored day

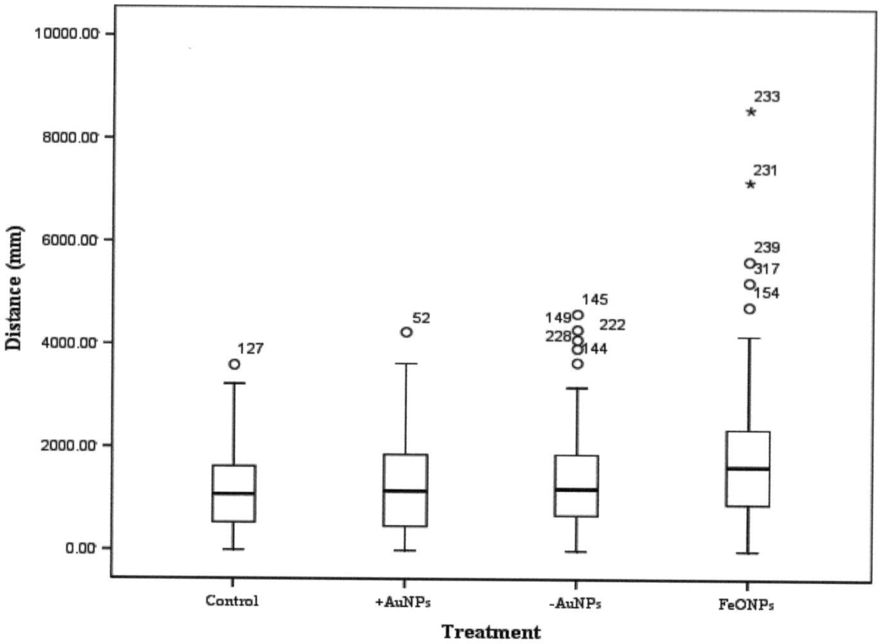

Fig. 3. Average roach walking distance. Drawn separately for each experimental group over the 2-month period: control group, +AuNPs, −AuNPs, and FeONPs. The FeONPs group shows higher mean walking distance. Numbers indicate outliners in statistical test.

to generate a plot of average walking distance vs. time as shown in Fig. 4. Categorize roach activity over time into three periods based on the plot (see Note 5).

7. Analyze the survival rates. Figure 5 shows an example of the recorded survival rates (see Note 6). The toxic effect of magnetic nanoparticles on insects may be partially due to magnetoreception (see Note 7).

8. Measure the weight of remaining food of each group once a week when cockroaches are fed, compare it with the original food weight (10 g). The amount of the consumed food should correlate with the distance measurements—more active cockroaches consume more food in order to support their hyperactivity. An example is shown in Fig. 6.

9. Transmission electron microscope analysis of in vivo interaction. In order to confirm the existence of NPs inside cockroaches' nervous cells, TEM imaging can be applied on cockroach's brain samples before and after NPs injection (see Note 8). Dissect cockroach's brain out and leave it in 2% (*v/v*) glutaraldehyde for 0.5 h at 37 °C, then go through TEM sample preparation process.

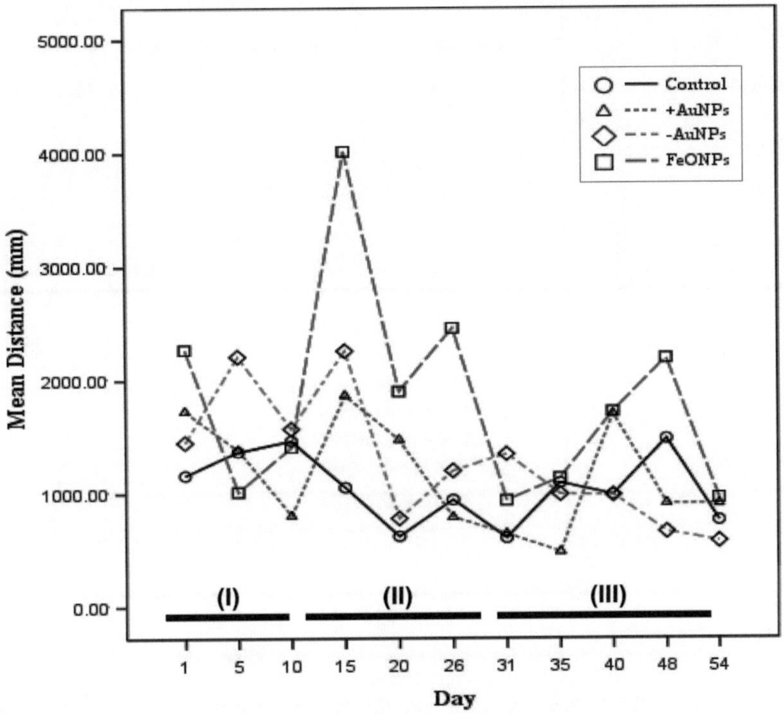

Fig. 4. Roach activity. Mean distance vs. time shows FeONPs group was hyperactive than the rest three groups (*circle* control, *triangle* AuNPs, *diamond* –AuNPs, and *square* FeONPs). Roach activity can be categorized into three periods: (1) Adjustment Period. Along this period, cockroaches adjust themselves to the injection in their nervous system. Their activity level is a reflection of trauma caused by syringe and the liquid (DI water or NPs solution) injected. (2) Stability Period. Cockroaches' activity level is a reflection of liquid content. In this period, there is a clear trend between different groups. Most likely, the NP-injected groups are more active than the control group. (3) Dying Period. Cockroaches start to die simultaneously after certain days. In this case, cockroach's activity levels were close to each other in Adjustment Period, which is between day 1 and day 10. Stability Period is from day 10 to day 30, NPs treated groups were more active than control group, and FeONPs treated group was the most active one. Cockroaches started to die at day 35, and dying period was between day 31and day 53.

4. Notes

1. Since surfactants might cause side effects on insects other than NPs, before NPs injection, remove extra surfactants in NPs solution by centrifugation and resuspend NPs in DI water.
2. The acid/alkaline property of injection solutions is another consideration. Select NPs in neutral solutions which are similar to the insect's neurons in an in vivo environment. Even though the solution should be neutralized/diluted by insect body fluids in a short period of time after injection, acid or alkaline solutions might do not match insect's in vivo environment. This could bring more factors in affecting insect's behavior.
3. It is important to minimize the sound and vibration around the laboratory during the recordings.

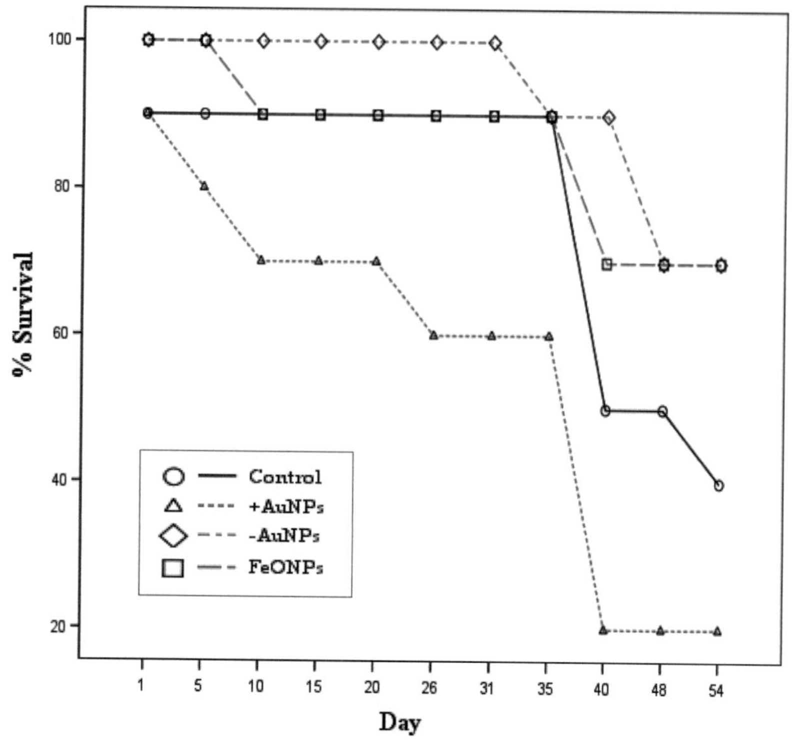

Fig. 5. Roach survival rate. Number of surviving roaches (in %) vs. time, where *circle* control, *triangle* +AuNPs, *diamond* –AuNPs, and *square* FeONPs.

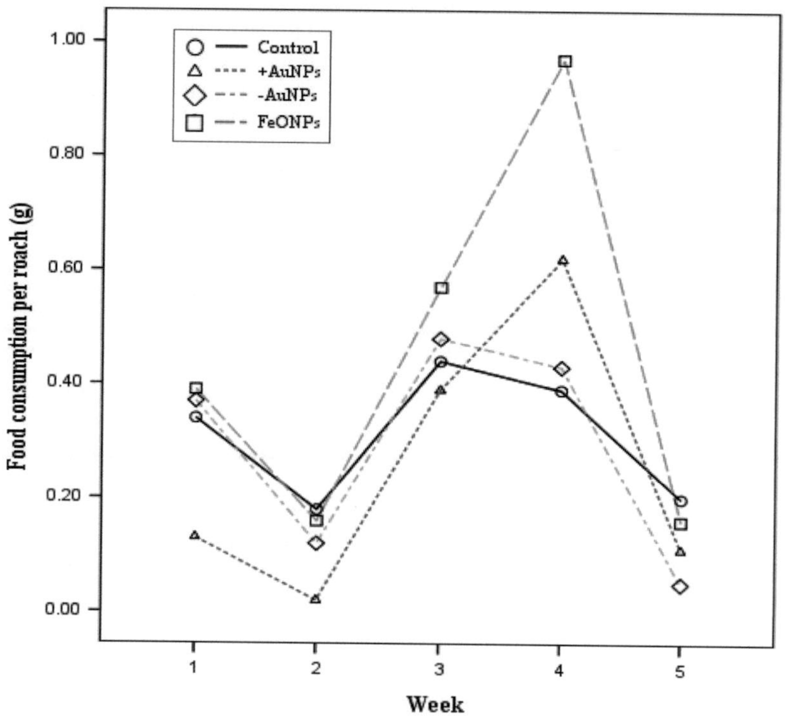

Fig. 6. Food consumption per roach on a weekly basis. In the second half of this experiment, FeONPs group consumed more food than the rest three groups, where *circle* control, *triangle* +AuNPs, *diamond* –AuNPs, and *square* FeONPs.

4. A suitable software for track recordings can be created based on GMimPro sequence processor software for Windows available free for academic users form (www.nimr.mrc.ac.uk/gmimpro). We recorded tracks manually.
5. If the lines in the plot grouped together over a period of time, NPs-injected groups show no evidence to be hyperactive compared with control group. If this happens at the beginning of the test when insects adjust themselves to the injection, this period can be categorized as Adjustment Period. If this happens at the end of the test when insects start to die, this period can be categorized as Dying Period. Stability Period is when insects' activity level is a reflection of liquid content. In this period, there is a clear trend between different groups and the plotted lines are generally separated
6. The survival rate reflects NPs toxicity on insects at organism level instead of nervous system level. Some NPs may be more toxic on insect organism, but less toxic on insect nervous system. For example, +AuNP show the most toxic effect at organism level supported by lowest survival rate. However, FeONP show the most toxic effect at insect nervous system level since FeONPs treated group is most active. The survival drops dramatically at a certain day, and this day usually lies in dying period.
7. The presence of nanoparticles inside the nervous system of discoid cockroaches has effects on the insects' behavior. Cockroaches have a tendency to be hyperactive and consume more food as a response to the NPs stimulus. Magnetic and nonmagnetic NPs have different effects on cockroaches' behavior. In our experiments, magnetic NPs are likely to have caused a neurotic disorder on magnetoreception. Magnetoreception has been found in honeybees (15–17) and fruit flies (18). Evidence suggests that roaches are equipped with a similar system (19, 20). It has been proposed that magnetoreceptor is composed of magnetite NPs and works together with mechano-sensitive structures. By receiving geomagnetic field in the form of torque, the information was then transmitted as electrochemical signals in nervous system (21). Recently, iron oxide NPs were isolated from the antennae of *Pachycondyl amarginata* ant, supporting the idea that the head and antennae of insect host magnetoreceptors (22, 23). It has also been reported that encapsulation of NPs was found inside the roach nervous system (24).
8. Smaller NPs have higher mobility inside neuron cells and within nervous system. The information on NPs aggregation and protein encapsulation can be gathered by TEM imaging.

Acknowledgments

This work was partially sponsored by the National Science Foundation (0515930), Texas Engineering Experimental Station, and the Texas A&M University. Assistance provided by Drs. Brad Vinson, Jorge Gonzelez, and Subrata Kundu was greatly appreciated.

References

1. Lewis LN (1993) Chemical catalysis by colloids and clusters. Chem Rev 93:2693–2730
2. Alivisatos AP (1996) Semiconductor clusters, nanocrystals, and quantum dots. Science 271:933–937
3. Henglein A (1989) Small-particle research: physicochemical properties of extremely small colloidal metal and semiconductor particles. Chem Rev 89:1861–1873
4. Taton T, Mirkin C, Letsinger R (2000) Scanometric DNA array detection with nanoparticle probes. Science 289:1757
5. Cao Y, Jin R, Mirkin C (2002) Nanoparticles with Raman spectroscopic fingerprints for DNA and RNA detection. Science 297:1536
6. Sandhu K, McIntosh C, Simard J, Smith S, Rotello V (2002) Gold nanoparticle-mediated transfection of mammalian cells. Bioconjug Chem 13:3–6
7. Loo C, Lowery A, Halas N, West J, Drezek R (2005) Immunotargeted nanoshells for integrated cancer imaging and therapy. Nano Lett 5:709–711
8. Menjoge AR, Kannan RM, Tomalia DA (2010) Dendrimer-based drug and imaging conjugates: design considerations for nanomedical applications. Drug Discov Today 15: 171–185
9. Ashcroft JM, Tsyboulski DA, Hartman KB, Zakharian TY, Marks JW, Weisman RB, Rosenblum MG, Wilson LJ (2006) Fullerene (C60) immunoconjugates: interaction of water-soluble C60 derivatives with the murine anti-gp240 melanoma antibody. Chem Commun (Camb) 28:3004–3006
10. Bianco A, Kostarelos K, Prato M (2005) Applications of carbon nanotubes in drug delivery. Curr Opin Chem Biol 9:674–679
11. Cho K, Wang X, Nie S, Chen ZG, Shin DM (2008) Therapeutic nanoparticles for drug delivery in cancer. Clin Cancer Res 14:1310
12. Bhavane R, Karathanasis E, Annapragada AV (2007) Triggered release of ciprofloxacin from nanostructured agglomerated vesicles. Int J Nanomedicine 2:407
13. Choi UB, Strop P, Vrljic M, Chu S, Brunger AT, Weninger KR (2010) Single-molecule FRET-derived model of the synaptotagmin 1-SNARE fusion complex. Nat Struct Mol Biol 17:318–324
14. Gupta AK, Gupta M (2005) Synthesis and surface engineering of iron oxide nanoparticles for biomedical applications. Biomaterials 26:3995–4021
15. Kirschvink J, Padmanabha S, Boyce C, Oglesby J (1997) Measurement of the threshold sensitivity of honeybees to weak, extremely low-frequency magnetic fields. J Exp Biol 200:1363
16. Kirschvink JL, Kirschvink AK (1991) Is geomagnetic sensitivity real? replication of the Walker-Bitterman magnetic conditioning experiment in honey bees. Am Zool 31:169
17. Walker MM, Bitterman M (1989) Short communication honeybees can be trained to respond to very small changes in geomagnetic field intensity. J Exp Biol 145:489
18. Phillips J, Sayeed O (1993) Wavelength-dependent effects of light on magnetic compass orientation in *Drosophila melanogaster*. J Comp Physiol A Neuroethol Sens Neural Behav Physiol 172:303–308
19. Vácha M (2006) Laboratory behavioural assay of insect magnetoreception: magnetosensitivity of *Periplaneta americana*. J Exp Biol 209:3882
20. Vácha M, Puzová T, Kvícalová M (2009) Radio frequency magnetic fields disrupt magnetoreception in American cockroach. J Exp Biol 212:3473
21. Wajnberg E, Acosta-Avalos D, Alves OC, de Oliveira JF, Srygley RB, Esquivel D (2010)

Magnetoreception in eusocial insects: an update. J R Soc Interface 7:S207

22. Abraçado L, Esquivel D, Wajnberg E (2008) Oriented magnetic material in head and antennae of *Solenopsis interrupta* ant. J Magn Magn Mater 320:e204–e206

23. De Oliveira JF, Wajnberg E, de Souza Esquivel DM, Weinkauf S, Winklhofer M, Hanzlik M (2010) Ant antennae: are they sites for magnetoreception? J R Soc Interface 7:143

24. Rocha A, Zhou Y, Kundu S, González JM, BradleighVinson S, Liang H (2011) In vivo observation of gold nanoparticles in the central nervous system of *Blaberus discoidalis*. J Nanobiotechnol 9:5

Chapter 36

Evaluation of Cytotoxicity of 3-Mercaptopropionic Acid-Modified Quantum Dots on *Medicago sativa* Cells and Tissues

Ana Raquel Santos, Ana Sofia Miguel, Pedro Fevereiro, and Abel Oliva

Abstract

Like most of the new technologies designed to interact with biological systems, the applications of nanomaterials needs a proper assessment for their potential impacts. It is only through addressing the issues raised by toxicological studies that nanotechnology will be able to acquire its full potential. Here, we describe the detailed protocols to study the responses of plant cells to their exposure to nanoparticles, including viability, oxidative stress detection, and reactive oxygen species enzymatic detoxification, as well as particle uptake.

Key words: Quantum dots, Cytotoxicity, ROS detection, Oxidative stress, Uptake, Antioxidant enzyme activity, Cell suspension cultures, *Medicago sativa*

1. Introduction

As a result of the rapid development of nanotechnology an emerging area of research is now focused on studying the short and medium term impacts of nanomaterials on the environment. This new discipline, named nano-ecotoxicology is now being developed to adequately evaluate the global impact of these new devices and their multiple applications (1). Cell-based in vitro studies play an essential role on meaningful toxicity testing. They allow the setting up of high-throughput systems for rapid and cost-effective screening of hazards, while targeting the biological responses under highly controlled conditions (2). Mechanisms of nanotoxicity should include the identification of the routes of uptake and intracellular accumulation of nanoparticles, the modes of its action, the damage to cellular structures, and the kinetics of toxi- and detoxification (3). The following protocols were developed to evaluate the plant

Mikhail Soloviev (ed.), *Nanoparticles in Biology and Medicine: Methods and Protocols*, Methods in Molecular Biology, vol. 906,
DOI 10.1007/978-1-61779-953-2_36,

cell responses when in contact with quantum dots (QDs), including cell viability, the histochemistry of reactive oxygen species (ROS) production and accumulation, the activity of ROS detoxification enzymes, and cell/plant–particle uptake.

Fluorescein diacetate (FDA) is a widely used viability test due to quickness and because it can be visualized within individual cells with the help of a fluorescence microscope and therefore not only it is used as a general viability assay but also facilitates the determination of the number of live cells in a population (4). FDA is converted into a fluorescent product if cell membrane is intact and therefore viable cells present a bright green fluorescence when excited at 488 nm (5).

Using light microscopy we can monitor ROS production and accumulation with ROS-specific tracer dye 3-3′ diaminobenzidine (DAB) for hydrogen peroxide (H_2O_2) and the nitroblue tetrazolium (NBT) for $O_2^{-\bullet}$. 2′,7′-Dichlorodihydrofluorescein diacetate (H_2DCFDA) was applied to measure oxidative responses in cell suspension cultures. This probe is oxidized due to the presence of ROS, and the emitted fluorescence intensity is directly proportional to the level of dichlorofluorescein (DCF) formed intracellularly (6).

There is much evidence obtained from various plant systems showing that the concentration and activity of enzymes involved in scavenging ROS are altered by different stresses. The measurement of the activity of three major antioxidative enzymes is described: superoxide dismutases (SOD), catalase (CAT), and glutathione reductase (GR) from the ascorbate-glutathione cycle.

2. Materials

2.1. Initiation of Medicago sativa Cell Suspension Cultures

1. Eight week dark grown callus from roots or petioles of *Medicago sativa*.
2. Murashige & Skoog (M&S) liquid medium supplemented with 0.5 mg/L of Dichlorophenoxyacetic acid and 0.5 mg/L of kinetin. pH of medium should be adjusted to 5.8 with 1 M NaOH before autoclaving (121 °C, 20 min). Growth regulators are filter sterilized (0.2 μm) and added to the cooled autoclaved medium.
3. Orbital shaker at 120 rpm in the dark, at 24 °C.

2.2. Determination of Cell Viability

1. Stock solution of CdSe nanocrystals filter sterilized (0.2 μm). To illustrate the methods we will be using Mercaptopropionic acid (MPA)-QDs nanocrystals coated with a multishell structure of ZnS (see Note 1).
2. Eight days old *M. sativa* cell suspension culture.

3. Stock solution of FDA: dissolve 10 mg of FDA in 2 mL of acetone and keep at −20 °C.
4. Fluorescence microscope with a filter set at 488 nm.
5. Murashige & Skoog (M&S) liquid medium supplemented with 0.5 mg/L of Dichlorophenoxyacetic acid and 0.5 mg/L of kinetin. pH of medium should be adjusted to 5.8 with 1 M NaOH before autoclaving (121 °C, 20 min). Filter-sterilize the growth regulators (0.2 μm) and add to the cooled autoclaved medium.

2.3. QDs Uptake by Cells of *M. sativa* in Suspension Cultures

1. Stock solution of CdSe nanocrystals filter sterilized (0.2 μm). To illustrate the methods we will be using MPA-QDs nanocrystals coated with a multishell structure of ZnS (see Note 1).
2. Eight days old *M. sativa* cell suspension culture.
3. Fluorescence microscope with a filter set at 588 nm.
4. Murashige & Skoog (M&S) liquid medium supplemented with 0.5 mg/L of Dichlorophenoxyacetic acid and 0.5 mg/L of kinetin. pH of medium should be adjusted to 5.8 with 1 M NaOH before autoclaving (121 °C, 20 min). Filter-sterilize the growth regulators (0.2 μm) and add to the cooled autoclaved medium.

2.4. In Vitro Uptake of QDs by Roots of *Medicago truncatula* Plantlets

1. Fourteen days old *Medicago truncatula* plantlets with roots of 1 cm length, grown in solid Murashige & Skoog (M&S) medium with 0.2 % (w/v) of gelrite.
2. Murashige & Skoog liquid medium supplemented with 200 mM of mannitol.
3. Four percent paraformaldehyde (PFA) solution in PBS: in a fume hood heat 20 mL of PBS to 60 °C in a water bath. Weigh 1 g of (PFA) and add to the PBS while continuously stirring and maintain on heating plate at 60 °C (do not allow water bath to go over 70 °C, or formaldehyde will vaporize!). Add one or two drops of 5 M NaOH with a Pasteur pipette. The solution should become almost clear fairly rapidly, but will still have some fine particles that will not disappear. Remove from heat and complete the volume up to 25 mL with PBS. Cool the solution to room temperature and adjust the pH to 7.2 with a solution of 1 M HCl. Filter, aliquot into tubes, and freeze at −20 °C.
4. Phytotron set to a 16 h photoperiod and a day/night temperature of 24/22 °C.
5. Fluorescence microscope with a filter set at 588 nm.
6. PBS: 8 g of NaCl, 0.2 g of KCl, 1.44 g of Na_2HPO_4, and 0.24 g of KH_2PO_4. Adjust pH to 7.4 with HCl and add distilled water to 1 L.
7. Vibratome.

2.5. ROS Detection

2.5.1. Hydrogen Peroxide Detection

1. Stock solution of CdSe nanocrystals filter sterilized (0.2 μm). To illustrate the methods we will be using MPA-QDs nanocrystals coated with a multishell structure of ZnS (see Note 1).
2. Three days old *M. sativa* cell suspension culture.
3. 10 mg/mL DAB stock solution: 20 mg of DAB in 2 mL DMSO. Vortex the solution and centrifuge because the DAB does not dissolve completely and filter with a 0.2 μm pore. Store in dark at 4 °C (see Note 2).
4. 0.5 M H_2O_2 solution: mix 102 μL of H_2O_2 and 2 mL of distilled water. Sterilize with a 0.2 μm filter and store in the dark at 4 °C. Prepare this solution fresh, just before running the assay.
5. Optical microscope.
6. Sterile 6-well plates.
7. Orbital shaker.

2.5.2. Superoxide Anion ($O_2^{-\bullet}$) Detection

1. Stock solution of CdSe nanocrystals filter sterilized (0.2 μm). To illustrate the methods we will be using MPA-QDs nanocrystals coated with a multishell structure of ZnS (see Note 1).
2. Three days old *M. sativa* cell suspension culture.
3. 12 mM NBT stock solution: dissolve 20 mg of NBT in 2 mL of distilled water. Vortex and sterilize with a 0.2 μm filter. Store at 4 °C.
4. Optical microscope.
5. Orbital shaker.

2.5.3. Oxidative Stress Detection

1. Stock solution of CdSe nanocrystals filter sterilized (0.2 μm). To illustrate the methods we will be using MPA-QDs nanocrystals coated with a multishell structure of ZnS (see Note 1).
2. Three days old *M. sativa* cell suspension culture.
3. 10 mM H_2DCFDA stock solution: dissolve 9.7 mg of H_2DCFDA in 2 mL of DMSO. Vortex and sterilize with a 0.2 μm filter. Store at −20 °C.
4. Fluorescence microscope with a filter set at 488 nm.
5. Orbital shaker.
6. Heated water bath.

2.5.4. Oxidative Stress Dose Response Assay

1. Stock solution of CdSe nanocrystals filter sterilized (0.2 μm). To illustrate the methods we will be using MPA-QDs nanocrystals coated with a multishell structure of ZnS (see Note 1).
2. Three days old *M. sativa* cell suspension culture.

3. 10 mM H_2DCFDA stock solution: dissolve 9.7 mg of H_2DCFDA in 2 mL of DMSO. Vortex and sterilize with a 0.2 μm filter. Store at −20 °C.
4. Fluorescence microscope with a filter set at 488 nm, with a color camera.
5. Open source software ImageJ (http://rsbweb.nih.gov/ij/) for quantification of the image intensity.

2.6. Antioxidant Enzymatic Activity

2.6.1. Enzyme Extraction

1. Stock solution of CdSe nanocrystals filter sterilized (0.2 μm). To illustrate the methods we will be using MPA-QDs nanocrystals coated with a multishell structure of ZnS (see Note 1).
2. Three days old *M. sativa* cell suspension culture.
3. 100 mM Tris–HCl extraction buffer, pH 7.5: for 100 mL add 1.22 g of Tris, 3 mM DTT, 1 mM EDTA, 0.2 % triton X-100. Adjust to pH 7.5 with 1 M HCl and store at 4 °C, use within 2–3 weeks.
4. Centrifuge capable of 12,000 × *g*.

2.6.2. Determination of SOD Activity

1. Extracts from cell suspension cultures.
2. 100 mM potassium phosphate buffer, pH 7.5. First make solution A by dissolving 13.61 g of KH_2PO_4 in 1 L of distilled water; and solution B by dissolving 14.2 g of Na_2HPO_4 in 1 L of distilled water. Add solution A to solution B until the required pH 7.5 is reached. Keep at 4 °C. The solution can be used for 2–3 weeks.
3. 0.5 mM xantine stock solution: dissolve 47.6 mg of xantine in 25 mL of potassium phosphate buffer, pH 7.5. Add 1 M NaOH dropwise with constant stirring, until a clear solution is obtained. Store at 4 °C, use within 2–3 weeks.
4. 0.1 mM EDTA stock solution: dissolve 23.27 mg of EDTA in 25 mL potassium phosphate buffer. Store at 4 °C, use within 2–3 weeks.
5. Solution A (superoxide production): dissolve 6.1 mg of Cytochrome C in 12.5 mL of potassium phosphate buffer, add 0.5 mL of EDTA stock solution, and 0.5 mL of xantine stock solution. Prepare fresh before use.
6. Solution X (xanthine oxidase): Prepare a diluted solution containing 0.72 U of xanthine oxidase in 800 μL of potassium phosphate buffer. Prepare fresh before use and keep on ice.
7. Spectrophotometer.
8. Disposable polystyrene cuvettes with 10 mm optical light path.

2.6.3. Determination of CAT Activity

1. Extracts from cell suspension cultures.
2. 10 mM H_2O_2 stock solution: 170 μL of H_2O_2 (30 %) in 50 mL potassium phosphate buffer. Prepare fresh before use.
3. UV/Vis spectrophotometer.
4. Quartz cuvette 10 mm optical light path.

2.6.4. Determination of GR Activity

1. Extracts from cell suspension cultures.
2. 100 mM potassium phosphate buffer, pH 7.5 with 1 mM EDTA: dissolve 372 mg of EDTA in 100 mL of potassium phosphate buffer. Store at 4 °C, use within 2–3 weeks.
3. 0.75 mM 5,5′-Dithiobis(2-nitrobenzoic acid) (DTNB) stock solution: dissolve 11.8 mg of DTNB in 10 mL potassium phosphate buffer. Prepare just before use, protect from direct light.
4. 0.1 mM NADPH stock solution: dissolve 3 mg of NADPH in 2 mL of potassium phosphate buffer. Prepare fresh before use.
5. 1 mM of Oxidized Glutathione (GSSG): dissolve 24.3 mg of GSSG in 2 mL of potassium phosphate buffer. Prepare fresh just before use.
6. Spectrophotometer.
7. Disposable polystyrene cuvettes with 10 mm optical light path.

3. Methods

All procedures should be carried out in a laminar flow chamber and with sterilized material unless otherwise specified. Unless mentioned otherwise, all cell suspension cultures are grown and maintained on the orbital shaker at 120 rpm in the dark, at 24 °C.

3.1. Initiation of *Medicago sativa* Cell Suspension Culture

1. Place two friable portions of 8 week grown callus in 50 mL of liquid M&S medium. Grow cells for 8 days. After 8 days, add 100 mL of fresh M&S liquid medium.
2. Grow cells for another 8 days.
3. Remove the cells from the shaker and allow the suspension to settle for 3–4 min and pipette 20 mL of cells to another flask with 100 mL of M&S liquid medium (see Note 3).
4. Grow cells until a density of 4×10^4 cells/mL is obtained. Change the medium every 8 or 9 days by transferring 20 mL of cell culture to 100 mL of fresh medium (in 250 mL Erlenmeyer flasks).

3.2. Determination of Cell Viability

1. Take 2 mL aliquots of 8-day-old cell suspension culture of *M. sativa* and subculture in a 50 mL Erlenmeyer flask containing 10 mL of fresh M&S liquid medium.
2. On the third day of culture determine the viability of cells. Prepare fresh diluted solution of FDA by adding 20 μL of the FDA stock solution and 1 mL of M&S liquid medium, then on a microscope slide mix 20 μL of suspension culture and 20 μL of diluted FDA and cover with a cover slip. Incubate 5 minutes at room temperature and perform the counts under the fluorescence microscope. FDA is converted into a fluorescent product if cell membrane is intact and therefore viable cells present a bright green fluorescence when excited at 488 nm. Count five random spots per slide and calculate the percentage of viability:

   $$\text{viability \%} = (\text{number of viable cells}/\text{total number cells}) * 100$$

3. After determining of the cellular viability add the MPA-QDs to the cell suspension cultures (use different concentrations ranging from 10 to 100 nM).
4. Grow cells for 5 days (until the eighth day of culture) and determine the cell culture viability as described above (see Note 4).

3.3. QDs Uptake by Cells of M. sativa in Suspension Cultures

MPA-QDs may accumulate intracellularly due to the hydrophilic character of MPA and the reduced size of the particles. The exact mechanisms of QDs intake by the plant cells are not known. It has been shown that MPA-QDs can accumulate in different cellular compartments when contacted with *M. sativa* cells in suspension cultures (7).

1. Inoculate 2 mL aliquot of 8 days old *M. sativa* cell suspension culture in 50 mL Erlenmeyer flasks containing 10 mL of fresh M&S liquid medium.
2. Culture the cells for 3 days.
3. Add the MPA-QDs to the cell suspension cultures (final concentration ranging from 10 to 100 nM).
4. Grow the cells for 48 h.
5. Visualize samples using a confocal and fluorescent microscope (Fig. 1).
6. Acquire thin time-course confocal optical sections (~2 μm thick) using <20 % laser intensity and operating in the mode 1,024 × 1,024, 400 Hz (~1/2 s per frame). A ×20 Plan Apo dry objective (NA = 0.75) (Leica) should be used.

3.4. In Vitro Uptake of QDs by Roots of Medicago truncatula Plantlets

1. Gently take out the 14-day-old plantlets from the agar using forceps (make sure that the roots are taken intact) and place it in 50 mL Falcon sterilized tubes with 200 μL of M&S liquid medium with mannitol. Cover with parafilm and incubate in a

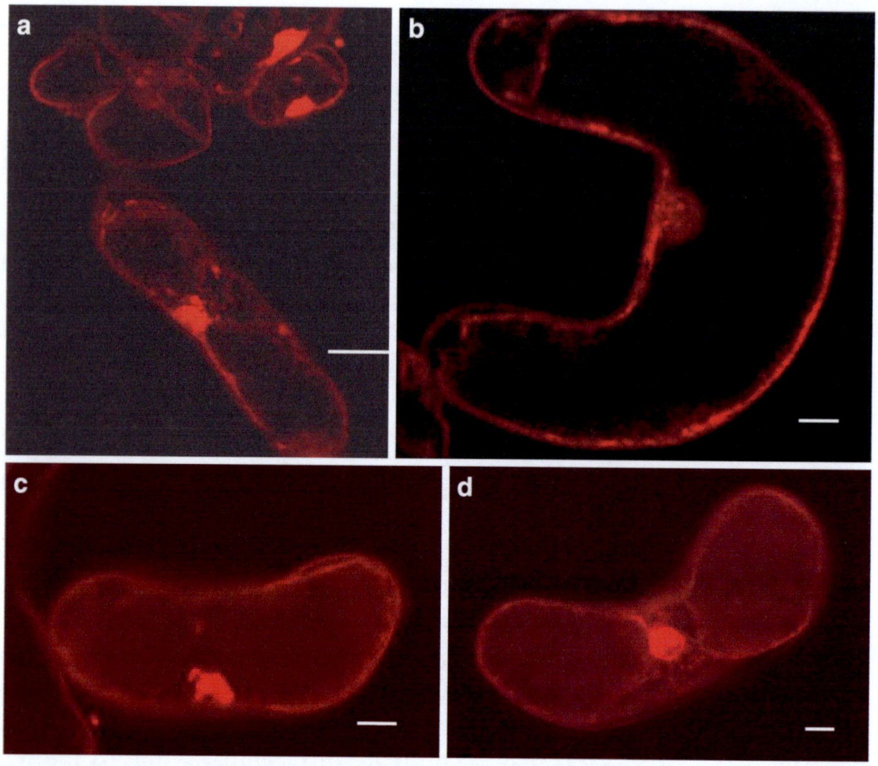

Fig. 1. Internalization of MPA-QDs in *Medicago sativa* cells. Confocal (**a**, **b**) and wide field (**c**, **d**) fluorescence images of *M. sativa* cells after 48 h of incubation with MPA-QDs. Scale bar = 20 μM. Reproduced from ref. (7).

Phytotron under standard culture conditions mentioned above.

2. After 48 h remove the plantlets from the tubes and place them in a petri dish with a drop of fixation solution. Cut the root into three pieces no larger than 1 cm: root cap, zone of maturation, and zone of differentiation.
3. Transfer the pieces into 15 mL Falcon tubes with 4 % PFA solution (approximately ten times the volume of the samples) at 4 °C. Then place the samples under vacuum for approximately 10 min (to facilitate the fixative penetration into the samples).
4. Let the fixation process run overnight (about 18 h) at 4 °C with smooth agitation.
5. Wash the samples three times 10 min in PBS and store it in 0.1 % PFA in PBS at 4 °C (see Note 5).
6. Make longitudinal cuts using a vibratome and visualize them on a fluorescence microscope under UV light.

3.5. ROS Detection

3.5.1. Hydrogen Peroxide Detection

1. Take seven 2 mL aliquots from 3-day-old cell suspension cultures of *M. sativa* and transfer into sterile 6-well plate. Add MPA-QDs to the cell suspension cultures to 0, 1, 5, 10, 20, 50, 100 nM final concentration.
2. Grow the cells for 48 h.
3. Add 0.1 mg/mL of DAB to all cell suspension cultures (see Note 6). As control, heat 2 mL of suspension culture at 45 °C for 20 min (heated in a water bath) and add 0.1 mg/mL of DAB (see Note 7), as a second control add 40 μL of H_2O_2 stock solution into 2 mL of suspension culture plus 0.1 mg/mL of DAB (see Note 8).
4. Incubate cells on the orbital shaker for 90 min and check cell cultures using a fluorescence microscope (see Note 9).

3.5.2. Superoxide Anion ($O_2^{-\bullet}$) Detection

1. Take seven 2 mL aliquots from 3-day-old cell suspension cultures of *M. sativa* and transfer into sterile 6-well plate. Add MPA-QDs to cell suspension cultures to 0, 1, 5, 10, 20, 50 and 100 nM final concentrations
2. Grow the cells for 48 h.
3. Add 10 μL of NBT stock solution to all cell suspension cultures (final concentration of 60 nM). As control, heat 2 mL of suspension culture at 45 °C for 20 min (heated in a water bath) and add 10 μL of NBT stock solution (final concentration of 60 nM).
4. Incubate cells on the orbital shaker for 4 h and visualize cells using a microscope under bright field illumination (see Note 10).

3.5.3. Oxidative Stress Detection

1. Take seven 2 mL aliquots from 3-day-old cell suspension cultures of *M. sativa* and transfer into sterile 6-well plate. Add MPA-QDs to cell suspension cultures to 0, 1, 5, 10, 20, 50 and 100 nM final concentrations.
2. Grow the cells for 48 h.
3. Add 1 μL of H_2DCFDA stock solution to all cell suspension cultures (to a final concentration of 5 μM). As control, heat 2 mL of suspension culture at 45 °C for 20 min (heated in a water bath), and add 1 μL of H_2DCFDA stock solution (to a final concentration of 5 μM).
4. Incubate cells on the orbital shaker for 1 h and visualize the samples using a fluorescence microscope (see Note 11).

3.5.4. Oxidative Stress Dose Response Assay

1. Take 27 2 mL aliquots from 3-day-old cell suspension cultures of *M. sativa* and add to sterilized 6 well plates. Aliquots of MPA-QDs are added to each well to obtain the following final concentrations: 1, 5, 10, 20, 40, 60, 100, 120, and 180 nM (all in triplicates). All plates are placed on an orbital shaker at 110 rpm in the dark at 24 °C.

2. After 48 h of incubation, 1 μL of H_2DCFDA is added to all treatments, to obtain a final concentration of 5 μM.
3. The fluorescence is quantified in terms of average pixel intensity, using the open source software ImageJ. Images are acquired 1 h after the H_2DCFDA addition, always with the same settings, and exposure time. For each treatment at least 15 pictures should be taken.
4. Using the ImageJ program split the channels of the pictures in red, blue, and green. Using the green channel histogram (to exclude the QDs fluorescence), multiply the number of pixels with their respective intensity value (0–255) and divide it for the total number of pixels of the image.

3.6. Antioxidant Enzyme Activity

3.6.1. Enzyme Extraction

1. Transfer four 10 mL aliquots from 3-day-old cell suspension cultures of *M. sativa* to 50 mL Erlenmeyer flasks. Add MPA-QDs to a final concentration of 0 (control), 10, 50, or 100 nM.
2. Grow the cells for 48 h.
3. Harvest the cells by filtration, wash for a few seconds with distilled water and make aliquots of about 500 mg and proceed with the enzyme extraction.
4. Prepare a positive control consisting of cell suspension culture heated at 50 °C for 20 min, and performing the same aliquoting procedure as described in previous step.
5. Homogenize each aliquot in a mortar with a pestle in 2 mL of 100 mM Tris–HCl buffer (pH 7.5) and 2 % (w/v) insoluble polyvinylpyrrolidone (PVP-40). Carry this procedure at 4 °C.
6. Pipette the homogenate into a microcentrifuge tube and centrifuge at 12,000 × *g* for 30 min at 4 °C.
7. Store the supernatant in small aliquots at −80 °C, until use (CAT, GR, SOD, and protein analyses) (see Note 12).

3.6.2. Determination of SOD Activity

1. Determine the enzyme activity using disposable cuvettes. Add 50 μL of the extract (supernatant obtained in Subheading 3.6.1) to a reaction mixture composed of 0.5 mM xanthine, 0.05 mM ferricytochrome-C, 0.1 mM EDTA, and 0.01 U of xanthine-oxidase to the final volume of 1 mL (see Table 1) (see Note 13). Prepare the negative control by heating the extract containing the enzyme (SOD) at 90 °C for 10 min.
2. Mix by inverting the cuvette several times and measure the increase in absorbance at 550 nm for 2 min (10 s interval).
3. Calculate the enzymatic activity (see Note 14).

$$\%\ \text{inhibition} = (\Delta\text{Abs control} - \Delta\text{Abs sample}) / \Delta\text{Abs control} * 100$$

$$\text{Units/mg protein} = (\%\ \text{inhibition}/50\ \%) * (1/v) / \text{mg of total protein}$$

Table 1
Reagents and proportions for SOD activity measurement

	Control sample (μL)	Sample (μL)
Solution A[a]	500	500
Phosphate buffer	487.5	437.5
Solution X[b]	12.5	12.5
Extract	–	50

[a] Solution A is defined in Subheading 2.6.2, item 5
[b] Solution X is defined in Subheading 2.6.2, item 6

Table 2
Reagents and proportions for CAT activity measurement

	Blank (μL)	Sample (μL)
Phosphate buffer	963	593
Extract	37	37
H_2O_2 stock solution	–	370

where 50 % = inhibition of cytochrome C reduction rate as per Unit definition; v is volume of extract in the cuvette ($v = 0.05$ mL, Table 1).

3.6.3. Determination of CAT Activity

1. In a quartz cuvette prepare 1 mL of solution containing the reagents in the order and proportions described in Table 2. Prepare a negative control by heating an extract with the enzyme (CAT) at 90 °C for 10 min.
2. Mix by inverting the cuvette several times and measure the decrease in absorbance at 240 nm for 2 min (10 s intervals) against the H_2O_2-free blank.
3. Calculate the CAT enzymatic activity (see Note 15):

$$\text{Units per mg of protein} = (\Delta\text{Abs}/\Delta t) * (1/\varepsilon) * (1/L) * (1/v) / \text{mg of total protein}$$

where ε is extinction coefficient of H_2O_2 (0.00394 μmol/mm); L is path length (10 mm); and v is volume of extract in the cuvette (0.037 mL, Table 2).

3.6.4. Determination of GR Activity

1. In a disposable cuvette prepare 1 mL of solution containing 1 mM EDTA in phosphate buffer, 0.75 mM DTNB, 0.1 mM NADPH, and 1 mM GSSG by pipetting the reagents in the

Table 3
Reagents and proportions for GR activity measurement

	Sample volume (μL)
Phosphate buffer + 1 mM EDTA stock solution	600
0.75 mM DTNB stock solution	250
0.1 mM NADPH stock solution	50
Extract	50
1 mM GSSG stock solution	50

order and proportions described in Table 3. Prepare a negative control by heating the extract with the enzyme (GR) at 90 °C for 10 min.

2. Mix by inverting the cuvette several times and measure the increase in absorbance at 412 nm (10 s interval during 2 min).
3. Calculate the activity of the enzyme (see Note 16):

$$\text{Units per mg of protein} = (\Delta\text{Abs}/\Delta T) * (1/\varepsilon) * (1/L) * (1/v)/\text{mg of total protein}$$

where ε is extinction coefficient of GSSG (0.62 mL μmol/mm); L is path length (10 mm); v is volume of extract in the cuvette (0.05 mL, Table 3).

4. Notes

1. *Preparation of* CdSe nanocrystals coated with a multishell structure of ZnS and their surface modification based on the successive ion layer adsorption and reaction (SILAR) method has been described elsewhere in this book. The first phase of the synthesis produces a hydrophobic surface consisting of a TOPO/HDA ligand mixture. This is followed by a surface modification of the Quantum Dots (QDs) with 3-mercapto-propionic acid (MPA) by the phase transfer method. The modified QDs become soluble in aqueous systems such as water or buffers for further biological applications.
2. DAB in solution can be kept at 4 °C for a maximum of 2 days; however, it is advisable to prepare a fresh solution for each set of experiments. DAB is difficult to dissolve in water unless the pH is reduced. An alternative procedure to dissolve DAB is to prepare a solution of water by adding HCl until pH 3.6 is

reached and then add the DAB. Shake the solution until all the DAB is dissolved (8).

3. For subculture use a pipette with an orifice narrow enough to allow single cells and small aggregates to pass through. The suspension culture is allowed to rest in order to settle down the large clumps, and the inoculum is taken from the upper part of the culture. Regular practice of this procedure should allow obtaining a fine cell suspension culture.
4. The presence of QDs can promote cell aggregation so be aware of this fact to have an accurate cell counting. The metabolism of FDA to fluorescein reflects the metabolic response of organisms when exposed to toxicants. Furthermore, the degree of fluorescence depends on the physical and metabolic state of the cell (5), therefore pay attention to the variations in the fluorescence.
5. Samples stored this way are stable for weeks.
6. The majority of studies investigate acute toxicity, generally 24 h, but longer incubation times are required to see the influence of nanoparticles (9). We also tested after 48 h of incubation with QDs, but extended periods can also be used. Note that the QDs should always be added at day 3 of culture (beginning of the exponential phase).
7. The treatment of cell suspension cultures at 45 °C for 20 min is a positive control to ensure a cellular stress.
8. The use of DAB technique requires the presence of endogenous peroxidase activity in order to ensure the polymerization, thus a simple test to check the required peroxidase activity is to expose cells to DAB and H_2O_2 (10).
9. DAB reacts rapidly with H_2O_2 in the presence of peroxidase, forming a brown polymerized product (10).
10. NBT is reduced by $O_2^{-\bullet}$ to form an insoluble blue formazan deposit (6).
11. H_2DCFDA is a nonfluorescent product that is converted by ROS into DCF (2′,7′-dichlorofluorescein), which can easily be visualized as a green fluorescence around 525 nm when excited at 488 nm (11). H_2DCFDA treated cells show a basal-level signal that accumulates in all cytoplasmic areas, including the nucleoplasm (12). DCF signal can suffer quenching and decrease the fluorescence within 5–10 min when continuously exposed to light.
12. To normalize the results the total cellular protein concentration must be determined. The Bradford method is commonly followed by using a protein assay kit, and BSA as a standard (13).
13. The extract volume can be adjusted according to the activity of the enzymes, e.g., if for a determined enzyme the activity is too

high you should use a smaller volume, but always maintaining the described proportions.

14. The enzymatic activity of SOD is defined as the quantity necessary for the inhibition of 50 % of cytochrome-C reduction per minute under the assay conditions (14).
15. The enzymatic activity of CAT is defined as the consumption of 1 μmol of H_2O_2 per minute per mL at room temperature, under the assay conditions (15).
16. The enzymatic activity of GR is expressed in U/mL mg protein wherein unit activity is the amount of enzyme which reduces 1 mM of GSSG per minute at 24 °C under assay conditions (16).

Acknowledgments

This work was supported by the project "Development of ultrasensitive detection methods and plant nano-vaccines for the fungi *Fusarium* spp. using nanotechnological devices" Iberian Capacitation Program in Nanotechnologies: Call 2006/2007. This work was also supported by Fundação para a Ciência e a Tecnologia through grant # PEst-OE/EQB/LA0004/2011.

References

1. Saeza G, Moreaua X, Jonga L, Thiérya A et al (2010) Development of new nano-tools: towards an integrative approach to address the societal question of nanotechnology? Nano Today 5:251–253
2. Auffan M, Rose J, Wiesner M, Bottero J (2009) Chemical stability of metallic nanoparticles: a parameter controlling their potential cellular toxicity in vitro. Environ Pollut 157:1127–1133
3. Kristen U (1997) Use of higher plants as screens for toxicity assessment. Toxicol In Vitro 11:181–191
4. Duncan D, Widholm J (1990) Measurements of viability suitable for plant tissue cultures. In: Pollard J, Walker M (eds) Plant cell and tissue culture, vol 6. The Humana Press, London, p 33
5. Regel R, Ferris J, Ganf G, Brookes J (2002) Algal esterase activity as a biomeasure of environmental degradation in a freshwater creek. Aquat Toxicol 59:209–223
6. Gerber I, Dubery A (2003) Fluorescence microplate assay for the detection of oxidative burst products in tobacco cell suspensions using 2′,7′-dichlorofluorescein. Methods Cell Sci 25:115–122
7. Santos A, Miguel A, Tomaz L, Malhó R, Maycock C, Patto M, Fevereiro P, Oliva A (2010) The impact of CdSe/ZnS quantum dots in cells of *Medicago sativa* in suspension culture. J Nanobiotechnology 8:24
8. Driever S, Fryer M, Mullineaux P, Baker N (2009) Imaging of reactive oxygen species in vivo. In: Pfannschmidt T (ed) Plant signal transduction, vol 479. Humana Press, London, p 115
9. Gormley A, Ghandehari H (2009) Evaluation of toxicity of nanostructures in biological systems. In: Sahu S, Casciano D (eds) Nanotoxicity from in vivo and in vitro models to health risks. Wiley, United Kingdom, p 137
10. Thordal-Christensen H, Zhang Z, Wei Y, Collinge D (1997) Subcellular localization of H_2O_2 in plants. H_2O_2 accumulation in papillae and hypersensitive response during the barley-powdery mildew interaction. Plant J 11:1187–1194
11. Ortega-Villasante C, Rellán-Alvarez R, del Campo F, Carpena-Ruiz R, Hernández L (2005) Cellular damage induced by cadmium and mercury in *Medicago sativa*. J Exp Bot 56:2239–2251

12. Ashtamker C, Kiss V, Sagi M, Davydov O, Fluhr R (2007) Diverse subcellular locations of cryptogein-induced reactive oxygen species production in tobacco bright yellow-2 cells. Plant Physiol 143:1817–1826
13. Ma Y (2009) In vitro models for nanotoxicity testing. In: Sahu S, Casciano D (eds) Nanotoxicity from in vivo and in vitro models to health risks. Wiley, United Kingdom, p 365
14. McCord J, Fridovich I (1969) Superoxide dismutase. An enzymic function for erythrocuprein (hemocuprein). J Biol Chem 244:6049–6055
15. Aebi E (1983) Catalase. In: Bergmeyer H (ed) Methods of enzymatic analysis, vol III, Oxidoreductases, transferases. Verlag Chemie, Germany, pp 273–277
16. Shanker A, Djanaguiraman M, Sudhagar R, Chandrashekar C, Pathmanabhan G (2004) Differential antioxidative response of ascorbate glutathione pathway enzymes and metabolites to chromium speciation stress in green gram (*Vigna radiata* (L.) R. Wilczek. cv CO 4) roots. Plant Sci 166: 1035–1043

Part VI

Nanoparticles Characterization

Chapter 37

Scanning Transmission Electron Microscopy Methods for the Analysis of Nanoparticles

Arturo Ponce, Sergio Mejía-Rosales, and Miguel José-Yacamán

Abstract

Here we review the scanning transmission electron microscopy (STEM) characterization technique and STEM imaging methods. We describe applications of STEM for studying inorganic nanoparticles, and other uses of STEM in biological and health sciences and discuss how to interpret STEM results. The STEM imaging mode has certain benefits compared with the broad-beam illumination mode; the main advantage is the collection of the information about the specimen using a high angular annular dark field (HAADF) detector, in which the images registered have different levels of contrast related to the chemical composition of the sample. Another advantage of its use in the analysis of biological samples is its contrast for thick stained sections, since HAADF images of samples with thickness of 100–120 nm have notoriously better contrast than those obtained by other techniques. Combining the HAADF-STEM imaging with the new aberration correction era, the STEM technique reaches a direct way to imaging the atomistic structure and composition of nanostructures at a sub-angstrom resolution. Thus, alloying in metallic nanoparticles is directly resolved at atomic scale by the HAADF-STEM imaging, and the comparison of the STEM images with results from simulations gives a very powerful way of analysis of structure and composition. The use of X-ray energy dispersive spectroscopy attached to the electron microscope for STEM mode is also described. In issues where characterization at the atomic scale of the interaction between metallic nanoparticles and biological systems is needed, all the associated techniques to STEM become powerful tools for the best understanding on how to use these particles in biomedical applications.

Key words: Scanning transmission electron microscopy, STEM, Inorganic nanoparticles, Biological systems, HAADF-STEM, Aberration-corrected microscopy

1. Introduction

Conventional transmission electron microscopy (TEM) is a beam characterization technique with almost 80 years since the development of the first microscope. In TEM, a large area of the specimen is illuminated, the magnification is performed by the lens system underneath the specimen, and subsequently the whole image is registered instantaneously. Conventional scanning electron

Mikhail Soloviev (ed.), *Nanoparticles in Biology and Medicine: Methods and Protocols*, Methods in Molecular Biology, vol. 906, DOI 10.1007/978-1-61779-953-2_37,

microscopy (SEM) was designed after the TEM and it has been widely used for surface analyses in biological and materials sciences. The scanning-transmission electron microscopy (STEM) combines both operational modes. In STEM mode, a convergent electron beam is focused to a small area of the sample. To register an image, the electron probe is raster-scanned and subsequently propagated through the sample. Due to the electron-matter interaction, the trajectory of the electrons is scattered away and different kinds of signals are registered in sync with the electron probe scanning. The resolution of the STEM is mostly determined by the size of the probe. The newest electron microscopes can work in both modes, STEM or TEM, so there is a tendency to discard the use of dedicated instruments. One of the main advantages of the STEM over the TEM is that the signal generated by the electrons scattered out to high angles on a high-angle annular dark field detector (HAADF), is chemically sensitive, and a sample with a definite crystalline arrangement is not necessarily a requirement. The capacity of STEM of generating these different levels of contrast is commonly known as Z-contrast; the contrast dependance goes approximately as Z^2, Z corresponding to the atomic weight of the element that caused the scattering of the electrons (1).

For nanostructures, TEM and STEM operational modes become necessary due to the fact that matter properties from bulk can result in different—sometimes even improved—properties at the nanoscale from those at bulk. This dependence in size is due to several reasons: quantum effects, surface effects, and modification of thermal behavior, among others (2, 3). In order to take advantage of the physical and chemical properties of the nanostructures by guiding the fine-tuning of these properties depending on the purposes for which the structures has been created, a deep understanding on the relation between shape, size, and function, is needed. The appropriate generation and interpretation of electron micrographs is crucial for this purpose. The interpretation of the electron micrographs is not always straightforward, since the intensity signal that correlates with the atomic positions depends not just in the length of the atomic columns parallel to the direction of the electron beam, but also on the chemical species, and on the microscope's parameters at which the micrograph was obtained (4). In this way, the problem of extracting a third dimension from the information contained in a strictly two-dimensional image requires the comparison of the observed images against simple models acquired by previous experience or inherited, and in many cases a theoretical model of the structure is necessary.

This chapter deals with the specific issues that need to be considered in order to obtain useful micrographs and to make an adequate interpretation of the micrograph of a nanostructure, with a particular emphasis in the imaging of nanoparticles used as good candidates to be incorporated in biomedical applications. We will concentrate in the use of STEM imaging. The chapter aims to be a guide for researchers interested in the characterization, at atomic scale, of metallic nanoparticles, and nanostructures. We will cover the working principles of STEM, the role of aberration correctors, and how the theory supporting the imaging of micrographs can be used to simulate the imaging process, to concentrate later on the practical issues concerning the comparison and interpretation of real and simulated electron micrographs of nanostructures.

2. Stem Imaging

In TEM, the information transfer in phase contrast imaging is determined by the objective lens and the size of it aperture, which are directly related to the temporal and spatial coherence of the electrons. In Fig. 1, a conventional high resolution transmission electron microscopy (HRTEM) micrograph of a bimetallic nanoparticle of 2 nm of diameter shows the interatomic distances; however, the contrast is not directly related with the different atomic positions in the bimetallic nanoparticle (5) proposed by the theoretical model shown in Fig. 1b. The optical arrangement of a conventional TEM compared with the STEM configuration is illustrated in the schematic diagrams shown in Fig. 2. The scattered electrons in STEM mode can be registered by three different detectors. The bright field (BF) detector collects the electrons transmitted in the path of the beam close to the optical axis, the BF electrons containing the total beam current. The annular dark field (ADF) and the HAADF detectors are used to record the electrons scattered out of the path of the beam. From the point of view of

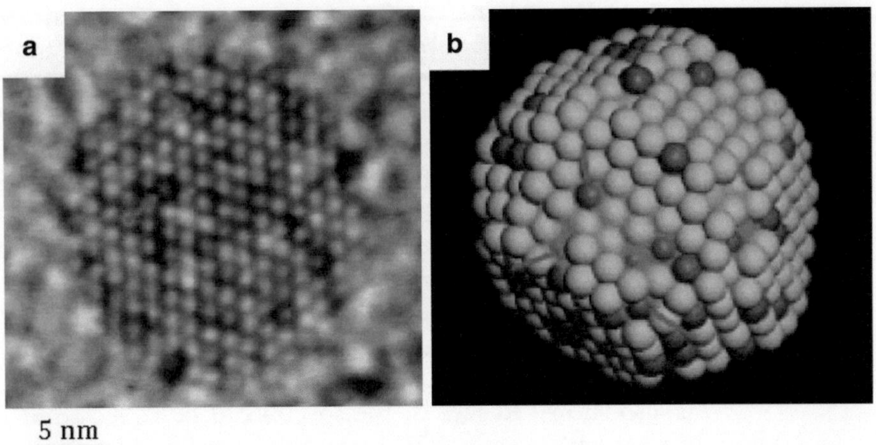

Fig. 1. Conventional high resolution transmission electron microscopy (HRTEM). (**a**) HRTEM image of a bimetallic gold-palladium nanoparticle. (**b**) Theoretical model of the bimetallic nanoparticle.

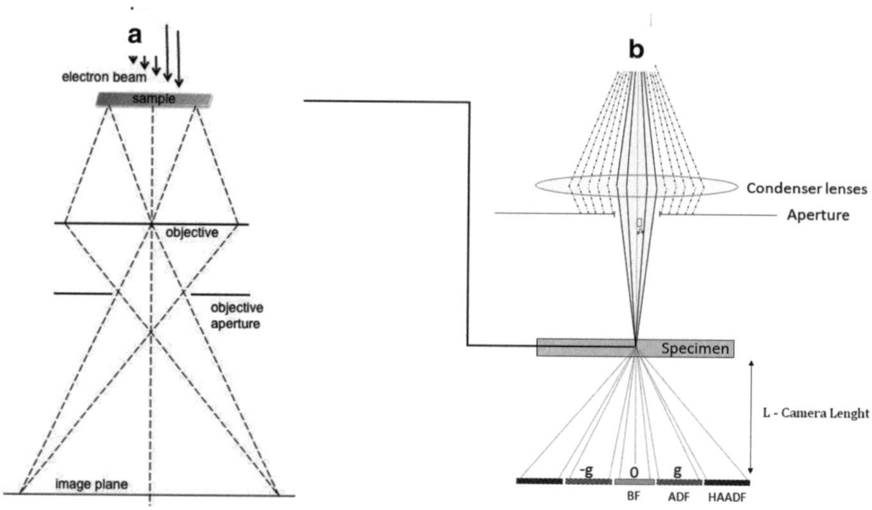

Fig. 2. Optical arrangement of a conventional transmission electron microscopy (TEM) and scanning transmission electron microscopy (STEM) configurations. (**a**) TEM and (**b**) STEM operational modes.

the reciprocity theorem, the STEM is optically equivalent to an inverted TEM, in the sense that if the source and the detector exchange positions, the electron ray paths remain the same (6). In the STEM, the objective lens—and all the relevant optics—are positioned before the specimen. The scattered angle of the ADF detector is set around ~40 mrad, and the outer angle is set around 60–200 mrad (see Fig. 2). STEM differs from conventional TEM in that the electron beam interacts only with a small section of the sample; the scanning process is the one in charge of generating the image as a whole, while in TEM, a broad beam is interacting instantaneously with the sample. X-ray energy dispersive spectroscopy (EDS) attached to the STEM mode provides with an elemental analysis directly from the point or line raster-scanned in the sample. Therefore, the imaging using the BF or HAADF detectors in a complementary way can be matched to the EDS information, and the elemental mapping at atomic resolution is obtained. Figure 3 shows the setup of the STEM mode including the detectors a set of the detectors typically used in STEM mode. Both the HAADF and BF detectors are shown in the picture. Examples of HAADF and BF images are included in Fig. 3 (on the right); both can be acquired simultaneously. *L* corresponds to the camera length, which is the effective distance or magnification between the specimen and the detector plane position. The X-ray detector is also shown in Fig. 3; the spectra can be collected at the same time together with STEM images. An example of the use of EDS and HAADF/BF-STEM is illustrated in Fig. 4, where a set of

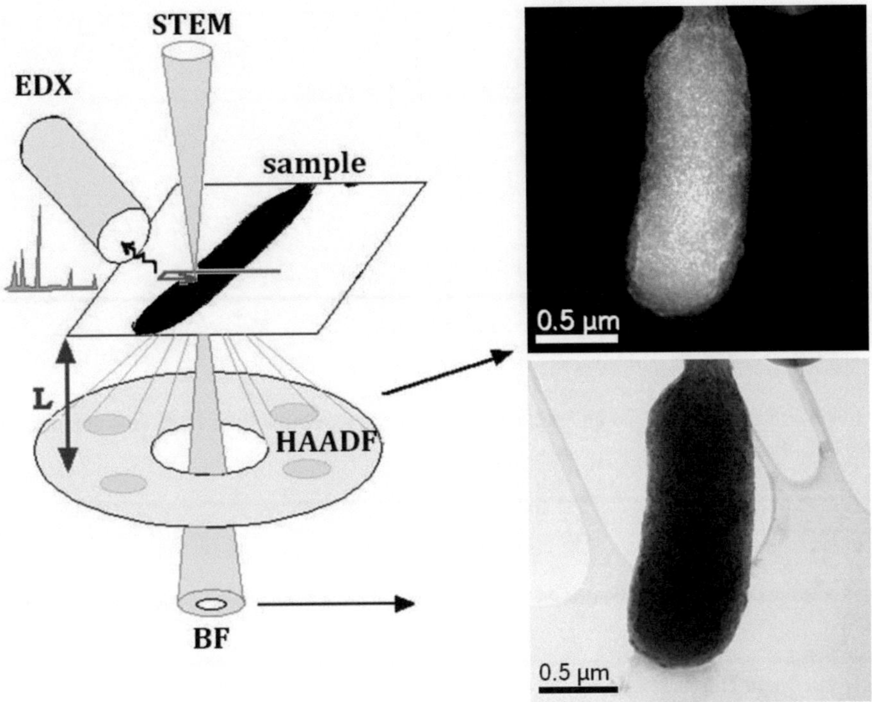

Fig. 3. STEM setup including the energy dispersive spectroscopy (EDS) spectrum detector.

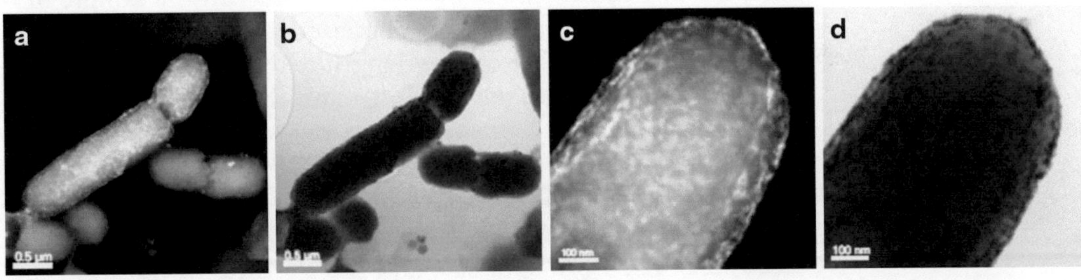

Fig. 4. STEM imaging of *Escherichia coli* bacteria. (**a**, **c**) high angular annular dark field (HAADF)-STEM, (**b**, **d**) BF-STEM.

STEM images from a bacterium show high contrast due to the presence of terbium inside and in the periphery of cells, which confirms the incorporation of Tb and indicates that TATA-binding protein (TBP) was mainly located in the periplasmic region. For HAADF images, the high contrast corresponds to the Tb incorporation in the bacteria. The EDS spectrum obtained in a transversal region of bacteria is shown in Fig. 5.

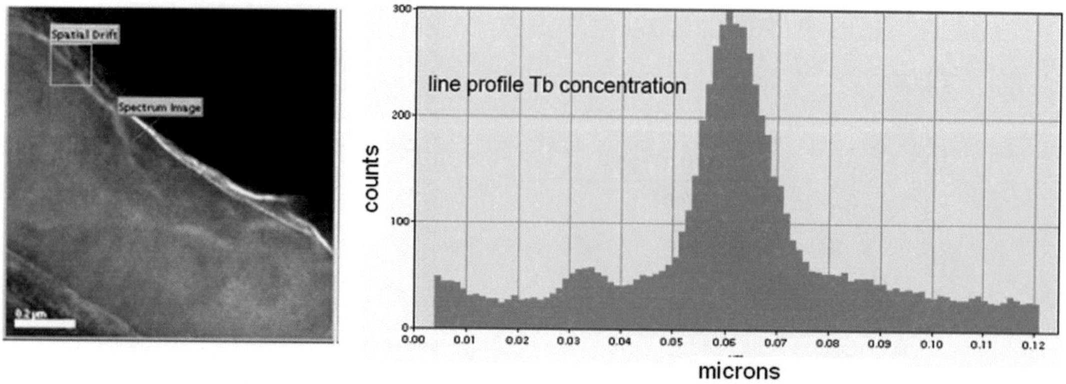

Fig. 5. Line profile EDS spectrum. STEM image showing transversal region analyzed (*left*). Tb line profile of transversal region analyzed (*right*).

3. Practical Considerations for Stem Imaging

The basic parameters described in this section are used to register high resolution STEM images. The process must be initialized in TEM mode for searching field of view and adjust the crystal orientation if necessary. The eucentric focus must be adjusted with the z-control. By following the TEM initial setup, the system can be switched to STEM. The illumination focused on the sample must be adjusted by the size of the condensed aperture, spot size, and camera length (*L*). The camera length is the effective distance or magnification between the specimen and the detector plane position (labeled in Fig. 2). The collection angle depends on the microscope camera length. Typical values for ADF and HAADF detectors vary from ~40 and 200 mrad, respectively.

The resolution in STEM depends upon the spot size of the electron beam. Both spot size and the aperture of the condenser lens have influence in the beam current density in the sample and as consequence, in the resolution of the instrument. The beam current density in the sample can be increased by a large aperture of the condenser lens or by using a lower spot size. Figure 6 shows two different settings for STEM. Increasing the spot size (spot number) leads to increased demagnification of the source and a decreased current in the beam. Increasing the spot number leads to a reduction of the beam current. Changing the size of the condenser lenses CL1 and CL2, produces that the beam current is also changed. In Fig. 7, the sketches for the rays using two different sizes of the condenser apertures are illustrated. The selected aperture must be centered in the Ronchigram (described in next section) as is shown in Fig. 7. Spot size and aperture of the

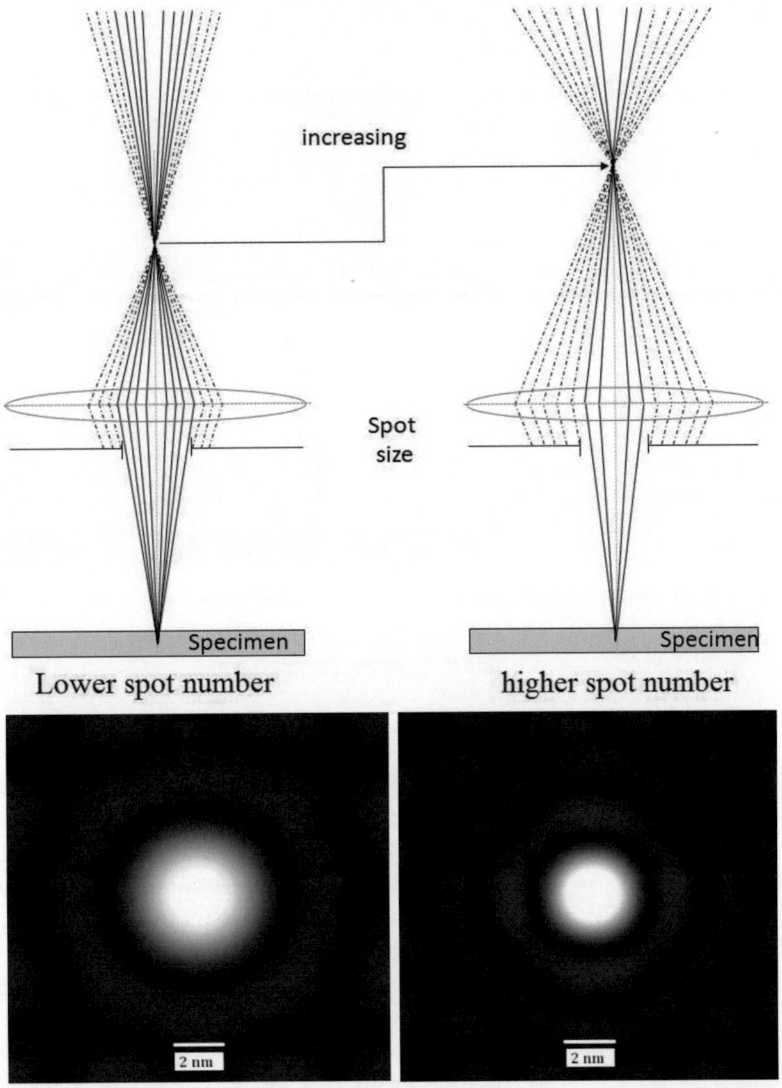

Fig. 6. Electron beam sketches varying the spot size and the potential projected simulated depending on the spot size. Smaller spot size (*left*), larger spot size (*right*).

condenser lens determine the resolution of the microscope in STEM mode. The spot size, measured in nanometers, can be simulated as a function of the parameters of the microscope (7).

4. The Ronchigram

A Ronchigram is actually a shadow image, referring to the features at the center of the convergent beam electron diffraction (CBED) pattern, and formed by a focused and stationary electron probe on

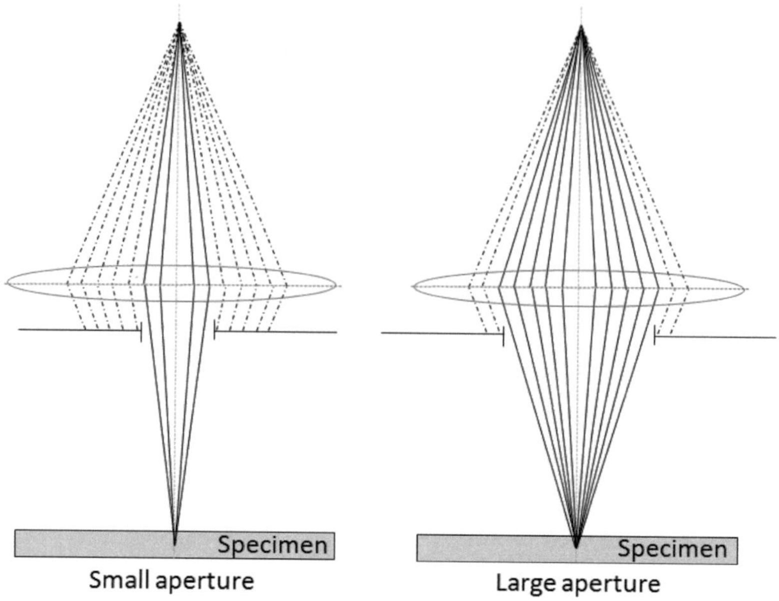

Fig. 7. Electron beam trajectories by varying the condenser lens apertures. Smaller aperture (*left*), larger aperture (*right*).

amorphous material. The quality and resolution of the STEM images depend directly on the proper alignment of the Ronchigram. It is easily observable in field emission gun (FEG) microscopes, but it might be difficult to obtain in the Lanthanum Hexaboride (LaB_6) filaments because the effective probe size is too large, and therefore both the spatial resolution and the brightness decrease dramatically, even using microscopes with voltages higher than 200 kV. All the parameters used to adjust different sets of contrast, spatial resolution, etc. are directly related with the Ronchigram. The Ronchigram can be registered in a CCD camera. The electrons coming out of the sample can be observed directly in a circle called the *Gabor hologram*, the *Ronchigram* or the *central zero-order disk* of the CBED pattern. The Ronchigram from an amorphous region should look as in the Fig. 8. The Ronchigram is actually a shadow image, referring to the feature within the central disk of a CBED pattern and formed by a small, focused, and stationary probe on thin amorphous region. The optimal defocus is useful to optimize the spatial resolution and it can be corrected from the Ronchigram. When a near-focused beam illuminated on the specimen, the Ronchigram changes with the illumination conditions. In under and overfocus, an image of the sample is observed. Overfocus, in-focus, and underfocus Ronchigrams and the beam incident on the sample are illustrated in Fig. 9. The shape of the Ronchigram can be simulated using the parameters of the microscope (7).

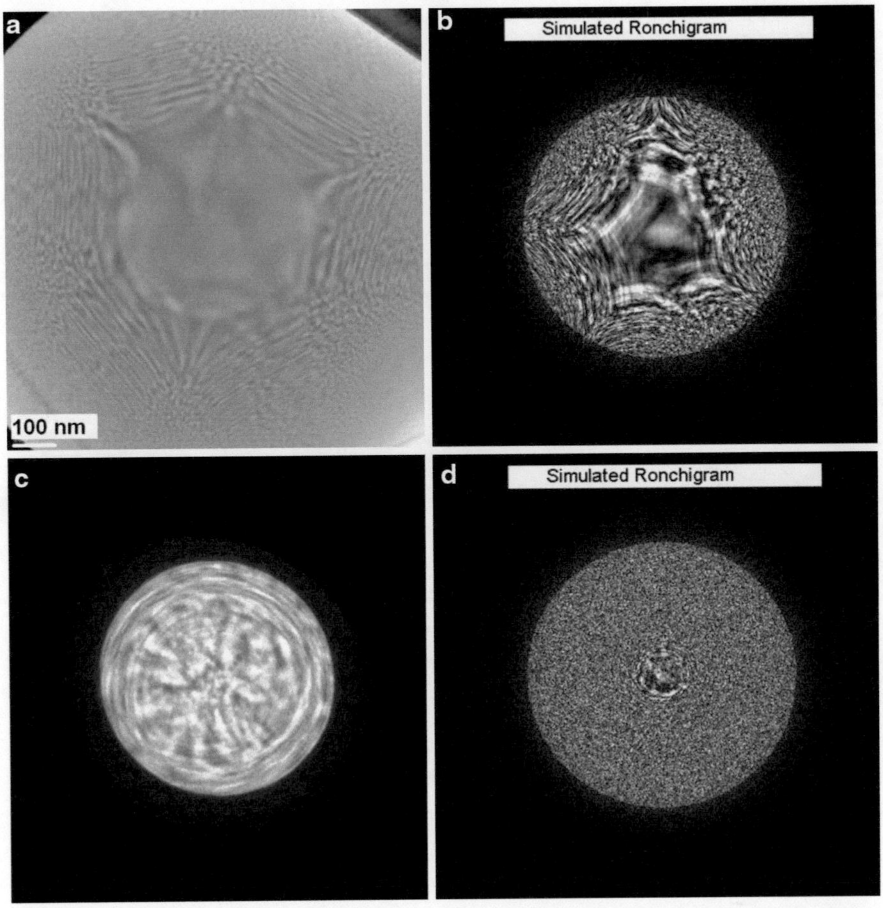

Fig. 8. Ronchigram images. Aberration-corrected experimental Ronchigram from an amorphous material (**a**); simulated Ronchigram (**b**) (7). Not corrected experimental Ronchigram from an amorphous material (**c**) and simulated Ronchigram (**d**) (7). The Ronchigram shown in (**a**) was obtained in an aberration-corrected microscope Jeol ARM200F, at the Kleberg Advance Microscopy Center at UTSA.

Electron microscopy has been improved greatly since 1990s with the correction of the spherical aberration and the chromatic aberration and is now capable of sub-angstrom resolution (1). Another important aberration to be corrected is the axial coma. The coma astigmatism occurs when the direction of illumination does not coincide with the direction of the true optical axis of the electron microscope. To reduce the effect of comma aberration in STEM, the beam must to be centered in the center in underfocus and overfocus conditions. The center of the Ronchigram must be positioned at the same point. Misalignment in *X* and *Y* direction can be observed in the Fig. 10. The corrected *X* and *Y* directions of the comma astigmatism are observed in the center of the Ronchigram illustrated in the Fig. 10. Astigmatism of the condenser lens must be aligned in *X* and *Y* directions. Non astigmatic Ronchigram and induced *X* and *Y* astigmatism are shown in Fig. 11.

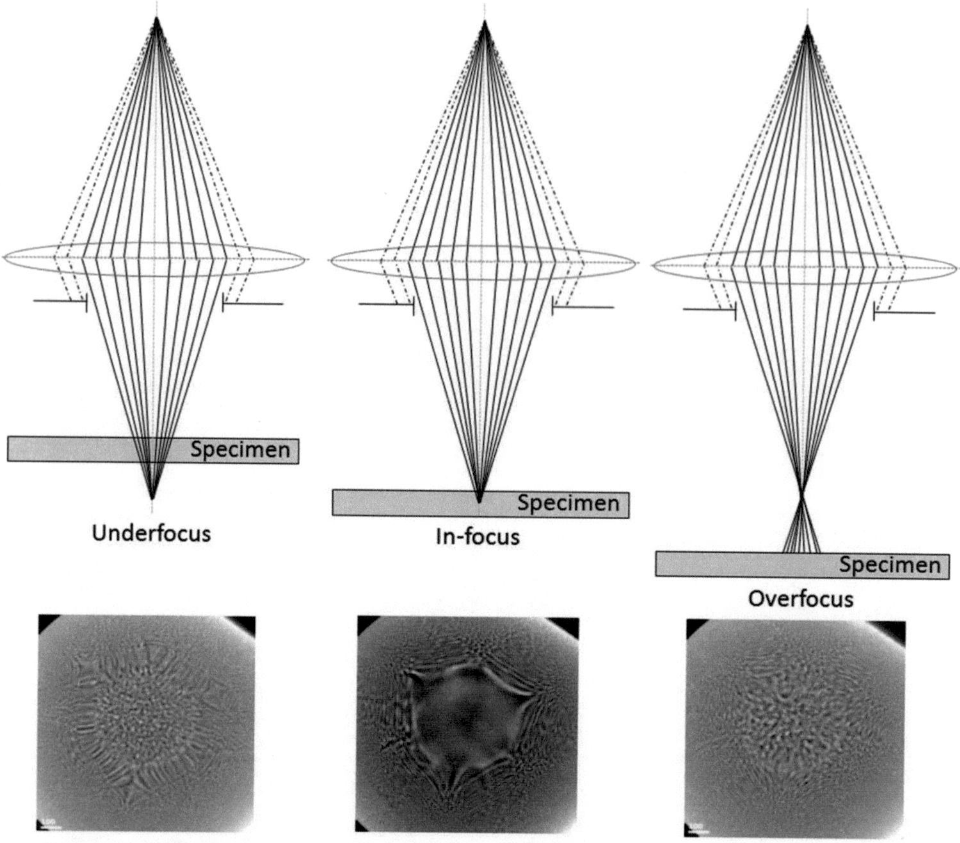

Fig. 9. Ronchigrams recorded with an underfocus, optimal defocus and overfocus electron probe on an amorphous specimen.

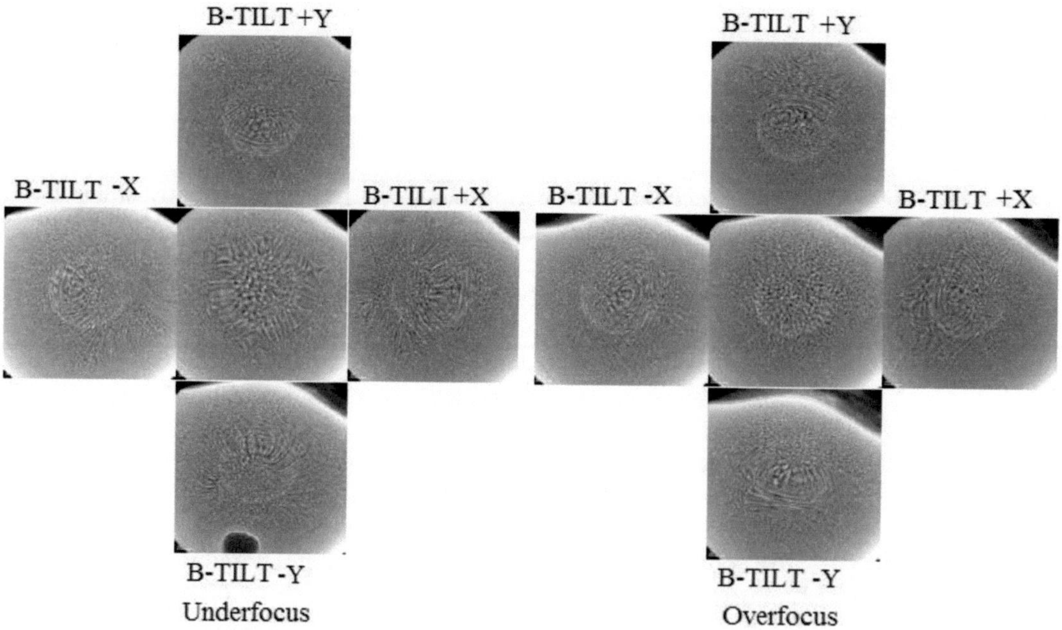

Fig. 10. Correction of the axial coma (beam tilt).

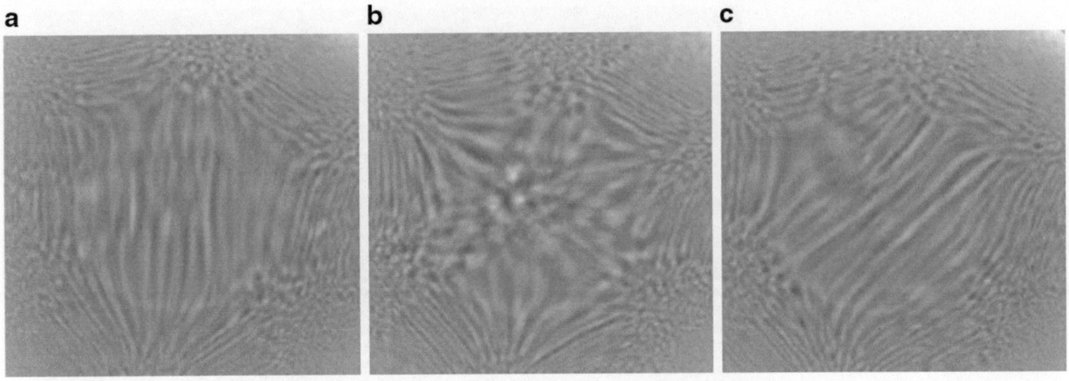

Fig. 11. Correction of the astigmatism in STEM. Induced astigmatism by *x* (**a**); no astigmatism (**b**); induced astigmatism by *y* (**c**).

5. Atomic Number (Z) Dependence in HAADF-STEM Imaging

HAADF image resolution is primarily determined by the size of the illumination probe, which is formed by the condenser system of the microscope. The best probe is formed at a balance of resolution (narrow illumination peak) and contrast (minimal tails), which means aperture semi-angle of $(4\lambda/Cs)^{1/4}$ and defocus of $-(\lambda/Cs)^{1/2}$. The resolution of incoherent STEM and TEM image will then be:

$$\text{STEM}: \rho = 0.43\,Cs^{1/4}\lambda^{3/4}$$
$$\text{TEM}: \rho = 0.65\,Cs^{1/4}\lambda^{3/4}$$

In this way, using the same energy in the microscope (same wavelength, λ) and the same spherical aberration (Cs), the STEM images are considerably better than TEM images, the spatial resolution is improved. It has been known for long time that the annular detector in a HAADF-STEM microscope will collect a large amount of the elastic scattered electrons, in such a way that the intensity of the signal collected by the detector will have a dependence on the scattering cross-section, and thus on the atomic number of the atoms in the sample (8). Pennycook mentions a dependence close to $Z^{3/2}$ (1). For a single atom, the distribution of electrons being scattered is more or less like a Gaussian function (Rutherford scattering). In order to confirm how the intensity signal generated in the HAADF detector depends on the atomic number of a column formed by just one atom of a specific element, we performed a simulation of HAADF intensities in 16 different atoms, which have been projected in a 2D plane (Fig. 12).

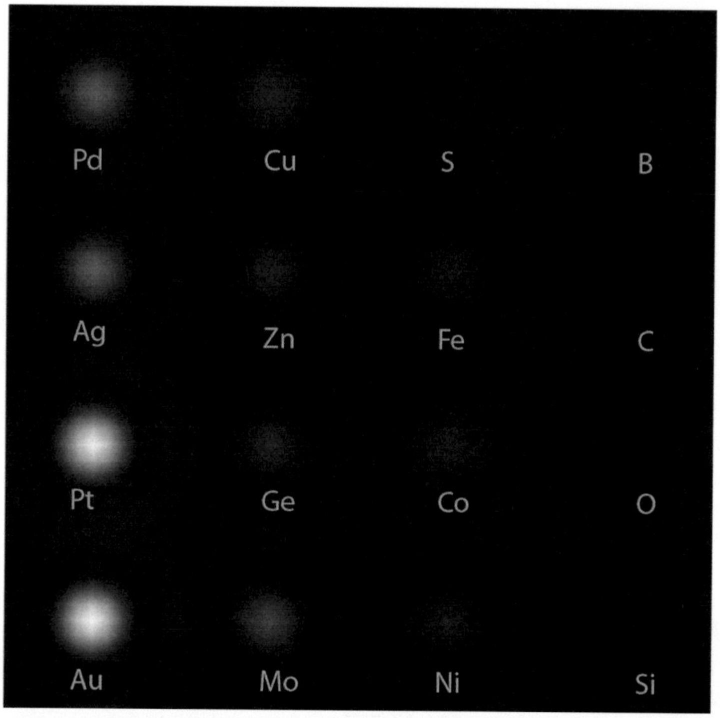

Fig. 12. Calculation of the relative HAADF-STEM contrast of different elements. Positions of the atoms are labeled by the name of their respective chemical element.

The line scan of the HAADF image intensity through the center of the atoms is shown in Fig. 13, where the high contrast between heavy and light atoms can easily be noted. The heights of the maxima from Fig. 13 are plotted against the relevant atomic numbers in Fig. 14. The extrapolation adjustment shows that the HAADF signal varies as approximately $Z^{1.46}$ in a good agreement with the previously reported $Z^{3/2}$ (1). STEM imaging simulation can be obtained by using the commercial version of the xHREM package, designed and implemented by Ishizuka (9). The simulation process can be carried out using the Multislice method. The basic idea underlaying the multislice method is that the potential of the sample can be approximated by defining a number of slices of thickness *dz*, and projecting the potential due to the atoms of a particular slice to the central plane of this slice. Thus, a solution to the electron wavefunction is obtained for one slice, and used as input for the calculation of the next slice. This means that the choice of very thick slices will speed up the calculations, but the potential function will be poorly approximated. On the other hand, the use of very thin slices will improve the calculation of the projected potential, but the errors due to approximations will accumulate and undermine the final result. The appropriate choice of the number of slices is an issue that requires a generous amount of

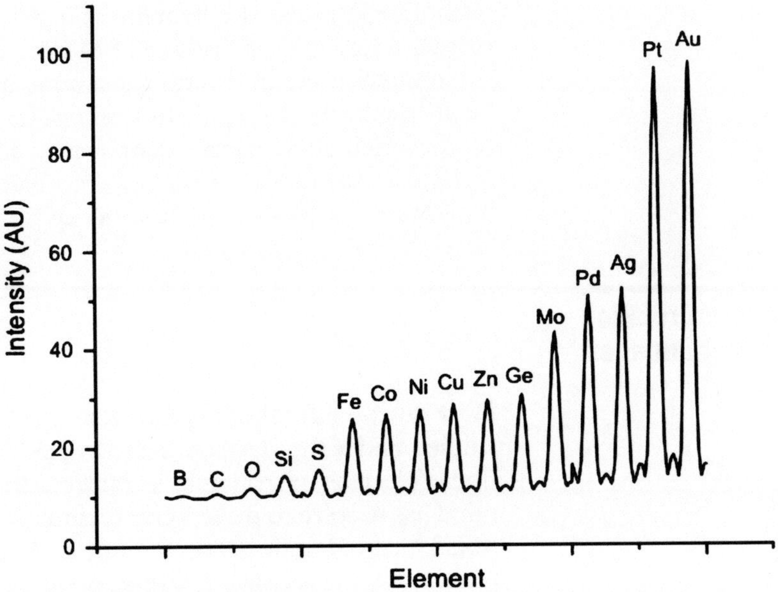

Fig. 13. HAADF-STEM Intensity signal profile relative to the chemical species, for the simulated STEM image shown in Fig. 12.

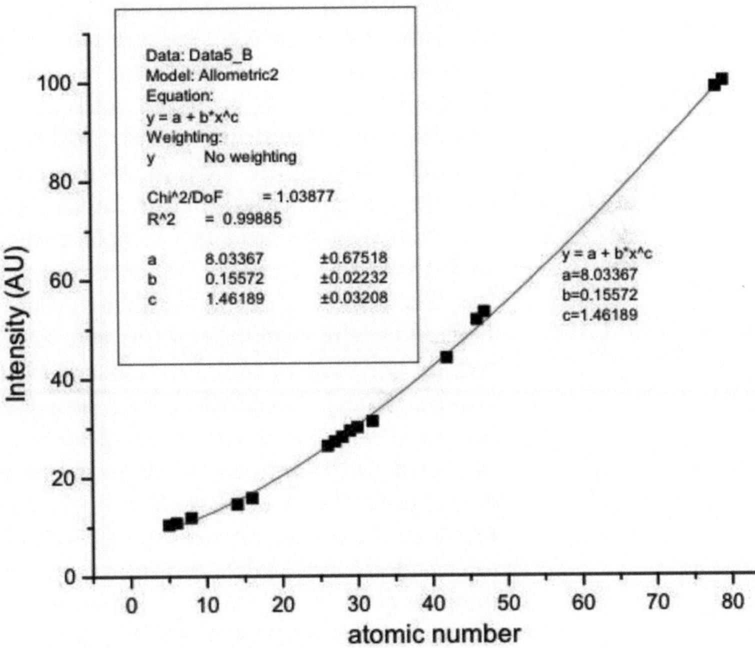

Fig. 14. Variation of the intensity of the contrast as a function of the atomic number.

effort and computing time, and the final decision must consider the benefits of a fast calculation (STEM simulations by the very nature of the technique are computationally expensive and the computing time depend both on the number of slices and the scan resolution) against the benefits of a very precise calculation (that may be critical in quantitative STEM). As pointed out by Ishizuka (9), the multislice method requires several approximations that could be critical in the interpretation of results for inclined illumination, so the orientation of the sample may also affect the final result. XHREM uses an algorithm based on Fast Fourier Transforms, and it allows the use of a large number of slices.

6. Interpretation of Stem Images

How can we infer the shapes, sizes, and composition of the objects studied under the electron microscope? The problem of interpretation of electron micrographs is similar to the problem that the human mind has to solve to make a correct interpretation of the two-dimensional images imprinted on the retina. As the linguist Steven Pinker puts it: "Optics is easy but inverse optics impossible" (10). The solution to this factual impossibility lays on the capacity of the human mind to make educated assumptions about the observed objects and their surroundings. The interpretation of the electron microscope images must be made under the same kind of assumptions: As a first approximation, one considers that the nanostructures are laying on an even surface (usually carbon), and that in inner regions of the nanostructure and far from twin boundaries, point-defects and dislocations, the atomic arrangement is close to that of the bulk.

As it was mentioned before, the approximations used in the different simulation techniques are strictly true only when the sample is thin enough for the linear assumption to be correct. In the multislice method the sample is dissected into several slices, each treated independently through an averaged projected potential. The selection of the number of slices is a tricky job, and it will depend on the thickness of the sample (the size of the nanostructure in the case of metal nanoparticles), the orientation, and the computational capabilities. The projected potential is calculated for each slice, and the corresponding output wavefunction is obtained. If the number of slices is increased, the number of projected potentials will increase as well. This will reflect on the final result of the image, as can be seen in Fig. 15, where a model of an icosahedral gold particle was used to generate two simulated STEM micrographs, being the only difference between the two number of slices, 10 slices (Fig. 15b), 20 slices (Fig. 15c). It is not a simple task to find out which of the images follows the experimental one in a better way, since the orientation and lattice parameters must be considered.

In the remaining of this chapter, for purposes of comparison we will restrict ourselves to use ten slices, irrespective of the model being simulated. We invite the reader to dedicate a generous amount of time to try out different possibilities.

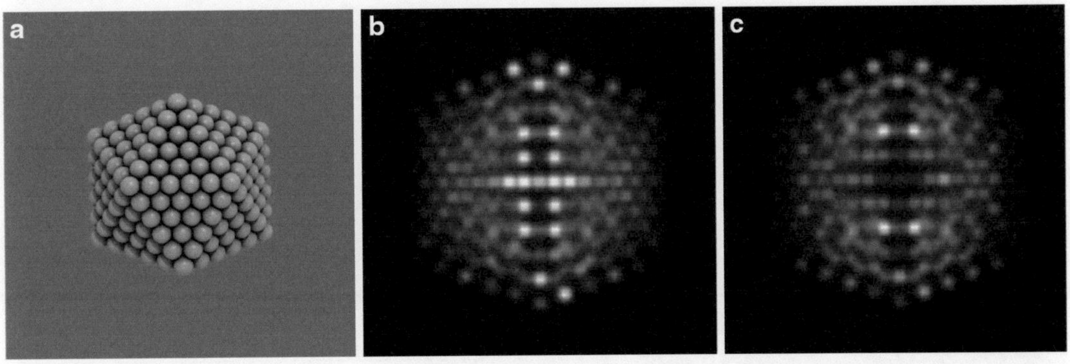

Fig. 15. Simulated STEM for icosahedral gold particle. Icosahedral gold particle model (**a**) and HAADF-STEM simulated images using 10 slices (**b**) or 20 slices (**c**) in the simulation.

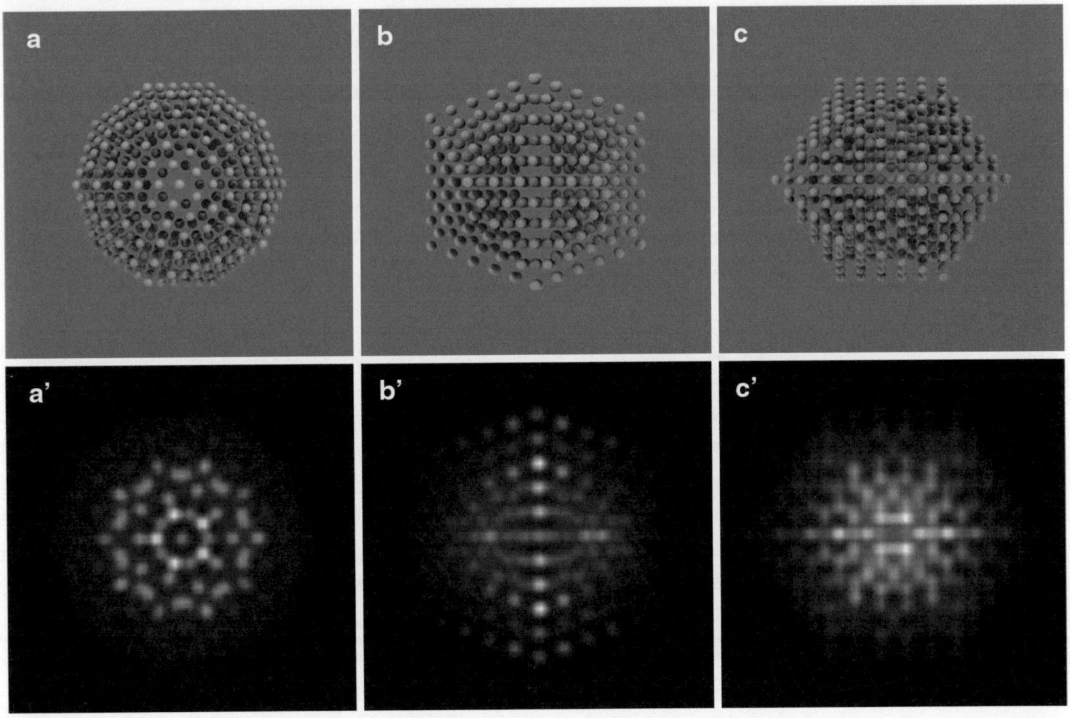

Fig. 16. Simulated STEM images of an icosahedral Au_{core}-Pd_{shell} particle. Core-shell structure model (**a**–**c**) and simulated STEM images (**a**′–**c**′), at three different orientations.

Returning back to the issue of the Z-contrast dependency on STEM, we take advantage of the previous figure to compare it against the simulated STEM micrographs of particles of the same size and shape, but with different composition. Figure 16 shows the STEM simulated images of an icosahedral Au_{core}-Pd_{shell} particle. The orientation shown in Fig. 16b is rotated 30° with respect to the orientation shown in Fig. 16a; the orientations shown in

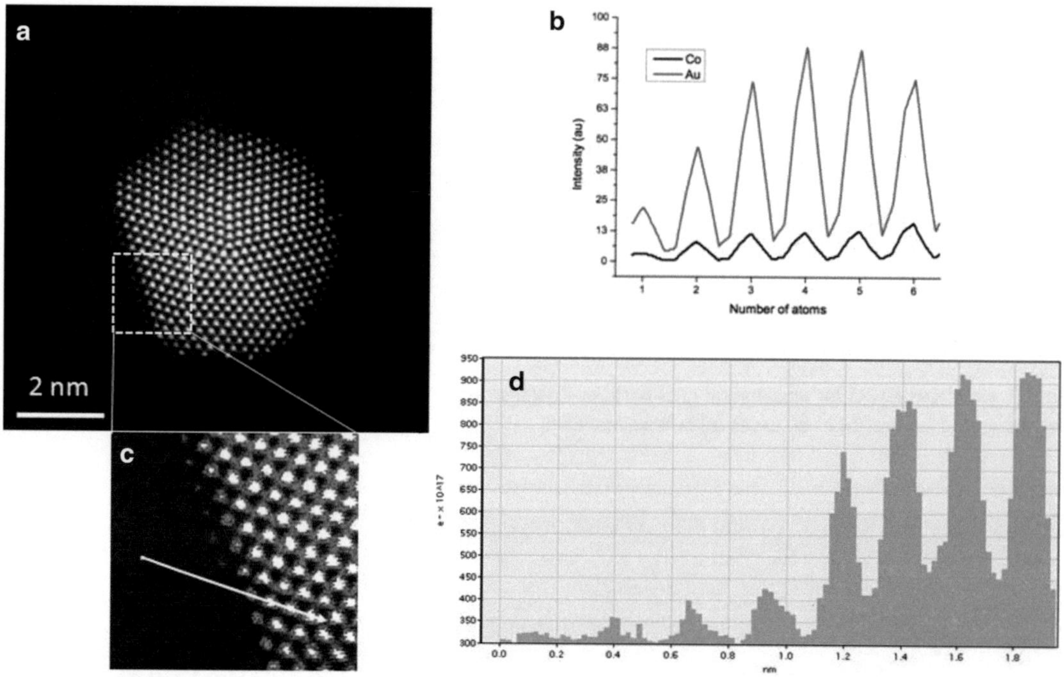

Fig. 17. Z-contrast capabilities of HAADF-STEM. Cs-corrected aberration HAADF-STEM image from Au/Co bimetallic nanoparticle (**a**). Calculated intensities profiles from Co and Au atoms (**b**). Enlarge Cs HAADF-STEM image (**c**) and experimental intensity profile including the Co and Au atomic spacing (**d**).

Fig. 16c is rotated 90° with respect to Fig. 16a. Figure 16a′–c′ show respective simulated images. The difference in intensity between core and shell are evident. How can the Z-contrast capabilities of HAADF-STEM be used in the investigation of real metal nanoparticles? As an example, consider the Au-Co particle presented in Fig. 17a. This particle was produced by the coalescence of a gold particle with a cobalt particle, both originally synthesized by inert gas condensation and deposited on a carbon substrate. After deposition, the system was subjected to a heating process. The particle in Fig. 17c shows an enlargement to remark the differences in intensity, as well as the intensity profile in Fig. 17d. The two intensity profiles plotted in Fig. 17b show the simulated intensity profiles, described in Fig. 14. This result exemplifies the use of STEM simulations for an approximate description of composition and shape of metal nanoparticles.

7. Examples of Stem-Imaging of Metallic Nanoparticles

Using lower aperture of the condenser lens, spot number higher than 6, the beam current density incident in the sample would be less than 30 pA. Under these conditions and the alignment

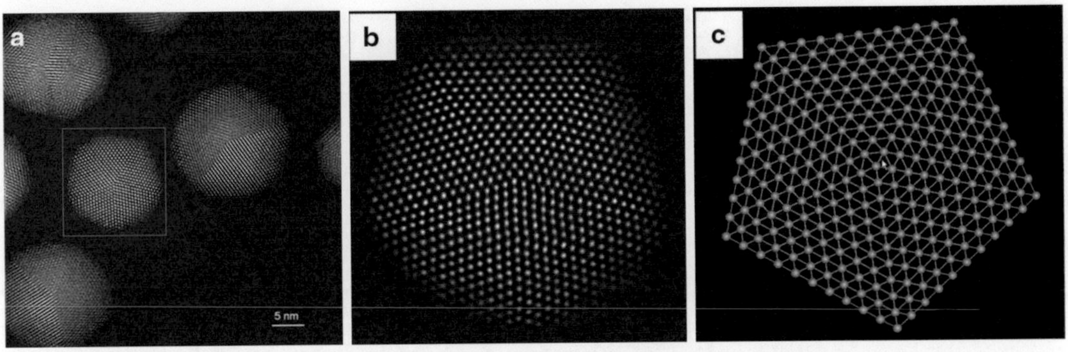

Fig. 18. Gold nanoparticles at atomic scale resolution. STEM image showing a few different gold nanoparticles (**a**); high magnification of a decahedral nanoparticle (**b**) and theoretical model of this decahedral nanoparticle (**c**).

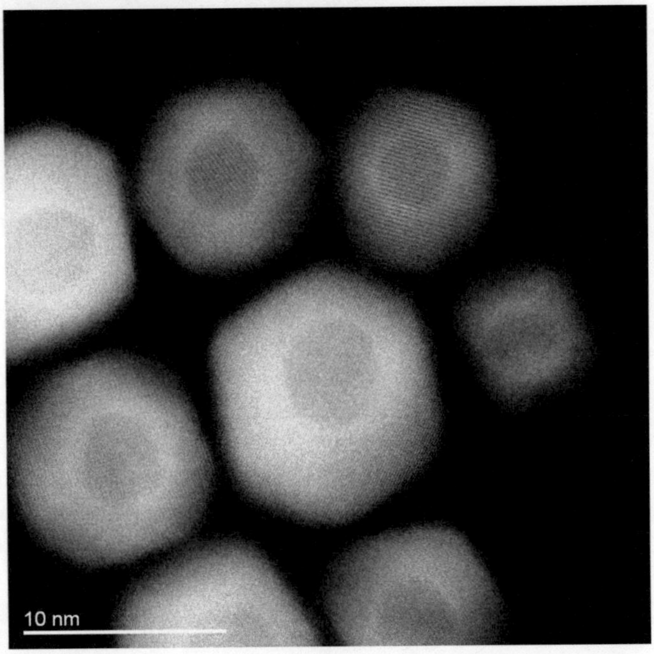

Fig. 19. Au/Pd core-shell structures obtained by HAADF-STEM.

mentioned in the previous sections, the high resolution STEM is guaranteed. Decahedra gold nanoparticles can be identified at atomic scale in the Fig. 18. Gold-palladium (Au/Pt) bimetallic Core-shell structures can be identified by differences in the contrast in the HAADF-STEM imaging in Fig. 19. The high contrast corresponds to the gold shell (Au with Z atomic weight 79) whereas the inner core showed lower contrast corresponding to a core of palladium (Pt with Z atomic weight 46).

The controlled incorporation of the metallic nanoparticles into biological systems is a challenge for the experimental scientist. The STEM imaging becomes as a critical tool to identify the incorporation and the influence of the multiple shapes into these biological

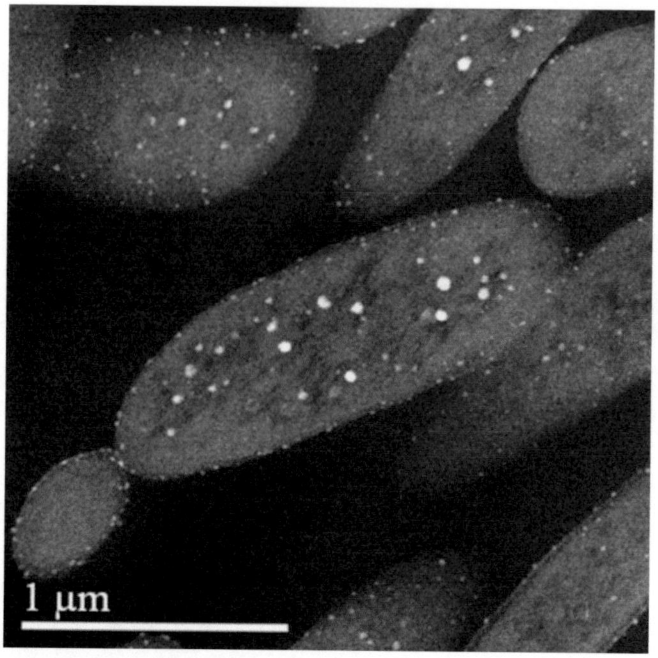

Fig. 20. HAADF-STEM image of silver nanoparticles incorporated into bacteria *E. coli.*

systems. Biologic systems contain mainly carbon, nitrogen, hydrogen as the matrix. Therefore, the contrast in HAADF-STEM imaging is lower compared with the contrast generated by metallic nanoparticles. In this way, the identification of metallic nanostructures embedded in biological tissue is relatively easy to observe. In the Fig. 20 silver nanoparticles are incorporated into bacteria and the particles can be clearly identified by contrast. HAADF-STEM imaging can avoid the staining procedure in comparison with the conventional TEM.

In summary, the scanning transmission electron microscope overcomes the barrier to atomic resolution, making it in one of the most powerful tools for characterization of nanomaterials. The new generation of aberration corrected microscopes, besides being a source of sophisticated characterization techniques, will certainly become a fundamental bridge between materials and health sciences. The capacity to control size, shape, and surface chemistry versatility of engineered nanoparticles has enabled their use into a plethora of biological and medical applications as drug delivery systems, therapeutics, diagnostics, and imaging contrast agents with different advanced functions and new properties. Depending of particular nanoparticle properties and mechanisms of interaction with biological systems, adsorption and uptake can produce dose-dependent decreases in cell viability and different metabolic and genetic alterations. These phenomena can be monitored with different well-established biochemical and molecular biology assays. The STEM methods mentioned in this chapter can be used to confirm attachment, internalization, and intracellular localization of these nanomaterials. As an example, cancer nanotechnology focuses on applications based on engineered nanoparticles designed for detection, diagnosis, targeting, and treatment of cancerous cells. Noble metal nanoparticles are biocompatible and non-toxic agents useful in all these biomedical applications at the cellular or molecular scale. Other important structural characteristics of nanoparticles for biological and biomedical applications that need to be determined are: size, shape, hydrodynamic radius, zeta potential, surface charge, defects, chemical composition, and atomic structure, that is highly related to functionality, stability, and reactivity of nanoparticles.

Acknowledgements

The authors would like to acknowledge THE WELCH FOUNDATION AGENCY PROJECT # AX-1615. "Controlling the Shape and Particles Using Wet Chemistry Methods and Its Application to Synthesis of Hollow Bimetallic Nanostructures." The authors would also like to acknowledge the NSF PREM Grant # DMR 0934218, Title: Oxide and Metal Nanoparticles—The Interface between life sciences and physical sciences. The authors would also like to acknowledge RCMI Center for Interdisciplinary Health Research CIHR. "The project described was supported by Award Number 2G12RR013646-11 from the National Center for Research Resources. The content is solely the responsibility of the authors and does not necessarily represent the official views of the National Center for Research Resources of the National Institutes of Health."

References

1. Pennycook S (1989) Z-contrast STEM for materials science. Ultramicroscopy 30:58–69
2. Ercolessi F, Andreoni W et al (1991) Melting of small gold particles: mechanism and size effects. Phys Rev Lett 66:911–914
3. Ferrando R, Jellinek J et al (2008) Nanoalloys: from theory to applications of alloy clusters and nanoparticles. Chem Rev 108:845–910
4. Carter CB, Williams DB (2009) Transmission electron microscopy: a textbook for materials science. Springer, New York
5. Fernández-Navarro C et al (2007) On the structure of Au/Pd bimetallic nanoparticles. J Phys Chem C 111:1256–1260
6. Kirkland EJ (1998) Advanced computing in electron microscopy. Springer, New York
7. Zuo JM, Olsen E (2011) http://cbed.matse.illinois.edu/JProbe/JProbe.html. Accessed 12 Aug 2011
8. Wall J, Langmore J, Isaacson M et al (1977) Scanning transmission electron microscopy at high resolution. Proc Natl Acad Sci 74:1802–1806
9. Ishizuka K (1998) Multislice implementation for inclined illumination and convergent-beam electron diffraction. In: Shiojiri M & Nishio K (eds) Proceedings of the International Symposium on Hybrid Analysis for Functional Nanostructure, Kyoto, Nakanishi Printing Co., pp 69–72
10. Pinker S (1997) How the mind works. W. Norton & Company, New York

Chapter 38

Total Internal Reflection Fluorescence (TIRF) Microscopy for Real-Time Imaging of Nanoparticle-Cell Plasma Membrane Interaction

Ladan Parhamifar and S. Moein Moghimi

Abstract

Nanoparticulate systems are widely used for site-specific drug and gene delivery as well as for medical imaging. The mode of nanoparticle-cell interaction may have a significant effect on the pathway of nanoparticle internalization and subsequent intracellular trafficking. Total internal reflection fluorescence (TIRF) microscopy allows for real-time monitoring of nanoparticle-membrane interaction events, which can provide vital information in relation to design and surface engineering of therapeutic nanoparticles for cell-specific targeting. In contrast to other microscopy techniques, the bleaching effect by lasers in TIRF microscopy is considerably less when using fluorescent nanoparticles and it reduces photo-induced cytotoxicity during visualization of live-cell events since it only illuminates the specific area near or at the plasma membrane.

Key words: TIRFM, Live-cell imaging, Cell surface, Widefield microscopy, Trafficking, Nano-particles

1. Introduction

Delineating the route of uptake and membrane interaction of various nanoparticulate systems with membrane barriers provides vital information in relation to design and surface engineering of nanoparticles for therapeutic and medical imaging interventions. The primary step in nanoparticle-mediated cellular drug delivery is nanoparticle interaction with the cell surface. Total internal reflection fluorescence (TIRF) microscopy, developed in the early 1980s, allows for the study of molecular interactions at, or very close to, the plasma membrane of cells (1, 2). TIRF therefore provides a unique opportunity for studying fluorescently-tagged nanoparticle adhesion processes to the plasma membrane in real-time. Unlike other microscopy techniques, such as confocal and multiphoton microscopy where the incident beam is shone directly (i.e., perpendicular to the specimen) with the light being refracted

Mikhail Soloviev (ed.), *Nanoparticles in Biology and Medicine: Methods and Protocols*, Methods in Molecular Biology, vol. 906, DOI 10.1007/978-1-61779-953-2_38, © Springer Science+Business Media, LLC 2012

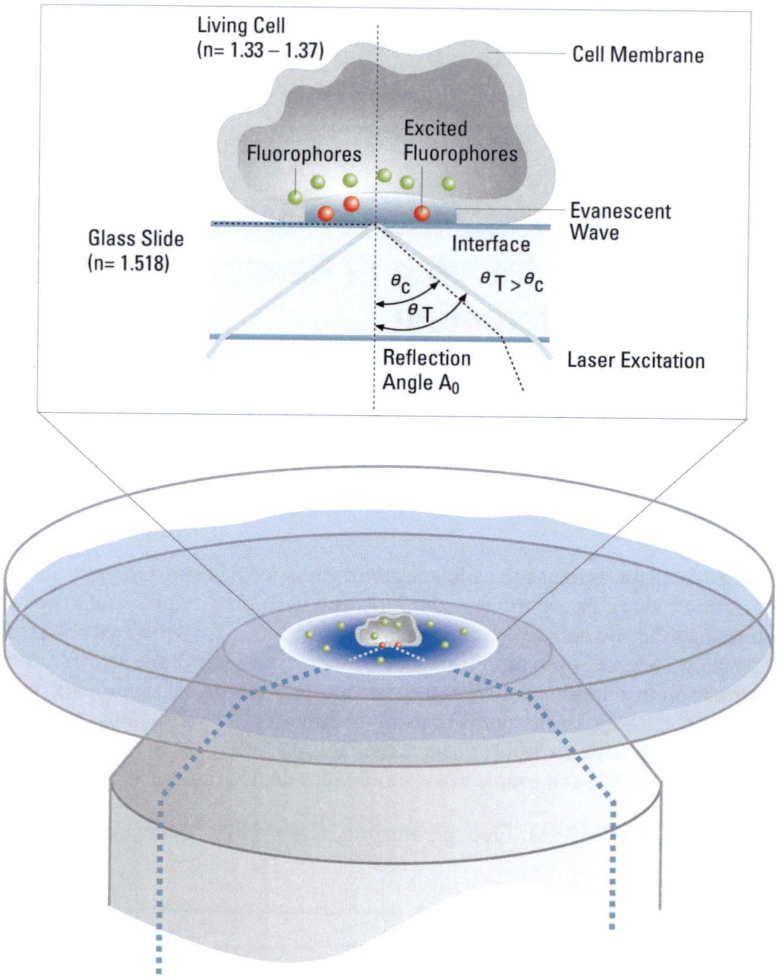

Fig. 1. TIRF microscopy principle. The "*n*" refers to refractive indices (10).

throughout the entire cell (3), in TIRF the angle of the incident beam is increased (it is no longer perpendicular). At a specific point, known as the critical angle, the vast majority of light is no longer refracted through the sample, but instead the incident light is totally internally reflected (2, 3). The reflection at the critical angle takes place at the glass-water (or buffer, when considering biological systems) interface, and is dictated by the refractive indices of the glass cover slip or plastic dish with coverslip base (for live-cell imaging) and the water/buffer. At this point, in addition to the light being reflected, an electromagnetic wave is created, called the evanescent wave or field (4), which occurs parallel to the glass-water interface and extends at short distance into the specimen, typically between 100 and 200 nm. Critically, the wavelength of the evanescent wave is the same as the incident light (thus being able to excite the same fluorochromes as the incident light would be able to), and in addition the wave undergoes exponential intensity decay with increasing distance from the surface, thus reducing the interference from fluorochromes not specifically at the plasma membrane region (3), Fig. 1.

TIRF microscopy has been applied in a multitude of different scenarios, for example studying cellular adhesion processes (5), drug-receptor interactions (6), and trafficking mechanisms of nanoparticles across the cell membrane (7). Such biological processes can be studied on fixed cells (8), but also to a greater degree on live cells (9) due to the fact that the incident light is not shone directly at the sample, therefore reducing the cellular damage, which would otherwise occur with other laser-based microscopy procedures. Due to the milder effects on cells, TIRF microscopy is becoming a routine procedure for studying biological plasma membrane processes in real-time. Here we describe protocols for nanoparticle-plasma membrane interaction and their tracking in live cells using TIRF microscope.

2. Materials

2.1. Preparing Cell Cultures and Their Exposure to Nanoparticles

1. We illustrate our methods using non-small lung cancer cells, H1299 (ATCC, CRL-5803) (see Note 1).
2. Trypsin-EDTA (5.0 g/L porcine trypsin and 2 g/L, EDTA·4Na in 0.9 % w/v sodium chloride, 10×).
3. Phosphate buffer saline for cell culture (PBS): KCl 0.2 g/L, KH_2PO_4 (anhydride) 0.2 g/L, NaCl 8 g/L, Na_2HPO_4 (anhydride) 1.15 g/L.
4. RMPI-1640 growth medium with 10 %v/v Fetal bovine serum, 100 mg/L streptomycin, 100 units/mL penicillin, 0.9 %w/v sodium chloride, and 2 mM L-glutamine (see Note 2).
5. Glass-bottom dishes/slides/plates with a thickness between 0.13 and 0.17 mm.
6. 15 μg/mL silica and Excitation nanoparticles (100 nm in diameter, Excitation 485 nm, Emission 510 nm) fluorescently tagged with fluorescein isothiocyanate (FITC). The choice of the fluorophore will depend on the laser and filters available.

2.2. TIRF System Alignment

1. Leica TIRF microscope system with the 488 nm laser. See Table 1 for available lasers for the Leica system.
2. TIRF microscope objectives (example ×63 and ×100, numerical aperture 1.47) (see Note 3).

2.3. TIRF Microscope Alignment

1. Leica TIRF microscope system with the 488 nm laser. See Table 1 for available lasers for the Leica system.

Table 1
Laser penetration depths using Leica LAS AF

Laser (nm)	Depth (nm)
405	70, 90, 110, 150, 200, 250, epi
488	70, 90, 110, 150, 200, 250, epi
561	90, 110, 130, 150, 200, 250, epi
632	90, 110, 130, 150, 200, 250, epi

2. TIRF microscope objectives (example ×63 and ×100, numerical aperture 1.47) (see Note 3).
3. Microscope immersion oil specific to the microscope used (always available and recommended by the microscope supplier). In our experiments we used Leica immersion oil type F.

2.4. TIRF Image Acquisition

1. Leica TIRF microscope system with the 488 nm laser. See Table 1 for available lasers for the Leica system.
2. TIRF microscope objectives (example ×63 and ×100, numerical aperture 1.47) (see Note 3).
3. Microscope immersion oil specific to the microscope used (always available and recommended by the microscope supplier). In our experiments we used Leica immersion oil type F.
4. Cells grown in a glass-bottom dishes, slides, or plates with a thickness between 0.13 and 0.17 mm.
5. Membrane dye (optional) (see Note 1).

3. Methods

3.1. Preparing Cell Cultures and their Exposure to Nanoparticles

1. To test nanoparticle detection with TIRF microscope add for example 10 μg/mL aqueous suspension of test nanoparticles (e.g., silica nanoparticles) to a glass-bottom dish containing growth medium. Incubate for 10 min, attempt nanoparticle detection with the TIRF microscope (see Subheading 3.4). Normally 2–3 depths can be chosen to check nanoparticle distribution in relation to the membrane. Moreover, a widefield channel and an epi-fluorescence channel can be used to demonstrate the difference between illuminating limited depths as compared to illuminating the entire sample, see Fig. 2.
2. Grow the cells (H1299 cell line) until 80% confluency (see Note 1).

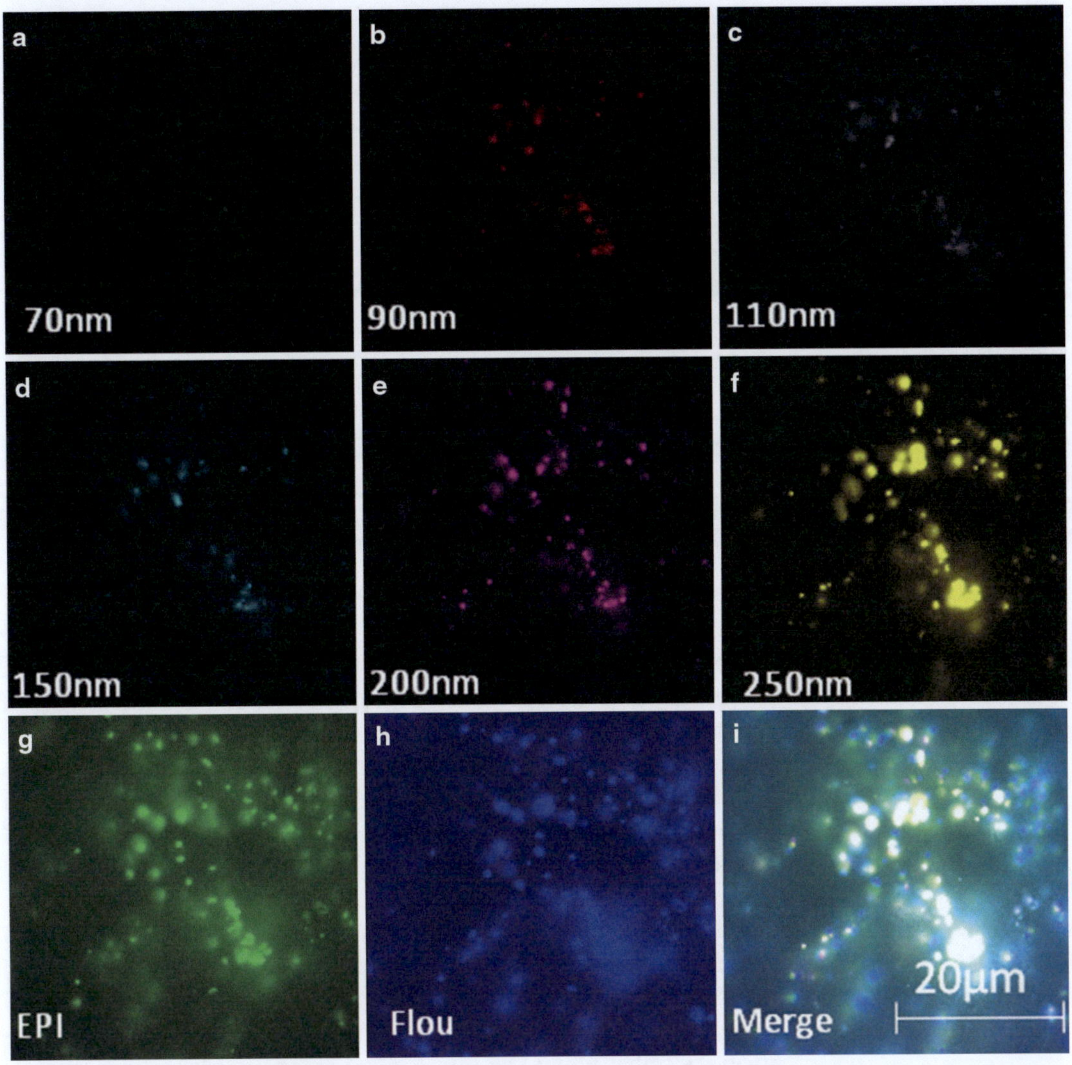

Fig. 2. Nanoparticle images with TIRFM using 488 nm Laser. Fluorescent nanoparticles (100 nm, 10 μg/mL) were imaged in a cell free environment using RPMI 1640 medium at 37 °C. Different pseudo colors in each channel show different depths: 70, 90, 110, 150, 200 and 250 nm (panels **a–f**), respectively. A widefield channel (**g**) and an epi-fluorescence channel (**h**) are also included to demonstrate the difference between illuminating limited depths compared with illuminating the entire sample. (**i**) Shows the merge image of (**g**) and (**h**).

3. Wash cells 2× in PBS and detach the cells using 1× trypsin-EDTA for 6 min at 37 °C. Resuspend the cells in growth medium, count the cells.
4. Add 25,000 cells in 2 mL of growth medium to each glass-bottom dish (see Note 4).
5. Grow cells until they reach 60–65% confluency.
6. Add fluorescent labeled nanoparticles directly to the growth medium in the glass-bottom dishes to the final concentration of 2.5 μg/mL and incubate 15 min at 37 °C, 5% CO_2.

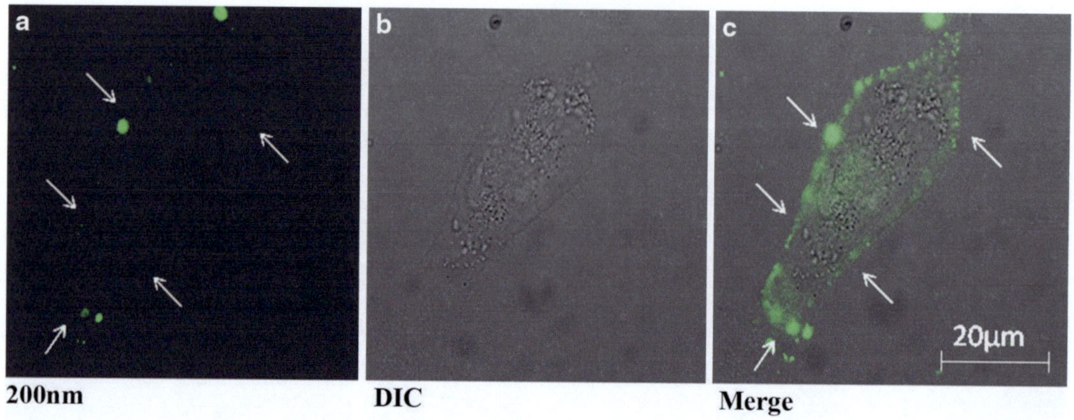

Fig. 3. Live-cell TIRF microscopy and DIC. H1229 cells were incubated with fluorescently-labeled nanoparticle (100 nm). Images were acquired 15 min after the incubation of cells with the nanoparticles. Final nanoparticle concentration was 2.5 μg/mL. Nanoparticles are shown on the plasma membrane (see *arrows*) in the TIRF channel 200 nm (**a**). The cell morphology is imaged in the DIC channel (**b**). (**c**) Shows the overlay of the TIRF and DIC channels.

7. Proceed with cell imaging using the TIRF microscope (see Subheadings 3.3 and 3.4). Figure 3 illustrates a TIRF image of particles using 200 nm depth combined with a differential interference contrast (DIC) overlay. The DIC image shows the cell and the overlay image (TIRF channel and the DIC channel overlayed) demonstrate that the particles are bound to the cell membrane (see also Note 5).

3.2. TIRF System Alignment (See Note 6)

Initially, the lasers have to be aligned in the LAS Configuration software (Leica TIRF System). Essentially this procedure ensures that the laser beam passes through the filter cube correctly (so all the light is passed through). This usually is performed by an application specialist when the microscope is set up. Otherwise the following procedure may be used:

1. Turn on the lasers, the computer and then the microscope.
2. Turn on LAS Configuration software (Leica TIRF System), make sure the LAS software is not running.
3. Select TIRF cube alignment in the configuration software.
4. Move out the transmitted light arm.
5. Press start to go into the "Align TIRF cube sequence."
6. Use a white sheet of paper and hold it horizontally above the objective to visualize the laser beam.
7. Position the laser in *X*, *Y* and focus in *Z* (making sure that the beam is passing strait through the objective from a 90° angle to the interphase.

8. Make a mark on the ceiling where the laser beam would hit at 90°. From now on all laser alignments performed before each TIRF experiment should be aligned using this mark on the ceiling, if one has not already been made by an application specialist (see Note 7).
9. This procedure should be repeated until all lasers have been aligned through the objective.
10. If there are several TIRF objectives in the system, the configuration software will ask you to "set the objective." This means that the laser alignments will now be set for the second objective.
11. Follow steps 8–10 to set the lasers for the second objective.
12. Once the system alignment has been completed, close down the Configuration Software.
13. Start the LAS AF software to proceed with TIRF alignment.

3.3. TIRF Microscope Alignment

This procedure should be performed before each new TIRF imaging experiment

1. Clean the objective using a piece of fresh special non-fibrous cleaning tissue.
2. Apply a small drop of microscope oil to the objective.
3. Clean the bottom of the glass dish with the cell culture using a non-fibrous tissue. Place the dish on the microscope stage making sure that it is flat; this is to ensure an even focal plane if the objective is moved to another location.
4. Slowly adjust the focal knob until the oil drop contacts the bottom of the dish. Continue to focus slowly until the "bottom" of the cells is observed in the objective; this is the glass-buffer interface.
5. Set the channel to TIRF mode, make sure the correct laser and filter are chosen (in the Leica system once the correct filter is chosen the correct laser is automatically set and the depths available for the laser are shown).
6. Select "auto-align" and follow the instructions on the screen. The computer will automatically align the objective for optimal viewing of the sample using the TIRF objective.
7. Click on the enable collimator.
8. Move out the transmitted light arm.
9. Focus the laser using the *Z* drive on the collimator centering and focusing the beam on the mark on the ceiling.
10. Move the transmitted light arm back to start the image acquisition experiment.

3.4. TIRF Image Acquisition (See Note 8)

A selection of angles for the evanescent field is available under the TIRF tab in the LAS software. This means that the angled laser beam can be directed at the objective from different angles. This parameter is rarely used and should in general not be changed. Available directions are represented as arrows and angles as follows: 0° (arrow left), 90° (arrow up), 180° (arrow right), 270° (arrow down).

1. Choose the depth of interest. A depth of 100–250 nm will show the area near and just within the membrane. Depths below this will show particles outside the cell.
2. A membrane dye can be used to visualize the interaction of the particle with the membrane (see Note 9).
3. A DIC channel should also be added to further enhance the imaging if this feature is available (Fig. 3).
4. Optionally, also add a channel with epi-fluorescence mode to highlight any differences between TIRF (restricted by depth) and normal epi fluorescence (no depth restriction is used). Epi-fluorescence should show a similar image to the widefield channel (Fluo) but with a higher intensity because of the use of a laser instead of the fluorescent lamp (Fig. 2).
5. To track the uptake of the particle in live mode use time laps (see Note 10).
6. *Z*-stacking is not used for TIRF imaging as the field illuminated is so narrow. However, to follow the movement of particles, several TIRF channels can be setup with various depths. This way, the entrance and exit of the particles in different depths can be tracked (see Notes 4 and 11) (an example is shown in Fig. 2, illustrating the case with 100 nm nanoparticles where different pseudo colors have been used in each channel to show the various depths).

4. Notes

1. Other cells or isolated organelles such as the cell nucleus can also be visualized under the microscope using TIRF. Here a nuclear membrane dye can be used instead. Fixed samples can also be used.
2. Other media may need to be used with other cell lines.
3. The higher the numerical aperture on the objectives, the better the resolution will be.
4. Particles can have a higher tendency of aggregating in serum containing medium. However, low serum concentration (1–5%) can dramatically affect the mode of nanoparticle-membrane interaction and subsequent uptake processes.

5. A nuclear staining in whole cells should not be visible in TIRF mode. That will ensure that the imaging is in TIRF mode. One of the advantages for studying nanoparticle uptake using TIRF over widefield or confocal imaging is lower degree of photo-bleaching achieved with TIRF.
6. This procedure should be performed when setting up the system for the first time or if the microscope is moved.
7. If a mark has been made previously by an application specialist but the microscope has been moved slightly from its original place, you have to adjust the mark again.
8. Leica system has the ability to use various critical angles to create evanescent fields of different depths (Table 1) for the available depths of each laser. These depths can be used either in automatic mode or expert mode. The difference is that in automatic mode several fixed depths are available to choose from, whilst in expert mode a slider is available to choose a depth. The expert mode is only recommended after substantial experience with TIRF microscope.
9. This can be useful if the edge of the cells (i.e., the membrane) is difficult to make out under DIC. A membrane dye (with a different fluorescence than your particles) for live imaging can be applied to the cells (incubation time and concentration varies and will be provided from the supplier) before the particles are added and the imaging is started.
10. If the microscope incubator has CO_2 supply a live acquisition can be programmed in the software under the time-laps section. The duration of imaging and the rate of acquisition should be entered into the software. If the microscope is not equipped with an appropriate incubator, a cell lab incubator can be used and acquisition should then be performed manually at specific time points.
11. A good indication of the TIRF mode is the "blinking" of labeled particles as they move in and out of focus.

Acknowledgements

Financial support from the Danish Agency for Science, Technology and Innovation (*Det frie forskningsråd for teknologi og production*), reference 274-08-0534, and Leica Microsystems (Ballerup, Denmark) are greatly acknowledged.

References

1. Toomre D, Bewersdorf J (2010) A new wave of cellular imaging. Annu Rev Cell Dev Biol 26:285–314
2. Axelrod D (1989) Total internal reflection fluorescence microscopy. Methods Cell Biol 30:245–270
3. Mattheyses AL, Simon SM, Rappoport JZ (2010) Imaging with total internal reflection fluorescence microscopy for the cell biologist. J Cell Sci 123:3621–3628
4. Stout AL, Axelrod D (1989) Evanescent field excitation of fluorescence by epi-illumination microscopy. Appl Opt 28:5237–5242
5. Weiger MC, Wang CC, Krajcovic M, Melvin AT, Rhoden JJ, Haugh JM (2009) Spontaneous phosphoinositide 3-kinase signaling dynamics drive spreading and random migration of fibroblasts. J Cell Sci 122:313–323
6. Murray TA, Liu Q, Whiteaker P, Wu J, Lukas RJ (2009) Nicotinic acetylcholine receptor alpha7 subunits with a C2 cytoplasmic loop yellow fluorescent protein insertion form functional receptors. Acta Pharmacol Sin 30:828–841
7. Aaron JS, Greene AC, Kotula PG, Bachand GD, Timlin JA (2010) Advanced optical imaging reveals the dependence of particle geometry on interactions between CdSe quantum dots and immune cells. Small 7:334–341
8. Ji W, Xu P, Li Z, Lu J, Liu L, Zhan Y, Chen Y, Hille B, Xu T, Chen L (2008) Functional stoichiometry of the unitary calcium-release-activated calcium channel. Proc Natl Acad Sci U S A 105:13668–13673
9. Jones JT, Myers JW, Ferrell JE, Meyer T (2004) Probing the precision of the mitotic clock with a live-cell fluorescent biosensor. Nat Biotechnol 22:306–312
10. http://www.leica-microsystems.com/products/light-microscopes/life-science-research/fluorescence-microscopes/details/product/leica-am-tirf-mc/

Chapter 39

Microscopy Imaging Methods for the Detection of Silver and Titanium Nanoparticles Within Cells

Robert Martin Zucker and Kaitlin Marie Daniel*

Abstract

Scientific evaluation of potential environmental hazards resulting from man-made nanomaterials has been hampered by the inability to optimally detect cell-associated nanoparticles. We have successfully imaged TiO_2 nanoparticles in ARPE-19 cells using different light microscope modalities commonly available to investigators including fluorescence, dark field, phase, interference, and confocal. In this report, we describe different optical and lighting conditions necessary for optimal nanoparticle imaging in ARPE-19 cells.

Microscopic examinations involved an E-800 Nikon microscope connected to a xenon light source along with special dark field objectives. For microscopy analyses, ARPE-19 cells were fixed in situ in cultured chamber slides or collected from T-25 flasks and then fixed in suspension. At the lowest concentrations of TiO_2 (0.1–0.3 μg/mL), it was possible to detect as few as 5–10 nanoparticles per cell due to intense light scattering by TiO_2. The degree of brightness detected indicated that the uptake of nanoparticles within ARPE-19 cells could be monitored using dark field microscopy. This report details how wide-field microscopy can be effectively used to detect nanoparticle uptake as well as to assess cellular health in ARPE-19 cell cultures.

Key words: Nanoparticle, Microscopy, Dark field, Toxicology, Nanotoxicity, Confocal microscopy

1. Introduction

Nanomaterials may have potential hazards to the environment, humans, or other organisms (1). Nanoparticles (NPs) can have health risks due to their composition, reactivity, and small size (2). It is difficult to detect nanoparticles in solution, tissues, or cells. Thus, the development of new methods to identify and characterize cells containing nanomaterials would greatly benefit the understanding of potential risks.

*Contractor to the USEPA, award # EP09D000042.

Mikhail Soloviev (ed.), *Nanoparticles in Biology and Medicine: Methods and Protocols*, Methods in Molecular Biology, vol. 906, DOI 10.1007/978-1-61779-953-2_39, © Springer Science+Business Media, LLC 2012

Different light microscopy methods are currently used to study NPs. However, the detection of certain NPs is beyond the resolution diffraction limit of conventional light microscopes (~200 nm). Therefore, transmission electron microscopy (TEM), scanning electron microscopy (SEM), and atomic force microscopy (AFM) have been used to study nanoparticles. Fluorescence microscopy (FM) and these electron microscopy (EM) techniques are useful methods to examine NP interactions with biological substrates. However, these methods have disadvantages that limit their usefulness for toxicological investigations. FM carries the risk of sample photo bleaching coupled with limited particle size detection. In addition, different nanoparticles may lack intrinsic fluorescent energy and therefore possibly not be detectable. EM is labor-intensive and time-consuming. Suspension of nanoparticles can also be assessed using dynamic light scattering (DLS) to yield a hydrodynamic diameter. While these techniques have been useful in determining the physical properties of nanoparticles, they provide limited quantitative information regarding cells interactions with nanoparticles and the rapid identification of the nanoparticles in the cells.

Different nano-sized metallic objects scatter light with relatively high intensity for their small size (3). The side scatter intensity values derived from a flow cytometry analysis relate to cellular mass and possibly protein content (3–5). It is hypothesized that high flow cytometry side scatter generated from different NPs may also be detected as bright objects using dark field and other microscopy contrast modes (6). Dark field microscopy (DFM) is essentially an illumination-based imaging technique used to enhance the contrast of different specimen for increased visibility (6, 7). Nanoparticles can best be detected using contrast enhancements in dark field modes that are below the normal diffraction wavelength limit of 200 nm. Dark field-captured NP images were approximately 100 times brighter compared to fluorescence-captured images.

Conventional DFM is a relatively straightforward illumination technique used in wide-field microscopy to optimize differences in contrast by capturing light scattered by the specimen. In brief, this is accomplished with the attachment of a specialized light condenser that incorporates a light stop comprised of an annulus with a narrow aperture to obliquely illuminate the specimen via a hollow cone of light (8–10). DFM detects scattered light that is refracted, diffracted, or reflected. This scattered light allows DFM to detect extremely small structures, offering a potential light microscopy tool to study cell-associated NPs. A DFM oil condenser (1.2–1.4 numerical aperture (NA)) is used; lenses with higher NAs may also be used to achieve greater optical resolution. In order to optimize DFM, the NA of the lens must be slightly below the NA of the condenser (6, 8–11).

The use of a shorter wavelength and a high intensity xenon light source has the greatest capacity to detect small objects. A xenon light source increases the resolution and detection, thus aids in distinguishing single nanoparticles from clumps using DFM. Clumps have a higher intensity and are larger in size. Single particles may be difficult to detect with dimmer light sources in DFM. Light may be reflected, refracted, and/or absorbed when using fluorescence, interference, dark field, and phase light sources. NPs are not normally detected by conventional light microscopy methods due to their small size (below optical resolution of 200 nm). However, if they are clumped together, they may be detected using interference, phase, or fluorescence microscopy.

A combined DFM and FM methodology adds the advantage of simultaneously visualizing cellular morphology together with nanoparticles. DFM-captured NP images were approximately 100 times brighter compared to fluorescence-captured images. Images (DFM and FM) can be combined using Nikon Elements software. DFM plus Fluorescence allows for the visualization of nanoparticles and cell structures. Other microscopy techniques may be used, but there appear to be problems associated with them. Interference contrast microscopy combined with fluorescence imaging may show a pixel shift. Bright fluorescent nanomaterials can be detected with fluorescent microscopy, but they are hard to visualize due to low signal from the small particle. Even phase contrast microscopy can be used, but it appears to be the least useful of these methods due to a lack of resolution. It is important to note that the application of DFM described in this report is limited in use as a detection method only, not a technique to directly observe single NPs. Thus, this method does not produce accurate size representations of NPs, but only their relative size based on the capacity of NPs to scatter light. Likewise, NP agglomerates, compared to single NPs, may be represented as a single point with a larger size and/or increased light scatter. Therefore, inferences regarding size, shape, and number of structures being examined are limited. In addition, factors including surface irregularity and reflectivity may also affect light scatter and signal strength in analyzing NPs using different DFM methods.

DFM has been interfaced with the power of point scanning confocal microscopy to resolve objects in the *z*-plane within cells (7). Confocal microscopy technique contributes increased resolution and optical sectioning of specimens for three-dimensional analyses, as well as the precise application of excitation wavelengths. The combination of these two techniques can yield images of nanoparticles in 3D space (7). To obtain the best dark field confocal microscopy images, it is necessary to attach a suitable DF condenser (NA 1.2–1.4), with an oil lens (×60 to ×100) containing an adjustable iris. With an inverted microscope, it is necessary to use an extender with the condenser. DF confocal microscopy can also be done with a conventional air condenser, but this will produce slightly inferior quality images compared to the higher NA oil condenser and oil objectives with an adjustable iris. These configurations allow for the production of a 3D image that can be enlarged and the nanoparticles can be located in 3D space of a cell or tissue. When using a point scanning confocal microscope, the light illuminating the specimen is focused by the objective and detected by a dark field condenser. Essentially, these instruments use a reverse light path from normal DF applications for DF detection in confocal mode. Consistent with normal DF illumination, the blockage of centralized light results in DF-CLSM reveals only structures of specimen that are capable of scattering the oblique rays of light towards the objective.

Wide-field confocal microscopy combined with dark field microscopy (DF-WFCM) can acquire images that can be used with deconvolution software to image nanoparticles within the cell in 3D space. In this operation, a stack of images is obtained and then combined into an image using extended depth focusing software. Using this approach the focal center of the nanoparticle can be obtained and combined with fluorescent images. A standard fluorescence dark field microscope can be used to create stack of images that can be enhanced using deconvolution software to combine the image stacks. DF-WFCM has the advantage of using an oil condenser to focus the dark field light and a lens with an iris to detect the scattered light yielding increased resolution. It is easy, accurate, and fast with thin samples and proper software.

DF-CLSM and DF-WFCM methodologies represent novel and workable options for many investigators searching for microscopy techniques that can be used to study different interactions between cells and NPs in 3D space. DFM can detect very small structures, and CLSM has the vertical sectioning capabilities to render specimens in 3D. By combining these two methodologies (DF-CSLM and DF-WFCM), solutions to NP toxicological studies can be provided.

2. Materials

2.1. Cell Culture in Chamber Slides

1. ARPE-19 cells (see Note 1).
2. RPMI and GlutaMax^{tm}-1 Medium with 25 mM HEPES supplemented with 10 % fetal bovine serum (FBS) (see Notes 2 and 3).
3. T25 tissue culture flasks.
4. TrypLEth express Stable Trypsin Replacement (see Note 3).
5. Incubator set at 37 °C and 5.0 % CO_2.

6. Dulbecco's phosphate buffered saline (PBS) without Calcium chloride and Magnesium Chloride: specify molecular composition here.
7. Centrifuge IECentra8R.
8. Prolong gold antifade reagent with DAPI (Invitrogen).
9. 22 × 22 mm × 1.5 mm microscope cover slips (see Note 4).
10. 25 × 75 × 1.0 mm microscope slides.
11. 1 mg/mL of TiO_2 primary particles NPs (10 nm, NanoAmor, Inc.) suspended in RPMI and GlutaMaxth-1 Medium with 25 mM HEPES (Gibco/Invitrogen) and 10 % FBS that has been sonicated with a probe sonicator operating at 4.5 W for 3 pulses, 2 s/pulse to disperse the NPs (see previous chapter for characterizing NPs by DLS) (see Note 5).
12. BD Flacon: 4 Chamber polystyrene vessel tissue culture treated glass slides with chamber removal accessory.
13. Misonix Microson Ultrasonic Cell Disrupter XL Probe sonicator.
14. Organelle, nuclei, and membrane staining probes (see Note 6).
15. 4 % paraformaldehyde.
16. Inverted Fluorescent microscope.

2.2. Creating Slides of Trypsinized Suspended Cells

1. Fixed ARPE-19 prepared for light scatter and nuclei assay (see previous chapter).
2. Items 9–11 from Subheading 2.1.

2.3. Creating NP Slides

1. Items 9 and 11 from Subheading 2.1.
2. 25 × 75 mm microscope slide cleaned with sonicator.
3. 22 × 22 mm 1.5 mm thickness cover slips (see Note 4).

2.4. Wide-Field Confocal Microscopy

1. Upright Nikon E800 Fluorescent with DF (see Note 7).
2. Oil immersion Dark field condenser (NA = 1.2–1.4) ×60 Plan Fluor oil immersion objective lens with adjustable iris permitting a Numerical Aperture (NA) ranging from 0.5 to 1.25.
3. Slides containing cells treated with NP.
4. Nikon NIS-Elements imaging software.
5. Immersion oil for microscopy type A. $n_D = 1.515$ (23 °C).
6. PC with Pentium 4, 3.2 GHz or higher CPU, 2 GB + RAM, Windows 7 or Vista, 600 MB + on hard disk, 1,280 × 1,024 True Color mode video.
7. A known test slide for checking optical configuration of microscope.
8. Andor Clara CCD microscope camera for taking images of cells and NPs.

9. 89 North Photo Fluor I light source for fluorescence.
10. Xenon light source for DFM (see Note 8).
11. GG 420 blockers eye pieces (see Note 9).

2.5. Leica SP1 Point Scanning Confocal and DIRBE Inverted Microscope

1. Leica DIRBE fluorescent microscope (see Note 10).
2. 1 mm dry Dark field condenser for ×63 Plan Apo lens (see Note 10).
3. Plan Apo ×63 with an adjustable iris diaphragm that was variable between 1.32 NA and 0.5 NA (see Note 10).
4. Slides containing cells treated with NP.
5. Leica SP1 software for collecting images.
6. BitPlane Imaris and Leica SPI software for postacquisition processing.

2.6. Nikon Ti

1. Nikon Eclipse C1si Spectral Confocal Imaging System.
2. Nikon Ti inverted microscope.
3. ×60 Plan Fluor oil immersion objective lens with adjustable iris permitting an Numerical Aperture (NA) ranging from 0.5 to 1.25 (see Notes 11 and 12).
4. Transmission detector (TD).
5. 404, 488, 561, and 633 nm lasers (see Notes 13 and 14).
6. Slides containing cells treated with NP.
7. Nikon E-Z-Cl software for collecting images.
8. Nikon NIS-Elements for postacquisition processing.
9. BitPlane Imaris for postacquisition analysis.

3. Methods

NPs contained within cells or solutions are studied using different types of light and confocal microscopes. The cells used for high resolution imaging are derived from chamber slides containing cells incubated with NPs, or from suspended cells derived from adherent cells growing in T-flask that were incubated with NPs. The NPs in solution are mounted on slides to make them immobile for 24 h.

For dark field, fluorescent, and wide-field confocal microscopy, we use the Nikon E800 upright microscope. For the point scanning confocal microscopy, a Nikon Ti inverted microscope is used with an oil condenser with the Nikon C1si confocal system and the Leica DMIRB inverted microscope is used with a dry condenser containing an appropriate dark field stop with the Leica SP1 confocal system. These microscopes were chosen because they were readily available for use in the laboratory. Other common models of microscopes and confocal microscopes are readily adaptable to dark field imaging.

3.1. Cell Culture in Chambered Slides

1. Grow ARPE-19 cells in T-flask to confluence with RPMI and GlutaMaxth-1 Medium containing 25 mM HEPES supplemented with 10 % FBS. Trypsinize and then dilute with media to 0.5 mL. Add 10 μL of cells and 1 mL of tissue culture media with serum into each of the 4 wells chambered slides (see Note 15). Incubate cells at 37 °C and 5.0 % CO_2.
2. After 24 h, add the suspended NPs in the experimental wells to produce the desired concentrations between 0.1 and 30 μg/mL of NPs within the well. The dose range used was determined in preliminary experiments to cause no toxicity and to show the ability to detect nanoparticles in cells by flow cytometry and microscopy. At 30 μg/mL, the cell is completely filled with nanoparticles and at 0.1 μg/mL an individual NP in cells can be observed. These NPs should be previously sonicated to ensure that the NPs are suspended homogeneously. Gently mix the NPs in the wells to disperse them equally before incubating (37 °C, 5 % CO_2) them for 24 h (see Note 16).
3. Stain the cells for organelles, nuclei, or membranes prior or after fixation depending on the specific stain. Incubate the cells and periodically check them with a fluorescent microscope (Nikon E800) to determine when they have become properly stained. Fix *cells* with a 1:1 dilution of 4 % paraformaldehyde (see Note 17).
4. Remove liquid from wells and remove the top with chamber accessory. Make sure the slide is relatively dry, if not remove the excess liquid. Add a drop of Prolong in each well and add a 22 × 50 mm 1.5 thickness cover slip. Let dry 24 h before observing with oil (Fig. 1).

3.2. Creating Slides of Trypsinized Suspended Round Cells

1. Using fixed cells, place 2 μL of cells on each microscope slide with one drop of Prolong.
2. Place the 1.5 mm cover slip on top. Set the slides to dry for 24 h (Fig. 2).

3.3. Creating NP Slides

1. Add 2 μL of 100 μg/mL of suspended TiO_2 NPs in serum to sonicated cleaned slide with a drop of particle-free mounting media.
2. Mix the NPs and the mounting media before placing the 1.5 mm thickness cover slip. Allow the sides to dry before use.
3. Under a wide-field confocal microscope, the sample will look like a star field (Fig. 3).

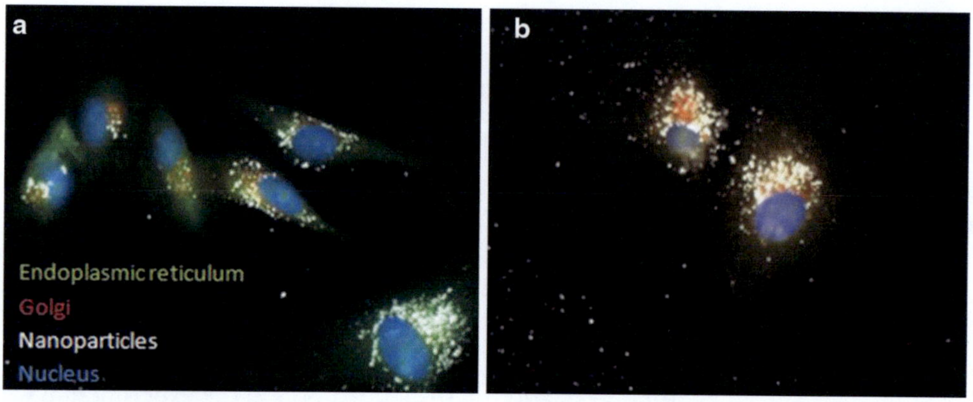

Fig. 1. Fate of TiO_2 nanoparticles (NP) in ARPE-19 cells exposed to NPs in a chamber slide. NPs (*white*) enter cells and are transported to the ER (*green*) and Golgi (*red*) and around the nuclei (*blue*). The NPs come together to form aggregates of different sizes. Magnification ×600. Panels (**a**) and (**b**) are similar APPE-19 cells with different staining conditions and digital photographic exposures.

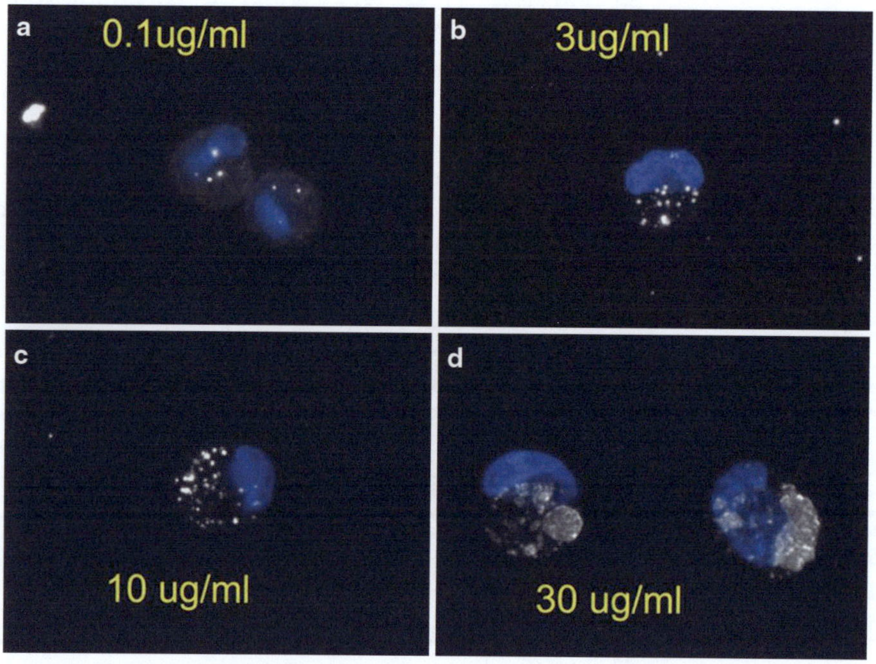

Fig. 2. Combined dark field and fluorescent image. Nanoparticles (NP) ((**a**): 0.1 μg/mL; (**b**): 3 μg/mL, (**c**): 10 μg/mL; (**d**): 30 μg/mL) were incorporated into ARPE-19 cells for 24 h in T-flasks. Cells were trypsinized from the T-flask for flow cytometry and then fixed in suspension with 4 % PF for a final concentration of 2 %. NPs are shown in *white* and nuclei were stained with DAPI and shown in *blue*. Magnification ×600. Plan Fluor ×60 lens with a NA between 1.25 and 0.55 was adjusted to about 0.70 NA to reduce background for dark field imaging images of fluorescent cells and dark field NPs were acquired sequentially and then combined using Nikon Elements software. Suspended cells from a light scatter and natural fluorescence assay will have more NP aggregates with increasing NP concentration. Reprinted with permission from ref. (6).

Fig. 3. TiO_2 nanoparticles (NP) in suspension at ×600. NP in solution was fixed to slides with particle-free adhesives. Dark field microscopy was used to detect particles. Magnification ×600.

3.4. Wide-Field Confocal Microscopy

1. Turn on Nikon E800 microscope. Turn off lights, as they will interfere with imaging.
2. In order to determine if the system is working, place a familiar test slide on the stage, add oil on the Dark field condenser. Take pictures of the test slide in fluorescence and Dark field to insure proper functioning of the microscope.
3. Place specimen slide onto the stages. To create a *z*-series, set the bottom and top focus and enter the optimal number of slices. Acquire the fluorescent *z*-series in the desired wavelengths (see Notes 18–20). Use Dark field to obtain the same *z*-series of the NPs (see Note 21).
4. Combine the fluorescent and DF *z*-series using Nikon Elements. Convert into an extended depth focus image. Deconvolute with software.

3.5. Leica SP1 Point Scanning Confocal and DIRBE Inverted Microscope

1. Turn on the computer, microscope, light sources, lasers, and software.
2. Run standard QA procedures to insure optical alignment and proper function of a confocal microscope.
3. Set up the Leica DIRBE microscope with a dry Dark field condenser, 1 mm condenser lens, and Plan Apo ×63 with an adjustable iris. Adjust the system for Kohler illumination. Set the Plan Apo objective to 0.5 NA which is the minimum value achievable. Adjust the iris to control the background and image quality. Adjust the sliders for proper fluorescent ranges and the interference optics to create images in dark field and fluorescence. Take DF-CLSM images of NPs using the lowest wavelength (488 nm)

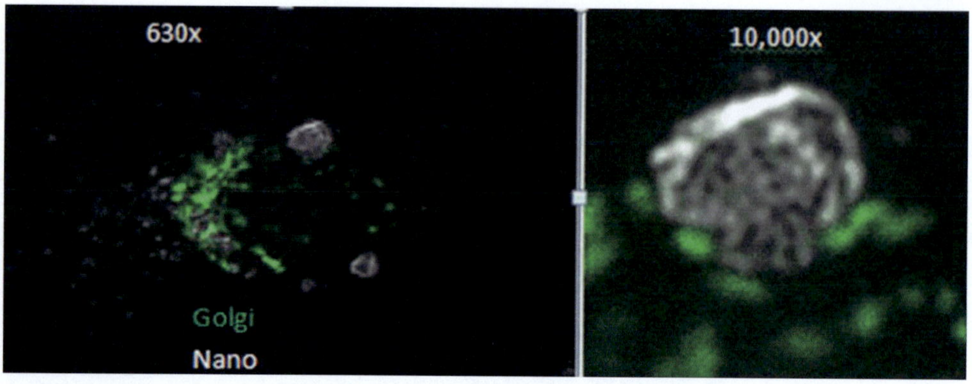

Fig. 4. Confocal microscopy analysis on nanoparticle (NP) clumps within cell. The NP clumps (*white*) embedded in the Golgi (*green*) were enlarged 10,000× with a zoom feature on a Leica SP1 confocal microscope.

laser light on the confocal. Use the fluorescence of organelle-specific probes to localize the organelles (see Note 6).

4. Collect images using Leica SP1. Next, use BitPlane Imaris software to conduct any postacquisition processing.
5. Create 2D and 3D images combining fluorescent (confocal) and transmission (dark field) images (Fig. 4).

3.6. Nikon Ti

1. Turn on the computer, microscope, light sources, and lasers.
2. Run standard QA procedures to insure optical alignment and proper function of a confocal microscope.
3. Set up the Nikon Ti microscope with an oil dark field condenser lens and Plan Fluor ×60 oil lens with an adjustable iris. Adjust microscope for Kohler illumination. Adjust the Plan Fluor objective to approximately 0.9 NA. Adjust the iris for best contrast balancing the background with image intensity. Use the fluorescence filters to create a proper image (see Notes 13 and 14).
4. Take DF-CLSM images of NPs using a low wavelength (405 or 488 nm) laser light. Use the fluorescence of cellular and organelle-specific probes in the cell to use Dark field to localize the probes.
5. Collect images using Nikon EZ-C1 software. Next, use Nikon NIS-Elements AR software to conduct any postacquisition processing (Fig. 4).
6. BitPlane Imaris software can be used to conduct any postacquisition processing.
7. Create 2D and 3D images combining fluorescent (confocal) and transmission (Dark field) images.

4. Notes

1. Other immortalized adherent cell lines can be substituted for the ARPE-19 cells.
2. Different tissue culture media may be substituted for different cell lines.
3. Live cells should be incubated with warm media and warm TrypLEth express to avoid shocking the cells.
4. A 1.5 mm cover slip is needed to achieve high-quality images with objectives. Other cover slip dimensions can be substituted depending on the area to be mounted such as 18 × 18 mm.
5. See Malvern instruction for nanoparticles detection and measurement (12). While working with NPs, it is also important to work in low-light conditions. Some NPs may be photosensitive and may change while exposed to light.
6. Different organelles dyes and GFP BacMam 2.0 transfections can be used to stain the mitochondria, peroxisomes Golgi complex cytoskeleton actin, and plasma. For nucleus staining, HO3342, YOPRO1, SYTOX green, or SYTOX orange can be used depending on the fluorescent wavelength desired. Check Molecular Probes catalog for suggested dye concentrations and spectral excitation/emission wavelengths. Usually an experiment will use unique cellular probes to yield a mix of blue, green, and red fluorescence emitting excitable probes.
7. Upright microscopes are superior and easier to use than inverted microscopes for dark field applications.
8. Xenon light is superior to halogen light as it contains primarily shorter wavelength (blue and UV vs. red) to observe NPs.
9. Add the GG 420 to eyepieces to block lower UV light from the Xenon light source. UV light can damage the eyes.
10. The Leica Confocal DFM configuration uses a 1 mm Working Distance (WD) condenser and a ×63 Plan Apo lens with an iris to observe NPs using the transmission detector. An approximate NA of about 0.5 or less must be used with the system as it has a dry condenser. Leica inverted microscopes do not have the possibility of using oil condensers. Oil condenser allows you to use a lens with an approximate 0.9 NA for better resolution.
11. Nikon Ti microscope uses same oil condenser (NA 1.2–1.4) as the E800 microscope and a ×60 Plan fluor oil lens with an iris.
12. Objectives that have low NA show a decrease in optical resolution and may introduce chromatic aberration into the image.

13. A single excitation wavelength for fluorescence and dark field is the best way to avoid chromatic aberration and fluorescent pixel colocalization errors. Chromatic aberration errors between the NPs derived from one wavelength and the fluorescent signals derived from the other laser lines will occur, so it is important to measure the NPs and the fluorescence with the same wavelength. In cases where multiple excitation wavelengths are needed, they should be used sequentially rather than simultaneously and the 404 line should not be used with the visible wavelengths (7). The shortest possible wavelength should be used for the acquisition of the DF signals for best resolution. It is important to use a wavelength in which the fluorescence structure is being identified and correlated to the NP.
14. UV and visible colocalization issues occur in CLSM (13, 14). It appears that the NPs accentuate these colocalization problems relative to submicron and micron particles as the lens were designed for particles above the diffraction limit of 200 nm. It is recommended that characterization of NPs observed at different wavelengths be made to address chromatic aberration and colocalization issues. Each imaging system and each lens used should be tested.
15. The height of the media influences how many NPs are absorbed into the cell. For example, the concentration of a flask with 10 mL of media will be half of that of a flask with 5 mL. In addition, it will take the NPs longer time to settle on the cells in a 10 mL than a 5 mL flask.
16. Cells can be exposed to NPs for different time periods between 4 and 72 h. Longer exposures involve nanoparticle utilization and clumping within the cell. It was found that 24 h incubation allowed most of the particles to settle onto the cells. Larger agglomerates sediment more quickly than smaller particles.
17. If with GFP BacMam 2.0 or RFP BacMam 2.0 organelle probes are used to transfect cells, it is important to have a good expression of the fluorescent probe prior to the addition of nanoparticles. Generally, in our experiments we transfected cells for 24 h prior to the addition of nanoparticles. It is useful to have a high power lens with an iris and have a lower power oil lens to locate the samples (i.e., ×20 MI oil and a ×60 with IRIS). GFP BacMam 2.0 or RFP BacMam 2.0 organelle probes are fixable.
18. Acquire fluorescent images using epi fluorescent mercury light or equivalent with the Xenon light off. Use the Xenon light to capture images of NP images in DF. Xenon light is much brighter than the mercury lamp and the exposure time is about a 100× faster.

19. Best image quality will occur when the RI (Reflexive Index) of the lens condenser and the material is matched and high NA lenses are used. This occurs using a Nikon oil condenser and a ×60 plan/fluor oil lens with the Nikon E-800 upright microscope or the Nikon Ti inverted microscope.
20. Always shut the light going to the stage by pulling the shutter out when not imaging. This will minimize photo bleaching of the sample.
21. If possible, capture a single image instead of using the *z*-series as it saves time. However, this may reduce the quality of the images because out of plane light will be acquired in addition to the focal plane.

Acknowledgements

Thanks are extended to Carl Blackman Eugene Gibbs, James Samet, Will Boyes, Laura Degn, Sarah Hutchinson, and Kristen Sanders for their helpful discussions on the research included in the protocol.

Government Disclaimer

Although the research described in this article has been supported by the United States Environmental Protection Agency, it has not been subjected to Agency technical review and therefore does not necessarily reflect the views of the Agency, and no official endorsement should be inferred. Mention of trade names or commercial products does not constitute endorsement or recommendation for use.

References

1. Auffan M, Rose J, Bottero JY et al (2009) Towards a definition of inorganic nanoparticles from an environmental, health and safety perspective. Nat Nanotechnology 4(10):634–641
2. Nohynek GJ, Lademann J, Ribaud C (2007) Grey goo on the skin? Nanotechnology, cosmetic and sunscreen safety. Crit Rev Toxicol 37(3):251–277
3. Salzman GC (2001) Light scatter: detection and usage. Curr Protoc Cytom Chapter 1:Unit 1.13
4. Shapiro HM (2003) Practical flow cytometry, 4th edn. Hoboken, New Jersey
5. Shapiro HM (2001) Optical measurements in cytometry: light scattering, extinction, absorption, and fluorescence. Methods Cell Biol 63:107–129
6. Zucker RM, Massaro EJ, Sanders KM et al (2010) Detection of TiO_2 nanoparticles in cells by flow cytometry. Cytometry 77:677–685
7. Gibbs-Flournoy EA, Bromberg PA, Hofer TPJ et al (2011) Dark field-confocal microscopy detection of nanoscale particle internalization by human lung cells. Part Fibre Toxicol 8(1):2
8. Murphy DB (2001) NetLibrary I: dark-field microscopy. In: Fundamentals of light microscopy and electronic imaging (electronic resource). Wiley-Liss, New York, pp 112–116

9. Spencer M (1982) Dark-field microscopy. In Fundamentals of light microscopy. Cambridge University Press, New York, pp 37–39
10. Wayne R (2009) Dark-field microscopy. In Light and video microscopy. Elsevier, Amsterdam, pp 95–97
11. Skebo JE, Grabinski CM, Schrand AM (2007) Assessment of metal nanoparticle agglomeration, uptake, and interaction using high-illuminating system. Int J Toxicol 26: 135–141
12. Murdock RC, Braydich-Stolle L, Schrand AM et al (2008) Characterization of nanomaterial dispersion in solution prior to in vitro exposure using dynamic light scattering technique. Toxicol Sci 101:239–253
13. Zucker RM (2006) Evaluation of confocal microscopy system performance. Methods Mol Biol 319:77–135
14. Zucker RM (2006) Quality assessment of confocal microscopy slide based systems: performance. Cytometry A 69:659–676

Chapter 40

Detection of TiO_2 Nanoparticles in Cells by Flow Cytometry

Robert Martin Zucker and Kaitlin Marie Daniel*

Abstract

Evaluation of the potential hazard of man-made nanomaterials has been hampered by a limited ability to observe and measure nanoparticles in cells. A FACSCalibur™ flow cytometer was used to measure changes in light scatter from cells after incubation with TiO_2 nanoparticle. Both the side scatter and forward scatter changed substantially in response to the TiO_2. Between 0.1 and 30 μg/mL TiO_2, the side scatter increased sequentially while the forward scatter decreased, presumably due to substantial light reflection by the TiO_2 particles. At the lowest concentrations of TiO_2 (0.1–0.3 μg/mL), the flow cytometer apparently could detect as few as 5–10 nanoparticles per cell as shown using dark field microscopy. The influence of nanoparticles on the cell cycle was detected by nonionic detergent lysis of nanoparticle-incubated cells. Viability of nanoparticle-treated cells was determined by PI exclusion.

These data suggest that the uptake of nanoparticles within cells can be monitored using flow cytometry and confirmed by dark field microscopy. This approach may help fill a critical need to assess the relationship between nanoparticle dose and cellular toxicity. Such experiments could potentially be performed quickly and easily using the flow cytometer to measure both nanoparticle uptake and cellular health.

Key words: Nanoparticles, Flow cytometer, Cytometry, Toxicology, Nanotoxicity

1. Introduction

Nanotechnology may have potential hazards on the environment, humans, and other organisms (1). Nanoparticles may pose unique health risks due to their composition, reactivity, and small size (2, 3). There are technical challenges in detecting and quantifying nanomaterials in environmental media, tissues, or cells, a step that is critical for evaluating the potential hazard of man-made nanomaterials. A suspension of nanoparticles can be assessed using dynamic light scattering (DLS) to yield an approximate

* Contractor to the USEPA, award # EP09D000042.

Mikhail Soloviev (ed.), *Nanoparticles in Biology and Medicine: Methods and Protocols*, Methods in Molecular Biology, vol. 906, DOI 10.1007/978-1-61779-953-2_40, © Springer Science+Business Media, LLC 2012

hydrodynamic diameter. However, this method has many limitations, one being that it cannot accurately distinguish the single nanoparticles from clumps of nanoparticles in suspension. Although a number of research groups have tried to detect the presence of these particles in the environment and in cells using a variety of techniques including transmission electron microscopy (TEM), scanning electron microscopy (SEM), and atomic force microscopy (AFM), there is no universally accepted method to monitor nanoparticles. While many of these techniques have been useful in determining the physical properties of nanoparticles, they provide limited quantitative information regarding cellular interactions. In addition, many of these techniques are expensive, labor-intensive, and time-consuming. In this communication, we demonstrate that the common technique of flow cytometry can be used to assess the quantity and the effects of nanoparticles within cells.

Flow cytometry has been used to measure both light scattering and fluorescence from particles or biological cells. Most flow cytometers measure both the small angle forward scatter (FSC) intensity (approximately 0–5°) with a photodiode and side scatter (SSC) intensity (approximately 85–95°) with a photomultiplier tube. Forward light scatter is usually proportional to cell volume as predicted by Mie theory (4–8), but can be influenced by strongly absorbing or reflecting materials, the cell nucleus, other organelles, irregular cell shape, or damage to the cell membrane. The presence of particles inside a cell can change its refractive index (RI), which will ultimately change the FSC intensity. Therefore, the FSC does not always reliably represent cell size (5). The SSC intensity is mainly thought to be determined by the inner structure of the cell. It is also related to cellular mass and possibly protein content as well (4–6). The SSC signal is affected by the RI of the cytoplasm and the number and type of organelles present in the cell. SSC has been used to show differences in the physical state of the cell, including mitosis, particle uptake, and sperm decondensation (8–11). Cells of interest can be defined by their light-scattering clusters in a cytogram and then the amount of fluorescence in various channels can be recorded.

Rayleigh scattering states that the intensity of light scatter in submicron-sized particles is proportional to the sixth power of the particle diameter (4–6). It appears that many metallic, nano-sized objects like Titanium dioxide (TiO_2) can scatter light with relatively high intensity for their size. TiO_2 nanomaterials occur in two crystalline structures: anatase and rutile. Both structures efficiently reflect visible and UV light. Rutile crystals scatter more light than anatase due to a higher refractive index. This property of rutile TiO_2 has been used by the sunscreen industry to produce effective UV blockers (1, 3) and also makes it possible to examine the particles through light-scattering techniques.

The rationale behind using flow cytometry to detect nanoparticles in cells was derived from the knowledge that 488-nm laser light has different scattering effects on various hematological cells and will have a high degree of scattering with granulocytes. Lymphocytes, monocytes, and granulocytes can be differentiated by their unique forward scatter (FSC) and side scatter (SSC) properties. Small lymphocytes scatter less light than larger monocytes. Granulocytes scatter the most light, primarily due to many granules located in the cytoplasm (4).

Flow cytometry was used to monitor the effects of nanoparticles on a human-derived line of retinal pigment epithelial cells (ARPE-19) and previously used with other cell types (12–14). The uptake and distribution of nanoparticles detected by flow cytometry was visually confirmed with dark field microscopy. These technologies could also provide a statistical basis to estimate nanoparticle uptake, cytotoxicity, and other biological effects. Methods for using flow cytometry to detect the effects of nanoparticles on cells are described in this communication.

2. Materials

2.1. Nanoparticles Suspensions

1. Resuspend TiO_2 primary Nanoparticles (10 nm, NanoAmor, Inc.), Cat #5485HT in RPMI and GlutaMax^{tm}-1 Medium, containing 25 mM HEPES supplemented with 10 % fetal bovine serum (FBS) (see Notes 1–3).
2. Dynamic light scatter (DLS, Malvern) Zetasizer Misonix Microson Ultrasonic Cell Disrupter XL Probe sonicator.
3. Cuvettes (Sarstedt 67.754 or equivalent) for the Zetasizer.
4. 50 mL centrifuge tube.
5. Probe sonicator (Misonix Microson Ultrasonic Cell Disrupter XL).

2.2. Cell Culturing and Nanoparticles Exposure

1. ARPE-19 cells (see Note 4).
2. 25 cm^2 Tissue Culture Flasks.
3. RPMI and GlutaMax^{tm}-1 Medium with 25 mM HEPES supplemented with 10 % FBS.
4. TrypLE Express Stable Trypsin Replacement to remove cells from plate.
5. Dulbecco's Phosphate Buffered Saline (PBS) (1×) without Calcium chloride and Magnesium Chloride (GIBCO #14190).
6. Incubator set at 37 °C and 5.0 % CO_2.

2.3. Flow Cytometry and Bead Alignment Test

1. BD FACSCalibur™ (BD Biosciences) flow cytometer containing a 488-nm laser, FSC diode detector, and photomultiplier tube SSC detector.
2. Polystyrene round bottom tubes #352008 (Becton Dickinson).
3. G4 Apple computer 1.2 MHz, operating system 9.2, 5 mL BD Cell Quest software.
4. Filtered water (cleaned with a 0.05 μm filter).
5. Duke 3.0-μm alignment beads (Duke Scientific Corp) (see Note 5).

2.4. Light Scatter and Fluorescence Assay

1. Items 2–6 from Subheading 2.2 and items 1–5 from Subheading 2.3.
2. ARPE-19 cells with Nanoparticle samples diluted with 0.5 mL media.
3. Microsoft Excel.
4. FSC Express 4.0 Flow Cytometry Analysis software (see Note 6).

Four percent paraformaldehyde (PF) in Dulbecco's PBS without Calcium chloride and Magnesium chloride.

2.5. Suspension Cell Viability Assay

1. All items from Subheading 2.4.
2. 40 μg/mL Propidium iodide in tissue culture media precleaned.
3. Microscope slides.
4. Microscope slide cover glasses—1.5 thickness 22 × 22 or 18 × 18
5. Nikon Eclipse C1Si Spectral Confocal imaging system.
6. Nikon E800 fluorescence upright Microscope with Dark field optics. Plan fluor 60× with iris (NA 0.55–1.25).

2.6. Cell Cycle Assay

1. All items from Subheading 2.5 except for item 2.
2. 40 μg/mL propidium iodide in 0.5 % IGEPAL CA-630 nonionic detergent (I-3021 Sigma) (see Note 7).
3. FSC Express Flow Cytometry Analysis software with MultiCycle (see Note 6).

3. Methods

The light scattering detects the effects that nanoparticles have on cell's biophysical parameters. The viability assay is used to determine if cell death occurred during incubation with nanoparticles. The cell cycle assay is used to determine how nanoparticles affect cell growth and division. Nonionic detergent lyses the cells and PI stains the nuclei to obtain the DNA histograms that will be used to quantify the distribution of cells in the G1, S, and G2 phases of the cell cycle.

To obtain reliable and reproducible results, it is important to process the cells as quickly as possible after removal from the T-flask. A long exposure time outside of the incubator may increase cell death or may physically and biochemically alter the cells. In addition, it is important to always warm the media and TrypLE express before using to avoid sending the cells into shock.

3.1. Nanoparticles Suspensions

1. Make a 10 mL of 1 mg/mL of TiO_2 10 nm nanoparticles in a 50 mL plastic centrifuge tube (see Note 8) by adding nanoparticles to RPMI and GlutaMaxth-1 Medium + 25 mM HEPES (Gibco/Invitrogen) supplemented with 10 % FBS.
2. Use a probe sonicator operating at 4.5 W for 3 pulses, 2 s/pulse to disperse the nanoparticles (see Note 8).
3. Run the 1 mg/mL nanoparticles in a DLS, Malvern Zetasizer Nano to determine the zeta potential (see Note 8). Suspensions with a zeta potential greater than 30 mV or less than −30 mV are considered stable. As the zeta potential approaches 0 mV, the particles aggregate and fall out of solution.

3.2. Cell Culturing and Nanoparticles Exposure

1. Passage the human-derived line of retinal pigment epithelial (ARPE-19) cells when approaching confluence with TrypLE express to create new cultures in T25 flasks. One T25 flask is required for each experimental data point. A 1/3 split will provide cultures that are approaching confluence after 24 h. Add 5 mL of RPMI and GlutaMaxth-1 Medium with 25 mM HEPES and 10 % FBS to each T25 flask (see Note 9). Incubate cells at 37 °C and 5.0 % CO_2.
2. Once cells approach confluence, add the suspended nanoparticles to the experimental T25 flasks to produce the desired concentrations of nanoparticles between 0.1 and 30 μg/mL in the flasks (see Note 10). Gently mix the nanoparticles with the treated cells before incubating for 24 h in a 37 °C CO_2 incubator. Cells can be exposed to NPs for different time periods between 4 and 72 h (see Note 11).
3. Following the incubation with nanoparticles, wash cells with PBS and remove cells from flask with TrypLE express Stable Trypsin Replacement (see Note 12). Transfer into centrifuge tubes and add medium with serum.
4. Centrifuge for 10 min at 900 rpm (about 200 g) using an IEC Centra-8R centrifuge with a 218 Rotor. Decant the supernatant and add 0.5 mL of media to resuspend the cells. Place the tubes on ice (see Note 13).

3.3. Flow Cytometry and Bead Alignment Test

1. Prepare the flow cytometer for operation.
2. Set up the work space to include a FSC vs. SSC cytogram, and histograms of FSC, SSC (and if necessary the fluorescent

channels on Cell Quest). Trigger on FSC. Set up the cytometer to measure SSC and fluorescence logarithmically and FSC linearly (see Note 14).

3. Test the fluidics of the system by checking the alignment of the system. To conduct a bead alignment test, add 0.5 mL of filtered water and drop Duke 3.0-μm alignment beads or equivalents into a flow cytometer tube. Insert tube into the flow cytometer and acquire the data in all channels. Low CVs and a symmetrical histogram are needed for proper performance (see Notes 14 and 15).

3.4. Light Scatter and Fluorescence Assay

1. Transfer 50–200 μL of sample (depended on cell concentration) into the flow cytometer tubes. Samples should be diluted to at least 300 μL with media to produce a desired count rate below approximately 500 at a low flow rate (12 μL/mL) setting of the flow cytometer (minimal sample volume of 300 μL).
2. For microscopy observations, set aside a portion of the sample (50–100 μL) in micro-centrifuge tubes. The sample will be fixed with an equal volume of paraformaldehyde (4 %) in PBS.
3. On the Cell Quest work space, create a FSC vs. SSC cytogram, and histograms of all PMTs. All PMTs should be set in Log scale, except for FSC which in linear (see Note 12).
4. Initially, run a sample with the highest concentration of nanoparticles and then run the control to set the PMTs voltage and get all the nanoparticles on scale. The cells with nanoparticles should appear around the middle of the spectrum (between 10^3 and 10^4 on a log scale) to set the dynamic scale. Nanoparticles incorporated into cells will increase the cellular scatter signal.
5. Run the different nanoparticle-treated samples. In between samples, clean the sample injection tube with a Kim Wipe and insert a flow tube filled with media.
6. Data Analysis. First, use FSC Express 4.0 to create a cytogram of FSC vs. SSC and gate the cell population(s) to exclude the debris. Use the gated cells to create histograms for FSC, SSC, and the fluorescent channels. Obtain the geometric mean from each of these histograms. Express data as an overlay of control and treated concentrations for Forward Scatter, Side Scatter and Fluorescence. Present the data as a ratio of geometric means (treated/control) using Microsoft Excel (see Figs. 1 and 2).

3.5. Suspended Cell Viability Assay

1. Transfer 100 μL of samples into flow cytometer tubes. Add 200 μL of 40 μg/mL Propidium iodide in media (see Note 16).
2. Check the cells using a fluorescent microscope to insure that the cells have been properly stained.

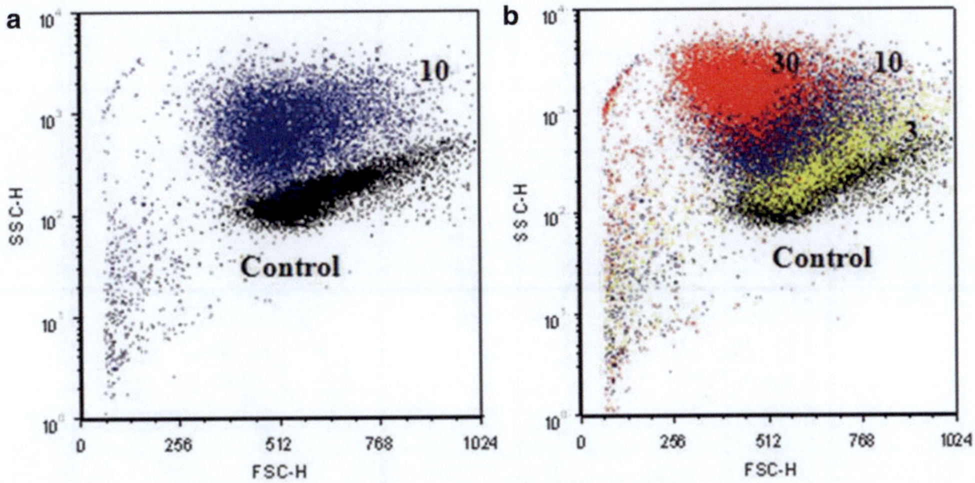

Fig. 1. Forward scatter (FSC-H) and side scatter (SSC-H) of ARPE PRE cells exposed to 0 (*black*), 3 (*yellow*), 10 (*blue*), and 30 (*red*) µg/mL of TiO_2. Charts indicate that side scatter increases and forward scatter decreases with TiO_2 dose. Reprinted with permission from ref. (12). (**a**) Cytogam showing for clarity, the difference between control (*black*) and 10 µg/mL (*blue*), (**b**) cytogram showing for completeness, 3 doses of TiO_2 (*yellow, blue,* and *red*) plus control (*black*).

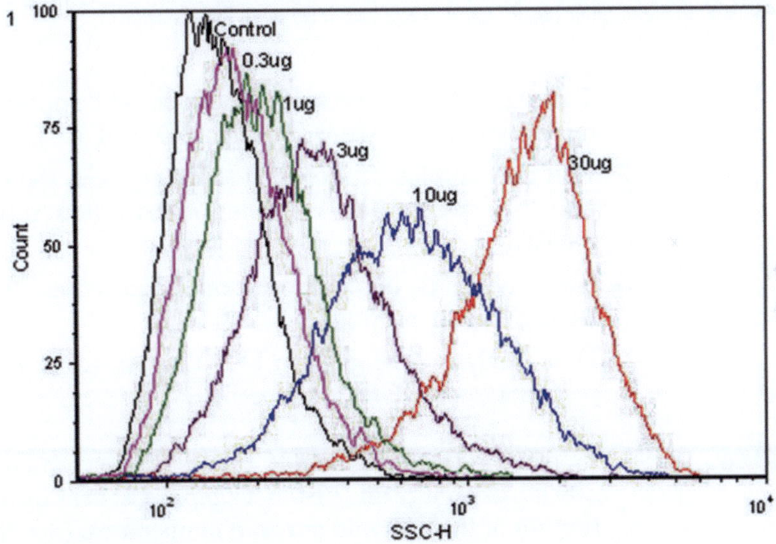

Fig. 2. Side scatter (SSC-H) histogram of ARPE-19 cells after exposure to 0 (*black*), 0.3 (*pink*), 1 (*green*), 3 (*purple*), 10 (*blue*), and 30 (*red*) µg/mL of TiO_2. Reprinted with permission from ref. (12).

3. On Cell Quest, create a cytogram of FSC vs. SSC and gate to remove the debris. Use another cytogram, PI (585/42) vs. SSC, to determine the percentages of labeled and unlabeled cells (see Fig. 3).
4. Initially, run a sample with the highest concentration of nanoparticles and then the control cells to set the PMTs on scale.

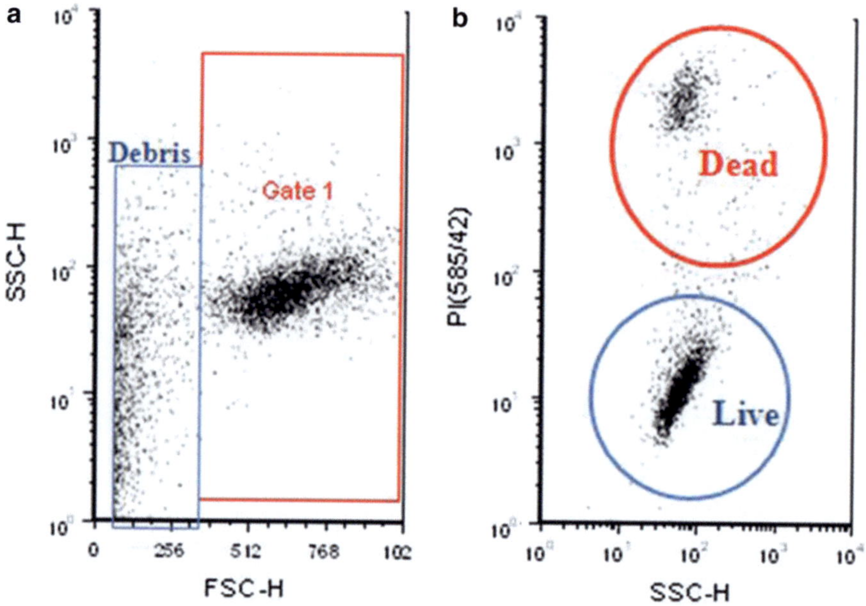

Fig. 3. ARPE-19 cells. (**a**) Cytogram of forward scatter (FSC-H) and side scatter (SSC-H) from ARPE-19 cells incubated for 24 h. The gated includes cells and excludes debris. (**b**) Viability cytogram uses cells gated from (**a**) and utilizing PI to mark dead cells. Cells with broken membranes absorb more PI and thus have more fluorescence.

The stained and nonstained cells should appear around the middle of the spectrum (between 10^1 and 10^3).

5. Run the samples. In between samples, clean the sample injection tube with a Kim Wipe and insert a flow tube filled with media.
6. Check viability on a fluorescent microscope (Nikon E800) using PI as an endpoint.
7. Data Analysis. Use gates in FCS Express 4.0 to determine the percentage of live and dead cells.

3.6. Cell Cycle Assay

1. Transfer 100 μL of sample into a flow cytometer tube with 200 μL of 40 μg/mL PI-0.5 % Igepal. Let cells incubate in solution for at least 30 min at room temperature (see Note 17).
2. Check cells under a microscope to monitor the membrane lysis of the cells. Has the membrane been removed and clean nuclei been produced? If not, note this observation.
3. The Cell Quest workspace for cell cycle should be set up to include a FSC (linear) vs. SSC (Log) cytogram (see Fig. 4), and a histogram for Fluorescence Area (564–606 nm) in linear should be obtained.
4. Initially, run a sample with the highest concentration of nanoparticles and then set the PMTs on scale on FL2 585/42 nm (area and height) and side scatter. The nuclei from the control cells should appear around the middle of the spectrum (between

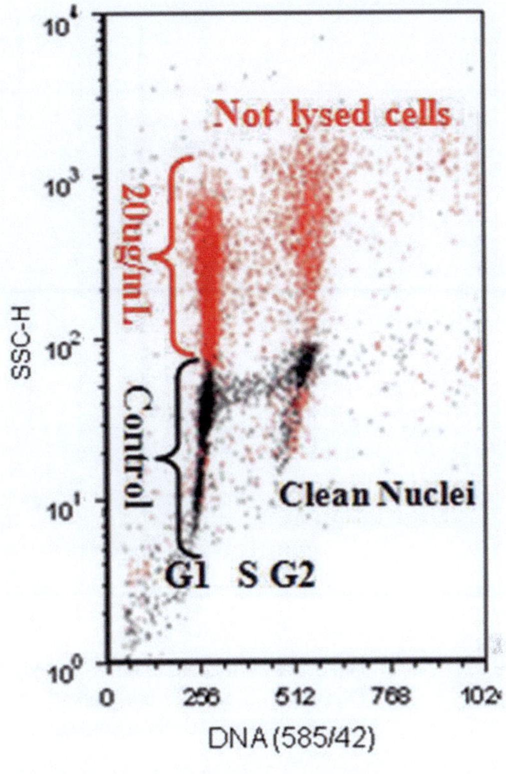

Fig. 4. Nuclei cytogram from ARPE-19 cells. The *black* represents control cells. *Red* was exposed to 20 mg/mL Ag nanoparticles. Cell within phase G1, S, and G2 are marked.

10^1 and 10^3) and the treated nuclei should appear between 10^2 and 10^4 (see Fig. 4). Run the samples. In between samples, clean the sample injection tube with a Kim Wipe and insert a flow tube filled with media.

5. Save and fix the leftover samples with 4 % PF for morphology (see next chapter).
6. Analyze the data with FSC express 4.0 and MultiCycle (see Note 9). Gate the nuclei in the DNA (585/42) vs. SSC cytogram (see Fig. 4). Then create a DNA histogram (see Fig. 5). Select the model that fits the data best. Report the percentage of cells in G1, S, and G2.

4. Notes

1. A variety of nanoparticles can be used i.e.: TiO_2 anatase 10 nm (catalog #44690, Alfa Aesar, Wardfill, MA), TiO_2 rutile 30–40 nm, 25 nm anatase/rutile (Aeroxide® TiO_2 P25, Degussa, Alpharetta, GA), 30–40 nm anatase/rutile (catalog #5485HT, NanoAmor, Houston, TX), 200 nm rutile (catalog

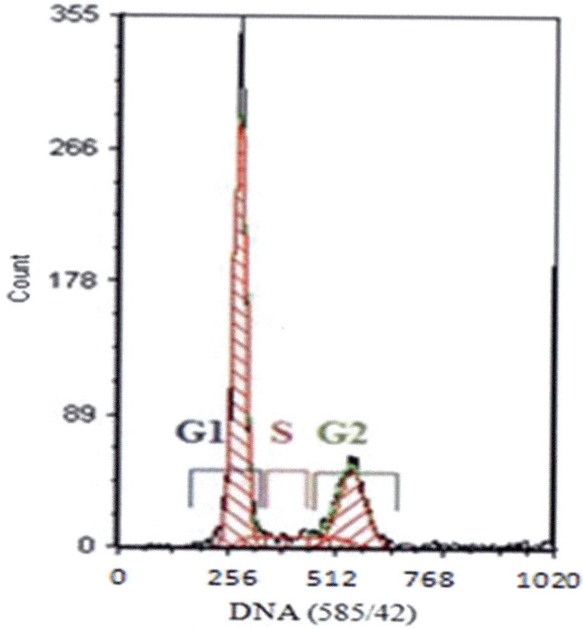

Fig. 5. DNA histogram of ARPE-19 cells derived from the detergent-treated lysis of cells. Phases G1, S, and G2 are marked. Percentages determined using MultiCycle (Phoenix Flow Systems and DeNovo Software).

#MKN-TiO_2-R250; Acros, Morris Plains, NJ), and 200–400 nm rutile (catalog #MKN-TiO_2 R250, Mknano, Mississauga, ON, Canada). Other sources such as Nanocomposix Inc., for silver nanoparticles, have been used successfully. Note that different types and sizes of nanoparticles have different degrees of effectiveness on cells. Uncoated TiO_2 nanoparticles tend to clump together in aqueous media creating a problem for proper dosimetry.

2. The presence of FBS in the dispersant was found to provide a relatively stable, monodispersed particle suspension on DLS evaluation.
3. Other tissue culture media can be used as long as it (a) contains 10 % FBS and (b) is compatible with cell line.
4. Other immortalized adherent cell lines have shown similar results. Other flow cytometers can be substituted for the FACSCalibur.
5. Other alignment beads can be used from companies such as: Thermo, SpheroTech, and Bangs Inc. Beads from other manufactures will have different characteristics and CV values. A homogeneous 2.5–3 μm bead with a small CV is desirable.
6. Flow Cytometry Analysis Program such as FlowJo, or Winlist can be substituted for FSC Express. Other cell cycle programs can be substituted for MultiCycle.

7. NP40 and other nonionic detergents can be substituted for Igepal CA-630.
8. Strictly follow the manufacturer's instructions for use of the Malvern Zetasizer Nano.
9. Height of media will determine the concentration of nanoparticles the cells are exposed to. The height of the media above the cells determines how many particles may land on the cells. For example, the nanoparticles in a T25 flask with 10 mL of media will have twice the concentration as a flask with only 5 mL of media. Thus, the 10 mL cells will have twice the concentration as on with 5 mL even though the number of nanoparticles per mL may be the same. In addition for short-term experiments, a 10 mL of Nano in a flask takes longer for the particles to settle compared to a 5 mL media in a flask.
10. These nanoparticles should be sonicated to ensure nanoparticles are suspended prior to addition.
11. Longer exposure involves nanoparticle utilization and clumping within the cell. It was found that 24 h incubation allowed most of the particles to settle onto the cells.
12. To include the cells that have detached due to treatment and are presumed to be dead collect the supernatant in a centrifuge tube. Centrifuge this supernatant before trypsin is added and concentrate the pellet in about ½ mL prior to analysis by flow cytometry and microscopy.
13. Cells can be put on ice or left at room temperature dependent on the experimental design.
14. QA should be on a flow cytometer (15). On digital machines, all parameters can be collected in Log scale. On analog scale, the side scatter and cellular fluorescent parameters should be acquired in log. The SSC increased and the FSC decreased in a dose-dependent manner after exposure to nanoparticles. Microscopic observations correlated the amount of particles in the cells with increased flow cytometry light scatter (12).
15. Forward light scatter is usually proportional to particle volume as predicted by Mie theory. However, this forward scatter parameter can be influenced by a number of other variables (reviewed by Shapiro practical flow cytometry V3, page 230–232) and it is not always directly related to cell volume. Larger beads will sometimes appear smaller in forward scatter. The presence of strongly absorbing or reflecting materials in cells tends to decrease the amplitude of the forward scatter while increasing the side scatter intensity (5, 12).
16. Cells with broken membrane will absorb the PI. The increased fluorescence of the broken cells differentiates between the two populations of live and dead cells. When some cells die during

the incubation, they no longer become adherent. To include the cells that have broken off, collect supernates before adding TrypLE express to the flask. Remove the adherent cells and add to the supernates.

17. The tubes may also be placed in a 37 °C water bath for 15 min to hasten the removal of the cell membrane. Do not leave in for more than 30 min. Put on ice after incubation.

Acknowledgements

Thanks are extended to Carl Blackman, Will Boyes, Laura Degn, Sarah Hutchinson, and Kristen Sanders for their helpful comments.

Government Disclaimer

Although the research described in this article has been supported by the United States Environmental Protection Agency, it has not been subjected to Agency review and therefore does not necessarily reflect the views of the Agency, and no official endorsement should be inferred. Mention of trade names or commercial products does not constitute endorsement or recommendation for use.

References

1. Maynard AD, Aitken RJ, Butz T et al (2006) Safe handeling of nanotechnology. Nature 444: 267–269
2. Auffan M, Rose J, Bottero JY et al (2009) Towards a definition of inorganic nanoparticles from an environmental, health and safety perspective. Nat Nanotechnol 4(10):634–641
3. Nohynek GJ, Lademann J, Ribaud C et al (2007) Grey goo on the skin? Nanotechnology, cosmetic and sunscreen safety. Crit Rev Toxicol 37(3):251–277
4. Salzman GC (2001) Light scatter: detection and usage. Curr Protoc Cytom Chapter 1:Unit 1.13
5. Shapiro HM (2003) Practical flow cytometry, 3rd edn. Wiley, Hoboken, NJ
6. Shapiro HM (2001) Optical measurements in cytometry: light scattering, extinction, absorption, and fluorescence. Methods Cell Biol 63:107–129
7. Steen HB (2004) Flow cytometer for measurement of the light scattering of viral and other submicroscopic particles. Cytometry A 57(2): 94–99
8. Zucker RM, Elstein KH, Easterling RE et al (1988) Flow cytometric discrimination of nuclei by right angle scatter. Cytometry 9:226–231
9. Nüsse M, Jülch M, Geido E et al (1989) Flow cytometric detection of mitotic cells using the bromodeoxyuridine/DNA technique in combination with 90 degrees and forward scatter measurements. Cytometry 10:312–319
10. Zucker RM, Perrault SD, Elstein KH (1992) Utility of light scatter in the morphological analysis of sperm. Cytometry 13:39–47
11. Giaretti W, Nüsse M (1994) Light scatter of isolated cell nuclei as a parameter discriminating the cell-cycle subcompartments. Methods Cell Biol 41:389–400
12. Zucker RM, Massaro EJ, Sanders KM et al (2010) Detection of TiO_2 nanoparticles in cells by flow cytometry. Cytometry A 77(7):677–685
13. Stringer B, Imrich A, Kobzik L (1995) Flow cytometric assay of lung macrophage uptake of environmental particulates. Cytometry 20(1): 23–32

14. Suzuki H, Toyooka T, Ibuki Y (2007) Simple and easy method to evaluate uptake potential of nanoparticles in mammalian cells using a flow cytometric light scatter analysis. Environ Sci Technol 41(8): 3018–3024
15. Zucker RM (2008) Flow cytometry quality assurance. Standardization and quality assurance in fluorescence measurements 2. Springer series on fluorescence. In: Resch-Genger U (volume editor), Wolfbeis OS (series editor) Methods and applications, Springer, Berlin, pp 343–370

Chapter 41

Nanoparticle Tracking Analysis for the Multiparameter Characterization and Counting of Nanoparticle Suspensions

Matthew Wright

Abstract

Nanoparticle Tracking Analysis (NTA) provides direct and real time visualization, sizing, and counting of particulate materials between 10 nm and 1 μm in liquid suspension. The technique works on a particle-by-particle basis, relating the degree of movement under Brownian motion to the sphere equivalent hydrodynamic diameter particle size, allowing high resolution particle size distributions to be obtained within minutes. Here we describe application of NTA to the analysis of 100 nm polystyrene nanospheres in liquid suspension. The technique can be adapted for use with nearly all particulate materials with sizes between 10 nm and 1 μm.

Key words: NTA, Nanoparticle tracking analysis, Nanoparticles, Sizing, Counting, Visualization, NanoSight, Light scattering

1. Introduction

The properties and potential that nanoparticulate materials provide make them attractive to both academic and industrial researchers in a wide range of applications. However, as the interest in these materials increases, so does the need to characterize their various properties. At this scale, a change in size of a few nanometers can have a major effect on their desired properties, for example, when looking at nanomaterials as a drug delivery vehicle, a few nanometers difference in size may have a significant impact on the bioavailability of the drug to the desired delivery site. Nanoparticle Tracking Analysis (NTA) offers researchers a multiparameter analytical technique for materials between 10 nm and 1 μm suspended in liquids, with real time visualization, sizing, and counting.

Current methodologies for the analysis of nanomaterials include dynamic light scattering (DLS), scanning electron microscopy (SEM), transmission electron microscopy (TEM), and Analytical

Mikhail Soloviev (ed.), *Nanoparticles in Biology and Medicine: Methods and Protocols*, Methods in Molecular Biology, vol. 906, DOI 10.1007/978-1-61779-953-2_41, © Springer Science+Business Media, LLC 2012

Disk Centrifugation; each of these techniques has advantages and limitations. For example, DLS is one of the most commonly used techniques for the analysis of nanoparticles in liquid suspension. This technique is ideally suited for the analysis of monodispersed systems. However, it has a limited capability to analyze polydisperse systems when trying to establish particle size distribution. It is unsuitable for determining concentration of nanoparticles. DLS works by using a digital correlator to determine the characteristic timescales of intensity fluctuations within the light scattered by a suspension of nanoparticles as a consequence of their Brownian motion. The average particle size and polydispersity index can be calculated by analyzing the resultant exponential autocorrelation function. However, relationship between the size of a particle and the amount of light that it scatters is complex, and in the nanoscale region, varies as a function of radius to the power of six. Therefore a small increase in the size of a particle can result in a large increase in the amount of light scattered from it. This intensity bias to larger particles in the population can effectively obscure the presence of smaller particles without the user being aware of their existence.

NTA is a new technology which offers the simultaneous, multi-parameter analysis of nanoparticles in liquid suspension. NTA offers direct and real time visualization, sizing, and counting of nanoparticles and provides information on the relative refractive index of nanoparticles, their zeta potential, and fluorescence (1–5).

1.1. Principle of Operation

NTA utilizes a laser beam of 635, 535 or 404 nm to illuminate particles in liquid suspension. The light scattered from the particles allows them to be visualized using a conventional microscope fitted with a ×20 microscope objective. As these particles are typically much smaller than the wavelength of light used, there is little information to be gained about the morphology of the particles; they simply act as light scattering points which allows visualization of their Brownian motion. Fitted to the microscope is an emCCD (Electron Multiplying Charged Coupled Device) camera, operating at 30 frames per second. This is used to capture a video of a field of view of particles within the illuminating laser beam of dimensions approximately 100×80 μm. Figure 1 shows a schematic of the laser and camera configuration used for the visualization of particles. The Brownian motion is tracked by the NTA software, which determines and tracks the center of each particle's movement in x and y on a frame-by-frame basis. By following each particle for a number of steps, the average distance moved is then calculated and from this is determined the diffusion coefficient (Dt) for each particle. It is also possible to estimate equivalent particle diameter using the Stokes-Einstein equation and the sample temperature T and solvent viscosity η.

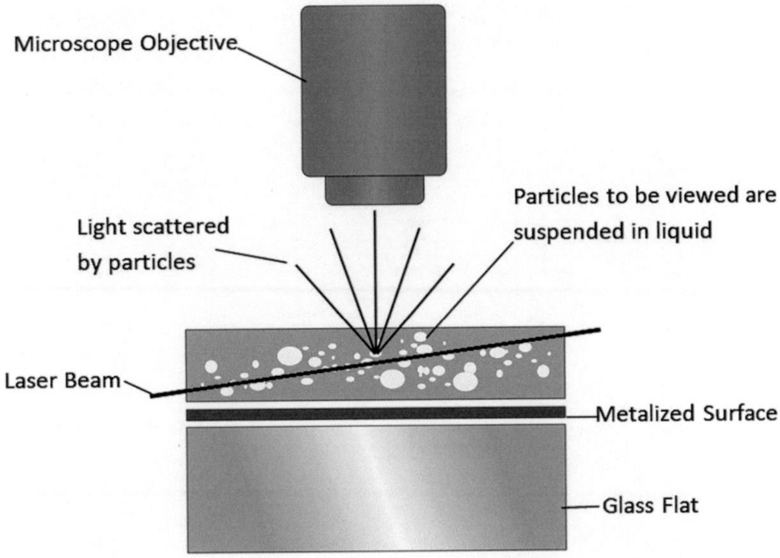

Fig. 1. Laser and camera configuration used in the Nanoparticle Tracking Analysis (NTA) methodology.

$$\frac{\overline{(x, y)}}{4} = Dt = \frac{K_B T}{3\pi\eta d_h}, \quad (1)$$

where K_b is Boltzmann's constant.

The user is then provided with a histogram which shows particle size in nm against concentration in terms of particles per mL. Figure 2 shows the results from an analysis of 100 nm particles. The additional black lines show tracks of particle movement.

1.2. Multiparameter Analysis of Particles with NTA

The NanoSight methodology allows multiparameter analysis of particles between 10 nm and 1 μm in liquid suspension. In addition to real time visualization, sizing, and counting of nanoparticles, NTA provides information on the relative refractive index of nanoparticles, their zeta potential, and fluorescence.

1.2.1. Relative Refractive Index

Although a measure of the amount of light scattered by a particle is not used in the sizing of materials, it is still recorded. Particles of different materials will scatter different amounts of light due to their differing refractive indexes. For example, if a particle population containing 50 % 100 nm gold particles and 50 % 100 nm polystyrene particles is to be analyzed, the population would simply appear as one peak of 100 nm on a particle size against concentration histogram as in Fig. 2. However, the relative refractive indexes of the particles involved will reveal the presence of two different populations of particles within the 100 nm peak.

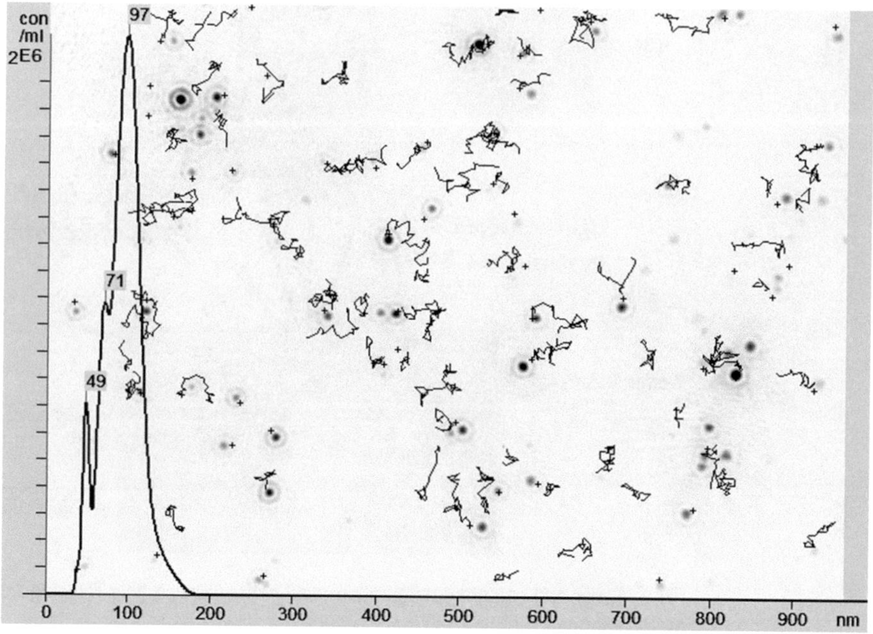

Fig. 2. Analysis of 100 nm particles using NTA. Horizontal axis—particle size in nm, vertical axis—particles concentration expressed as number of particles per mL. *Thin black lines* show nanoparticle tracks.

Figure 3 shows the results obtained from a sample containing two particle populations. The modal particle size is 111 nm, although it is apparent that there may be another population present as the particle size distribution shows a slight shoulder at around 80 nm. Taking into account the relative refractive indexes, the two populations of the nanoparticles can be easily identified (Fig. 4). This view of the data clarifies the fact that there are clearly two particle populations present, which may have been overlooked had the user relied solely upon the particle size against concentration histogram.

1.2.2. Fluorescence

Few of the existing light-scattering technologies allow the analysis of fluorescently labeled, or inherently fluorescent nanoparticles. NTA allows users the ability to directly visualize, size, and count fluorescent nanoparticles. Three different laser wavelengths are available for NTA systems, 404 nm (Blue/Violet), 532 nm (Green), and 638 nm (Red), to excite fluorescence within particles. A filter is placed in the optical path of the instrumentation that removes the excitation wavelength, allowing only the emitted light through, therefore only the fluorescent particles are visible to the system. These filters cut out light below 430 nm (for Blue/Violet excitation), 565 nm (for Green excitation), and 650 nm (for Red excitation). By analyzing the same sample twice, once with the filter in place and once without, one is able to investigate which particle

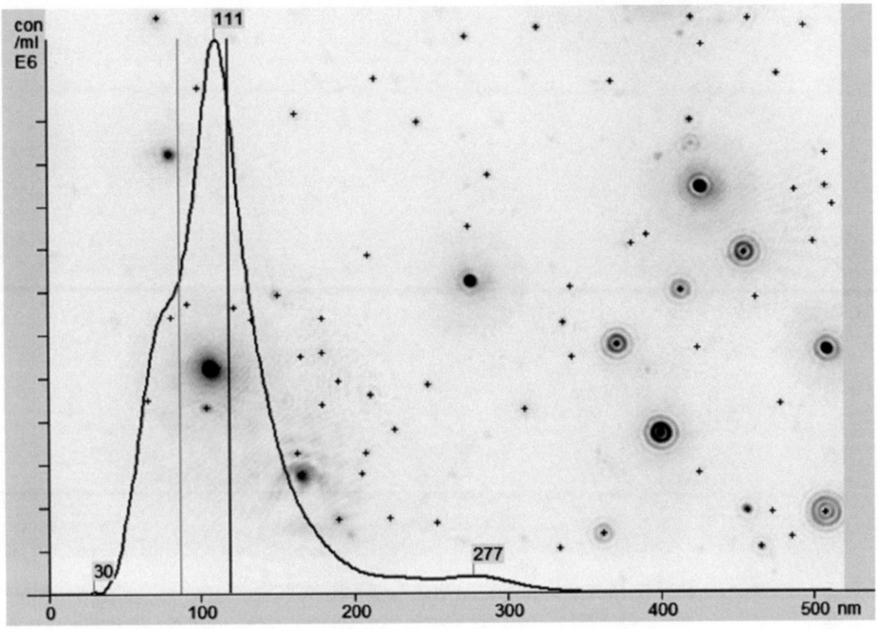

Fig. 3. Analysis of a bimodal system using NTA. Horizontal axis—particle size in nm, vertical axis—particles concentration expressed as number of particles per mL. *Thin black lines* show nanoparticle tracks.

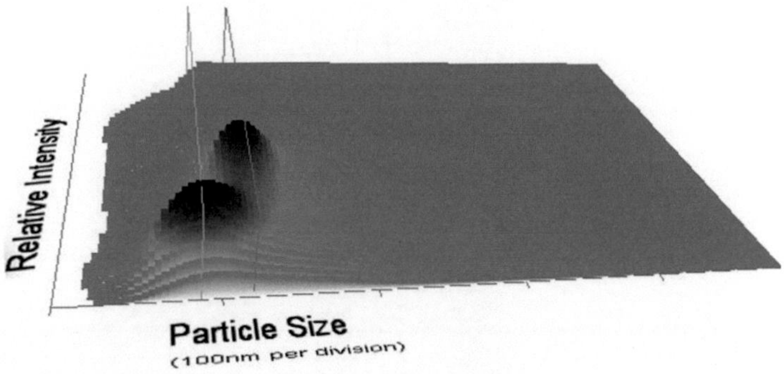

Fig. 4. Particle size against concentration against relative intensity. Particle size is shown along the *x* axis, concentration on the *y*, and relative intensity on the *z*.

populations exhibit fluorescence and which do not. Figures 5 and 6 show the results obtained from two different run-throughs of the same bimodal sample. Figure 5 shows the sample under normal light scatter mode. Here it is clear that two populations are present, one at 50 nm and one at 200 nm, but it is not clear which of these is fluorescently labeled. Figure 6 shows the results obtained from the same sample with a fluorescent filter in place. It is now clear that the population of 50 nm particles is labeled.

Samples with a high background material, such as blood plasma, may not be suitable for the analysis using the traditional

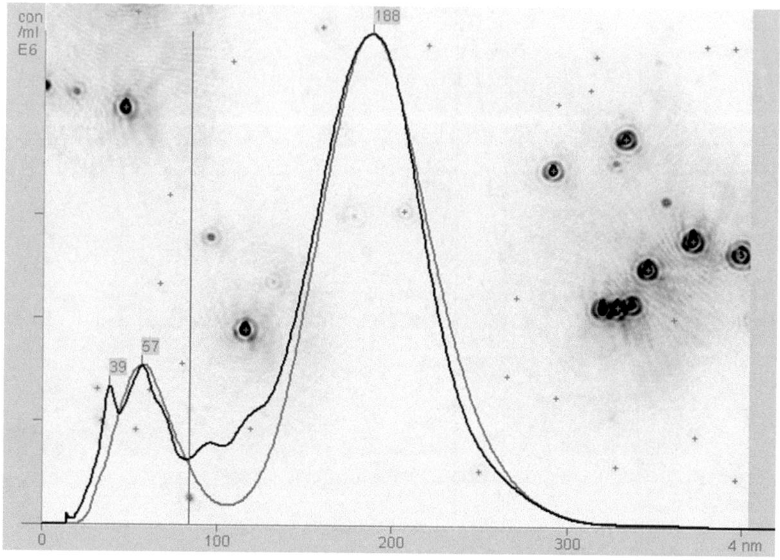

Fig. 5. Sample containing two populations of particles under light scatter conditions. Horizontal axis—particle size in nm, vertical axis—particle concentration expressed as number of particles per mL.

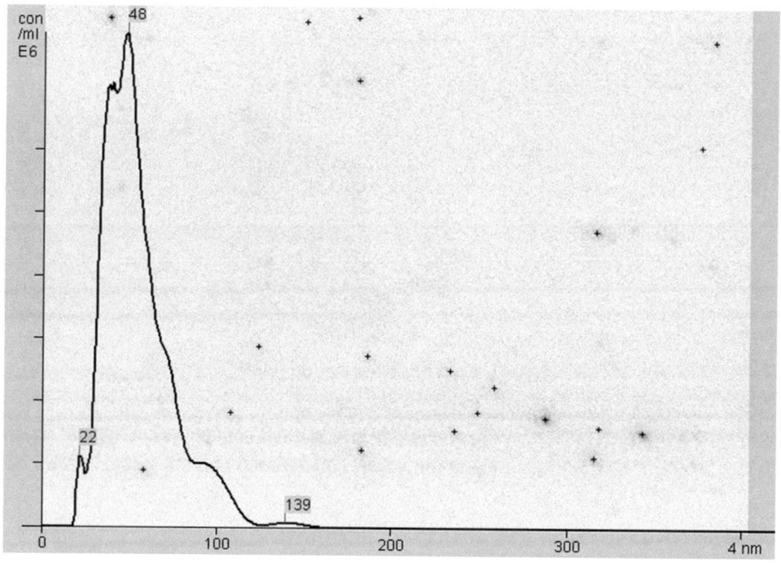

Fig. 6. Sample containing two populations of particles studies in fluorescence mode. Horizontal axis—particle size in nm, vertical axis—particle concentration expressed as number of particles per mL.

light-scattering mode of NTA, as other materials in suspension would also be sized and counted. However, by utilizing the fluorescence capability, and a flurophore that binds specifically to the material of interest, it is possible to analyze these materials.

To minimize the problem of flurophore bleaching, NTA utilizes a peristaltic sample delivery system to slowly move their sample through the field of view. This ensures that particles are not within the field of view for too long, and the fresh particles are introduced to the laser at a constant rate. The motion of particles caused by this pumping does not affect the particle sizing, as the NTA software has a drift correction functionality which removes the drift component from the tracked particle movement before the movement is related to size.

1.2.3. Zeta Potential

Understanding zeta potential is of critical importance in controlling dispersion and determining the stability of a nanoparticle suspension, i.e., to what degree aggregation will occur over time. The zeta potential is the measure of the electric potential at the slip plane between the bound layer of diluent molecules surrounding the particle and the bulk solution. Zeta Nanoparticle Tracking Analysis (z-NTA) system includes a sample chamber with platinum electrodes, which allows a variable electric field to be applied across the nanoparticles in liquid suspension. The z-NTA technique records the apparent drift velocity for each tracked particle, which will be a superposition of these two motions. By observing the total velocity at different depths within the sample chamber, it is possible to separate these components and obtain a measurement of the electro-phoretic velocity (due to the force impinged directly on the particles) and hence the zeta potential of the particles.

2. Materials

2.1. Sample Introduction and Data Capture

1. NanoSight LM10 Series Nanoparticle Analysis Instrument consisting of a laser block containing sample chamber, optical microscope, emCCD camera (see Note 1), and computer with NTA 2.2 software installed (see Note 2).
2. As the NTA methodology requires samples to be fairly dilute (see Note 3), an amount of solvent may be required for dilution. A variety of solvents may be used, including deionized water. Solvents should be noncorrosive, of low enough viscosity that it does not hinder Brownian motion, and also such that it does not dissolve the particles in question or cause them to aggregate.
3. 100 nm polystyrene nanospheres (Thermo Fischer Scientific).
4. A 2 mL disposable syringe.

2.2. Data Analysis

1. A PC with NTA 2.2 Software installed. The minimum requirements for the PC are outlined in Note 2.

2.3. Cleaning and Maintenance

1. Cleaning solution: 70 % ethanol.
2. Canister of Compressed Air Duster.

3. Methods

3.1. Sample Introduction and Data Capture

1. Dilute polystyrene nanosphere sample with pure water to create 0.5 mL of solution at a concentration of approximately 10^8 particles per mL of solution (see Notes 4–8).
2. Using a 2 mL syringe, add the diluted sample to the sample chamber taking care to ensure that air bubbles are not introduced to the system. The syringe should be left in place in order to maintain the seal.
3. Place the sample chamber on the stage of the microscope. The point at which the laser beam reaches the glass-solvent interface is found manually using the *x*, *y*, and *z* controls of the microscope and visualized using the oculars. This region is identified by a characteristic diffraction pattern of concentric ovals.
4. Follow the beam from this point until the diffraction pattern is no longer within the field of view. This is the optimum region for analysis. Slide the ocular/camera slider to the camera position and the NanoSight NTA2.2 software can then be used to visualize the particles on screen.
5. Once the optimal region for analysis has been found, record a video file of the particles moving under Brownian motion. The camera level must be set appropriately. This setting controls both the shutter speed of the camera and the artificial gain. Adjust the setting so that no information is lost (i.e., the smaller or dimmer particles) by having it too low and that the image is not oversaturated, which again results in a loss of information.
6. The length of time that the system will record for is then established. The NTA software provides a lookup table that uses the response to two questions in order to establish the length of video. By telling the system if the sample is of "low," "medium," or "optimal" concentration and whether the sample is "low," "medium," or "highly" polydisperse, the software can assess how long to record for. For a sample such as this, of optimal concentration and low polydispersity, a video of 30 s is recommended (see Note 9).
7. Start recording a video file of the particles moving under Brownian motion. The system will then require the user to input the recorded temperature, which is read from the thermocouple included in the NanoSight system (see Note 10).

3.2. Data Analysis

1. The captured video file is loaded automatically into the software for analysis. Parameters for analysis must be selected. The first of these is the detection threshold (see Note 11). This parameter can be set by the user; however, in order to analyze

a simple monodispersed system such as 100 nm polystyrene nanospheres, the automatic setting should ideally be used (to select this option click the box to the left of the slider bar).

2. The amount of blur on the image is then determined. It is again recommended to use the automatic function of the software by clicking the box to the left of the slider bar (see Note 12).
3. The "Minimum Track Length" is then set (see Note 13). It is recommended that the user activates the automatic Minimum Track Length. In this mode the program automatically selects and updates the optimum Minimum Track Length based on its calculated size at that instant.
4. The Minimum expected particle size setting should always be slightly below that which is expected from the results; therefore the minimum expected particle size setting should be adjusted to 80 nm (see Note 14).
5. The software will now analyze the recorded video using the parameters outlined after capture. This process takes the same amount of time as the video took to record. When this step is completed, the data is shown in the form of a histogram displaying particle size in nm on the *x* axis and concentration in terms of particles per mL on the *y* axis, as in Figs. 2, 3, 5, and 6. The concentration values shown are for the diluted sample rather than the original concentration. The data can also be exported in the form of a .csv file. The mean average particle size and modal average can also be seen in the lower left hand corner of the screen.

3.3. Cleaning and Maintenance

1. The sample is extracted using the syringe and the sample chamber is dissembled for cleaning. The glass flat and top plate cleaned using 70 % ethanol.
2. Carryover of fluid in the luer fittings may result in a slight error in concentration reading, as this fluid will combine with any new sample introduced, diluting it slightly. In order to circumvent this problem, compressed air is passed through the fittings and flow cell to dry them (see Note 15).
3. Check that the *o*-ring is placed within the specified recess in the top plate, and that it has not swelled or degraded due to the use of an inappropriate solvent before reassembling the sample chamber (see Note 16).

4. Notes

1. Other instrumentation utilizing the NTA methodology is available, namely the NS500, the NS200, and the LM20 models. These all operate on the same principles as outlined in the

methodology; however are not built around a microscope, and so visualization is only available through the use of the camera. However, these instruments offer other benefits, such as automated sample delivery and dilution as well as automated stage control.

2. The minimum requirements for the PC used for capture and analysis are as follows: Processor: An Intel Core 2 Duo or i5 processor in excess of 1.8 GHz, 3 GHz is recommended; RAM: 2 GB DDR2 RAM; Hard drive: A 7,200 rpm S/ATA hard drive with the minimum recommended capacity of 80 GB. The size is dependent on the expected usage of the system. Note that one run of a typical (60 s) analysis requires approximately 0.6 GB of hard disk space; Screen: recommended resolution 1,680 × 1,050 (minimum 1,280 × 1,024); Operating System: 32-bit Windows XP and 32-bit Windows 7 are supported. The software cannot be used to capture data on a 64-bit operating system; 3 Free USB 2.0 ports; IEEE 1394 "Firewire" Port. The PC should have a "Firewire A" port. This can either be 4pin or 6pin; 256 Mb graphics card is recommended but not crucial. Conflicts due to antivirus software and networks cannot be supported and programs that run in the background on the PC should be limited.
3. NTA works best at concentrations between 10^7 and 10^{10} particles per mL. This generally requires the sample to be "water clear" before analysis. Below this value, there are too few particles visible within the field of view and therefore the data obtained would not be statistically viable. Above this value, there are too many particles within the field of view and so the software is unable to track particles for a satisfactory amount of time before they cross behind another particle, therefore it is no longer possible to determine their trajectory. An optimum dilution would present around 100 particles in the field of view when inserted into the sample chamber and visualized.
4. It is important to ensure that solvents used are particle-free. In order to do this, take a small amount of solvent (≈0.5 mL) in a disposable syringe and introduce this, via the luer fitting, to the LM10 sample chamber. Load the sample chamber onto the viewing platform and visualize the particles using the technique described in Subheading 3.1, step 1. If particles can be seen, then these can be assumed to be from the solvent, and a filtering technique must be used or a new solvent sourced.
5. It is often advantageous to allow the sample to settle over a suitable time period, during which larger particles are allowed to settle under gravity. Allowing the sample to stand undisturbed for a period of a few hours or, for instance, overnight can often result in larger, potentially interfering particles settling to the bottom of the sample chamber, allowing the user

to extract an aliquot for analysis from the upper layer of the sample container. While any particle which settles out rapidly (e.g., within an hour) represents a particle size which is probably too large for the NanoSight technique to analyze with accuracy and should be allowed to clear from the sample, it should be borne in mind that, over extended periods of settling, increasingly smaller particles will settle and the region from which a sample aliquot is extracted from the sample container will reflect such changes in particle population.

6. NanoSight is ideally suited for the analysis of particles between 10 nm and 1 μm. When particles exceed this size, the rate at which they move under Brownian motion is negligible, and the vast amount of light that they scatter relative to the smaller particles makes it difficult for the software to use Gaussian profiling in order to find the center of the particle. As a result of this, it may be necessary to remove larger particles from a system before analysis. The most common way to do this would be through the use of filters. However, it is worth noting that nanoscale filter sizes refer to only the mean pore size, so initially they may allow particles larger than the stated size through, so as the pores clog up, particles smaller than the mean pore size will be extracted from the system. Incidentally, the checking of blocking states and rates of filters is another application for the NTA methodology.

7. Should the sample contain detergents, it is recommended that it is introduced slowly to prevent the formation of bubbles or foam in the sample chamber, which may prevent visualization of particles.

8. Choice of diluents is important and must not change the nature of the nanoparticle suspension. Figure 7 shows the effect of purity of water on a sample of calibration quality gold nanoparticles in which it can be seen that only ultrapure water does not cause some aggregation.

9. If the video file is too short in length, then it is unlikely that enough data will be gathered to present a statistically representative data set. The lower the polydispersity the shorter the time and the lower the concentration the longer the time. For example, a sample of low polydispersity and optimal concentration need only be assessed for 10 s, whereas one of high polydispersity and low concentration must be analyzed for 90 s in order to find a statistically relevant set of results.

10. Temperature is one of the few parameters that the NTA methodology requires users to input. However, it is of significance as for every 1 °C error in temperature input, the measured particle size will deviate by 2.4 %. Temperature may also be automatically recorded when using the NS500, LM20, or NS200 versions of the NTA instrumentation.

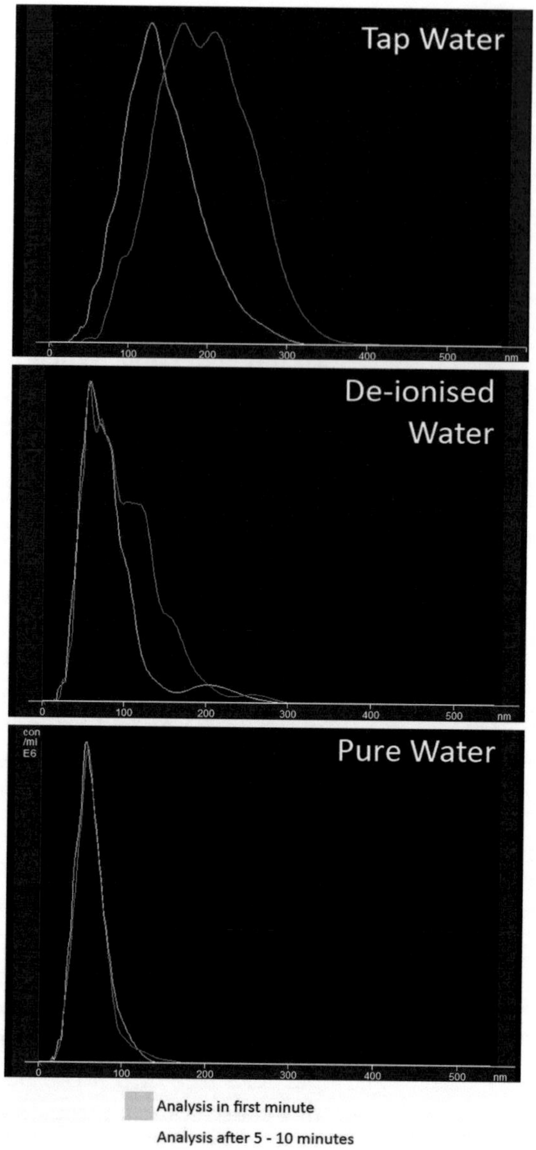

Fig. 7. Aggregation of gold nanoparticles in different solvents. Horizontal axis—particle size in nm, vertical axis—particle concentration expressed as number of particles per mL. *Gray line*—recording within the first minute, *white line*—analysis after 5–10 min.

11. The detection threshold determines the minimum gray scale value of any particle image necessary for it to qualify as a particle to be tracked for analysis. If this threshold is set too high, this will result in viable particles not being tracked by the software. If set too low, the result will be that noise from the camera may affect the result given.
12. Blur acts as a smoothing function to eliminate noise (such as diffraction rings surrounding larger particles) from around or

within a particle image. This value is set by either adjusting the settings until all "false centers" (more than one tracking point on a particle) are removed, or by using the automatic blur function which determines the amount of blur required for each particle and alters parameters accordingly. A 3×3 blur integrates intensities over a 3×3 matrix. Higher blur values (5×5, 7×7 or 9×9) increase smoothing, but become computationally expensive and cause the rate of analysis to slow. Furthermore, excessive blur may cause smaller particles to be smoothed to the point they fall below the detection limit.

13. The "Minimum Track Length" setting refers to the number of frames a particle must have been tracked for before the movement can be considered to be representative of the behavior of the particle. Because NTA 2.2 estimates the size of any given particle from measuring the average distance the particle moves, clearly the longer the time the particle is tracked for (the higher the number of steps taken), the better will be the statistical accuracy of the average and therefore of the particle size estimation. However, the time any given particle remains in the small scattering volume is limited, particularly so for small particles whose Brownian motion is rapid.
14. Even the best quality "standards" may have a slight breadth in size distribution, and it is best to assume slightly under the size expected.
15. Users wishing to clean the sample chamber by flushing through with an appropriate solvent should beware that the current design of the flow cell is such that flushing will not always displace the previous sample without excessive throughput.
16. Occasionally, after reassembling the sample chamber, users may find that during the next analysis the sample "drifts" somewhat (i.e., all the particles move in the same direction owing to an outside force). This is not usually a problem, as the NTA software can compensate for this and remove any errors it may cause from the reading. However, should the rate of movement be too great for the software to negate, the cause of this problem is often found to be due to placement of the o-ring within the sample chamber.

References

1. Carr R, Hole P, Malloy A, Smith J, Weld A, Warren J (2008) The real-time, simultaneous analysis of nanoparticle size, zeta potential, count, asymmetry and fluorescence in liquids. NSTI Nanotech Technical Proc 1:866–870
2. Warren J, Carr R (2007) Application of nanoparticle tracking analysis to BioNano systems. In: 2nd international congress of nanobiotechnology & nanomedicine (NanoBio2007), San Francisco, USA, 19 June 2007
3. Filipe V, Hawe A, Jiskoot W (2010) Critical evaluation of nanoparticle tracking analysis (NTA) by NanoSight for the measurement of nanoparticles and protein aggregates. Pharm Res 27(5):796–810. doi:10.1007/s11095-010-0073-2

4. Saveyn H, De Baets B, Thas O, Hole P, Smith J, Van der Meeren P (2010) Accurate particle size distribution determination by nanoparticle tracking analysis based on 2-D Brownian dynamics simulation. J Colloid Interface Sci 352:593–600

5. NanoSight Application Note. Nanoparticle tracking analysis (NTA) and dynamic light scattering (DLS)—a comparison. http://www.nanosight.com/technology/comparative-technologies. Accessed 1 Aug 2011

Chapter 42

Relevance of Two-Dimensional Brownian Motion Dynamics in Applying Nanoparticle Tracking Analysis

Paul Van der Meeren, Marios Kasinos, and Hans Saveyn

Abstract

Nanoparticle Tracking Analysis (NTA) is an image analysis-based technique that deduces the particle size from the changing position of scattering diffusing particles in a sequence of images. It is shown that a basic understanding of the underlying physical principles largely helps to prevent artifacts. In fact, an inappropriate selection of both software parameters, such as maximum particle jump or minimum track length, as well as sample preparation aspects (such as degree of dilution) may give rise to inaccurate or even erroneous results.

Key words: Nanoparticle tracking analysis, Photon correlation spectroscopy, Nanoparticles observation, Image processing, Nanoparticles diffusion, Nanoparticles size analysis, Nanoparticles size distribution, Two-dimensional nanoparticles diffusion, Nanoparticles aggregation

1. Introduction

The interest in the functional properties of nanoparticles has largely increased during the last decades. Hereby, particle size has been shown to be a decisive factor not only for the desired functionality, but also for adverse effects, such as health or ecotoxicological effects (1–3). Hence, accurate determination of the particle size distribution is a major issue in nanoparticle technology.

Lately, nanoparticle tracking analysis (NTA) has been developed as a new technique, which can image the light scattered from diffusing nanoparticles in liquid suspension (4–8). Particles in liquid suspension move under Brownian motion, the speed of which can be used to calculate the particle size through the application of the Stokes–Einstein equation. Both NTA and photon correlation spectroscopy (PCS) measure Brownian motion and relate this

Mikhail Soloviev (ed.), *Nanoparticles in Biology and Medicine: Methods and Protocols*, Methods in Molecular Biology, vol. 906, DOI 10.1007/978-1-61779-953-2_42, © Springer Science+Business Media, LLC 2012

movement to an equivalent hydrodynamic diameter. NTA measures this movement through image analysis tracking the movement of the particles on a particle-by-particle basis. PCS observes the time-dependent fluctuations in scattering intensity caused by constructive and destructive interference resulting from the relative Brownian movements of the particles within a sample. NTA provides a unique ability to directly visualize and size nanoparticles in a liquid suspension. The visualization of the particles allows each particle to be simultaneously and individually sized, overcoming inherent problems associated with techniques such as PCS, which is also referred to as dynamic light scattering (DLS). The intensity of light scattered from a nanoparticle as a function of particle radius follows a power law and increases with the sixth power for a Rayleigh particle. Hence, the average particle size produced by PCS (which measures total light scattered from an ensemble of particles) is heavily weighted towards small numbers of large, perhaps contaminant, particles. Electron microscopy, on the other hand, requires time-consuming sample preparation and imaging and only views a small area, thus risking a nonrepresentative analysis of the sample as a whole. NTA does not use the intensity of light scattered as a measure, and hence, there is no requirement for knowledge of the refractive index of the solvent and of the particles in the calculation.

In this contribution, the physical background of the NTA technique is presented, and its relevance for accurate particle size analysis is stressed. Hereby, the theoretical concepts are illustrated by NTA results obtained using a nanosphere size standard sample.

2. Mathematics of Two-Dimensional Diffusion

2.1. Two-Dimensional Diffusion Distance Distribution

Diffusion kinetics are generally simulated using the random walk approach. Hereby, diffusion is approximated by a succession of a number of subsequent steps of identical distance but random direction. For particles in Brownian motion, the apparent number of steps n is proportional to the diffusion time. Considering diffusion in the microscopic field of view, only two dimensions are involved. The probability of arriving at a distance x from the origin after some time t in a two-dimensional field (as is the case in microscopic observation) is largely determined by the particles' diffusion coefficient D (9, 10).

$$P(x,t)\mathrm{d}x = \frac{2x\mathrm{d}x}{4Dt}\exp\left(-\frac{x^2}{4Dt}\right) \qquad (1)$$

For spherical particles diffusing at temperature T (in Kelvin) in a dilute suspension of viscosity μ (in Pa.s), the Stokes–Einstein

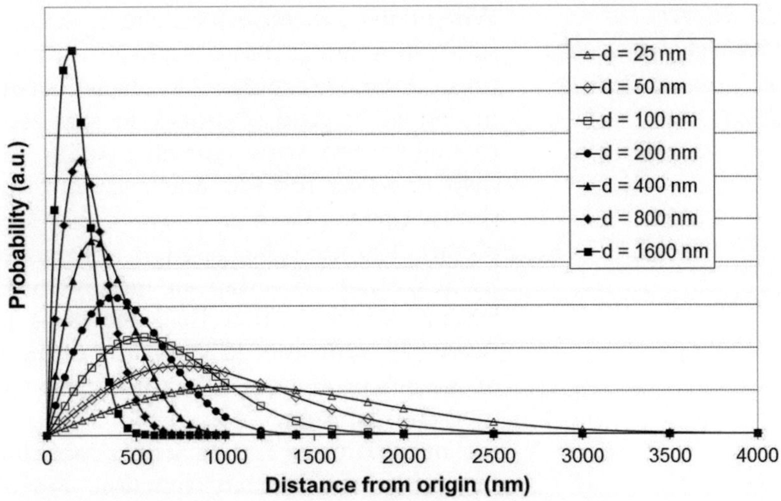

Fig. 1. Particle diffusion in water. Probability distribution function for the distance covered by diffusing colloidal particles in water at 20°C (1 mPa.s viscosity) after 33 ms as a function of their diameter.

equation relates the diffusion coefficient to the particle diameter d through the Boltzmann constant k:

$$D = \frac{kT}{3\pi\mu d} \tag{2}$$

Figure 1 reveals that the distance over which particles are displaced by diffusion strongly depends on their size. Considering a 166 nm pixel size, 50 nm particles can move over about 10 pixels in 33 ms, whereas 800 nm particles can hardly be displaced over 3 pixels under the same conditions.

Equation (1) corresponds to a Rayleigh distribution of the form: $R(x,\sigma) = x/\sigma^2 \exp\{-x^2/(2\sigma^2)\}$, where the σ^2-parameter of the Rayleigh distribution is equal to $2Dt$, and corresponds to the modal diameter squared. The mean and variance of a Rayleigh distribution correspond to:

$$\begin{aligned} \text{mean} &= \sigma\sqrt{\frac{\pi}{2}} \approx 1.253\ \sigma = \sqrt{\pi Dt} \\ \text{variance} &= \frac{4-\pi}{2}\sigma^2 \approx 0.429\ \sigma^2 = (4-\pi)Dt \end{aligned} \tag{3}$$

As both the mean and the standard deviation are proportional to the square root of both time and diffusion coefficient, it follows that neither the time period between subsequent frames (i.e., the step time), nor the diffusion coefficient has any influence on the sharpness of the step length distribution.

2.2. Relevance of Diffusion Mathematics on Maximum Particle Jump

When trying to reconstruct the track of individual particles from a set of subsequent photographs, it is important to define the maximum distance (expressed in pixels) between the particle positions in two subsequent photographs that are still considered as being caused by the same particle. If this so-called maximum jump distance is set too low, tracks cannot be found because particles diffuse outside the area in which they are searched for in the next picture. On the other hand, if this parameter is set too large, the particle observed within the next picture may actually be a neighboring particle, rather than the same particle which is slightly moved by diffusion. Whereas (3) enables to estimate the average displacement, the question arises up to what distance particles may move by diffusion.

Integration of the Rayleigh probability density function yields the cumulative density function $F(x)$, which corresponds to: $F(x) = 1 - \exp\{-x^2/(2\sigma^2)\}$. Based on this equation, it can be shown that the x-value that yields a cumulative frequency $F(x)$ may be calculated:

$$x = \sigma\sqrt{-2\ln(1 - F(x))} \quad (4)$$

From (4), it follows that 99% of all particles will move over a distance between 0 and 3σ, whereas only 1% of the particles will move further than this distance. Some numerical data are included in Table 1. The fifth column shows that small particles move over a

Table 1
Diffusion characteristics of spherical particles[a]

d (nm)	D (10^{-12} m^2/s)	σ (nm)	Mean/diameter	Distance$_{99\%}$ (pixels)	φ_v at $L = 10\sigma$	Dilution from 1%
25	17.18	1,065	53.4	19.2	1.3e-8	772,920
50	8.59	753	18.9	13.6	2.9e-7	34,159
100	4.30	532	6.7	9.6	6.6e-6	1510
200	2.15	377	2.4	6.8	1.5e-4	68
400	1.07	266	0.8	4.8	3.4e-3	3
800	0.54	188	0.3	3.4	7.7e-2	–
1,600	0.27	133	0.1	2.4	>1	–

[a]Diffusion characteristics of spherical particles of diameter d (and hence diffusion coefficient D) in a dilute aqueous suspension at 20°C (i.e., 1 mPa.s viscosity). From the sigma factor σ of the Rayleigh distribution, the mean diffusion displacement (relative to the particle diameter), the maximum distance over which 99% of all particles diffuse (relative to a pixel size of 166 nm), the volume fraction at an interparticle distance (L) of 10σ, and the corresponding dilution factor starting from a 1% (v/v) nanosuspension are calculated

large distance and hence require a large value for the maximum jump distance, whereas micron-sized particles cannot be probed based on diffusion as they hardly move.

2.3. Relevance of Diffusion Mathematics on Sample Dilution

In order to avoid that erroneous tracks are constructed by connecting nearby particles in subsequent frames, the (average) interparticle distance should be (much) larger than the diffusion length (which may be approximated by 3σ). According to the Wigner–Seitz approach, the average separation L between noninteracting particles is twice their Wigner–Seitz radius r_s.

$$L = 2r_s = 2\sqrt[3]{\frac{3V}{4\pi N}} \approx 1.24/\sqrt[3]{N/V} \tag{5}$$

The latter corresponds to the radius of a sphere having a volume that corresponds to the suspension's volume V divided by the number of incorporated particles N. Hereby, N/V represents the particle density n, from which the volume fraction φ_v may be obtained by multiplication by the particle volume: $\varphi_v = (N/V)\pi d^3/6$.

From the density function expressing the probability to find a nearest neighbor (noninteracting) particle at a distance r (11, 12), it follows that 2% of the particles are closer than 30% of the average distance. Hence, it follows that particles should be at least 10σ apart to limit the chance that connected positions in two subsequent frames are due to different nearby particles (rather than the same displaced particle) to 2%, whereas particles should be 12σ apart to limit this chance to 1%. The last column but one in Table 1 mentions the calculated volume fraction at an average interparticle distance L which corresponds to 10σ, whereas the last column mentions the required degree of dilution to reach this concentration starting from a 1% (v/v) nanosuspension. From Table 1, it is obvious that moderate dilution is sufficient to guarantee accurate tracking based on subsequent photographs for particles of several hundreds of nm in diameter, whereas nanoparticle (e.g., 25 nm diameter) suspensions require a high degree of dilution to prevent erroneous tracking.

2.4. Relevance of Diffusion Mathematics on Particle Size Analysis

In practice, particles are not just observed during one single step, but during a sequence of subsequent steps, which is indicated as a track. Hereby, the average step length SL follows from the track length (TL), divided by the number of steps (N) by which the track has been formed:

$$\mathrm{SL} = \mathrm{TL}/N \tag{6}$$

From a statistical point of view, the accumulated distance after N steps (i.e., the track length) may be approximated (provided that N is sufficiently large) by a normal distribution for which the average value is equal to N times the average value of the Rayleigh distribution, whereas the variance σ^2 is N times the σ^2 of the Rayleigh distribution for 1 step:

$$\begin{aligned} \text{mean(TL)} &= N\sqrt{\pi Dt}, & \text{mean(SL)} &= \sqrt{\pi Dt} \\ \text{variance(TL)} &= N(4-\pi)Dt, & \text{variance(SL)} &= (4-\pi)Dt/N \end{aligned} \tag{7}$$

From (7), it follows that the particle diameter may be derived from the average displacement $<x>$ (i.e., the step length, SL) between two subsequent frames (e.g., taken every 33 ms):

$$\begin{aligned} \langle x \rangle &= \text{SL} = \sqrt{\pi Dt} \\ D &= \frac{\text{SL}^2}{\pi t} = \frac{kT}{3\pi \mu d} \\ d &= \frac{kTt}{3\mu \text{SL}^2} \end{aligned} \tag{8}$$

As (7) indicates that a step length distribution is obtained even if all particles are exactly the same in size, it follows that direct transformation of the step length (distribution) to a corresponding diameter will result in an apparent particle size distribution with largely overestimated particle size distribution width. As (7) shows that the variance of the step length distribution is inversely proportional to the track length (N), it follows that longer tracks will lead to narrower particle size distributions. Saveyn et al. demonstrated that two-dimensional Brownian dynamics simulation may at least partly overcome this artificial particle size distribution broadening (13).

3. Experimental Verification

In order to evaluate the accuracy of NTA, the particle size distribution of a nanosphere standard sample of 125 nm nominal diameter was measured by NTA, as well as DLS. This suspension, containing 1% (w/w) polystyrene microspheres in water, originated from Duke Scientific Corporation (Cat. no. 3125A; lot no. 22865). According to the manufacturer, the mean diameter of this sample is 126 ± 3 nm, whereas the standard deviation of the diameter distribution is 4.4 nm. Samples were prepared by dilution in 0.02% NaN_3 solution to prevent microbial growth. NTA was performed using an LM10 System (Nanosight, UK), characterized by a pixel size of 166 nm. Video images of the particles' movement under Brownian motion were recorded every 33 ms and analyzed by the NTA, version 1.3 (B196), image analysis software (Nanosight). Each video clip was captured during 30 s.

For DLS results (obtained on a 200 times diluted sample), cumulant analysis yielded a *z*-average diameter of 125.6 ± 4.7 and

124.4 ± 0.7 nm at a scattering angle of 90° and 150°, respectively, which was very similar to the mean diameter specified by the manufacturer. For NTA, a suitable dilution factor was chosen to obtain accurate results within a reasonable time span. The nanosphere standard sample was first diluted twice by a factor of 10, and subsequently six times by a factor of 2 with 0.02% NaN_3 solution.

3.1. Effect of Maximum Particle Jump in NTA Particle Size Analysis

The choice of the parameters for the NTA software during processing video images seemed to play an important role in the achievement of accurate particle size distributions. Figure 2 reveals that the accuracy of the results obtained by NTA may largely depend on the selected value for the maximum particle jump parameter. Considering the sample that was diluted 12,800 times, the number-weighted mean diameter strongly depended on the maximum particle jump at values below 10, but became rather unaffected by this parameter at larger values. As discussed in Subheading 2.1, Brownian motion is a statistical phenomenon, whereby the distance traveled by a 125 nm particle within the time period between two subsequent photographs (i.e., 33 ms) ranges between 0 and about 1.4 μm according to (4). As one pixel corresponds to 166 nm, it follows that a particle can be displaced over up to 9 pixels in subsequent photographs. Hence, selecting a maximum particle jump value below 9 caused that only tracks with shorter distances between subsequent photographs were retained, which gave rise to an overestimation of the average diameter. The latter is a logical consequence from the fact that larger particles diffuse over smaller distances. Selecting a maximum particle jump value of at least 10 prevented this pitfall.

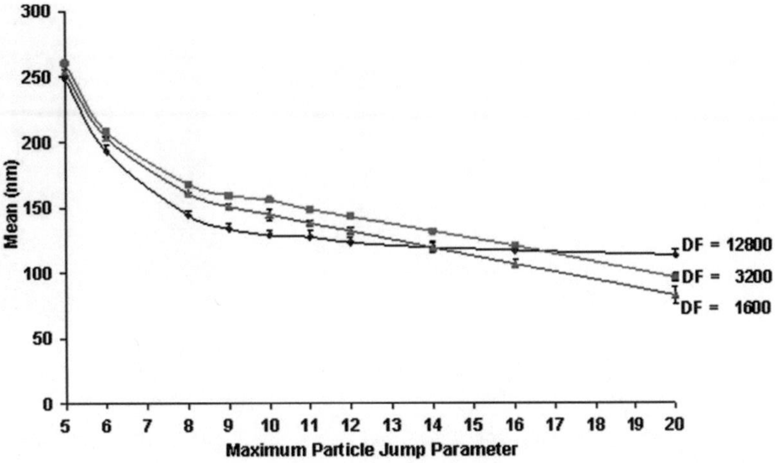

Fig. 2. Estimated mean diameter of a 125 nm latex sample. Influence of the value of the maximum particle jump parameter on the number-weighted average diameter for 125 nm nanosphere latex particles upon 1,600 (*triangle*), 3,200 (*square*), and 12,800 (*circle*) fold dilution; the minimum track length was fixed at 10.

3.2. Effect of Dilution in NTA Particle Size Analysis

The effect of insufficient dilution is clearly observed in Fig. 2 at large values of the maximum particle jump for the data related to samples with dilution factor 1,600 and 3,200. As a maximum particle jump of 20 pixels corresponds to 3.3 μm, and the average distance between particles is only 6.8 μm at 1,600-fold dilution according to (5), it follows that two particle positions in subsequent frames within this maximum distance may be due to two different particles, rather than due to the same particle that moved over this distance. Hence, apparent tracks of large step size are obtained, which are translated by the software to small particles, which in turn leads to an underestimation of the average particle diameter. As the interparticle distance increases at increasing dilution, the chance to have this disturbing effect decreases. Whereas the overestimation of the mean diameter at too low maximum particle jump values is not affected by dilution, its underestimation at high maximum particle jump values becomes less pronounced at higher dilution. From Fig. 2, it was concluded that a maximum particle jump of at least 10 and a dilution factor of 12,800 gave rise to an accurate estimation of the mean particle size which was not significantly affected by small variations in parameter values.

3.3. Particle Size Distribution Width in NTA

Figure 3b shows that the distribution width is largely overestimated. According to the manufacturer, the size distribution has a standard deviation of 4.4 nm, from which it can be calculated that the *D*10 and *D*90 should correspond to 119 and 131 nm, respectively.

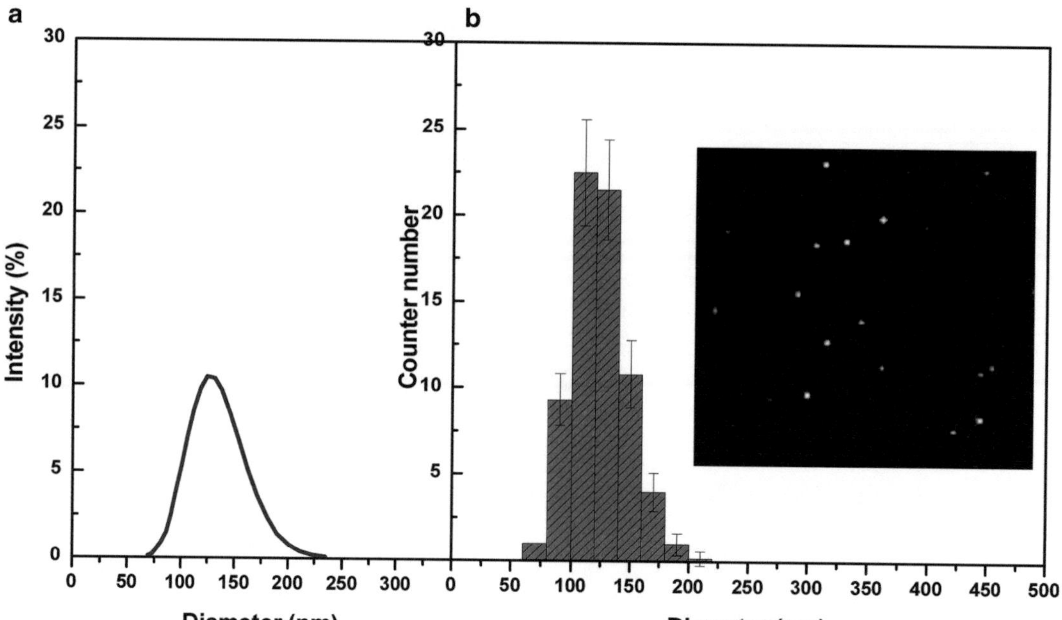

Fig. 3. Particle size distribution of a 125 nm latex sample. Intensity-weighted and number-weighted particle size distribution, and picture captured from a video of a 12,800 times diluted suspension of 125 nm latex spheres, as measured by DLS (**a**) and NTA (**b**); both the maximum particle jump and the minimum track length were assigned a value of 10.

Table 2
NTA results obtained using a 6,400 times diluted 125 nm latex at 22°C and maximum particle jump 10 as a function of the minimum track length (*N*); the standard deviation was obtained based on six measurements

Minimum track length	*D*10 (nm)	*D*50 (nm)	*D*90 (nm)	Span (–)	Mode (nm)	Mean (nm)
5	90 ± 3	121 ± 3	165 ± 7	0.62	119 ± 5	134 ± 6
10	93 ± 2	121 ± 3	159 ± 5	0.55	119 ± 5	129 ± 4
20	97 ± 2	121 ± 3	152 ± 5	0.45	120 ± 5	126 ± 3
40	101 ± 2	120 ± 5	145 ± 8	0.37	121 ± 6	124 ± 4

Hereby, *D*10 and *D*90 represent the particle diameters for which 10 and 90% of the particles are larger in diameter, respectively. As discussed in Subheading 2.4, this effect is expected to be largely affected by the track length, as sharper step length and hence particle diameter distributions are obtained from longer tracks. Table 2 clearly shows that both the median diameter (*D*50) and the modal diameter are hardly affected by the minimum track length, whereas the distribution span ($\{D90-D10\}/D50$) is gradually decreased as the shorter tracks are progressively eliminated from the data set. Whereas the overestimation of the particle size distribution width is most pronounced for short tracks (i.e., 5 frames), even considering tracks of at least 40 steps, the obtained particle size distribution is still broader than specified by the manufacturer. Keeping account of the inverse square root dependency of the standard deviation of the step length distribution on the number of steps in the tracks, it is obvious that especially tracks consisting of less than 10 steps are largely broadened. Hence, a minimum track length of 10 steps is thought to be appropriate to minimize artificial distribution broadening and to ensure a sufficiently large number of tracks.

Upon further increasing this parameter, the width is only slightly reduced, whereas the number of tracks that fulfills this criterion becomes largely reduced, which has a bad effect on the reliability of the results. Comparing Fig. 3a, b shows that a rather similar degree of overestimation of the distribution width is obtained by both NTA and PCS.

4. Conclusions

The above theoretical discussion and the experimental data revealed that NTA can provide accurate particle size data, provided that the sample dilution and image processing parameters are adjusted to prevent artifacts. Under these conditions, NTA has the clear benefit that visual observation is possible, from which aggregation can be easily observed.

Acknowledgments

Patrick Hole (Nanosight Ltd, UK) is kindly acknowledged for the stimulating discussions related to the NTA theoretical background.

References

1. Buzea C, Pacheco I, Robbie K (2007) Nanomaterials and nanoparticles: sources and toxicity. Biointerphases 2(4):MR17–MR71
2. Gill S, Lobenberg R, Ku T et al (2007) Nanoparticles: characteristics, mechanisms of action, and toxicity in pulmonary drug delivery—a review. J Biomed Nanotechnol 3:107–119
3. Grassian VH (2008) When size really matters: size-dependent properties and surface chemistry of metal and metal oxide nanoparticles in gas and liquid phase environments. J Phys Chem C 112:18303–18313
4. Finder C, Wohlgemuth M, Mayer C (2004) Analysis of particle size distribution by particle tracking. Part Part Syst Charact 21:372–378
5. Xu CS, Cang H, Montiel D, Yang H (2007) Rapid and quantitative sizing of nanoparticles using three-dimensional single-particle tracking. J Phys Chem C 111:32–35
6. Malloy A, Carr B (2006) Nanoparticle tracking analysis—the halo (TM) system. Part Part Syst Charact 23:197–204
7. Ryu JH, Bang SY, Yoon JW et al (2007) Pulsed laser induced synthesis of scheelite-type colloidal nanoparticles in liquid and the size distribution by nanoparticle tracking analysis. Appl Surf Sci 253:8408–8414
8. Kusaka Y, Adachi Y (2007) Determination of hydrodynamic diameter of small flocs by means of direct movie analysis of Brownian motion. Colloids Surf A Physicochem Eng Asp 306:166–170
9. Hiemenz PC (1986) Principles of colloid and surface chemistry, 2nd edn. Marcel Dekker, New York
10. Saxton MJ (1997) Single-particle tracking: the distribution of diffusion coefficients. Biophys J 72:1744–1753
11. Hertz P (1909) Uber den gegenseitigen durchschnittlichen Abstand von Punkten, die mit bekannter mittlerer Dichte im Raume angeordnet sind. Math Ann 67:387–398
12. Torquato S, Lu B, Rubinstein J (1990) Nearest-neighbour distribution functions in many-body systems. Phys Rev A 41:2059–2075
13. Saveyn H, De Baets B, Thas O et al (2010) Accurate particle size distribution determination by nanoparticle tracking analysis based on 2-D Brownian dynamics simulation. J Colloid Interface Sci 352:593–600

Chapter 43

Nanoparticle Bridges for Studying Electrical Properties of Organic Molecules

Klaus Leifer, Ken Welch, Syed Hassan Mujtaba Jafri, and Tobias Blom

Abstract

The use of single molecules as building blocks for practical electronic devices and sensors has high potential for novel applications due to the versatility of electronic properties of the molecules. Nano-sized molecules offer great potential for further miniaturization of electronic devices. We describe a method where such molecules are used to bridge a nanoparticles-nanoelectrode interface and thus determine the electrical properties of such a junction. We describe in detail the fabrication of the platform, its functionalization with molecules, and the basics of the electrical measurements. This platform has been shown to guide electrical current through a few molecules. The versatility of such nanoparticle-molecule-nanoelectrode heterojunctions makes this platform suitable for both basic molecular electronics measurements and also for molecular sensing devices in biological and medical applications.

Key words: Nanoparticles, Nanoparticle bridge, Molecular electronics, Functionalization, Nanoelectrode, Dielectrophoretic trapping

1. Introduction

The establishment of molecular junctions between metallic electrodes has attracted increasing interest in recent years due to possible applications in microelectronic devices and as molecular sensors. When a current passes through these electrical junctions containing molecules, the measured electrical signature can be used to identify molecules with relevance in biological and medical applications such as proteins and DNA (1). Thus, such electrical measurements can be used to sense biomolecules. Since molecules typically have lengths in the nanometer range, electrodes with corresponding separation distances are required to directly connect single molecules. There are several approaches to fabricate such electrodes (2–5), but to date specialized laboratory setups are

Mikhail Soloviev (ed.), *Nanoparticles in Biology and Medicine: Methods and Protocols*, Methods in Molecular Biology, vol. 906,
DOI 10.1007/978-1-61779-953-2_43,

needed to directly connect the molecules to the electrodes and form the molecular junctions. One strategy to obtain single molecular electronics devices that are useful for both fundamental measurements and applications consists in the connection of molecules to the nanoelectrodes via nanoparticle (NP) bridges. One advantage of such a platform is its stability, i.e., once the nanoparticle-molecule-nanoelectrodes arrangement is established, it is structurally and electrically stable. Another benefit is that conventional functionalization schemes can be applied to the electrodes or nanoparticles, allowing a large variety of molecules to be utilized and studied. Although the electrode and nanoparticle material in this platform can be varied, here we concentrate on the use of gold for both.

The creation of the molecular electronics platform starts from nanoelectrodes such as shown in Fig. 1. Then, the nanoelectrode surface is functionalized by molecules containing thiol groups. The nanoparticles can also be functionalized prior to their trapping. Since the detailed description of such functionalization schemes would go beyond the frame of this chapter, the reader is referred to refs. (6, 7) for more information.

The formation of the molecular junctions based on such type of platform can be achieved by several methods (8). The method used in this work to form nano-sized junctions is to dielectrophoretically trap nanoparticles in between functionalized nanoelectrodes (2, 9), see Fig. 2. After the trapping procedure, the molecules are linked on both ends to the gold surfaces, either through chemical bonds or Van-der-Waals interactions. The electrical response of this system is thus determined by the band structure of electrodes and nanoparticles, the electronic structure of the molecules, and the electronic properties of the gold-molecule junctions. We have used such a platform to form and electrically characterize junctions with different molecular systems (9, 10).

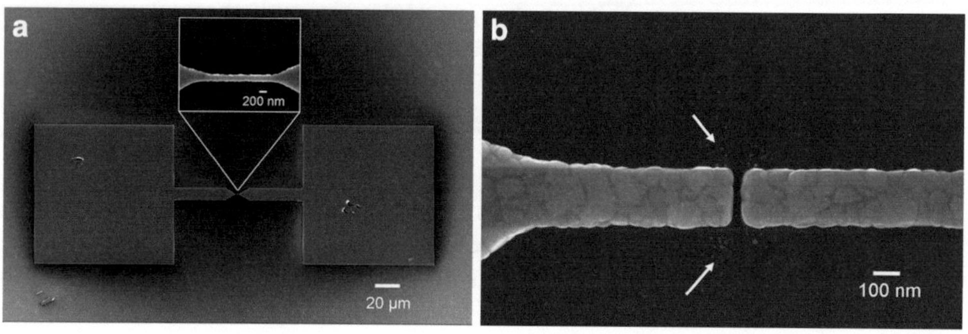

Fig. 1. SEM images of nanoelectrodes. (**a**) A nanoelectrode platform including contact pads with an *inset* that shows the 200 nm wide uncut wire and (**b**) a 26 nm nanogap created by FIB. Particles arising from backsputtering in the FIB can be observed on both sides of the electrodes and are marked with *arrows*. Reproduced from ref. (2) with permission from IOP Publishing Ltd.

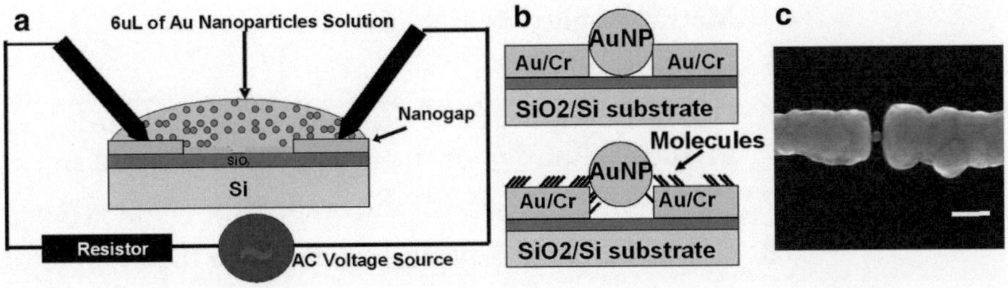

Fig. 2. Dielectrophoretic trapping of nanoparticles between functionalized nanoelectrodes. (**a**) Schematic of the nanoelectrode setup during the trapping experiment. (**b**) Schematic of cross-section through the nanoparticle-molecule-nanoelectrode bridge platform with clean AuNP and molecule functionalized electrodes, respectively. (**c**) Single nanoparticle trapped between the electrodes. The SEM scale bar is 100 nm. Reproduced from ref. (9) with permission from IOP Publishing Ltd.

2. Materials

Prepare all suspensions using purified deionized water. If suspensions cannot be immediately used, they must be stored in the fridge and kept away from light. Carefully follow all waste disposal regulations when disposing waste materials.

2.1. Gold Nanoparticle Suspensions

Commercially available gold nanoparticles (AuNP) can be used in preparing the nanoelectrode/nanoparticle bridge platform described below. If more control over the AuNP shells and solutions is needed, fresh gold nanoparticle suspensions might be prepared in-house. AuNP suspensions are typically available in polar solvents (e.g. water) or nonpolar organic solvents (e.g., toluene) (see Notes 1–4).

1. AuNP: 30 nm AuNP colloid solution stabilized by citrates, dispersed in deionized H_2O, e.g., EM.GC30 (BBInternational) or similar.
2. For AuNP manufacturing, prepare 5 mM solution of $HAuCl_4$ by dissolving 0.0985 g of $HAuCl_4$ in 50 mL of deionized H_2O.
3. Prepare 25 mM of sodium salt of l-glutamic acid by dissolving 0.211 g of monosodium glutamate ($C_5H_8NNaO_4$) in 50 mL of deionized H_2O.
4. Hot plate and magnetic stirrer.

2.2. Nanoelectrode Geometry and Materials

1. Field emission scanning electron microscope FEI xl30 ESEM equipped with electron beam lithography (EBL) software from (www.jcnabity.com) (see Note 5).
2. Poly(methyl methacrylate) (PMMA).
3. Thermally grown 1 μm silicon dioxide (SiO_2) on Si wafers.

4. Methyl isobutyl ketone (MIBK).
5. Isopropanol (IPA).
6. Commercially available evaporator, e.g., Edwards III Auto 306 FL400 (Edwards) for the deposition of Cr and Au.
7. Chromium for use with the evaporator (99.9 % purity).
8. Gold wire (99.9 % purity) for use with the evaporator.
9. A Karl Süss MA6 mask aligner was used for photo lithography.
10. FEI Strata DB235 dual beam focused ion beam (FIB) microscope.
11. UV-ozone photoreactor PR-100 (UVP Inc.).

2.3. Molecular Self-Assembly on Gold Electrodes

Self-assembled monolayers of the organic molecules such as alkanes (1-octanethiol) or cyclohexanes (4,4′-biphenyldithiol) are fabricated. Such molecules must have at least one thiolated end in order to enable covalent binding to the gold nanoelectrode surfaces and/or the bridging AuNPs (see Note 6).

1. Prepare 0.1 M solution of 1-octanethiol in ethanol by mixing 1.463 g of 1-octanethiol in 100 mL of ethanol.
2. 1 mM toluene solutions of 4,4′-biphenyldithiol: dissolve 0.0218 g of 4,4′-biphenyldithiol in 100 mL of toluene (see Note 7).

2.4. Dielectrophoretic Trapping

1. Pure deionized water.
2. Nitrogen gas for blow drying.
3. Karl Suss probe station with an optical microscope with micromanipulators including tungsten probe needles.
4. Function generator Tektronix AFG 3102 which can apply voltages in the 1–10 $V_{\text{peak–peak}}$ range at frequencies in the 1 MHz range.

2.5. Electrical Characterization of the Molecular Junctions

1. Electrical characterization should be carried out at ambient conditions by using the same probe station as mentioned in Subheading 2.4. The probe station is connected to Agilent B1500A semiconductor parameter analyzer source meters (see Note 8).
2. Field emission scanning electron microscope equipped with an in-lens detector, e.g., Zeiss 1550 or similar.

3. Methods

3.1. Gold Nanoparticle Suspensions

Gold nanoparticles that are stabilized electrostatically by ions (charges) in solution can be purchased or alternatively, if i.e., better size control is wished be fabricated in the laboratory. Gold nanoparticles in water are prepared by reduction of 2 mL of 5 mM solution gold (III) chloride in 100 mL of deionized water with 3 mL of 25 mM sodium phosphate at 100 °C. Further heating and stirring will change the color of the solution from blue (bigger nanoparticles) to bright red (smaller) AuNPs (see Notes 9–11). For the cleaning procedure of AuNPs, a UV-ozone photoreactor can be used.

1. Add 2 mL of 5 mM HAuCl to 100 mL of deionized H_2O.
2. Heat the solution, while stirring it, until it starts boiling.
3. Upon boiling, add 3 mL of 25 mM sodium glutamate solution.
4. After a few seconds, the solution will turn pale red indicating an onset of formation of gold nanoparticles.
5. Continue heating, until the solution turns bright red.
6. Transfer the solution quickly to ice water bath to stop the reaction, then transfer flask to room temperature.

3.2. Nanoelectrode Geometry and Materials

1. Perform EBL on using PMMA in order to obtain 2 μm long by 200 nm wide lines. Acceleration voltage of the EBL pattern is typically 30 kV. Dissolve the exposed area, i.e., lines in the PMMA, using a 1:3 MIBK:IPA solution.
2. After developing the EBL patterns, resistive evaporation is used to deposit a 5–10 nm adhesion layer of chromium (see Note 12) followed by a 80–120 nm layer of gold onto lines generated by EBL (see Note 13).
3. Use photolithography to fabricate 300 nm thick gold contact pads having lateral dimensions 100 μm by 100 μm and which connect to the EBL patterned lines. See Fig. 1a.
4. A 20–30 nm wide nanogap is milled in the EBL line using a FIB of Ga^+ ions at an acceleration voltage of 30 kV (see Note 14). See Fig. 1b.
5. Remove any residual organic contamination on the electrodes from the fabrication process by a 2 min oxygen plasma treatment, or alternatively by cleaning the sample for 10–30 min in a UV-ozone photoreactor (see Notes 15 and 16).
6. Place and store the cleaned nanoelectrode samples in pure ethanol immediately after the above-described cleaning procedure (see Notes 17 and 18).
7. If nonfunctionalized (i.e., clean and uncoated) AuNPs are trapped between the nanoelectrodes and metallic contact with the AuNP is difficult to achieve, irradiate the AuNP suspension prior to trapping with 254 nm UV light for 30 min to remove contamination from the surface of the AuNPs (see Note 19).

3.3. Molecular Self-Assembly on Gold Electrodes and on AuNPs

Nanoelectrodes are functionalized by creating a self-assembled monolayer (SAM) of the desired molecule on gold surfaces (see Note 20). Samples are placed in a solution of the molecules for a specified time interval. While placing the nanoelectrode samples in the solvents mentioned here, Au surface is wetted.

1. To functionalize nanoelectrodes with 1-octanethiol (OT), place the nanoelectrode samples in a 0.1 M of 1-octanethiol in ethanol or toluene for 24 h.
2. To prepare nanoelectrodes with a SAM of 1,4-biphenyldithiol (BPDT), place the nanoelectrode samples in a 1 mM solution of BPDT in ethanol (99.5 % pure) under Argon atmosphere for 48 h.
3. Immediately after SAM formation, take out the samples from the solutions and blow dry them with nitrogen.

3.4. Dielectrophoretic Trapping

1. To trap AuNPs on the nonfunctionalized electrodes, rinse the sample surface with deionized water prior to trapping and blow dry with a nitrogen flow (see Note 21).
2. For the experiments using electrodes functionalized with 1-octanethiol or 4,4-SAMs, immediately prior to trapping, remove the nanoelectrodes from their solutions, rinse with deionized water, and blow dry with a nitrogen flow.
3. Make electrical connection to the contact pads of the nanoelectrodes using tungsten probe needles in a probe station (see Note 22), see Fig. 2a.
4. Place a 6–10 μL drop of AuNP suspension with a concentration of roughly 10^{11} particles/mL on the nanogap. See Fig. 2a.
5. Apply an alternating voltage of 1–2 $V_{peak-peak}$ at a frequency of 1 MHz for 30–60 s (see Notes 23 and 24).
6. Immediately after the trapping, rinse the nanoelectrodes with deionized water to remove excessive AuNP from the solution and blow dry with a nitrogen flow (see Note 25). See Fig. 2b, c.

3.5. Electrical Characterization of the Molecular Junctions

Electrical characterization is considered here as a standard procedure where the tungsten probe needles in the probe station are placed on the contact pads connecting to the nanoelectrodes. The applied voltage is scanned, starting from 0 V, up to a certain value and (e.g., 1 or 2 V) across the nanogap between the electrodes while measuring the corresponding current. The current–voltage (*I*–*V*) curves obtained from the measurement give information about the resistance of the nanogap. If the *I*–*V* curve is linear, the ohmic resistance can be extracted from the slope of the curve.

1. Perform electrical characterization as soon as possible after the trapping of the AuNPs using a source meter in which voltages can be sourced and currents measured (see Notes 26 and 27).

2. Ground the tungsten needle before placing them on the contact pads (see Note 28).
3. Make both positive and negative voltage sweeps and record the corresponding current (see Note 29), see Fig. 3a. In order to minimize the possibility of fusing of trapped nanoparticles due to high currents, initially perform current–voltage (*I–V*) sweeps in 25 μV steps to a maximum applied voltage limit of ±1 mV, see Fig. 3b. If low resistances (in the order of 100 Ω or below) are measured, then the nanoelectrode devices are discarded (see Note 30). If metallic contact (i.e., high currents) is not observed, increase the maximum applied voltage to ±2–3 V with step sizes of 10–25 mV, see Fig. 3c (see Note 31). Other measurements are also possible (see Note 32).
4. Carry out SEM imaging to correlate *I–V* response with the actual number of nanoparticle(s) trapped between the nanoelectrodes (see Note 33).
5. After recording the *I–V* responses, use SEM to characterize the nanoelectrode-nanoparticle device itself. Use acceleration voltage of 15 kV and a working distance of 3 mm (see Note 34).

4. Notes

1. Advantages of purchasing commercially available AuNPs are convenience and narrow size distribution of nanoparticles. However, we found that better control over particle size, size distribution, and purity of suspension can be achieved by production of AuNPs in house.
2. The AuNP suspension turning black (typically for small AuNPs under 10 nm) or black particles precipitating out of the suspension would indicate that then AuNPs have agglomerated and the suspension has become unstable. This occurs more frequently with electrostatically stabilized AuNPs in polar solvents. In such cases, use a fresh AuNP suspension.
3. Always store AuNP suspension in fridge to protect from heat and light, but do not let the suspension freeze as freezing will lead to destabilization of AuNPs in solution.
4. Electrostatically stabilized AuNPs, such as with water suspensions, cannot be resuspended once agglomerated. However, in nonpolar organic solvents, AuNPs are stabilized by steric hindrance (normally they have a functionalized surface), and as such, they can be taken out of suspension by, e.g., centrifugation, washed, and resuspended.
5. The EBL patterning used here aims for the fabrication of 200 nm large wires. Thus, a scanning electron microscope

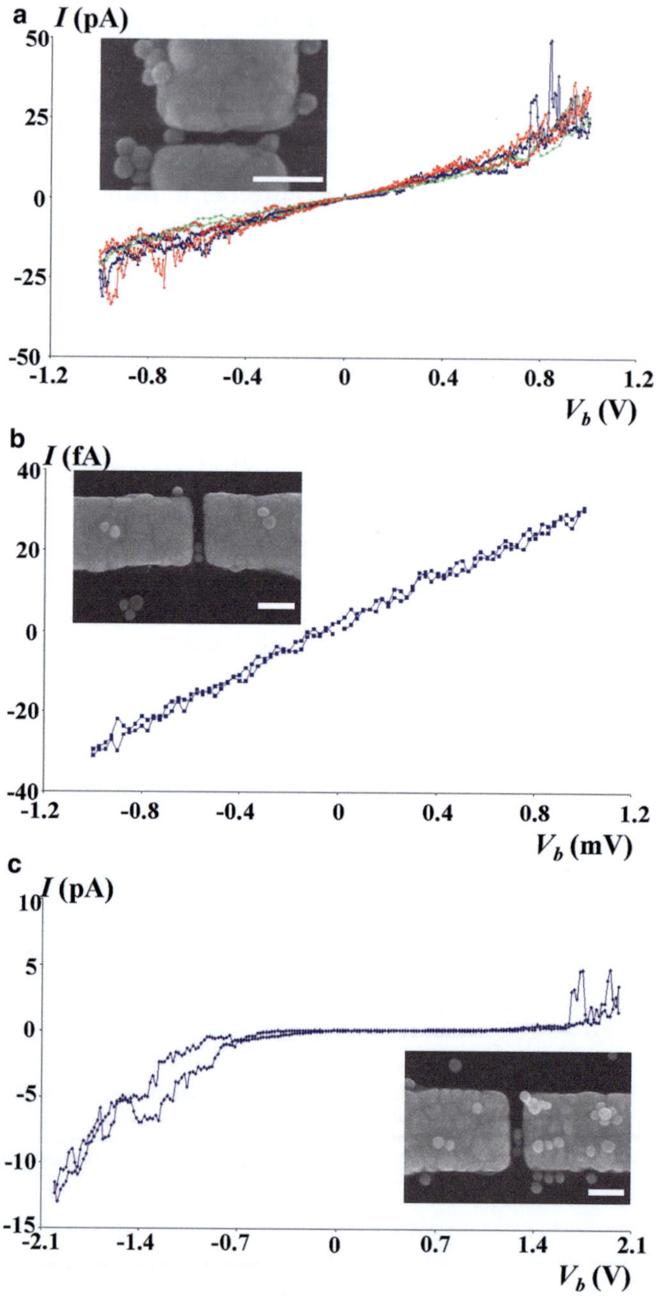

Fig. 3. Electrical characterization of the molecular junctions. (**a**) Three successive *I–V* measurements on platform containing trapped nanoparticles between octanethiol-coated nanoelectrodes (*inset*: SEM image of device). The three close to identical *I–V* curves indicate that the platform remains stable upon electrical measurements. (**b**) High resolution (25 μV steps) *I–V* scan of the trapped AuNPs between OT-coated nanoelectrodes (*inset*: SEM image of device). (**c**) *I–V* characteristics of trapped AuNPs (*inset*: SEM of device) between OT-coated nanoelectrodes at applied voltages up to 2 V. All scale bars are 100 nm. Reproduced from ref. (9) with permission from IOP Publishing Ltd.

equipped with a beam blanker and an external electron beam patterner are sufficient to obtain these structures. There are a range of external electron beam patterners. Dedicated EBL equipment can be used, but is not required for this process step.

6. Thiols can form disulphide bonds (i.e., the molecules can bond to each other) as well as reacts with oxygen to form oxides; both of these situations prevent molecules from forming the desired gold-thiol bond.
7. To prevent formation of disulphide bonds and oxides, keep the solution in Argon atmosphere (i.e., work in a glove box).
8. Spurious electrical disturbances can be minimized by placing the probe station in an electrically shielded enclosure that also kept out light and mechanical vibrations were reduced by placing the test setup on a stone table. With proper isolation and measurement equipment, current levels down to 10^{-17} A can be measured. For making molecular electronics measurements in such devices, a current sensitivity of at least 10^{-12} A is recommended.
9. The specific size of AuNPs (10–100 nm) can be obtained by stopping the reaction at an appropriate point during the synthesis (indicated by the color), then quenching and thermalizing the AuNP solution to room temperature.
10. The size of AuNPs can also be controlled by varying the concentration of gold salt and reduction agent.
11. Direct reduction of gold salts in the presence of the molecules in organic solvent often results in AuNPs less than 10 nm in size due to reaction dynamics. Larger AuNP suspensions in organic solvents are usually prepared by phase transfer. In order to accomplish the phase transfer, molecules having gold affinity are added to the AuNP suspension in water (electrostatically stabilized AuNPs). These molecules spontaneously attach to the AuNP surface, after which the suspension can be mixed with an organic solvent such as chloroform, which results in phase transfer of the AuNP to the organic solvent.
12. Alternatively, titanium can be used as an adhesion layer.
13. Prior to the evaporation, residual PMMA in the patterns can be removed by gentle oxygen plasma treatment. However this step can cause an increase in the size of the patterns as well as roughening of the PMMA.
14. Nanoelectrodes composed of Au can change their geometry when they are stored for some months. Sharp edges tend to become more rounded with time. Therefore, it is recommended to use the nanoelectrodes for experiments shortly after the structuring.

15. Both of these cleaning procedures remove organic contamination by essentially the same oxidative cleaning mechanism, but oxygen plasma treatment is generally more aggressive. Sufficient treatment/cleaning times can be checked by checking the sample substrate surface for degree of hydrophilicity: a surface free from contamination will be very hydrophilic and a drop of ultrapure water will wet the surface completely.
16. These cleaning processes only remove organic contamination. Inorganic and other particulate contamination must be first removed by other chemical cleaning steps.
17. Placing the nanoelectrodes in ethanol immediately after the oxygen plasma or UV-ozone treatment will remove (reduce) the gold oxide that forms during the strong oxidizing conditions and also reduce the possibility of unknown surface contamination. This ensures a reproducible procedure for gold surface preparation.
18. Because the substrate becomes very hydrophilic after the cleaning process, current measurements across the nanogap can be significantly affected due to leakage currents that run via the substrate surface. This is an important consideration if making sensitive measurements of high impedance samples that bridge the nanoelectrodes since the sample current may be masked by the large leakage currents. The amount of leakage current will depend on environmental conditions such as humidity. The substrate surface can be rendered hydrophobic by placing the sample in a vacuum and thus greatly reduce leakage currents on the substrate surface.
19. In our experience, it was found that this treatment resulted in more frequent metallic contact between the AuNPs and the electrodes. The AuNPs may have acquired a nonconductive contamination layer during storage.
20. It is important to note that the AuNPs can also be functionalized. This is not treated here since such functionalization involves a description of such methods that would go beyond the scope of this chapter.
21. Rinsing with water will help remove any accumulation of net charge on the surface, which otherwise may result in nanoelectrode damage by electric discharge when probe needles are connected to the contact pads.
22. Oxide formation on tungsten needles can result in poor contact with the contact pads. The oxide layer can be removed by soaking the probe needles in a concentrated NaOH solution. Other contamination can be easily removed by solvents such as ethanol and acetone, or even mechanically by piercing aluminum foil with the probe needle. Normally our final cleaning step is to rinse with deionized water and blow dry in a nitrogen flow.

23. Use low trapping voltages to avoid fusing of nanoparticles, but if trapping attempts are not successful, increase the applied AC voltage up to 2 V. In our experience, the risk of fusing increases greatly with voltages higher than 2 V. Increase the trapping time from 30 to 60 s as required. Length of trapping time may be limited by evaporation of AuNP suspension. AuNPs in more volatile solvents will evaporate much quicker than water, limiting the time for trapping. If the drop dries during the trapping, AuNPs will concentrate on the surface along with dried solvent residues.
24. Use a 10 MΩ resistor in series to reduce excessive currents. If metallic contact occurs during the trapping process, the impedance of the nanoparticle bridge may drop, allowing high currents that in turn may fuse the junction. Placing a high impedance resistor in series will limit the current to some degree in this situation, but not completely because of the AC voltage used and the associated capacitance in the setup.
25. It is difficult to achieve a reproducible number of nanoparticles in the gap. However, success rate for trapping after adjusting parameters can vary between 50 and 75 %. Also of interest is the fact that trapping can be induced in nearby electrode gaps that are not directly connected to the AC voltage, but that are close enough so that the AuNP drop covers the electrodes. Induced electric fields in the AuNP suspension sometimes results in single AuNP trapping of adjacent nanoelectrodes.
26. Clean the needles again as mentioned before to avoid any contamination on needle surfaces that might occur during trapping process.
27. Ideally using two source meters increases the sensitivity of electrical measurements. One source meter sources the voltage to one electrode, while the second source meter held at ground potential measures the current that flows through the nanogap. This measurement strategy removes the stray current contribution that may flow from the source nanoelectrodes and not through the nanogap.
28. Grounding of needles will remove accumulation of charges and subsequently will reduce the chance of electrical discharge which may damage the nanoparticle/nanoelectrode junction. As well, wear a wrist ground strap to reduce operator-induced static discharges.
29. It is generally better to start applying voltage from 0 V and record current in both rising and falling voltage steps. Any hysteresis between rising and falling loop may indicate voltage-induced structural changes in the metal-molecule junction.
30. Low resistances indicate metallic contact between the nanoparticles and the electrodes with no molecules involved.

31. In general, $I–V$ curves over more than roughly ten nanoparticle-molecule-nanoelectrode bridges are needed to obtain converging results from $I–V$ curves. Individual $I–V$ curves may also be of interest, particularly when nonlinear effects or hysteresis are observed.
32. This platform can be used for several other types of electrical measurements such as differential current measurements and the electrical properties of the molecules in this platform can be evaluated both at low temperatures (i.e., 4 K) and room temperature.
33. It is possible to estimate the junction resistance if the exact number of nanoparticles in junction is known and resistance histograms (showing the number of measurements with certain resistances) can be normalized. However, this is difficult if more than a few AuNPs are imaged as all the trapped AuNPs may not be visible in the SEM image.
34. Since the SEM imaging could possibly damage the molecules or result in the deposition of amorphous carbon by the electron beam, it was performed after the $I–V$ measurements.

References

1. Perkins FK, Tender LM, Fertig SJ et al (2002) Sensing macromolecules with microelectronics. Proc SPIE 4608:251–265
2. Blom T, Welch K, Stromme M et al (2007) Fabrication and characterization of highly reproducible, high resistance nanogaps made by focused ion beam milling. Nanotechnology 18:285301
3. Fischbein MD, Drndic M (2006) Nanogaps by direct lithography for high-resolution imaging and electronic characterization of nanostructures. Appl Phys Lett 88:063116
4. Morpurgo AF, Marcus CM, Robinson DB (1999) Controlled fabrication of metallic electrodes with atomic separation. Appl Phys Lett 74:2084–2086
5. Park H, Lim AKL, Alivisatos AP et al (1999) Fabrication of metallic electrodes with nanometer separation by electromigration. Appl Phys Lett 75:301–303
6. Sastry M, Kumar A, Mukherjee P (2001) Phase transfer of aqueous colloidal gold particles into organic solutions containing fatty amine molecules. Colloid Surf A 181:255–259
7. Zhang P, Chu AYC, Sham TK et al (2009) Chemical synthesis and structural studies of thiol-capped gold nanoparticles. Can J Chem 87:335–340
8. Liao JH, Mangold MA, Grunder S et al (2008) Interlinking Au nanoparticles in 2D arrays via conjugated dithiolated molecules. New J Phys 10:065019
9. Jafri SHM, Blom T, Leifer K et al (2010) Assessment of a nanoparticle bridge platform for molecular electronics measurements. Nanotechnology 21:435204
10. Welch K, Blom T, Leifer K, Stromme M (2011) Enabling measurements of low-conductance single molecules using gold nanoelectrodes. Nanotechnology 22:125707

Index

Mikhail Soloviev (ed.), *Nanoparticles in Biology and Medicine: Methods and Protocols*, Methods in Molecular Biology, vol. 906, DOI 10.1007/978-1-61779-953-2, © Springer Science+Business Media, LLC 2012

D

MIX
Papier aus verantwortungsvollen Quellen
Paper from responsible sources
FSC® C105338

If you have any concerns about our products,
you can contact us on
ProductSafety@springernature.com

In case Publisher is established outside the EU,
the EU authorized representative is:
Springer Nature Customer Service Center GmbH
Europaplatz 3, 69115 Heidelberg, Germany

Printed by Libri Plureos GmbH
in Hamburg, Germany